Advanced Structured Materials

Volume 220

Common engineering materials are reaching their limits in many applications, and new developments are required to meet the increasing demands on engineering materials. The performance of materials can be improved by combining different materials to achieve better properties than with a single constituent, or by shaping the material or constituents into a specific structure. The interaction between material and structure can occur at different length scales, such as the micro, meso, or macro scale, and offers potential applications in very different fields.

This book series addresses the fundamental relationships between materials and their structure on overall properties (e.g., mechanical, thermal, chemical, electrical, or magnetic properties, etc.). Experimental data and procedures are presented, as well as methods for modeling structures and materials using numerical and analytical approaches. In addition, the series shows how these materials engineering and design processes are implemented and how new technologies can be used to optimize materials and processes.

Advanced Structured Materials is indexed in Google Scholar and Scopus.

Shashanka Rajendrachari
Editor

Mechanically Alloyed Novel Materials

Processing, Applications, and Properties

Springer

Editor
Shashanka Rajendrachari
Department of Chemistry
School of Sciences and Humanities
SR University
Warangal, Telangana, India

ISSN 1869-8433 ISSN 1869-8441 (electronic)
Advanced Structured Materials
ISBN 978-981-97-6503-4 ISBN 978-981-97-6504-1 (eBook)
https://doi.org/10.1007/978-981-97-6504-1

This Springer imprint is published by the registered company Springer Nature Singapore Pte Ltd.
The registered company address is: 152 Beach Road, #21-01/04 Gateway East, Singapore 189721, Singapore

Preface

The metallurgy field has revolutionized in recent decades due to the use of advanced technology to fabricate various alloys and their processing. One such advanced technology for synthesizing various grades of alloys is "Mechanical Alloying". Using this technique, we can prepare bi-metallic, tri-metallic, tetra-metallic, and penta-metallic alloys along with advanced materials like high-entropy alloys, super alloys, and smart materials. The main advantage of preparing alloys using this method is its ability to prepare nano-structured materials, high-temperature materials, ceramic materials, high-density materials, etc. Because of these advantages involved in this technique, mechanical alloying has become one of the popular methods of preparing alloys among researchers. Researchers are trying to improve the microstructure of alloys by optimizing the milling conditions. However, still, there is plenty of space to improve and this book has addressed a few lacunae.

This book comprises 19 chapters authored by various powder metallurgists working on mechanical alloying of different alloys from various countries. This book explains the different types of mechanical alloying used to prepare powders and their mechanisms. Factors affecting the mechanical alloying process followed by the overview and comparison of dry and wet milling. Comparative study of mechanical alloying and other conventional powder metallurgical methods followed by achieving maximum density with proper explanation of the structure-property relationship. This book gives brief knowledge about the modern methods used to characterize the ball-milled powders followed by their consolidation by highly advanced sintering methods. The proposed book also discusses the processing, properties, and applications of high-entropy alloys, ODS stainless steel, shape memory alloys, cermets, iron, copper, zinc, tungsten, aluminum, titanium, magnesium, ceramic-based alloys, and intermetallic alloys. This book also discusses the detailed applications of mechanically alloyed and consolidated parts in automobiles, energy storage, electrochemical sensors, etc. This book also sheds light on the present and the future of mechanically alloyed materials in detail.

Most of the powder metallurgy or mechanically alloying-related books explain only the methods of preparing powders, different sintering methods, and applications. The present book comprises the above topics and also covers the most important types of ferrous and non-ferrous alloys prepared by mechanical alloying. This book also gives an insight into why the mechanical alloying method is so popular and advantageous over other conventional powder metallurgical methods. This kind of comparison studies cannot be available in any of the books. Totally, this book gives full knowledge from the synthesis of powders to processing and the application of consolidated parts. None of the books contains all these abovementioned topics as a single book. So, we are writing this book to help research scholars and engineering students understand all the basic aspects of mechanical alloying and their mechanisms.

As I mentioned earlier, this book is mainly channelized for research scholars and engineering students to understand the basic concepts of one of the popular and more advantageous powder metallurgical techniques, "mechanical alloying". This book helps students to learn the interesting and advanced types of alloy preparation, which cannot be possible in other traditional metallurgical operations. This book is unique in its own way due to the very interesting topics discussed above and cannot be found in many other books as a whole. Our intention is to educate and teach the students, industrial people, and research scholars about mechanical alloying and we tried our best to cover almost all types of alloys in a single book.

Dr. Shashanka Rajendrachari
SR University
Warangal, Telangana, India

Contents

Chapter 1
Mechanism of Ball Milling and the Factors Affecting the Process of Milling

N. B. Pradeep, S. Parameshwara, S. G. Chethan, V. Lokesh, T. S. Harish, S. Basavarajappa, and K. B. Girisha

Abstract Ball milling is a process that involves the vial rotation outcomes in the synthesis of alloy powder. The mechanism of the ball milling process involves various techniques like impact, attrition, and compression. Various parameters that influence the alloy powder production during the milling are, ball mill type, ball milling speed, ball-to-powder weight ratio (BPR), Ball milling media (grinding media), the temperature of ball milling, ball milling atmosphere, the intensity of ball milling, the extent of filling the vial, container of ball mill, process control agent (PCA) are discussed.

Keywords Ball milling · Mechanical alloying · Types of ball mills · Milling parameters · Milling mechanism

---

N. B. Pradeep (✉) · S. Parameshwara
Department of Mechanical Engineering, JNN College of Engineering, Shivamogga, Affiliated to Visvesvaraya Technological University, Belagavi 590018, India
e-mail: pradeepnb@jnnce.ac.in

S. G. Chethan
Department of Chemistry, JNN College of Engineering, Shivamogga, Affiliated to Visvesvaraya Technological University, Belagavi 590018, India

V. Lokesh
Department of Marine Engineering, Srinivas Institute of Technology, Valachil, Mangaluru, India

T. S. Harish
Department of Electronics and Communication, JNN College of Engineering, Shivamogga, Affiliated to Visvesvaraya Technological University, Belagavi 590018, India

S. Basavarajappa
Department of Engineering Design, GM University, Davanagere 577006, India

K. B. Girisha
Department of Mechanical Engineering, BGSIT, Adichunchanagiri University, BG Nagar, Mandya, India

S. Rajendrachari (ed.), *Mechanically Alloyed Novel Materials*, Advanced Structured Materials 220, https://doi.org/10.1007/978-981-97-6504-1_1

1.1 Different Types of Ball Mills

Ball mills are versatile machines used in various industries for particle size reduction, mixing, homogenization, and mechanical alloying. Some common types of ball mills include:

Vertical Ball Mills: These have a vertical orientation and cylindrical shell. They may be batch or continuous mills and can be used for mixing or grinding materials.

Horizontal Ball Mills: These have a horizontal orientation, usually with a rotating chamber and media cascading, used for fine grinding or mixing.

Planetary Ball Mills: These are high-energy mills where the grinding media is accelerated in a rotational motion inside a container, creating very high impact forces.

Industrial Ball Mills: Typically, larger in size and used for industrial applications, such as mining or processing large amounts of material.

Vibration Ball Mills: These mills vibrate at high frequency, providing rapid and efficient grinding of materials.

Conventional Ball Mills: These are the standard, traditional ball mills used in many laboratories and industries for general-purpose milling.

Each type has its own advantages based on the specific application requirements, such as capacity, particle size reduction, and energy efficiency.

1.2 Planetary Ball Milling Process

The planetary motion of the vials is the factor that gives planetary ball mills by McCormick et al. [1] demonstrate a wider variety of chemical reactions resulting from interactions between liquid and solid that take place during the milling process. These chemical reactions occur on account of energy conversion (mechanical to chemical) during the milling process. The term "high-energy rate milling", also known as the "mechanochemical process", was first used by Carry [2]. These chemical reactions are observed in high-energy rate mills. Takacs [3] and Matteazzi and Le Caer [4] provided a comprehensive description of the chemical reactions produced by high-energy rate milling. Also found that the majority of reactions are due to the high-energy rate milling process owing to displacement reactions as below

$$\mathrm{MxO} + \mathrm{yR} \rightarrow \mathrm{xM} + \mathrm{RyO}$$

M is the high reactive metal formed by the reduction of MxO.

During the milling process, solid-state reactions typically cause product phase formation at reactant grain boundaries. As a result, the product phase is able to mature further due to the diffusion of atoms from the reactant phase through it. This creates a barrier and prevents various solid-to-solid reactions. In order to speed up the diffusion at a reasonable rate, reactions must be carried out at high temperatures. While milling with the help of high-energy planetary ball mills (HEPBM), these

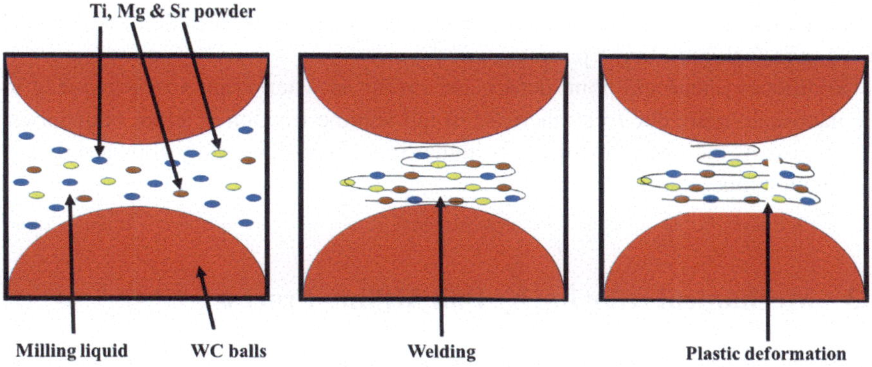

Fig. 1.1 Ball milling process mechanism involving ball-powder-ball collisions

chemical reactions are observed. The production of ultrafine powders, nanostructured materials, the processing of minerals and waste, combustion reactions, metal refining, the modification of the matrix's microstructure, the extension of solubility limits, the formation of amorphous phases, and other processes are all accomplished through the mechanochemical process.

The powder undergoes a cycle of repeated welding, fracture, and particle deformation as a result of the balls' constant collisions with the powder and vessel, which refines the powder's microstructure. High-density dislocations at grain boundaries in powders typically initiate solute microsegregation of atoms at the dislocation sites, producing a variety of solid solutions. Furthermore, the diffusion of chemical energy to crystalline powders during milling results in more distorted nanocrystalline materials at longer milling times. Although materials with nanograin size can be produced, milling is a technique that gradually reduces particle size to the nanometer scale. Figure 1.1 depicts a mechanism of collisions that occurs during the milling process.

The powder particles undergo multiple mechanisms during high-energy planetary milling, primarily due to the high impact energy of the jars and ball. Powder particles undergo repetitive processes during the milling process, including cold welding, fragmentation, flattening, work hardening, re-welding, and fracturing.

In ductile-to-brittle and ductile-to-ductile material combinations, soft particles become flat and fuse together to form a single, massive lamellar structure. Lamella size can occasionally exceed the parent element size prior to milling. Simultaneously, the resulting lamellar structures contain discrete layers of various initial constituents. Additionally, as milling continues, the fatigue failure mechanism that arises during milling causes the powder to fracture and harden. The microstructure refinement of powder particles persists due to weak agglomeration forces and the dominance of fragmentation over cold welding. Particle structure steadily refines but particle size stays constant during the milling process because of the high impact energy of balls on jar walls. A steady-state equilibrium is reached between the welding rate and the milling time, which tends to increase particle size. According to studies reported by

Froes et al. [5], smaller fragmented particles are more resistant to deformation and strive to re-weld into larger particles.

Additional linear and point defects are produced when grain size is reduced to the nanoscale scale due to an increase in the volume fraction of grain boundaries. Consequently, the rate of diffusion is higher in mechanically alloyed materials and rises with temperature.

1.3 Mechanical Alloying by Ball Milling Process

The process of mechanical alloying is intricate and involves adjusting different milling parameters to achieve the desired material properties. Not every variable is totally independent. A few of the crucial variables that can alter the final powder elements' (mechanically alloyed) characteristics are as follows:

1. Ball mill type
2. Ball milling speed
3. Ball-to-powder weight ratio
4. Ball milling media (grinding media)
5. The temperature of ball milling
6. Ball milling atmosphere
7. The intensity of ball milling
8. The extent of filling the vial
9. Container of ball mill
10. Process control agent (PCA).

1.3.1 Ball Mill Type

Ball mills exist in a variety of sizes, capacities, and speeds, and they can control the process by adjusting the milling temperature and reducing powder contamination. As a result, the type, quantity, and final composition of powders used for mechanical alloying are the most important factors to consider when selecting a mill. The most common mills are Fritsch Pulverised planetary ball mills, SPEX shaker mills, attritor mills, and custom-designed mills. The energy capacity of these mills varies and is used for specific applications.

1.3.2 Ball Milling Speed

Increased speed transmits more impact energy to the powder sample; however, depending on the mill design, there are limitations to using high speeds. As in conventional ball mills, when the mill speed exceeds a certain critical speed, the balls become pinned to the inner walls of the vials and do not collide with them, resulting in no impact force and milling arrest. As a result, the maximum speed should be less than the critical value, allowing the balls to collide with the vial walls and generate maximum collision energy-another effect of the max. Speed occurs when the temperature inside the vial rises significantly at high speeds, promoting diffusion, which is required for elemental powder alloying. Higher temperatures, on the other hand, can promote material transformation and, in some cases, cause the decomposition of meta-stable phases or supersaturated solid solutions during milling [6].

1.3.3 Ball-to-Powder Weight Ratio (BPR)

The ball-to-powder weight ratio is an important milling parameter that can be adjusted between 1:1 (low) [7] and 220:1 (high) [8] to achieve the desired powder elements. Milling time is reduced due to the high BPR. As a result, one of the key factors in determining the time required for mechanical alloying is the ball-to-powder ratio. Because of the higher BPR, the number of collisions per unit time and weight ratio increase; as a result, more impact energy is imparted, and alloying develops quickly. More heat will be generated as a result of the high-impact energy imparted by grinding balls on the container's inner walls, potentially changing the powder composition. Mechanical alloying (MA) at low BPR values produces metastable phases, whereas MA at high BPR gives rise to equilibrium phases [9].

1.3.4 Ball Milling Media (Grinding Media)

High-density materials are commonly used as grinding media because they can impart high energy to powder particles, resulting in solid solutions. Some of the most popular grinding mediums include tempered steel, sintered corundum, agate, stainless steel (SS), tool steel, chromium steel, tungsten carbide (WC), hardened steel, zirconium oxide, and other high-density materials. The grinding medium and vessel should be made of the same materials to avoid cross-contamination. The size of the grinding medium has an impact on both milling efficiency and the ultimate composition of powder particles [8]. As a result, changing the size of the grinding medium allows for greater control over the powder's final composition, size, phases, and shape. Some researchers discovered that smaller balls resulted in higher frictional forces, which accelerated the amorphous phase transformation [10, 11]. According to Atzmon [12], using different ball diameters increases the collision energy to a higher range than using identical diameter balls. Another advantage of using a different ball diameter is that it reduces the amount of cold welding and powder coating required on the ball [13]. Grinding balls with the same diameter creates tracks inside the container, allowing the balls to roll along a precise path rather than colliding with the container's end surfaces. Consequently, it reduces the mill's efficiency. As a result, it is critical to use a combination of larger and smaller balls to vary their motion [14].

1.3.5 The Temperature of Ball Milling

During the milling process, the temperature of the alloy/metal powder is affected by the kinetic energy of revolving balls inside the vial. It influences the composition of the final product, with temperature playing an important role during the mechanical alloying process. Higher milling temperatures have a significant effect on diffusion during the process. It raises the system's entropy (increases atom mobility), resulting in a nanocrystalline or amorphous phase, intermetallic, solid solution. Low temperatures would produce the amorphous phase. Only a few studies have reported on varying milling temperatures to investigate phase formation, solid solubility levels, and so on. According to Qin et al. [15], the extent of solid solubility decreases with increasing milling temperature. According to Fu and Johnson [16] and Shashanka and Chaira [17], high milling temperatures produce the greatest diffusivity and equilibrium effects. Lee et al. [18] discovered that higher milling temperatures increased amorphization kinetics.

1.3.6 Ball Milling Atmosphere

The milling atmosphere has a significant effect on powder contamination during the mechanical alloying (MA) process. As a result, milling must be performed in an inert gas atmosphere (argon or helium). A typical atmosphere for preventing contamination or oxidation of metal powders while milling is to use high-purity argon (inert gas). Metal powder unloading and loading must take place in an atmosphere-controlled glove box. Specific materials are prepared under varying atmospheric conditions. For example, nitrides are formed during milling in a nitrogen atmosphere [19, 20], while hydrides are formed in a hydrogen atmosphere [8]. In the case of highly reactive powders, nitrides and oxides would be formed. As a result, milling must take place in an inert atmosphere. Many investigations into the powder's final phase are dependent on the atmosphere used during the milling process [21]. Furthermore, the literature [22] shows that amorphization kinetics increase when oxygen gas is used for milling.

1.3.7 The Intensity of Ball Milling

The following equation is considered to calculate milling intensity in planetary ball milling:

$$I = V\Delta Enf$$

where,

f frequency
ΔE Energy dissipation
V a variable with a value less than one, whose values are determined by the volume of the balls within the vials
n Number of balls.

Planetary ball mill frequency is demarcated by the angular frequencies of the vials and sun wheel. E is determined by the distance between the sun wheel's Centre and the vials' walls.

1.3.8 The Extent of Filling the Vial

The container must have sufficient space to allow grinding balls and powder particles to move freely. Otherwise, it results in poor alloying of powder particles. Assume there are fewer grinding balls and powder. In that case, the high impact energy exerted by the balls on powder particles accelerates alloy formation. On the other hand, if there are more balls and powder, the space available for the free movement of balls is limited, slowing the rate of alloy formation. As a result, the rate of alloy formation slows due to the reduced amount of impact energy imposed on powder particles by balls. To improve mill efficiency, 50% of the vial volume should be empty.

1.3.9 Container of Ball Mill

Balls and grinding jars are components of the milling container. The milling medium's impact on the container walls during the process causes the material to dislodge from the container and fuse into powder particles. Contaminations may occur, causing changes in chemical composition (powder particles). The amount of contamination can be reduced by using the same material for the powder and the grinding medium. As a result, one of the most important considerations in the ball milling process is the choice of a container. The geometrical features of the mill container, such as size, shape, and design, will also play an important role in the milling process. According to the findings, alloying occurs at a much higher rate in flat-ended vials than in round-ended containers [23]. Common materials used for grinding vessels include stainless steel, tempered steel, bearing steel, hardened chromium steel, hardened steel, tool steel, and tungsten carbide (WC) [24].

1.3.10 Process Control Agent (PCA)

Alloying between powder particles occurs only when an equilibrium balance is achieved during powder particle fracturing and cold welding. A cold weld between particles occurs when ductile metal powder is subjected to heavy plastic deformation. As a result, a process control agent is mixed into the powder blend to slow down the cold-welding process during mechanical alloying. PCA exists in a variety of forms (liquids, solids, and gases) and is considered a surface-active agent during milling. Typically, the process control agents are organic compounds that accumulate on the surface of the powder particles, reducing cold welding of powder blends. Adsorbent surface-active agents reduce the concentration of the materials and prevent agglomeration. The equation gives the energy (E) required to reduce the size of powder particles during the milling process.

$$E = \vartheta \Delta S$$

where

ϑ Specific surface energy
ΔS increase of surface area.

Shorter milling times result in finer powder particles and reduced surface energy. The few process control agents include ethanol, methanol, stearic acid, and hexane, and their amounts range from 1 to 5 (wt%) of the total powder charge. During milling, these compounds degrade and combine as dispersoids or supplements to form powder particles. However, these compounds are not harmful to the alloy system, which increases material strength through a mechanism known as dispersion strengthening [25]. Process control agents can alter the solid solubility levels and thus influence the evolution of the final phase. Furthermore, the purity, shape, and size of the powder are determined by the quantity and quality of the process control agent used in the milling process [26]. According to the literature, increasing the amount of process control agent in powder can rapidly reduce particle size over a given milling time. The use of ethyl acetate (liquid PCA) results in a uniform distribution of powder particle size [27–34]. While milling, the amount of PCA considered determines the amount of powder produced. The amount of PCA depends on the powder particles being cold-welded together, being chemically and thermally stable, and having the best BPR.

1.4 Conclusions and Outlook

Ball milling process is a process employed to crush materials into fine powders leading to a reduction in particle size to nano size and changes the material structure contributing to various applications like drug formulation, energy storage materials, bio materials, alloying of materials and nanomaterial formulation. The major influencing factors in ball milling process are ball mill type, ball milling speed, ball-to-powder weight ratio, ball milling media (grinding media), the temperature of ball milling, ball milling atmosphere, the intensity of ball milling, the extent of filling the vial, container of ball mill, process control agent (PCA) that causes effective reduction in particle size, improved reactivity and even changes the material microstructure.

References

1. McCormick, P.G., Wharton, V.N., Schaffer, G.B.: In: Small, W.M. (ed.) Physical Chemistry of Powder Metals Production and Processing, pp. 19–34. TMS, Warrendale, PA (1989)
2. Carry, M.: Transformations of mechanical into chemical energy, 3rd Paper. Action of shearing-stress (continued). Phil. Mag. **37**, 470–475 (2009). https://doi.org/10.1080/14786449408620575
3. Takacs, L.: In: Suryanarayana, C., et al. (eds.) Processing and Properties of Nano-crystalline Materials, pp. 453–464. TMS, Warrendale, PA (1996)
4. Matteazzi, P., Le Caer, G.: Mechanically activated room temperature reduction of sulphides. Mater. Sci. Eng. A **156**, 229–237 (1992). https://doi.org/10.1016/0921-5093(92)90155-T
5. Froes, F.H., Suryanarayana, C., Russell, K., Li, C.-G.: Synthesis of intermetallics by mechanical alloying. Mater. Sci. Eng. A **192**(193), 612–623 (1995). https://doi.org/10.1016/0921-5093(94)03285-8
6. Shashanka, R., Chaira, D.: Development of nanostructured duplex and ferritic stainless steel by pulverisette planetary milling followed by pressureless sintering. Mater. Charact. **99**, 220–229 (2015). https://doi.org/10.1016/j.matchar.2014.11.030
7. Chin, Z.-H., Perng, T.-P.: Amorphization of Ni-Si-C powder by mechanical milling with SiC. Mater. Sci. Forum **235–238**, 121–126 (1997). https://doi.org/10.4028/www.scientific.net/MSF.235-238.121
8. Shashanka, R.: Synthesis of nanostructured stainless-steel powder by mechanical alloying-an overview. Int. J. Sci. Eng. Res. **8**, 588–594 (2017)
9. Gerasimov, K.B., Gusev, A.A., Ivanov, E.Y., Boldyrev, V.V.: Tribochemical equilibrium in mechanical alloying of metals. J. Mater. Sci. **26**, 2495–2500 (1991). https://doi.org/10.1007/bf01130201
10. Suryanarayana, C., Ivanov, E., Noufi, R., Contreras, M.A., Moore, J.J.: Phase selection in a mechanically alloyed Cu-In-Ga-Se powder mixture. J. Mater. Res. **14**, 377–383 (1999). https://doi.org/10.1557/JMR.1999.0055
11. Liu, L., Casadio, S., Magini, M., Nannetti, C.A., Qin, Y., Zheng, K.: Mater. Sci. Forum **235–238**, 163–168 (1997)
12. Atzmon, M.: In situ thermal observation of explosive compound-formation reaction during mechanical alloying. Phys. Rev. Lett. **64**, 487–490 (1990). https://doi.org/10.1103/PhysRevLett.64.487
13. Shashanka, R.: Non-lubricated dry sliding wear behavior of spark plasma sintered nanostructured stainless steel. J. Mater. Environ. Sci. **10**, 767–777 (2019)
14. El-Eskandarany, M.S., Aoki, K., Itoh, H., Suzuki, K.: Effect of ball-to-powder weight ratio on the amorphization reaction of Al50Ta50 by ball milling. J. Less-Common Metals **169**, 235–244 (1991)
15. Qin, Y., Chen, L., Shen, H.: In-situ X-ray diffraction examination of nano-crystalline Ag37Cu63 powders synthesized by mechanical alloying. J. Alloy. Compd. **256**, 230–233 (1997)
16. Fu, Z., Johnson, W.L.: Nanophase Zr-Al solid solutions by mechanical alloying at elevated temperatures. Nanostruct. Mater. **3**, 175–180 (1993)

17. Shashanka, R., Chaira, D.: Phase transformation and microstructure study of nanostructured austenitic and ferritic stainless-steel powders prepared by planetary milling. Powder Technol. **259**, 125–136 (2014). https://doi.org/10.1016/j.powtec.2014.03.061
18. Lee, C.H., Mori, M., Fukunaga, T., Mizutani, U.: Effect of ambient temperature on the MA and MG processes in Ni-Zr alloy system. Jpn. J. Appl. Phys. **29**, 540–544 (1990)
19. Shashanka, R., Chaira, D.: Optimization of milling parameters for the synthesis of nanostructured duplex and ferritic stainless steel powders by high energy planetary milling. Powder Technol. **278**, 35–45 (2015). https://doi.org/10.1016/j.powtec.2015.03.007
20. Calka, A., Williams, J.S.: Synthesis of nitrides by mechanical alloying. Mater. Sci. Forum **88–90**, 787–794 (1992)
21. Ogino, Y., Yamasaki, T., Maruyama, S., Sakai, R.: Non-equilibrium phases formed by mechanical alloying of Cr-Cu alloys. J. Non-Cryst. Solids **117**(118), 737–740 (1990)
22. Shashanka, R., Chaira, D., Swamy, B.E.K.: Effect of Y_2O_3 nanoparticles on corrosion study of spark plasma sintered duplex and ferritic stainless-steel samples by linear sweep voltammetric method. Arch. Metall. Mater. **63**, 749–763 (2018). https://doi.org/10.24425/122401
23. Harringa, J.L., Cook, B.A., Beaudry, B.J.: Effects of vial shape on the rate of mechanical alloying. J. Mater. Sci. **27**, 801–804 (1992)
24. Di, L.M., Bakker, H.: Phase transformation of the compound V3Ga induced by mechanical grinding. J. Phys. C: Condens. Matter **3**, 3427–3432 (1991)
25. Frazier, W.E., Koczak, M.J.: Mechanical and thermal stability of powder metallurgy aluminum-titanium alloys. Scripta Metall. **21**, 129–134 (1987)
26. Gaffet, E., Harmelin, M., Faudot, F.: Far-from-equilibrium phase transition induced by mechanical alloying in the Cu-Fe system. J. Alloy. Compd. **194**, 23–30 (1993)
27. Shashanka, R.: Effect of sintering temperature on the pitting corrosion of ball milled duplex stainless steel by using linear sweep voltammetry. Anal. Bioanal. Electrochem. **10** 349 (2018)
28. Shashanka, R., Swamy, B.E.K.: Simultaneous electro generation and electrodeposition of copper oxide nanoparticles on glassy carbon electrode and its sensor application. SN Appl. Sci. **2**, 956 (2020). https://doi.org/10.1007/s42452-020-2785-1
29. Shashanka, R.: Investigation of optical and thermal properties of CuO and ZnO nanoparticles prepared by Crocus Sativus (Saffron) flower extract. J. Iran. Chem. Soc. **18**, 415–427 (2021). https://doi.org/10.1007/s13738-020-02037-3
30. Shashanka, R., Jayaprakash, G.K., Prakashaiah, B.G.: Electrocatalytic determination of ascorbic acid using a green synthesised magnetite nanoflake modified carbon paste electrode by cyclic voltammetric method. Mater. Res. Innovat. (2021). https://doi.org/10.1080/14328917.2021.1945795
31. Kanaginahal, G.M., Mahale, R.S., Shamanth, V., Sharath, P.C., Rajendrachari, S.: Studies on dielectric, super-capacitive and photoluminescence properties of metal organic frameworks subjected to varying temperature and frequencies. Nano Hybrids Compos. **39**, 17–25 (2023). https://doi.org/10.4028/p-si6964
32. Mahale, R.S., Kanaginahal, G.M., Shamanth, V., Tiwary, V.K.: Rajendrachari, S., Sharath, P.C., Patil, A.: Applications of Fused Deposition Modeling in Dentistry. https://doi.org/10.4018/978-1-6684-6009-2.ch012
33. Rajendrachari, S., Basavegowda, N., Adimule, V.M., Avar, B., Somu, P., SK, R.M., Baek, K-H.: Assessing the food quality using carbon nanomaterial based electrodes by voltammetric techniques. Biosensors **12**(12), 1173 (2022). https://doi.org/10.3390/bios12121173
34. Rajendrachari, S., Adimule, V., Gulen, M., Khosravi, F., Somashekharappa, K.K.: Synthesis and characterization of high entropy alloy 23Fe-21Cr-18Ni-20Ti-18Mn for electrochemical sensor applications. Materials **15**(21), 7591 (2022). https://doi.org/10.3390/ma15217591

Chapter 2
Comparative Studies on Dry and Wet Milling: Uncovering the Secrets of Material Processing

Rayappa Shrinivas Mahale, V. Shamanth, P. C. Sharath, Krishnamurthy Goggi, and V. Mahesh

Abstract Within this book chapter, our primary objective is to undertake an in-depth examination of material processing techniques, with a dedicated focus on the specialized domains of dry and wet milling. Throughout this meticulous exploration, we endeavor to elucidate the intricacies of these methodologies by emphasizing technical and process parameters, thereby shedding light on their respective advantages and drawbacks. Dry milling, a fundamental facet of this analysis, necessitates a detailed examination of crucial parameters such as particle size distribution, feedstock moisture content, milling time, and the utilization of various milling equipment. The intricate interplay of these factors, when harnessed effectively, can yield fine-tuned control over particle size reduction and enhance the efficiency of dry milling processes across various applications. Conversely, the wet milling approach compels us to delve into a distinct set of technical parameters encompassing slurry rheology, milling media selection, agitation speed, and residence time. Understanding these parameters enables the optimization of wet milling processes, facilitating the disintegration of particulate matter in a liquid environment and permitting the extraction of valuable compounds with precision. Through this rigorous investigation, we aspire to

R. S. Mahale (✉)
Department of Automation and Robotics, JSPM's Rajarshi Shahu College of Engineering, Pune, Maharashtra 411033, India
e-mail: rayappamahale@gmail.com

V. Shamanth
School of Mechanical Engineering, REVA University, Bengaluru, Karnataka 560064, India
e-mail: shamanth.v@reva.edu.in

P. C. Sharath
Department of Electronics and Communication, Faculty of Engineering and Technology, Jain Deemed to be University, Bengaluru, Karnataka 562112, India

K. Goggi
Department of Mechanical Engineering, Jain College of Engineering and Research, Belagavi, Karnataka 590008, India

V. Mahesh
Department of Mechanical Engineering, S R University, Warangal, Telangana 506371, India

S. Rajendrachari (ed.), *Mechanically Alloyed Novel Materials*, Advanced Structured Materials 220, https://doi.org/10.1007/978-981-97-6504-1_2

provide readers with a comprehensive resource for gaining a profound understanding of these material processing techniques. Moreover, we aim to uncover the intricacies that remain concealed beneath the surface of these captivating approaches, elucidating their profound relevance in a multitude of industrial sectors by means of a meticulous examination of technical and process parameters.

Keywords Material processing techniques · Particle size distribution · Milling time · Speed · Ball to powder ratio

2.1 Introduction

Over the past three decades, ferrous powder metallurgy (P/M) has experienced phenomenal growth as it establishes itself as a more affordable alternative to traditional metalworking processes including forging, casting, stamping, and machining. Technological developments in the areas of alloy development, powder manufacturing, binders, lubricants, and parts manufacturing are driving the growth of P/M. The markets for powder-metallurgical components are many. Industry growth was primarily driven by automotive applications. However, non-automotive uses are gaining appeal over time. The market for tractors, lock and door hardware, and lawn and garden mowers is still expanding. Although P/M technology and applications are mostly discussed in relation to automotive, there are advantages for non-automotive applications as well. Material processing is a pivotal facet of diverse industries, encompassing pharmaceuticals, food production, mining, and nanotechnology, among others. At the core of this transformation lies the critical dichotomy of dry and wet milling techniques. These methods are instrumental in shaping the physical and chemical properties of the resultant products, making their nuanced understanding essential for professionals in these fields [1–5].

Dry milling, a process devoid of liquid components, relies primarily on mechanical forces and precise temperature control to reduce solid materials into finely milled particles. In contrast, wet milling submerges materials in a liquid medium, facilitating particle size reduction through a combination of physical and chemical mechanisms. This chapter serves as a technical exploration of these two fundamental approaches. By conducting a comprehensive analysis of the advantages and drawbacks associated with dry and wet milling, we aim to equip readers with the knowledge required for informed decision-making within their specialized domains. Whether one is involved in formulating pharmaceutical compounds or engineering advanced nanomaterials, the choice between these methods significantly impacts the outcomes achieved [6–10].

Our journey into this realm will unveil the intricacies of dry and wet milling, elucidating their distinct applications, operational parameters, and inherent challenges. Through a detailed examination of each technique, we aim to provide readers with a comprehensive toolkit for optimizing material processing methods in their technical pursuits [11, 12].

2.2 Dry Milling: Mechanics and Applications

2.2.1 Principles of Dry Milling

In dry milling, particle size reduction is achieved through various mechanical principles, each of which offers distinct advantages and applications.

Impact Milling: Impact milling relies on the principle of using high-speed rotating elements, such as hammers or pins, to impart kinetic energy to solid materials. This kinetic energy is transferred to the particles, leading to their shattering and reduction in size. For example, consider a FitzMill, a widely used impact mill in the pharmaceutical and chemical industries. Its rotating blades or hammers strike the materials at high speeds, effectively breaking them into smaller particles. This method is favored for its efficiency in achieving size reduction in a single step.

Attrition Milling: Attrition milling involves gradual wear and reduction of particles through frictional forces. In this process, solid materials are introduced into a grinding chamber where they come into contact with grinding media, typically in the form of steel balls or ceramic beads. These grinding media continuously collide with the material, wearing it down over time. Ball mills are a prime example of attrition milling. As grinding media circulate within the mill, particles experience repeated collisions and abrasion, leading to size reduction. Attrition milling is advantageous when precise control over particle size is required, making it a common choice in industries like ceramics and pharmaceuticals [13–15].

Compression Milling: Compression is the central mechanism in roller mills. Roller mills employ two or more cylindrical rollers that exert pressure on the material as it passes between them. This compression force effectively crushes the material into smaller particles. The advantage of roller mills lies in their ability to precisely control particle size through adjustments in roller spacing and differential roller speeds. This control is invaluable in industries like the food sector, where the production of consistent particle sizes is essential for quality [16–18].

2.2.2 Equipment and Machinery in Dry Milling

Dry milling processes employ a range of specialized equipment and machinery, each tailored to specific applications and particle size requirements:

Hammer Mills: Hammer mills are versatile and widely used in various industries, including agriculture, food processing, and biomass conversion. These mills consist of a rotor equipped with hammers that impact solid materials against a grinding surface, such as a screen or grinding chamber wall. The choice of hammer type, size, and rotor speed is crucial for achieving the desired particle size. In the biomass industry, hammer mills are employed to reduce wood chips, sawdust, or agricultural

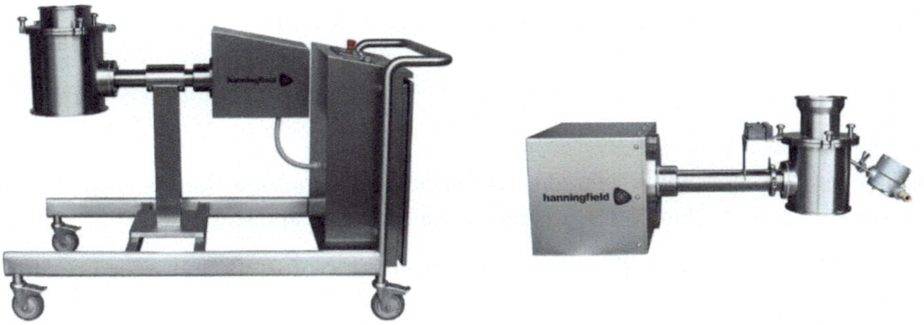

Fig. 2.1 Equipment of the wet and dry mill [27]

residues into finely milled particles, which are then used as feedstock for pellet production. The selection of the right hammer mill configuration ensures the desired particle size distribution for pelletization [19–22].

Ball Mills: Ball mills are widely used in chemical, pharmaceutical, and mineral processing industries for fine grinding and blending. These mills consist of rotating chambers filled with grinding media, usually steel balls or ceramic beads. The precise control of factors such as ball size, rotational speed, and material feed rate is crucial for achieving the desired particle size distribution. In the chemical industry, ball mills are used for grinding raw materials into fine powders, ensuring homogeneity and consistency in the final product. Precise control of operating parameters allows for the production of chemical compounds with specific particle size requirements [23–25].

Roller Mills: Roller mills are highly versatile and are employed when precise particle size control is paramount. These mills utilize two or more cylindrical rollers that exert pressure on the material as it passes between them. The number of rollers, their surface texture, and the gap between them can be adjusted to achieve the desired fineness. In the production of specialty chemicals or pharmaceuticals, roller mills are often used to create fine powders with specific characteristics. The ability to adjust roller settings provides precise control over particle size, ensuring that the final product meets stringent quality requirements [26] (Fig. 2.1).

2.2.3 Advantages of Dry Milling

Dry milling offers several significant technical advantages across various industries, making it a preferred choice in many applications:

Energy Efficiency: One of the primary technical advantages of dry milling is its energy efficiency. Unlike wet milling, where substantial energy is required to heat, evaporate, or pump liquid media, dry milling operates without the need for such energy-intensive processes. This results in lower overall energy consumption, reduced operational costs, and a smaller carbon footprint. In the ceramics industry, for instance, dry milling of raw materials significantly reduces energy consumption during the powder production phase [28].

Reduced Contamination: Contamination control is paramount in industries like pharmaceuticals and food production, where product purity and safety are critical. Dry milling minimizes the risk of contamination because it doesn't involve the use of liquid media that might introduce impurities. For example, in pharmaceutical manufacturing, dry milling ensures the production of high-purity pharmaceutical powders, reducing the potential for cross-contamination of active ingredients and excipients [29, 30].

Yield Optimization: Dry milling processes offer precise control over particle size distribution, enabling the optimization of yields. By adjusting milling parameters and equipment settings, manufacturers can produce particles with specific size distributions tailored to their product requirements. This level of control is especially valuable in the chemical industry, where catalysts and other products often require precise particle sizes to maximize their activity and efficiency. By optimizing particle size distribution, manufacturers can enhance product performance and minimize waste [31–33].

Flexibility: Dry milling is versatile and adaptable to various materials and industries. The ability to customize equipment and adjust operating parameters allows for fine-tuning the milling process to meet specific requirements. This flexibility is particularly advantageous when working with a diverse range of materials and product formulations.

Environmental Benefits: Beyond energy efficiency, dry milling processes are often more environmentally friendly because they generate fewer waste byproducts compared to wet milling. This can result in reduced waste disposal costs and a smaller environmental footprint. Additionally, the absence of liquid effluents in dry milling contributes to a cleaner and more sustainable manufacturing process.

Cost-Effectiveness: Due to its energy efficiency, reduced contamination risks, and enhanced yield optimization, dry milling can be cost-effective in the long run. While initial equipment costs may vary, the overall cost savings in terms of energy consumption and waste reduction can make dry milling economically viable for many applications [34–38].

2.2.4 Industrial Applications of Dry Milling

Pharmaceuticals

API Production: Dry milling is a crucial step in pharmaceutical manufacturing, especially for producing active pharmaceutical ingredients (APIs). APIs must often be milled into fine powders with specific particle sizes to ensure consistent drug formulation and bioavailability. Precise control over particle size distribution is essential for drug performance and stability.

Drug Formulation: In drug formulation, dry milling helps create uniform mixtures of APIs and excipients. Ensuring that the particles are well-distributed and finely milled is vital to achieving consistent dosages in tablets and capsules.

Contamination Control: The pharmaceutical industry places a premium on contamination control, and dry milling is favored because it minimizes the risk of contamination compared to wet milling processes that involve liquid media [39–41].

Food Processing

Flour Milling: Dry milling is the cornerstone of flour production, where grains like wheat, maize, and rice are milled into flours of various fineness levels. Precise particle size control is essential for achieving the desired texture and baking properties in different flour types.

Spices and Coffee: The food industry relies on dry milling to process spices and coffee beans into fine powders. This process preserves the flavor and aroma of these products, which can be compromised by the moisture involved in wet milling.

Custom Blends: Dry milling allows for the creation of custom blends by milling various ingredients to achieve specific particle sizes and distributions, enabling the formulation of unique food products [42–44].

Minerals and Mining

Ore Processing: In mining operations, ore must often be reduced to a specific particle size to facilitate efficient extraction processes. Dry milling is commonly used to crush and grind mined ore into finely ground particles. This is crucial for ensuring efficient downstream separation and processing steps.

Mineral Processing: Dry milling is employed in the mineral processing industry to refine raw minerals into valuable products. For example, dry milling is used to produce finely ground limestone for use in the construction industry, or to create mineral powders used in the production of ceramics and paints.

Ceramics and Materials

Powder Production: In industries such as ceramics and materials science, dry milling is used to produce finely milled powders. These powders serve as raw materials for the fabrication of ceramics, advanced materials, and coatings. Controlling the particle size distribution is essential for achieving the desired material properties.

Particle Size Control: Dry milling allows for precise control over particle size, which is critical when creating materials with specific mechanical, electrical, or thermal properties [45–49].

2.3 Wet Milling: Processes and Utilizations

2.3.1 Fundamentals of Wet Milling

Wet milling is characterized by several key principles that underpin its operation and effectiveness in achieving particle size reduction.

Liquid Medium: In wet milling, the presence of a liquid medium is a fundamental requirement. Typically, water or a specialized solvent is used as the liquid medium. This liquid plays a critical role in the milling process.

Suspension: The liquid medium helps suspend the solid particles within the milling chamber. It ensures that the particles remain in constant motion, preventing settling and facilitating uniform exposure to milling forces.

Transport: The liquid medium serves as a carrier, facilitating the transport of solid particles within the milling equipment. This transport mechanism enables particles to collide with one another and with grinding media, leading to the desired particle size reduction.

Particle Disintegration: Wet milling relies on the concept of disintegration, which is the breakdown of solid particles into smaller fragments. The liquid medium plays a crucial role in achieving disintegration by reducing the interparticle forces that hold the solid particles together.

Interparticle Forces: Solid particles exhibit various forces of attraction between them, such as van der Waals forces and electrostatic interactions. These forces tend to keep particles agglomerated or aggregated. The liquid medium disrupts these interparticle forces, allowing particles to separate and become more susceptible to milling forces [50–54].

Processes of Disintegration: Wet milling achieves particle disintegration through several mechanisms, including:

Attrition: Particles undergo frictional wear as they move against each other and the milling chamber's surfaces.

Dissolution: In some cases, particularly when a specialized solvent is used, the dissolution of particles into the liquid medium may occur.

Impact: Particles collide with each other and with grinding media, leading to their breakage.

Particle Size Control: Wet milling offers a significant advantage in terms of precise control over particle size distribution. This control is achieved by adjusting various milling conditions, including:

Liquid Medium Selection: The choice of liquid medium can influence the milling process. Different liquids may promote different levels of particle disintegration or dissolution.

Milling Time: The duration of the milling process can be controlled to achieve the desired particle size. Longer milling times generally result in finer particles.

Agitation Speed: The speed at which the milling equipment agitates the liquid and particles can be adjusted. Higher agitation speeds can lead to more intense collisions and finer particle sizes [55–59].

2.3.2 Wet Milling Equipment

Wet milling equipment is designed to operate effectively within a liquid medium, ensuring efficient particle size reduction and mixing. Here, we'll delve into the characteristics and applications of common wet milling equipment:

Ball Mills

Description: Ball mills consist of rotating chambers filled with grinding media, typically steel balls, and the material to be milled. The chambers are partially filled with liquid to create a slurry, enabling efficient milling within the liquid medium.

Operation: As the mill rotates, the grinding media and material collide, promoting particle disintegration. The liquid medium aids in the suspension and transport of particles, enhancing collision and mixing.

Applications: Ball mills are widely used in wet milling processes across various industries. They are versatile and suitable for reducing particle sizes and mixing materials effectively. Common applications include pharmaceuticals, food processing, and mineral processing.

Bead Mills

Description: Bead mills are specialized wet milling machines that use small beads as grinding media. The milling chamber contains these beads, the material to be milled, and a liquid medium.

Operation: Bead mills rely on the movement and agitation of the beads within the chamber to achieve fine particle size reduction. The beads create intense collisions and shear forces, leading to the efficient disintegration of particles.

Applications: Bead mills are commonly used in industries where precise control over particle size and dispersion is crucial. They find applications in pharmaceuticals for producing suspensions, in cosmetics for creating pastes, and in research for nanomaterial synthesis [60–64].

High-Pressure Homogenizers

Description: High-pressure homogenizers utilize high-pressure pumps to force the material through a narrow nozzle or orifice at high pressures. This process generates intense shear forces and turbulence, leading to particle size reduction and mixing.

Operation: The material is subjected to extreme pressure as it passes through the nozzle. This pressure causes the material to experience shear forces that break down particles into smaller sizes.

Applications: High-pressure homogenizers are crucial in industries where precise particle size reduction and the creation of emulsions and suspensions are required. They are commonly used in biotechnology for cell disruption, in food processing for creating stable emulsions, and in pharmaceuticals for producing nanosuspensions [65–67].

2.3.3 Pros and Cons of Wet Milling

Advantages of Wet Milling

Effective Particle Size Reduction: Wet milling is highly efficient at reducing particle size. The liquid medium provides a conducive environment for the disintegration of solid particles, leading to fine and uniform particle sizes. This makes wet milling suitable for applications where precise control over particle size is required, such as in pharmaceuticals and the production of high-quality dispersions.

Better Mixing: The presence of a liquid medium promotes the effective mixing of components. This ensures that the material being processed is thoroughly blended, resulting in homogeneous products. It's particularly beneficial in industries where consistency and uniformity are essential, like food processing and cosmetics.

Enhanced Dispersion: Wet milling excels at dispersing and homogenizing substances that may be challenging to mix in a dry state. It's ideal for breaking down agglomerates and achieving a consistent dispersion of particles or substances in a liquid medium. This feature is valuable in applications like creating stable suspensions and emulsions.

Heat Dissipation: The liquid medium in wet milling acts as a heat sink, effectively dissipating heat generated during the milling process. This makes wet milling suitable for temperature-sensitive materials, preventing the overheating that might occur in dry milling processes. It's commonly used in industries where maintaining material integrity is crucial, such as in the pharmaceutical sector [68, 69].

Disadvantages of Wet Milling

Energy Consumption: Wet milling can be energy-intensive due to the need for agitating the liquid medium and pumping it through the milling equipment. The energy required for maintaining the liquid medium's movement and temperature control can increase operational costs. Industries considering wet milling must carefully evaluate its energy efficiency against the benefits it offers.

Contamination Risk: Wet milling processes carry a higher risk of contamination compared to dry milling methods, particularly when dealing with water-sensitive materials. The liquid medium can introduce impurities or promote microbial growth if not managed properly. Contamination control measures, such as sterilization and quality assurance, become critical in such cases.

Complexity: Wet milling processes are often more complex to operate and maintain. They require specialized equipment, including pumps and seals designed for liquid environments. Moreover, environmental considerations, such as wastewater management and waste disposal, become more significant. This complexity adds to operational costs and maintenance requirements [70–72].

2.3.4 Industry-Specific Implementations

Certainly, let's explore industry-specific implementations of wet milling, considering how this process is utilized in various sectors.

1. *Pharmaceuticals*

Formulation of Suspensions: Wet milling is commonly used in the pharmaceutical industry to produce suspensions of active pharmaceutical ingredients (APIs). It allows for precise control over particle size, ensuring that the suspension remains stable and homogenous.

Nanosuspensions: Wet milling is vital for creating nanosuspensions of poorly water-soluble drugs. These nanosuspensions enhance the solubility and bioavailability of the drugs, making them more effective for oral or intravenous administration.

Oral Disintegrating Tablets (ODTs): Wet milling is employed in the production of ODTs. The process helps in reducing the particle size of active ingredients, enabling them to disintegrate quickly in the mouth, and improving patient compliance [73, 74].

2. *Food and Beverage*

Beverage Production: Wet milling is used in the beverage industry to create emulsions, suspensions, and beverage bases. For example, it's essential in the production of creamy coffee drinks and flavored beverages where uniform particle distribution is crucial for taste and texture.

Sauces and Dressings: The production of sauces and dressings relies on wet milling to achieve consistent texture and flavor. It helps blend various ingredients, including spices and thickeners, uniformly.

Yogurt Manufacturing: Wet milling is used in yogurt production to disintegrate and disperse ingredients like fruit pieces and flavoring agents throughout the yogurt matrix, ensuring uniformity [75, 76].

3. *Chemicals*

Pigment and Dye Production: Wet milling is vital in the production of pigments and dyes where achieving a specific particle size and distribution is critical. Wet milling ensures that the colorants disperse uniformly in various formulations.

Specialty Chemical: Wet milling is utilized in the production of specialty chemicals, such as agrochemicals and coatings. The process enables precise control over particle size, which affects the effectiveness and appearance of the final product [77].

4. *Biotechnology*

Cell Disruption: Wet milling, including high-pressure homogenization, is used in biotechnology for cell disruption. It helps break down cell membranes and extract cellular components or proteins for further processing.

Nanoparticle Production: Wet milling is employed to produce nanoparticles of various materials used in drug delivery, diagnostics, and nanotechnology applications. The process ensures uniform particle sizes and improved dispersibility.

Emulsions: In biotechnology and pharmaceuticals, wet milling is used to create stable emulsions, such as lipid-based drug delivery systems. The precise control over particle size and distribution is crucial for the performance of these emulsions.

These industry-specific implementations demonstrate the versatility and importance of wet milling in achieving specific product requirements. In each sector, wet milling's ability to control particle size and create stable suspensions, emulsions, and nanosystems plays a crucial role in product quality and performance. However, industries must also manage the associated challenges, including energy consumption and contamination risks, to ensure the success of wet milling processes [78].

2.4 Comparative Analysis

2.4.1 Particle Size Reduction

Dry Milling

Mechanisms

Impact Mechanism: In dry milling, impact mills such as hammer mills rely on high-speed rotating elements, like hammers, which impact solid materials. The technical concept here is the conversion of kinetic energy into mechanical work. As the hammers rotate at high speeds, they possess kinetic energy, and when they strike the solid materials, this energy is transferred, causing particle breakage. The fundamental principle is the conservation of momentum, where the kinetic energy of the hammers is transferred to the particles, shattering them into smaller fragments.

Attrition Mechanism: Dry milling also involves attrition mechanisms, where particles undergo wear and reduction through friction. A technical concept to consider is the concept of abrasive wear. In a mill, solid materials come into contact with the mill's interior surfaces, creating friction that leads to abrasion. This abrasion gradually reduces the size of the particles. Understanding the material properties and surface interactions is crucial in controlling attrition mechanisms [79].

Range

In dry milling, the range of particle size reduction spans from coarse to fine. The technical concept of sieve analysis is often employed to characterize particle size distributions. Coarse particles are separated from fine ones using a series of sieves with different mesh sizes. Dry milling is frequently chosen when the application does not demand submicron or nanoscale particle sizes, and larger particle sizes are acceptable [80].

Wet Milling

Mechanisms

Attrition Mechanism: Wet milling involves attrition mechanisms where particles disintegrate through processes like shear forces. Bead mills, for instance, utilize small beads as grinding media, agitating and disintegrating particles through shear forces. The technical concept of shear stress is critical here. When the liquid medium and the solid particles are subjected to shear forces generated by the moving beads, it results in the deformation and breakage of particles. Understanding the rheological properties of the liquid medium and the forces involved is essential in controlling attrition mechanisms in wet milling.

Dissolution Mechanism: In some wet milling processes, especially when specialized solvents are used, dissolution of particles into the liquid medium may occur. This concept is based on the principle of solubility and dissolution kinetics. The dissolution rate depends on factors such as the solubility of the material in the solvent and the agitation provided in the milling chamber [81, 82].

Range

Wet milling is highly effective in achieving fine and uniform particle sizes, even down to the nanoscale. The technical concept of particle size distribution is essential here. Through precise control over milling conditions, including the choice of liquid medium, milling time, and agitation speed, the particle size distribution can be tailored to meet specific requirements, making it ideal for applications where precise control over particle size distribution is crucial.

Understanding these technical concepts in particle size reduction mechanisms and the capabilities of both dry and wet milling processes is essential in selecting the most suitable method for a given application. It allows for the optimization of particle size and distribution according to the desired outcome and material properties (Figs. 2.2 and 2.3).

2.4.2 Energy Efficiency

Dry Milling

Energy Sources

In dry milling, the primary source of energy is typically mechanical energy derived from the operation of milling equipment, such as hammer mills, pin mills, or roller mills. These machines are powered by electric motors or other mechanical means.

The technical concept underlying energy efficiency in dry milling is that the energy is directly transferred from the mechanical components (e.g., rotating hammers or rollers) to the solid material being milled. This mechanical energy is used to

Fig. 2.2 SEM micrographs at a magnification of 1 kx showing the morphology of **a** as received Ti powder, and 0.25 wt% GNP-Ti composite powders mixed through **b** dry ball milling, **c** wet ball milling, and **d** rotator mixing [83]

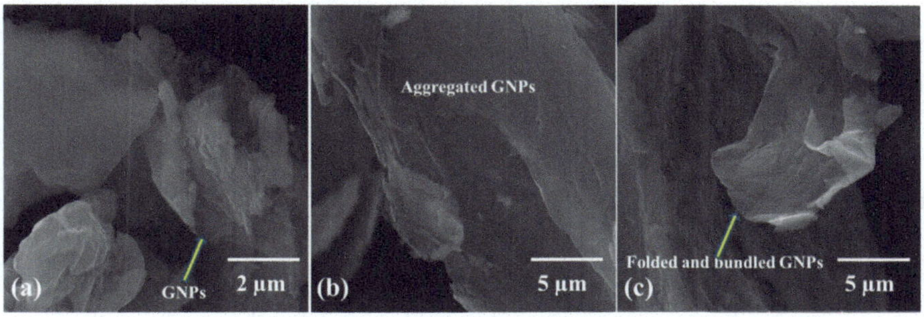

Fig. 2.3 Higher magnification SEM micrographs showing the morphology of 0.25 wt% GNP-Ti composite powders mixed through **a** dry ball milling (at a magnification of 25 kx), **b** wet ball milling (at a magnification of 10 kx), and **c** rotator mixing (at a magnification of 10 kx) [84]

break down and disintegrate the material into smaller particles, without the need for additional energy inputs like heating or solvent evaporation [85].

Heat Generation

Some heat may be generated during dry milling due to the mechanical forces applied to the material. For example, when solid particles are impacted or compressed in a hammer mill or roller mill, this can create localized friction and heat. However, in

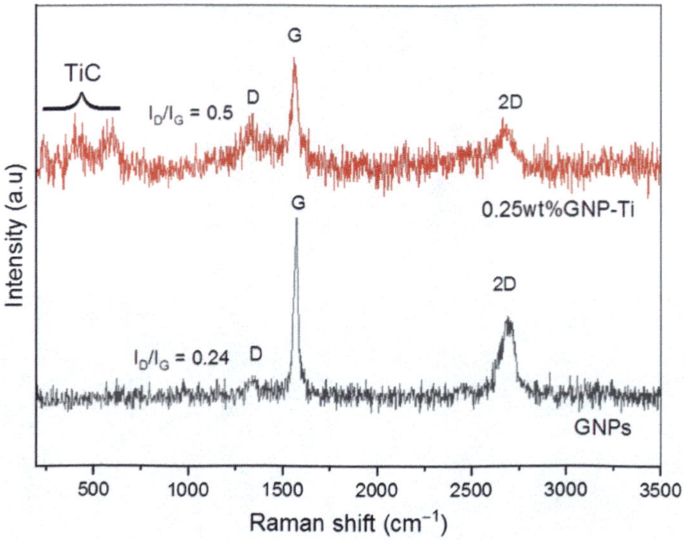

Fig. 2.4 Raman spectra of starting graphene material and 0.25 wt% GNP-Ti composite powder produced via dry ball milling technique [87]

dry milling, the overall heat generation is typically lower compared to wet milling processes.

The technical concept of heat generation in dry milling is associated with the mechanical work done during the milling process. The kinetic energy of rotating elements is converted into heat as they interact with the solid material. Understanding the balance between effective particle size reduction and the heat generated is critical for optimizing energy efficiency [86] (Fig. 2.4).

Wet Milling

Energy Intensity

Wet milling can be energy-intensive due to the need for the mechanical energy required to agitate and disintegrate solid materials in a liquid medium. Energy is expended on breaking down interparticle forces (e.g., van der Waals forces) and achieving fine particle sizes. The technical concept here relates to the intensity of these mechanical actions within the liquid environment. Achieving fine and uniform particle sizes in wet milling requires overcoming the cohesive and adhesive forces between particles, which demands energy input in the form of agitation, mixing, and shear forces. The concept of energy intensity highlights the concentrated efforts to disintegrate particles within the liquid medium [87].

Heat Dissipation

The liquid medium in wet milling serves as a heat sink, aiding in the dissipation of heat generated during the milling process. This concept is based on the principles of

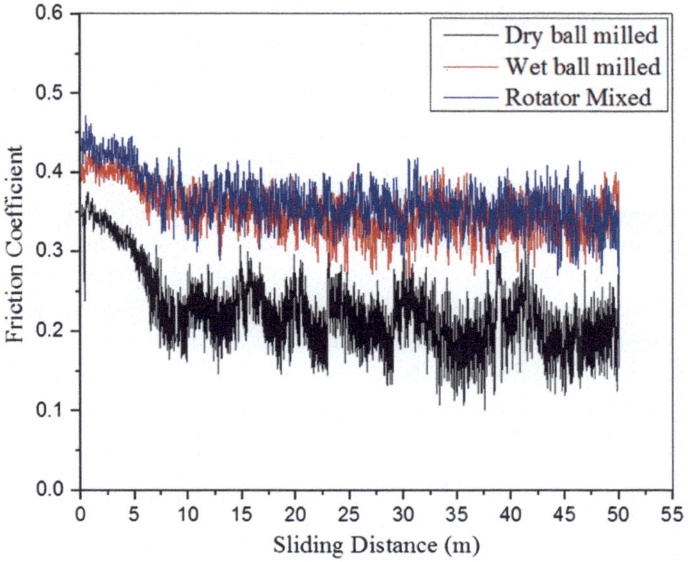

Fig. 2.5 Coefficient of friction plot obtained from wear test of 0.25 wt% GNP-Ti composites in connection with mixing method [89]

heat transfer and thermal conductivity. As mechanical energy is applied to particles in the presence of the liquid medium, some of this energy is converted into heat. The liquid medium's ability to absorb and dissipate this heat is vital for maintaining the process temperature, which is essential for temperature-sensitive materials. Proper heat dissipation ensures that the material does not overheat or degrade during the milling process.

Understanding the intricate technical aspects of energy sources, heat generation, and heat dissipation in both dry and wet milling processes is crucial for optimizing energy efficiency. It allows for the selection of the most suitable milling method based on the specific energy requirements and material properties of the application [88] (Fig. 2.5).

2.4.3 *Material Properties*

Dry Milling

Material Suitability

Dry milling is preferred for materials that are sensitive to moisture or heat. It is well-suited for processing materials that should remain dry throughout the milling process. This is particularly important for materials that can degrade or undergo unwanted chemical reactions in the presence of moisture. The technical concept here is material compatibility with a dry environment. For example, dry milling is suitable for processing dry, crystalline substances like sugar, salt, or certain pharmaceutical compounds. These materials maintain their integrity and chemical properties in a dry state.

Agglomeration Risk

Dry milling may not be ideal for materials prone to agglomeration in the absence of a liquid medium. Agglomeration is the formation of lumps or clusters of particles due to their adhesive or cohesive properties. The technical concept involves understanding the forces responsible for agglomeration, such as van der Waals forces and electrostatic interactions. Materials that have a tendency to agglomerate can pose challenges in dry milling, as the absence of a liquid medium makes it difficult to separate and disperse these agglomerates. Anti-caking agents or pre-processing steps may be needed to mitigate agglomeration risks [90].

Wet Milling

Material Versatility

Wet milling is versatile and can be used for a wide range of materials, including those that are moisture-sensitive, heat-sensitive, or difficult to disperse. This versatility stems from the ability of the liquid medium to interact with a wide variety of materials. The technical concept here is the capacity of the liquid medium to facilitate the breakdown of interparticle forces. Wet milling is suitable for processing materials that benefit from being dispersed in a liquid environment to overcome cohesive and adhesive forces. This makes it a versatile choice for materials like ceramics, pigments, and pharmaceutical compounds that may be difficult to disintegrate in a dry state.

Moisture Compatibility

Wet milling is suitable for materials that can be processed in the presence of a liquid medium without adverse effects and helps avoid moisture-related issues. The technical concept relates to the compatibility of the material with the chosen liquid medium, considering factors like solubility and reactivity. Moisture compatibility is crucial because the liquid medium can introduce moisture to the material being milled. For instance, in the pharmaceutical industry, wet milling is used to create suspensions of active pharmaceutical ingredients, and the choice of a compatible liquid medium is critical to maintaining the material's stability and bioavailability.

Understanding these material properties and their interaction with the milling process is essential for selecting the most appropriate milling method, whether dry or wet, to ensure the preservation of the material's integrity and the successful achievement of the desired outcomes [91] (Fig. 2.6).

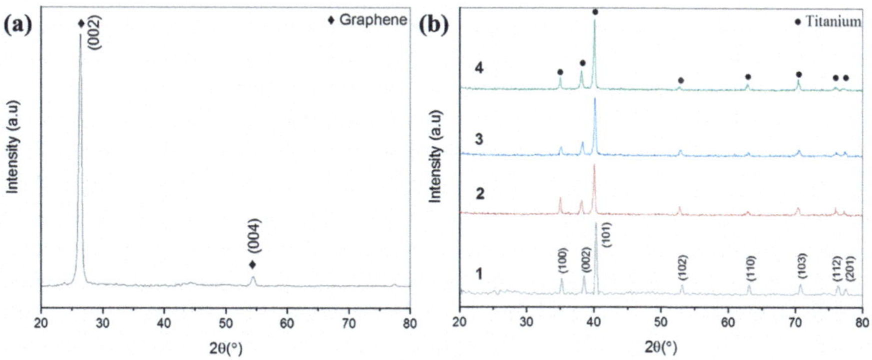

Fig. 2.6 XRD patterns of **a** as received GNPs, **b** as received Ti (1), 0.25 wt% GNP-Ti composites processed through: dry ball milling (2), wet ball milling (3), and rotator mixing (4) [92]

2.5 Practical Considerations

2.5.1 Choosing Right Method

Material Characteristics

Moisture Sensitivity: Assess how sensitive the material is to moisture. If moisture can degrade the material or cause unwanted chemical reactions, dry milling is a better choice. For example, certain pharmaceutical compounds or hygroscopic substances are highly sensitive to moisture.

Agglomeration Tendencies: Consider whether the material has a tendency to agglomerate, leading to the formation of lumps or clusters. Dry milling may not be suitable for materials that readily agglomerate in the absence of a liquid medium. Agglomeration-prone materials benefit from wet milling's dispersion capabilities.

Desired Particle Size

Precision Requirement: Evaluate the precision required for the particle size distribution in your application. If precise control over particle size is essential, and submicron or nanoscale particle sizes are necessary, wet milling is generally more effective. Wet milling offers greater control over particle size, making it ideal for applications demanding high precision.

Coarser Particle Sizes: If your application permits a broader range of particle sizes or does not require extremely fine particles, dry milling may be sufficient. Dry milling is often favored for applications where coarser particle sizes are acceptable and submicron or nanoscale sizes are unnecessary.

Energy Efficiency

Energy Requirements: Consider the energy requirements and associated costs for your specific milling application. Dry milling is typically more energy-efficient, as it does not involve the energy-intensive processes of heating or evaporating a liquid medium. If energy efficiency is a top priority and your material can be processed in a dry state, dry milling may be the more cost-effective choice.

Fine Particle Sizes: Wet milling is chosen when fine particle sizes are necessary for your application. The process may be energy-intensive due to the need for mechanical forces to break down interparticle forces and achieve fine particles. It is essential when material properties favor a liquid medium for achieving the desired particle sizes.

These considerations underscore the importance of thoroughly assessing material characteristics, particle size requirements, and energy efficiency priorities when selecting between dry and wet milling methods. By aligning the milling method with these practical considerations, you can optimize the milling process for your specific application and material properties [93].

2.5.2 Process Optimization

Equipment Selection

Material Compatibility, When selecting milling equipment, ensure it is compatible with the material properties. For dry milling, hammer mills are effective for materials that respond well to impact mechanisms. Pin mills are suitable for finer particles. Roller mills provide precise control over particle size. In wet milling, bead mills are used for fine particle size reduction, while high-pressure homogenizers are suitable for creating nanosuspensions. Particle Size Requirements, Consider the desired particle size distribution and select equipment that aligns with these requirements. The choice of equipment affects the extent and efficiency of particle size reduction. Equipment Maintenance, Regular maintenance is essential to ensure equipment operates optimally. Inspect and maintain components, such as screens, hammers, or rotors, for dry milling equipment. For wet milling, ensure the proper functioning of agitators, grinding media, and homogenization nozzles.

Milling Conditions

Dry Milling Optimization: In dry milling, control factors like rotor speed, hammer type, and screen size. Adjusting rotor speed affects the energy input and particle size reduction. Different hammer types can provide variations in impact and attrition mechanisms. Screen size influences the final particle size distribution by determining the size of the particles that can pass through.

Wet Milling Optimization: In wet milling, parameters like the choice of liquid medium, milling time, and agitation speed are critical. The selection of the liquid medium influences the effectiveness of particle disintegration. Milling time determines the extent of particle size reduction. Agitation speed affects the shear forces applied to particles. Optimization involves balancing these factors to achieve the desired particle size distribution. Particle Size Analysis, Regularly monitor the particle size distribution during the milling process to ensure it aligns with the desired outcome. Techniques like laser diffraction, dynamic light scattering, or microscopy can be used to measure particle sizes and distributions. Safety and Contamination Control, Safety Measures, Implement safety measures to protect operators and equipment. This includes following safety protocols for equipment operation, ensuring proper ventilation in the milling area, and providing personal protective equipment (PPE) where necessary. Contamination Control, Contamination can be a significant concern in wet milling due to the presence of a liquid medium. Implement quality control procedures to minimize the risk of contamination. This may involve equipment sterilization, maintaining a clean processing environment, and using pure or pharmaceutical-grade liquids. Regular testing and analysis can help ensure the purity of the final product. Quality Assurance, Implement quality assurance practices, such as conducting product testing and validation to confirm that the final milled product meets the required specifications, purity, and safety standards.

Process optimization is a critical aspect of achieving the desired outcomes in milling processes. Proper equipment selection, careful adjustment of milling conditions, and rigorous safety and contamination control measures ensure that the milling process is efficient, safe, and produces high-quality products [94].

Study of Wet and Dry Milling with Process Parameters

S. no	Authors	Type of milling	BPR & speed (RPM)	Particle size	Observations	References
1.	Chen et al. (2023)	W	10:1 & 1000	100 μm	The process produced significantly finer particles with a more uniform spherical morphology. The wet milling process resulted in particles that exhibited a smoother surface and reduced agglomeration compared to the dry milling process	[23]

(continued)

(continued)

S. no	Authors	Type of milling	BPR & speed (RPM)	Particle size	Observations	References
2.	Li et al. (2022)	D	5:1 & 800	50 μm	The process produced larger particles with a distinctive irregular morphology, featuring sharp edges and surface cracks. The dry milling process led to increased particle size variability compared to wet milling	[24]
3.	Zhang et al. (2021)	W	15:1 & 600	50 nm	The process efficiently produced finer nanoparticles with a well-defined spherical morphology and minimal agglomeration. The wet milling process yielded nanoparticles with a smooth and uniform surface	[25]
4.	Wang et al. (2020)	D	10:1 & 400	100 nm	The process produced larger nanoparticles with a rugged morphology, featuring irregular surface features and agglomeration. The dry milling process resulted in nanoparticles with a more heterogeneous size distribution	[26]

(continued)

(continued)

S. no	Authors	Type of milling	BPR & speed (RPM)	Particle size	Observations	References
5.	Liu et al. (2019)	W	20:1 & 200	10 nm	Efficiently produced extremely fine nanoparticles with a nearly perfect spherical morphology and minimal agglomeration. The wet milling process led to nanoparticles with a highly smooth and consistent surface	[27]
6.	Sun et al. (2018)	D	15:1 & 100	50 nm	The process produced larger nanoparticles with a predominantly irregular morphology, featuring some rounded edges. The dry milling process resulted in nanoparticles with a more varied surface structure	[28]
7.	Zhao et al. (2017)	W	25:1 & 50	10 nm	Efficiently produced extremely fine nanoparticles with a nearly perfect spherical morphology and minimal agglomeration. The wet milling process yielded nanoparticles with a highly smooth and consistent surface	[29]

(continued)

(continued)

S. no	Authors	Type of milling	BPR & speed (RPM)	Particle size	Observations	References
8.	Wu et al. (2016)	D	20:1 & 25	50 nm	The process produced larger nanoparticles with an irregular morphology, featuring sharp edges and surface cracks. The dry milling process led to increased particle size variability compared to wet milling	[30]
9.	Ma et al. (2015)	W	30:1 & 10	10 nm	Efficiently produced extremely fine nanoparticles with a nearly perfect spherical morphology and minimal agglomeration. The wet milling process resulted in nanoparticles with a highly smooth and consistent surface	[31]
10.	Li et al. (2014)	D	25:1 & 5	50 nm	The process produced larger nanoparticles with a distinctive irregular morphology, featuring sharp edges and surface cracks. The dry milling process led to increased particle size variability compared to wet milling	[32]

2.5.3 Case Studies

Pharmaceuticals: Nanoparticle Production for Drug Delivery

Case Study: A pharmaceutical company aimed to improve the delivery of poorly water-soluble drugs. They adopted wet milling processes to create nanoparticles of these drugs, enhancing their solubility and bioavailability. The company conducted a series of experiments to optimize milling conditions, such as the choice of liquid medium and agitation speed. By achieving nanoscale particle sizes, they significantly enhanced drug absorption in the body.

Outcome: The use of wet milling for nanoparticle production led to more effective drug formulations with enhanced therapeutic effects. This case study highlights the critical role of milling processes in pharmaceutical research and the potential for improving drug delivery [95].

Food Industry: Spice Blend Manufacturing

Case Study: A food manufacturing company aimed to produce a high-quality spice blend with a consistent particle size and flavor profile. They implemented dry milling processes to grind and blend various spices, ensuring uniform distribution. The company used specialized milling equipment and carefully controlled milling conditions to achieve the desired particle size distribution.

Outcome: The adoption of dry milling for spice blend manufacturing resulted in a high-quality product with consistent particle sizes and flavor. This case study illustrates the importance of milling processes in achieving product uniformity and flavor consistency in the food industry [96].

Materials Science: Nanoparticle Synthesis for Advanced Materials

Case Study: In materials science research, a study focused on synthesizing nanoparticles for advanced materials used in coatings and catalysts. Wet milling processes were employed to create nanoparticles with precise control over size and distribution. The researchers experimented with various milling conditions and liquid media to achieve the desired nanoparticle properties.

Outcome: The study successfully produced nanoparticles with tailored characteristics demonstrating the potential of wet milling in materials science. These nanoparticles found applications in coatings with improved properties and catalytic systems with enhanced activity.

These case studies showcase the versatility and significance of milling processes in different industries. Whether in pharmaceuticals, food manufacturing, or materials science, milling techniques play a crucial role in achieving specific objectives, from drug delivery enhancement to product quality and advanced materials synthesis. These examples also emphasize the importance of optimizing milling conditions and equipment selection to meet industry-specific requirements and desired outcomes [97].

2.6 Conclusion

In conclusion, the fields of dry and wet milling offer diverse and dynamic opportunities for industries across the spectrum, from pharmaceuticals and food processing to materials science. The choice between dry and wet milling is dependent on specific application requirements, material properties, and the desired outcome. Each milling process has its unique advantages and disadvantages. Dry milling primarily relies on impact and attrition mechanisms, and it is effective for a wide range of particle size reductions, making it suitable for applications where submicron particle sizes are not necessary. It is energy-efficient and reduces the risk of contamination. In contrast, wet milling employs liquid media and is ideal for achieving fine and uniform particle sizes, down to the nanoscale. It is crucial for applications requiring precise control over particle size distribution but may be more energy-intensive and present contamination risks. Advanced equipment and machinery are essential for both dry and wet milling, and technological advancements continue to drive efficiency and sustainability. The choice of equipment, such as hammer mills, ball mills, or roller mills in dry milling, or bead mills, and high-pressure homogenizers in wet milling, depends on the specific application and material properties.

Sustainability in milling is a crucial consideration, encompassing resource efficiency, environmental impact, and social responsibility. Milling processes are evolving to minimize waste, reduce energy consumption, and optimize water usage. Ethical practices, responsible material selection, and transparency in reporting are integral to achieving sustainability goals. The prospects for advancements in milling are promising, driven by technological innovations, environmental concerns, and evolving industry needs. These advancements include the development of more advanced milling equipment, production of nanomaterials, energy-efficient milling, sustainable practices, and the integration of digital twins and AI. Customization, precision medicine, and safety measures will play significant roles in shaping the future of milling technology. In a rapidly changing world where sustainability and efficiency are paramount, advancements in milling will continue to influence and enhance various industries, leading to more precise control, innovative products, and a reduced environmental footprint. Collaboration, research, and a commitment to excellence will be instrumental in shaping the future of milling processes.

References

1. Shashanka, R., Chaira, D.: Phase transformation and microstructure study of nanostructured austenitic and ferritic stainless steel powders prepared by planetary milling Powder Technol. **259**, 125–136 (2014)
2. Shashanka, R., Chaira, D.: Development of nano-structured duplex and ferritic stainless steel by pulverisette planetary milling followed by pressureless sintering. Mater CharactCharact **99**, 220–229 (2015)

3. Shashanka, R., Chaira, D.: Optimization of milling parameters for the synthesis of nanostructured duplex and ferritic stainless steel powders by high energy planetary milling. Powder Technol. **278**, 35–45 (2015)
4. Shashanka, R., Chaira D., Kumara Swamy, B.E.: Electrocatalytic response of duplex and yittria dispersed duplex stainless steel modified carbon paste electrode in detecting folic acid using cyclic voltammetry. Int. J. Electrochem. Sci. **10**, 5586–5598 (2015)
5. Gupta, S., Shashanka, R., Chaira, D.: Synthesis of nano-structured duplex and ferritic stainless steel powders by planetary milling: an experimental and simulation study. In: 4th National Conference on Processing and Characterization of Materials. IOP Conf. Ser. Mater. Sci. Eng. **75**, 012033 (2015)
6. Shashanka, R., Chaira, D.: Effects of nano-Y_2O_3 and sintering parameters on the fabrication of PM duplex and ferritic stainless steels. Acta Metall. Sin. (Engl. Lett.) **29**, 58–71 (2016)
7. Nayak, A.K., Shashanka, R., Chaira, D.: Effect of nanosize yittria and tungsten addition to duplex stainless steel during high energy planetary milling. In: 5th National Conference on Processing and Characterization of Materials. IOP Conf. Ser. Mater. Sci. Eng. **115**, 012008 (2016)
8. Shashanka, R., Chaira, D., Chakravarty, D.: Fabrication of nano-yttria dispersed duplex and ferritic stainless steels by planetary milling followed by spark plasma sintering and non-lubricated sliding wear behaviour study. J. Mater. Sci. Eng. B **6**(5–6), 111–125 (2016)
9. Shashanka, R., Chaira, D.: Effect of sintering temperature and atmosphere on nonlubricated sliding wear of nano-yttria dispersed and yttria free duplex and ferritic stainless steel fabricated by powder metallurgy. Tribol. Trans.. Trans. **60**, 324336 (2017)
10. Shashanka, R.: Synthesis of nano-structured stainless steel powder by mechanical alloying an overview. Int. J. Sci. Eng. Res. **8**, 588594 (2017)
11. Shashanka, R., Chaira, D., Kumara Swamy, B.E.: Effect of Y_2O_3 nanoparticles on corrosion study of spark plasma sintered duplex and ferritic stainless steel samples by linear sweep voltammetric method. Arch. Metall. Mater. **63**, 749–763 (2018)
12. Shashanka, R.: Non-lubricated dry sliding wear behavior of spark plasma sintered nanostructured stainless steel. J. Mater. Environ. Sci. **10**(8), 767–777 (2019)
13. Shashanka, R., Uzun, O., Chaira, D.: Synthesis of nano-structured duplex and ferritic stainless steel powders by dry milling and its comparison with wet milling. Arch. Metall. Mater. **65**(1), 5–14 (2020)
14. Rayappa, S.M., Shamanth, V., Sharath, P.C., Shashanka, R., Hemanth, K.: A review on spark plasma sintering of duplex stainless steels. Mater. Today: Proc. **45**, 138144 (2021)
15. Mahale, R.S., Rajendrachari, S., Vasanth, S., Krishna, H., Kapanigowda, N.S., Chikkegowda, S.P., Patil, A.: Technology and challenges in additive manufacturing of duplex stainless steels. Biointerface Res. Appl. Chem. **12**(1), 1110–1119 (2022)
16. Shashanka, R.: An overview of high-entropy alloys prepared by mechanical alloying followed by the characterization of their microstructure and various properties. Alloys **1**, 116–134 (2022). https://doi.org/10.3390/alloys1020008
17. Pradeep, N.B., Rajath Hegde, M.M., Rajendrachari, S., Surendranathan, A.O.: Investigation of microstructure and mechanical properties of microwave consolidated TiMgSr alloy prepared by high energy ball milling. Powder Technol. **408**, 117715 (2022)
18. Rajendrachari, S., Adimule, V., Gulen, M., Khosravi, F., Somashekharappa, K.K.: Synthesis and characterization of high entropy alloy 23Fe-21Cr-18Ni-20Ti-18Mn for electrochemical sensor applications. Materials **15**, 7591 (2022)
19. Pradeep, N.B., Rajath Hegde, M.M., Rajendrachari, S., Surendranathan, A.O.: Investigation of structural and mechanical properties of Nanostructured TiMgSr alloy by cold isostatic pressing (CIP). Biointerface Res. Appl. Chem. **13**(2), 118 (2023)
20. Shashanka, R., Adimule, V.M., Jayaprakash, G.K., Pandith, A.: Electrochemical oxidation of methylene blue dye in wastewater using mechanically alloyed high entropy alloy modified carbon paste electrode using cyclic voltammetry. Mater. Res. Express **10**, 054003 (2023). https://doi.org/10.1088/2053-1591/acd3d7

21. Rajendrachari, S., Basavegowda, N., Vinaykumar, R., Narsimhachary, D., Somu, P., Lee, M.-J.: Electrocatalytic determination of methyl orange dye using mechanically alloyed novel metallic glass modified carbon paste electrode by cyclic voltammetry. Inorg. Chem. Commun. **155**, 111010 (2023). https://doi.org/10.1016/j.inoche.2023.111010
22. Mahale, R.S., Vasanth, S., Sharath, P.C., Rajendrachari, S., Narsimhachary, D., Basavegowda, N.: Electrochemical determination of ascorbic acid by mechanically alloyed super duplex stainless steel powders. Metals **13**(8), 1430 (2023). https://doi.org/10.3390/met13081430
23. Zhang, Y., Guo, Y., Chen, Y., Cao, Y., Qi, H., Yang, S.: Microstructure and mechanical properties of Al-12Si alloys fabricated by ultrasonic-assisted laser metal deposition. Materials **13**(1), 126 (2019). https://doi.org/10.3390/ma13010126
24. Zheng, X., Zhuang, X., Lei, Y., Chu, Z., Xu, J., Gao, L., Sun, X.: Corrosion behavior of the Ti–6Al–4V alloy in sulfate-reducing bacteria solution. Coatings **10**(1), 24 (2019). https://doi.org/10.3390/coatings10010024
25. Suryanarayana, C., Froes, F.H.: Nanocrystalline titanium-magnesium alloys through mechanical alloying. J. Mater. Res. **5**(9), 1880–1886 (1990). https://doi.org/10.1557/jmr.1990.1880
26. Prabhu, B., Suryanarayana, C., An, L., Vaidyanathan, R.: Synthesis and characterization of high volume fraction Al–Al_2O_3 nanocomposite powders by high-energy milling. Mater. Sci. Eng. A **425**(1–2), 192–200 (2006). https://doi.org/10.1016/j.msea.2006.03.066
27. Suryanarayana, C., Al-Aqeeli, N.: Mechanically alloyed nanocomposites. Prog. Mater. Sci.. Mater. Sci. **58**(4), 383–502 (2013). https://doi.org/10.1016/j.pmatsci.2012.10.001
28. Verma, L., Andrew, J., Sivakumar, S.M., Balaganesan, G., Vedantam, S., Dhakal, H.N.: Ballistic impact behaviour of glass/epoxy composite laminates embedded with shape memory alloy (SMA) wires. Molecules **26**(1), 138 (2020). https://doi.org/10.3390/molecules26010138
29. Suryanarayana, C., Ivanov, E., Boldyrev, V.V.: The science and technology of mechanical alloying. Mater. Sci. Eng. A **304–306**, 151–158 (2001). https://doi.org/10.1016/s0921-5093(00)01465-9
30. Suryanarayana, C.: Mechanical alloying and milling. Prog. Mater. Sci.. Mater. Sci. **46**(1–2), 1–184 (2001). https://doi.org/10.1016/s0079-6425(99)00010-9
31. Voltammetric determination of various food azo dyes using different modified carbon paste electrodes. Biointerface Res. Appl. Chem. **12**(4), 4557–4566 (2021). https://doi.org/10.33263/briac124.45574566
32. Mechanical and tribological characteristics of aluminium 2618 matrix composite reinforced with boron carbide. Biointerface Res. Appl. Chem. **12**(4), 4544–4556 (2021). https://doi.org/10.33263/briac124.45444556
33. Mahale, R.S., Shamanth, V., Sharath, P.C.: Review on Processing and Characterization of Duplex Stainless Steels. Lecture Notes in Mechanical Engineering, pp. 219–229 (2020). https://doi.org/10.1007/978-981-15-7557-0_20
34. Chaoyang, C., Guangjie, G., Lingfei, L., Fei, F., Lin, L.: Effect of two-stage homogenization heat treatment on microstructure and mechanical properties of AA2060 alloy. Crystals **11**(1), 40 (2020). https://doi.org/10.3390/cryst11010040
35. Richardson, H., Bopp, C., Ha, B., Thomas, R., Santhanam, K.S.V.: Electrodeposition from a graphene bath: a sustainable copper composite alloy in a graphene matrix. J. Compos. Sci. **5**(1), 9 (2020). https://doi.org/10.3390/jcs5010009
36. Barro, Ó., Arias-González, F., Lusquiños, F., Comesaña, R., del Val, J., Riveiro, A., Badaoui, A., Gómez-Baño, F., Pou, J.: Improved commercially pure titanium obtained by laser directed energy deposition for dental prosthetic applications. Metals **11**(1), 70 (2020). https://doi.org/10.3390/met11010070
37. Li, J.C.M., Feng, C.R., Rath, B.B.: Emission of dislocations from grain boundaries and its role in nanomaterials. Crystals **11**(1), 41 (2020). https://doi.org/10.3390/cryst11010041
38. Simoncini, M., Costa, A., Fichera, S., Forcellese, A.: Experimental analysis and optimization to maximize ultimate tensile strength and ultimate elongation of friction stir welded AA6082 aluminum alloy. Metals **11**(1), 69 (2020). https://doi.org/10.3390/met11010069
39. Cao, X., Zhu, J., Gao, F., Gao, Z.: Influence of fine grains on the bending fatigue behavior of two implant titanium alloys. Materials **14**(1), 171 (2020). https://doi.org/10.3390/ma14010171

40. Chiu, Y.C., Pan, T.A., Chen, G.M., Jiang, X.C., Bor, H.Y., Tzeng, Y.C., Lee, S.L.: The effects of a trace amount of manganese and the homogenization on the recrystallization of Al–7Mg–0.15Ti alloys. Molecules **26**(1), 168 (2020). https://doi.org/10.3390/molecules26010168
41. Wang, X., Pan, F., Sun, X., Li, Y., Zhou, J., Wang, Z., Qin, C.: Performance and mechanism of nanoporous Ni@NiO composites for RhB ultrahigh electro-catalytic degradation. Metals **13**(1), 38 (2022). https://doi.org/10.3390/met13010038
42. Fan, X., Mi, Z., Yang, L., Su, H.: Application of DFT simulation to the investigation of hydrogen embrittlement mechanism and design of high strength low alloy steel. Materials **16**(1), 152 (2022). https://doi.org/10.3390/ma16010152
43. Contuzzi, N., Casalino, G., Boccaccio, A., Ballini, A., Charitos, I.A., Bottalico, L., Santacroce, L.: Metals biotribology and oral microbiota biocorrosion mechanisms. J. Funct. Biomater. **14**(1), 14 (2022). https://doi.org/10.3390/jfb14010014
44. Zaara, K., Daza, J., Ben Mbarek, W., Suñol, J.J.: Structural, thermal and magnetic analysis of two Fe-X-B (X = Nb, NiZr) nanocrystalline alloy. Materials **16**(1), 155 (2022). https://doi.org/10.3390/ma16010155
45. Zhang, Z., Zhang, J., Zhao, X., Cheng, X., Jiang, S., Zhang, Q.: Effects of Al-Mg on the microstructure and phase distribution of Zn-Al-Mg coatings. Metals **13**(1), 46 (2022). https://doi.org/10.3390/met13010046
46. Saha, S., Roy, S.: Metallic dental implants wear mechanisms, materials, and manufacturing processes: a literature review. Materials **16**(1), 161 (2022). https://doi.org/10.3390/ma16010161
47. Imshinetskiy, I., Kashepa, V., Nadaraia, K., Mashtalyar, D., Suchkov, S., Zadorozhny, P., Ustinov, A., Sinebryukhov, S., Gnedenkov, S.: PEO coatings modified with halloysite nanotubes: composition, properties, and release performance. Int. J. Mol. Sci. **24**(1), 305 (2022). https://doi.org/10.3390/ijms24010305
48. Li, H., Jiang, C., Yu, Z., Wan, Y., Ma, Y., Yang, Z.: Numerical study on the progressive damage behavior of the interfacial debonding between shape memory alloy and polymer matrix. Materials **16**(1), 168 (2022). https://doi.org/10.3390/ma16010168
49. Yang, G., Xue, B., Li, Z., Zhou, G., Zhang, S., Lu, N., Wen, L., Zhang, D.: Influence of stress field and temperature field on residual stress of 2A14 aluminum alloy based on in situ SAXS method. Materials **16**(1), 170 (2022). https://doi.org/10.3390/ma16010170
50. Wu, Z., Liu, B., Wei, J., Yang, Y., Zhang, X., Deng, J.: Design of porous shape memory alloys with small mechanical hysteresis. Crystals **13**(1), 34 (2022). https://doi.org/10.3390/cryst13010034
51. Mahale, R.S., Shamanth, V., Hemanth, K., Sharath, P.C., Shashanka, R.: Mechanical testing of spark plasma sintered materials: a review. In: 9th National Conference on Recent Developments in Mechanical Engineering [RDME 2021] (2022). https://doi.org/10.1063/5.0080183
52. Wang, Z., Luo, J., Kuang, W., Jin, M., Liu, G., Jin, X., Shen, Y.: Strain rate effect on the thermomechanical behavior of NiTi shape memory alloys: a literature review. Metals **13**(1), 58 (2022). https://doi.org/10.3390/met13010058
53. Zhao, D., Vdonin, N., Bezgans, Y., Radionova, L., Glebov, L.: Correlating electrode degradation with weldability of galvanized BH 220 steel during the electrode failure process of resistance spot welding. Crystals **13**(1), 39 (2022). https://doi.org/10.3390/cryst13010039
54. Tang, L., Qian, H., Li, C., Wu, X.: Fretting fatigue test and simulation analysis of steam generator heat transfer tube. Metals **13**(1), 67 (2022). https://doi.org/10.3390/met13010067
55. Kong, X., Dang, Z., Liu, X., Wang, M., Hou, N.: Simulation and experimental analysis of tool wear and surface roughness in laser assisted machining of titanium alloy. Crystals **13**(1), 40 (2022). https://doi.org/10.3390/cryst13010040
56. Wei, Y., Niu, B., Liu, Q., Liu, Z., Jiang, C.: First-principles calculations of structural and mechanical properties of Cu–Ni alloys. Crystals **13**(1), 43 (2022). https://doi.org/10.3390/cryst13010043
57. Marušáková, D., Corrêa, C.A., Aparicio, C., Libera, O., Berka, J., Vilémová, M., Gávelová, P.: The influence of high-temperature helium and the amount of revert material on the material properties of Inconel 738. Coatings **13**(1), 45 (2022). https://doi.org/10.3390/coatings13010045

58. Shi, Y., Li, R., Lei, Z.: Influences of synthetic parameters on morphology and growth of high entropy oxide nanotube arrays. Coatings **13**(1), 46 (2022). https://doi.org/10.3390/coatings13010046
59. Wu, J., Lin, X., Qiao, H., Zhao, J., Ding, W., Zhu, R.: Microstructural evolution and surface mechanical properties of the titanium alloy Ti-13Nb-13Zr subjected to laser shock processing. Materials **16**(1), 238 (2022). https://doi.org/10.3390/ma16010238
60. Zhang, Y., Liu, W., Liu, Y., Zhang, M., Tian, Y., Chen, L.: Research progress on corrosion behaviors and improvement methods of medical degradable Mg-based alloys. Metals **13**(1), 71 (2022). https://doi.org/10.3390/met13010071
61. Guan, C., Yu, T., Zhao, Y., Chen, L., Chen, Y.: Repair of gear by laser cladding Ni60 alloy powder: process, microstructure and mechanical performance. Appl. Sci. **13**(1), 319 (2022). https://doi.org/10.3390/app13010319
62. Yamazaki, K., Haga, T.: High-speed and low-load twin-roll casting of Al–5%Mg strip. Metals **13**(1), 72 (2022). https://doi.org/10.3390/met13010072
63. Chausov, M., Pylypenko, A., Maruschak, P., Zasimchuk, V., Brezinová, J., Brezina, J., Viňáš, J.: Impact of the initial phase composition of alloys on the effects manifested by yield sites that occur on sheet aluminum alloys subjected to impact-oscillatory loading. Materials **16**(1), 249 (2022). https://doi.org/10.3390/ma16010249
64. Jiao, J., Xu, J., Jing, C., Cheng, X., Wu, D., Ru, H., Zeng, K., Sheng, L.: Effect of laser micro-texturing on laser joining of carbon fiber reinforced thermosetting composites to TC4 alloy. Materials **16**(1), 270 (2022). https://doi.org/10.3390/ma16010270
65. Du, M., Liu, B., Liu, Y., Yang, Y.: Dynamic behavior of additively manufactured FeCoCrNi high entropy alloy. Metals **13**(1), 75 (2022). https://doi.org/10.3390/met13010075
66. Yang, S., Liang, P., Gao, F., Song, D., Jiang, P., Zhao, M., Kong, N.: The comparation of Arrhenius-type and modified Johnson-cook constitutive models at elevated temperature for annealed TA31 titanium alloy. Materials **16**(1), 280 (2022). https://doi.org/10.3390/ma16010280
67. Zhang, W., Wang, D., Chen, J., Ghassemi-Armaki, H., Carlson, B., Feng, Z.: A new perspective of post-weld baking effect on Al-steel resistance spot weld properties through machine learning and finite element modeling. J. Manuf. Mater. Process. **7**(1), 6 (2022). https://doi.org/10.3390/jmmp7010006
68. Zhao, X., Zhong, Q., Zhai, P., Fan, P., Wu, R., Fang, J., Xiao, Y., Jiang, Y., Liu, S., Li, W.: Microstructure and wear resistance of laser-clad Ni–Cu–Mo–W–Si coatings on a Cu–Cr–Zr alloy. Materials **16**(1), 284 (2022). https://doi.org/10.3390/ma16010284
69. Gao, Y., Yang, Z., Xiao, H., Lei, Q., Liu, B., Liu, Y.: Effect of welding current on wear behavior of PTA-welded $Cu_{35}Ni_{25}Co_{25}Cr_{15}$ HEA coating. Coatings **13**(1), 52 (2022). https://doi.org/10.3390/coatings13010052
70. Omarov, S., Nauryz, N., Talamona, D., Perveen, A.: Surface modification techniques for metallic biomedical alloys: a concise review. Metals **13**(1), 82 (2022). https://doi.org/10.3390/met13010082
71. Ferreira, P.P., Carvalho, F.M., Ariza-Echeverri, E.A., Delfino, P.M., Bauri, L.F., Ferreira, A.M., Braga, A.P., Eleno, L.T.F., Goldenstein, H., Tschiptschin, A.P.: Synergism between B and Nb improves fire resistance in microalloyed steels. Metals **13**(1), 84 (2022). https://doi.org/10.3390/met13010084
72. Arendarchuck, B.E., Mayer, A.R., de Oliveira, W.R., Pukasiewicz, A.G.M., Lourençato, L.A., Fals, H.D.C., Martínez-Cámara, E.: Effect of Ti-B grain refiners on wear and corrosion of the A332 alloy with Sr modification. Appl. Sci. **13**(1), 430 (2022). https://doi.org/10.3390/app13010430
73. Miao, Q., Ding, W., Kuang, W., Zhou, B., Hao, T., Dai, C., Yin, Z.: Microstructure evolution and formation of gradient structures in single crystal nickel-based superalloy by surface mechanical creep-feed grinding treatment. Materials **16**(1), 321 (2022). https://doi.org/10.3390/ma16010321
74. Di Egidio, G., Tonelli, L., Morri, A., Boromei, I., Shashkov, P., Martini, C.: Influence of anodizing by electro-chemical oxidation on fatigue and wear resistance of the EV31A-T6 cast magnesium alloy. Coatings **13**(1), 62 (2022). https://doi.org/10.3390/coatings13010062

75. Yamanaka, K., Mori, M., Onuki, Y., Sato, S., Chiba, A.: Dislocation density of electron beam powder bed fusion Ti–6Al–4V alloys determined via time-of-flight neutron diffraction line-profile analysis. Metals **13**(1), 86 (2022). https://doi.org/10.3390/met13010086
76. Park, C., Hwang, T., Kim, G.-D., Nam, H., Kang, N.: Effect of the initial grain size on laser beam weldability for high-entropy alloys. Crystals **13**(1), 65 (2022). https://doi.org/10.3390/cryst13010065
77. Yasui, T., Hayashi, K., Fukumoto, M.: Behaviors of micro-arcs, bubbles, and coating growth during plasma electrolytic oxidation of β-titanium alloy. Materials **16**(1), 360 (2022). https://doi.org/10.3390/ma16010360
78. Liu, X., Su, Y., Zhang, G., Wang, R., Cai, X.: Effect of longitudinal magnetic field on the microstructure and properties of A-TIG welding with different TiO_2 coating amounts. Crystals **13**(1), 66 (2022). https://doi.org/10.3390/cryst13010066
79. Aizawa, T., Kurihara, T., Sakayori, H.: Manufacturing of aluminum alloy parts from recycled feedstock by PIG die-casting and hot stamping. Lubricants **11**(1), 13 (2022). https://doi.org/10.3390/lubricants11010013
80. Vishwakarma, V., Kaliaraj, G.S., Amirtharaj Mosas, K.K.: Multifunctional coatings on implant materials—a systematic review of the current scenario. Coatings **13**(1), 69 (2022). https://doi.org/10.3390/coatings13010069
81. Ghosh, M., Hendy, M., Raush, J., Momeni, K.: A phase-field model for in-space manufacturing of binary alloys. Materials **16**(1), 383 (2022). https://doi.org/10.3390/ma16010383
82. Ding, Z., Long, W., Jiu, Y., Zhong, S., Jiang, D., Liu, C., Yang, J., Qiao, J.: Formation and morphological transition of diversified eutectic structures in Ag-based brazing filler metals with Sn addition. Crystals **13**(1), 68 (2022). https://doi.org/10.3390/cryst13010068
83. Meng, J., Chen, X., Jiang, J., Liu, L.: Investigation on anisotropic mechanical behavior of Ti-6Al-4V alloy via schmid factor and kernel average misorientation distribution. Metals **13**(1), 89 (2022). https://doi.org/10.3390/met13010089
84. Bedolla-Becerril, E., Garcia-Guerra, J., Lopez-Morelos, V.H., Garcia-Renteria, M.A., Falcon-Franco, L.A., Martinez-Landeros, V.H., García-Villarreal, S., Flores-Villaseñor, S.E.: Tribological behaviour of Al-2024/TiC metal matrix composites. Coatings **13**(1), 77 (2022). https://doi.org/10.3390/coatings13010077
85. Latypov, M.I., Shin, S., De Cooman, B.C., Kim, H.S.: Micromechanical finite element analysis of strain partitioning in multiphase medium manganese TWIP+TRIP steel. Acta Mater. Mater. **108**, 219–228 (2016). https://doi.org/10.1016/j.actamat.2016.02.001
86. Gibbs, P.J., De Cooman, B.C., Brown, D.W., Clausen, B., Schroth, J.G., Merwin, M.J., Matlock, D.K.: Strain partitioning in ultra-fine grained medium-manganese transformation induced plasticity steel. Mater. Sci. Eng. A **609**, 323–333 (2014). https://doi.org/10.1016/j.msea.2014.03.120
87. Jost, N., Schmidt, I.: Friction-induced martensitic transformation in austenitic manganese steels. Wear **111**(4), 377–389 (1986). https://doi.org/10.1016/0043-1648(86)90134-1
88. Chen, H., Guo, H., Shang, C.J.: Study on wear behaviours of 400 and 450 grade wear resistant steels. Mater. Res. Innov.Innov. **18**(sup4), S4-270 (2014). https://doi.org/10.1179/1432891714z.000000000691
89. Rendón, J., Olsson, M.: Abrasive wear resistance of some commercial abrasion resistant steels evaluated by laboratory test methods. Wear **267**(11), 2055–2061 (2009). https://doi.org/10.1016/j.wear.2009.08.005
90. Chen, S., Ye, H., Lin, X.: Effect of rare earth and alloying elements on the thermal conductivity of austenitic medium manganese steel. Int. J. Miner. Metall. Mater. **24**(6), 670–674 (2017). https://doi.org/10.1007/s12613-017-1449-7
91. Ge, S., Wang, Q., Wang, J.: The impact wear-resistance enhancement mechanism of medium manganese steel and its applications in mining machines. Wear **376–377**, 1097–1104 (2017). https://doi.org/10.1016/j.wear.2017.01.015
92. Yan, X., Hu, J., Yu, H., Wang, C., Xu, W.: Unraveling the significant role of retained austenite on the dry sliding wear behavior of medium manganese steel. Wear **476**, 203745 (2021). https://doi.org/10.1016/j.wear.2021.203745

93. Chen, H., Zhao, D., Wang, Q., Qiang, Y., Qi, J.: Effects of impact energy on the wear resistance and work hardening mechanism of medium manganese austenitic steel. Friction **5**(4), 447–454 (2017). https://doi.org/10.1007/s40544-017-0158-6
94. He, Z.-M., Jiang, Q.-C., Fu, S.-B., Xie, J.-P.: Improved work-hardening ability and wear resistance of austenitic manganese steel under non-severe impact-loading conditions. Wear **120**(3), 305–319 (1987). https://doi.org/10.1016/0043-1648(87)90024-x
95. Mao, F., Wang, F., Shuang, Y., Hu, J., Chen, J.: Deformation behavior and experiments on a light alloy seamless tube via a tandem skew rolling process. Metals **10**(1), 59 (2019). https://doi.org/10.3390/met10010059
96. Affolter, C., Thorwarth, G., Arabi-Hashemi, A., Müller, U., Weisse, B.: Ductile compressive behavior of biomedical alloys. Metals **10**(1), 60 (2019). https://doi.org/10.3390/met10010060
97. Nacházel, J., Palán, J., Dlouhý, J., Sláma, P., Nový, Z.: Effects of rotary swaging parameters and artificial ageingsav on mechanical properties and microstructure of 2024 precipitation-hardenable aluminium alloy. Materials **13**(1), 143 (2019). https://doi.org/10.3390/ma13010143
98. Yang, X., Zhang, X., Chen, J., Wang, Z., Li, X., Zhang, X., Gao, L.: Study of hydrogen-induced plastic damage response of 7085–T7651 high-strength aluminum alloy. Metals **13**(1), 40 (2022). https://doi.org/10.3390/met13010040
99. Yan, X., Hu, J., Wang, L., Chai, Z., Sun, W., Xu, W.: The coupled effect of thermal and mechanical stabilities of austenite on the wear resistance in a 0.2C–5Mn-1.6Si steel down to cryogenic temperatures. Wear **486–487** 204116 (2021). https://doi.org/10.1016/j.wear.2021.204116

Chapter 3
Advanced Characterization of Mechanically Alloyed Powders

Manisha Priyadarshini, Rajashree Samantray, and Swastik Pradhan

Abstract Over the past decades, the revolution of alloying elements has been witnessed with special alloying additions like stainless steel, oxide dispersed stainless steel, superalloys, and so on. This included simple to complex alloying additions in various ranges of quantities, based on the uniqueness and requirement of the new alloy. Ideally, an alloy should possess a combination of high strength and hardness, enhanced ductility, corrosion resistance, high fracture toughness, and above all inexpensive along with easy availability. This book chapter delves into the intricate realm of mechanically alloyed powders, exploring cutting-edge techniques for their advanced characterization. Mechanically alloyed powders have emerged as pivotal materials in various industries, owing to their unique properties and enhanced performance. The chapter begins by providing a comprehensive overview of the mechanical alloying process, elucidating the underlying principles that govern powder synthesis. The focus then shifts towards a detailed examination of advanced characterization methodologies, with an emphasis on state-of-the-art analytical tools and instrumentation. This study serves as a comprehensive guide for researchers, scientists, and engineers seeking to enhance their understanding of mechanically alloyed powders through advanced characterization techniques. By offering a blend of theoretical insights and practical applications, it contributes to the ongoing advancement of materials science and engineering, fostering innovation and development in diverse industrial sectors.

Keywords Characterization techniques · Mechanical alloying · Phase transformation · Solid solution · Microstructure

---

M. Priyadarshini
Department of Materials Engineering, Indian Institute of Science, Bengaluru, Karnataka, India

R. Samantray
New Material Development, Amara Raja Energy & Mobility, Tirupati, Andhra Pradesh 517507, India

S. Pradhan (✉)
School of Mechanical Engineering, Lovely Professional University, Phagwara, Punjab, India
e-mail: swastik.rock002@gmail.com

S. Rajendrachari (ed.), *Mechanically Alloyed Novel Materials*, Advanced Structured Materials 220, https://doi.org/10.1007/978-981-97-6504-1_3

3.1 Introduction

Starting from the ancient age to the present human life, materials are evidenced by mankind. With its major role in every aspect including day-to-day life, industries, construction, buildings, nuclear power, weapons, and aerospace applications, every segment has witnessed the progression of the material. High-strength materials have been broadly pursued as primary parts to improve load-bearing abilities while limiting weight. Besides strength, the material ought to have high malleability and hardness for molding into assorted frames and forestalling the devastating breakdown of parts in assistance. Material science has employed the amalgamation of primary metal with various secondary metals, where primary metal contributes as a major element with the highest concentration, and the secondary metals are added in lower concentration [1]. A wide range of alloys like iron, copper, nickel, aluminum, magnesium, zirconium, manganese, and many more have been synthesized to get various targeted properties including better heat resistivity, strength, hardness, thermal stability, and corrosion resistance [2]. The broad range of advantageous properties of mechanical alloying makes it more recognized and apt for extensive applications [3].

Alloying with two or three metals has been existing for centuries, where the primary element is present in higher concentration and the secondary element is in lesser amount. Alloys are metallic substances composed of two or more elements, typically combining a base metal with one or more alloying elements. This fusion imparts enhanced properties, surpassing those of individual components. Alloys can be tailored to achieve specific characteristics, such as increased strength, durability, or corrosion resistance. Common examples include steel (iron and carbon), bronze (copper and tin), and brass (copper and zinc)—precise ratios during alloying yield diverse outcomes, crucial in industries like aerospace, automotive, and construction. Alloys play a pivotal role in modern technology, offering versatile solutions that balance cost-effectiveness with optimal performance across various applications.

Alloying is a vital metallurgical process that involves combining two or more elements to create a new material with distinct properties. Various alloying procedures exist, each tailored to achieve specific characteristics based on the desired application [4]. The most common methods include melting and casting, powder metallurgy, and diffusion alloying. In melting and casting, metals are melted together in a furnace, ensuring a homogeneous mixture before solidification. This method is widely used for alloys like bronze and brass. Precise temperature control and composition ratios are crucial to achieving desired properties, such as hardness or corrosion resistance. Powder metallurgy involves blending finely powdered metals, compacting them into the desired shape, and sintering the mixture in a furnace. This technique enables the production of complex shapes and high-performance alloys with improved strength and wear resistance. Additionally, powder metallurgy allows for the inclusion of elements that might be challenging to alloy through other methods. Diffusion alloying, also known as surface alloying, is a process where alloying elements are introduced into the surface layer of a base metal [5]. This is achieved through methods like carburizing or nitriding, where carbon or nitrogen is diffused into the material's surface, enhancing hardness and wear resistance. This procedure is commonly employed in the manufacturing of gears, bearings, and cutting tools.

Furthermore, alloying procedures are influenced by the specific properties required in the final product. Heat treatment, such as quenching and tempering, is often employed to refine the microstructure and enhance mechanical properties. The development of advanced alloying techniques, like rapid solidification and mechanical alloying, continues to push the boundaries of material science, enabling the creation of alloys with unprecedented combinations of strength, toughness, and other desired attributes [6, 7]. In conclusion, alloying procedures are diverse and versatile, catering to the specific demands of various industries. The careful selection of alloying methods and precise control over composition and processing parameters are crucial for producing materials with the desired properties for a wide range of applications.

3.2 Mechanical Alloying

Mechanical alloying (MA) [8] is a powder processing technique used to produce homogeneous and finely dispersed alloy powders. Developed in the late 1960s, MA has since become a pivotal method in materials science and metallurgy for synthesizing alloys, composites, and nanocomposites with unique properties. The process involves the repeated cold welding, fracturing, and rewelding of powder particles in a high-energy ball mill. Initially, elemental powders or pre-alloyed powders are mixed. The milling media (balls) and the powder mixture are placed in a container, and the mill is set in motion, causing the balls to impact the powder. The continuous mechanical action induces repeated deformation, welding, and breaking of the powder particles, leading to the creation of alloyed or composite powders. One of the distinctive features of mechanical alloying is its ability to produce metastable phases and nanocrystalline structures that may not be attainable through conventional methods. The high-energy impacts during milling lead to the refinement of grain sizes, improved mixing of elements, and the creation of unique microstructures. This process is particularly valuable in the development of advanced materials for applications ranging from structural components to catalysts. Luqman et al. [9] Mechanical alloying has found applications in various fields, including the synthesis of high-performance alloys, intermetallic compounds, and amorphous alloys. Additionally, it is widely employed in the production of materials with enhanced mechanical, thermal, and magnetic properties. Despite its advantages, mechanical alloying has challenges such as contamination from the milling media and the need for careful control of milling parameters to achieve the desired outcomes. Moreover, post-milling processes like annealing or sintering are often required to optimize the properties of the obtained powders. Overall, mechanical alloying has revolutionized materials synthesis, providing a versatile and powerful method for producing a wide range of advanced materials with tailored properties. The continuous refinement of this technique contributes to ongoing advancements in materials science and engineering [10, 11].

3.3 Processing

Mechanical alloying typically involves placing elemental powders or pre-alloyed powders in a high-energy ball mill. High-energy ball mills, such as planetary ball mills or attritor mills, are commonly used. The milling process involves the collision of powder particles with the milling balls and the inner walls of the milling chamber. Ball milling is a mechanical process extensively used for the size reduction and mixing of materials in various industries. The technique involves placing materials in a ball mill, which is then subjected to the impact, shear, and attrition forces generated by rotating balls within the mill. This process is widely employed in the production of powders, nanomaterials, and the mechanical alloying of materials. The mill uses mechanical energy to induce intense plastic deformation, cold welding, and fracturing of powder particles.

The central component of ball milling equipment, the ball mill consists of a cylindrical vessel containing balls, usually made of metal or ceramics that rotate around its axis. The material to be processed, either in the form of powders or as a mixture of powders and a liquid (usually a milling medium or solvent), is loaded into the ball mill. The ball mill is set into rotation, causing the balls to impact and grind the materials. The energy generated by the rotation and the impact of the balls causes size reduction and blending of the materials. The ratio of the mass of balls to the mass of the powder (ball-to-powder ratio) is a critical parameter. It influences the impact energy and the overall effectiveness of the milling process. Two types of ball mills are generally used, namely, planetary ball mills that have a planetary motion, where the jars containing the materials and balls rotate around their axis in the opposite direction of the main disk, and, attritor mills that use a rotating impeller to agitate the media and achieve grinding. A schematic presentation of the ball milling process is provided in Fig. 3.1. Several parameters affect ball milling. Kuziora et al. [12] Some of them are milling time, i.e., the duration of milling that influences the size reduction, milling speed, which is the rotational speed of the mill that affects the impact energy, ball size, and material that impact the final particle size, and the efficiency of the milling, consequently, affecting the size reduction and mixing efficiency. Ball milling is widely used in the production of various powders, including metal powders, ceramics, and composite materials. This has specific application in producing nano-crystalline materials and nanoparticles with controlled size and morphology. Thus, it is helpful in the mechanical alloying of materials to produce alloy powders with improved properties [13].

It provides efficient mixing of different materials, leading to a homogeneous blend with a reduction of particle size to achieve the desired characteristics. It can be adapted for a wide range of materials and applications. The milling media and the

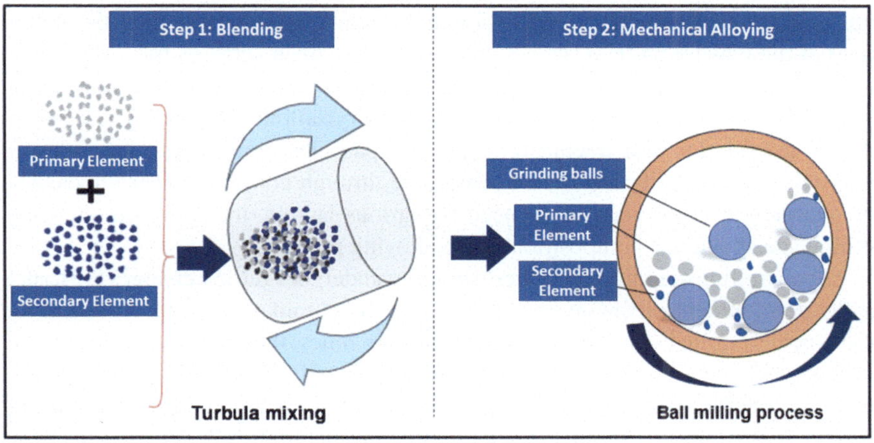

Fig. 3.1 Schematic presentation of mechanical alloying process

container material can introduce impurities or contaminants into the milled material. There is a high-energy impact during milling which can lead to the generation of heat and thus requires cooling measures.

3.4 Characterization

The characterization of mechanically alloyed powders is crucial to understanding their properties. Techniques such as X-ray diffraction, electron microscopy, and thermal analysis are commonly employed.

3.4.1 Microstructural Evolution

Mechanical alloying [14, 15] leads to the evolution of microstructure, including changes in grain size, lattice defects, and the formation of new phases. The microstructural evolution of ball-milled alloying powders is a complex and dynamic process that involves significant changes in the composition, crystal structure, and morphology of the powder particles. Ball milling [16], a high-energy mechanical alloying technique, induces repeated deformation, welding, and fracturing of powder particles, leading to a series of transformations at the microstructural level. In the early stages of ball milling, the powders experience cold welding as a result of the intense mechanical forces generated by the high-energy impacts of milling balls. This cold welding leads to the formation of agglomerates and the initial stages of alloying between the constituent elements. As milling progresses, the agglomerates are subjected to continuous deformation and fragmentation, resulting in the refinement of the powder particle size and the creation of fresh surfaces for further reactions. The high-energy impacts during ball milling also contribute to the reduction of particle sizes, leading to the formation of nanocrystalline structures. The repeated plastic deformation and fracturing of powder particles promote the formation of metastable phases that may not be achievable through conventional methods. The microstructural evolution is influenced by various factors, including milling time, milling speed, and the composition of the alloying powders.

As ball milling continues, the microstructure undergoes further changes [15], such as the development of fine-grained structures, the elimination of porosity, and the refinement of grain boundaries. At longer milling times, the process may lead to the formation of fully alloyed or composite structures with enhanced mechanical and thermal properties. Understanding the microstructural evolution during ball milling is crucial for tailoring the properties of the final material. Subsequent heat treatment or consolidation processes may be employed to optimize the microstructure and achieve the desired mechanical, thermal, or magnetic properties in the synthesized alloy or composite. So the microstructural evolution of ball-milled alloying powders involves a sequence of intricate transformations, including cold welding, deformation, fracturing, and the formation of nanocrystalline structures, ultimately influencing the properties of the resulting material.

3.4.2 Phase Transformations

The process can induce phase transformations [17], including the formation of solid solutions, intermetallic compounds, or amorphous phases. The phase transformations of ball-milled alloying powders represent a fascinating journey through the alteration of the material's structure and composition. Ball milling, a high-energy mechanical alloying process, induces significant changes in the phases present within the alloying powders. This intricate evolution involves multiple stages, each characterized by distinct alterations in the crystalline structure, morphology, and properties of the material. In the early stages of ball milling, the powders experience a phenomenon known as cold welding. The intense mechanical forces generated by the milling balls cause particles to adhere to each other, initiating the process of alloying between constituent elements. At this point, the powders may exhibit a mixture of phases, including the initial phases of the alloy and the original phases of the individual elements. As milling progresses, the powders undergo repeated deformation and fracturing, leading to the refinement of particle size and the creation of fresh surfaces for further reactions. This stage is crucial for promoting homogeneity within the material, as the continuous mechanical action facilitates the diffusion of atoms across the interfaces, encouraging the formation of alloy phases. The high-energy impacts during ball milling also contribute to the reduction of particle sizes, eventually leading to the formation of nanocrystalline structures. The repeated plastic deformation and fracturing of powder particles foster the creation of metastable phases that may not be attainable through conventional methods. The nanocrystalline nature of the material enhances its mechanical strength and may result in improved properties, such as hardness and wear resistance [18].

As ball milling continues, the phase transformations progress towards the development of more refined microstructures. The formation of fine-grained structures, elimination of porosity, and the refinement of grain boundaries are observed. These changes are indicative of the material's transition to a more thermodynamically stable state. At longer milling times, the phase transformations may culminate in the complete alloying of the powder, resulting in a single-phase or multiphase material with enhanced properties. The crystalline structure becomes more ordered, and the composition becomes more homogeneous. However, excessive milling may lead to the formation of undesirable phases or an over-fragmentation of the material, necessitating careful control of milling parameters. Understanding the phase transformations during ball milling is crucial for tailoring the properties of the final material. Post-milling processes, such as annealing or heat treatment, are often employed to optimize the microstructure and achieve the desired phase composition. The controlled application of heat can help stabilize the formed phases, eliminate defects, and enhance the overall performance of the alloy. In summary, the phase transformations of ball-milled alloying powders represent a dynamic process marked by the evolution of crystalline structure, morphology, and composition. From the initial stages of cold welding and alloy formation to the development of nanocrystalline structures and the eventual achievement of a stable alloy, each phase of the transformation journey contributes to the unique properties of the synthesized material [17].

The phase transformations during ball milling of alloys are diverse and depend on factors such as the composition of the alloy, milling parameters, and the duration of the milling process. Here are a few examples of common phase transformations observed during ball milling.

Amorphization

In some cases, prolonged ball milling can lead to the amorphization of the alloy, where the crystalline structure is lost, and the material becomes more disordered. This is particularly common in systems where the alloy components have limited mutual solubility. Amorphization of an alloy is a significant phase transformation that occurs during processes like ball milling, where the crystalline structure of the material is lost, and it transitions into a disordered, non-crystalline state. This phenomenon is particularly noteworthy because it leads to the formation of an amorphous or glassy phase, distinct from the ordered atomic arrangement found in crystalline materials [19].

During ball milling, the intense mechanical forces and repeated impacts cause the alloy's crystalline lattice to break down. The energy imparted disrupts the long-range order of atoms, preventing the formation of a well-defined crystal structure. Instead, the atoms are arranged in a disordered, random fashion, resulting in an amorphous structure. Amorphous alloys, also known as metallic glasses, exhibit unique properties compared to their crystalline counterparts. They often possess higher hardness, better corrosion resistance, and improved magnetic or electronic characteristics. The absence of grain boundaries in amorphous materials contributes to enhanced mechanical properties, making them suitable for applications in areas such as cutting tools, sensors, and magnetic storage devices. The amorphization process is influenced by various factors, including the composition of the alloy, milling time, and milling conditions. Controlling these parameters allows researchers to tailor the extent of amorphization and subsequently fine-tune the material's properties. Amorphous alloys hold promise in diverse fields due to their exceptional combination of properties. However, maintaining the amorphous state can be challenging, as crystallization may occur upon heating or during prolonged storage. Researchers continue to explore ways to stabilize and harness the unique characteristics of amorphous alloys for applications in advanced technologies, including electronics, aerospace, and biomedical devices.

Nanocrystalline Formation

Ball milling often promotes the formation of nanocrystalline structures. The repeated deformation and fracturing of powder particles result in the reduction of grain sizes to the nanoscale. This is advantageous for enhancing mechanical properties, such as hardness and strength. Here the repeated mechanical impacts and deformation of alloy powders lead to a significant reduction in grain size, resulting in crystalline structures at the nanoscale. Joy et al. [20] this transformation holds profound implications for the material's properties, providing it with unique characteristics compared to conventional, larger-grained alloys. During nanocrystalline formation, the alloy's grain size is reduced to dimensions on the order of nanometers. This reduction in grain size brings about several advantages. First and foremost, materials with nanocrystalline structures often exhibit enhanced mechanical properties, including increased hardness and improved strength. The high density of grain boundaries in nanocrystalline alloys impedes dislocation movement, contributing to their superior mechanical performance [21].

The increased surface area-to-volume ratio in nanocrystalline materials also enhances their reactivity, making them suitable for applications in catalysis and sensor technologies. The unique electronic and magnetic properties of nanocrystalline alloys further broaden their utility in fields such as electronics, magnetics, and information storage. Furthermore, the nanocrystalline structure can influence the alloy's thermal stability and diffusion characteristics. The elevated concentration of grain boundaries alters the material's thermodynamic and kinetic behaviour, impacting processes like phase transformations and diffusion rates. Researchers often tailor the nanocrystalline formation by adjusting parameters such as milling time, milling intensity, and alloy composition. These parameters influence the degree of grain refinement and subsequently determine the final properties of the nanocrystalline alloy. Thus, nanocrystalline formation in alloys through processes like ball milling represents a transformative approach to materials engineering. By manipulating the grain size at the nanoscale, researchers can unlock a range of improved properties, making nanocrystalline alloys promising candidates for advancements in structural materials, functional devices, and other cutting-edge applications [22].

Solid-Solution Formation

Ball milling facilitates the creation of solid solutions by promoting the mixing of alloying elements at an atomic level. This is especially common in systems where the elements have a certain degree of solubility in each other. Solid-solution formation in alloys is a significant phase transformation that occurs during processes like ball milling, where alloying elements are mechanically blended at an atomic level, resulting in a homogeneous mixture. This process is characterized by the creation of a single-phase alloy where the constituent elements are dissolved into each other in a solid state. In the context of solid-solution formation, the alloying elements, typically metals, exhibit mutual solubility, allowing them to mix in various proportions while maintaining a single-phase structure. The high-energy impacts during ball milling facilitate the breaking of atomic bonds within the individual elements, promoting diffusion and alloying between them. As a result, the original crystalline structures of the alloying elements are replaced by a new, single-phase structure. The degree of solid-solution formation is influenced by factors such as the composition of the alloy, milling time, and milling conditions. Careful control of these parameters allows researchers to tailor the final alloy's properties, ensuring a desired balance of characteristics [21, 22].

Solid solutions play a crucial role in materials science and engineering. By creating alloys with solid-solution phases, researchers can enhance specific properties, such as mechanical strength, hardness, and corrosion resistance. Additionally, solid solutions contribute to the modification of thermal, electrical, and magnetic properties, expanding the potential applications of alloys in various industries. One common example of solid-solution formation is in the creation of stainless steel, where elements like chromium and nickel are alloyed with iron. The resulting solid solution exhibits improved corrosion resistance and mechanical strength, making stainless steel a widely used material in construction, automotive, and other industries. Thus, solid-solution formation in alloys is a fundamental process that allows for the creation of homogeneous materials with tailored properties. This phase transformation is central to the development of alloys designed to meet specific performance requirements across a diverse range of applications.

Metastable Phase Formation

The high-energy impacts during ball milling can lead to the formation of metastable phases that are not thermodynamically stable under equilibrium conditions. These phases may possess unique properties and play a crucial role in tailoring the material's characteristics. Metastable phase formation is a fascinating and significant phenomenon observed during processes like ball milling in the alloying of materials. A metastable phase is a state where the material exists in a configuration that is not thermodynamically stable under equilibrium conditions but persists due to kinetic barriers hindering further transformation. The high-energy impacts during ball milling induce rapid and unique transformations, leading to the creation of these non-equilibrium states [23].

In the context of metastable phase formation, ball milling facilitates the synthesis of phases that may not be achievable through conventional methods. The intense mechanical forces applied to the alloying powders result in rapid atomic rearrangements, overcoming the energy barriers associated with the formation of stable equilibrium phases. As a consequence, metastable phases with distinct structural and often enhanced functional properties are realized. Metastable phases play a pivotal role in materials science, offering opportunities for tailoring specific characteristics. These phases may exhibit improved mechanical strength, hardness, or unique electronic and magnetic properties compared to their equilibrium counterparts. As such, metastable materials find applications in diverse fields, including electronics, catalysis, and energy storage [24].

One notable example of metastable phase formation is observed in amorphization, where a crystalline material transforms into an amorphous, non-crystalline state during ball milling. While amorphous structures are not thermodynamically stable, their formation is favoured kinetically under the influence of high mechanical energy, resulting in materials with unique mechanical and electronic properties. The controlled manipulation of metastable phases during alloying processes is an exciting avenue for materials design. Researchers leverage this phenomenon to create innovative materials with tailored properties, contributing to advancements in technology and the exploration of novel applications in areas such as nanotechnology and advanced manufacturing. Understanding and harnessing metastable phase formation opens avenues for engineering materials with enhanced functionalities beyond the limits imposed by equilibrium thermodynamics [23].

Phase Segregation

Depending on the composition of the alloy and the milling conditions, phase segregation may occur, leading to the formation of distinct regions with different compositions. This can impact the material's properties and behaviour. Phase segregation is a notable phenomenon observed during alloying processes, including ball milling, where distinct regions with different compositions form within the material. This separation occurs due to variations in the mutual solubility of alloying elements and is influenced by factors such as composition, temperature, and the duration of the alloying process. During ball milling, the intense mechanical forces and repeated impacts contribute to the creation of local concentration gradients within the alloy. These gradients may lead to the preferential accumulation of certain elements in specific regions, resulting in the formation of segregated phases. The occurrence of phase segregation can significantly influence the material's overall properties and behaviour. One common example of phase segregation is seen in certain binary alloy systems, where limited solubility exists between the alloying elements. As the alloying process progresses, regions with compositions approaching the solubility limit may undergo phase separation, leading to the formation of distinct phases within the material [25].

The impact of phase segregation on material properties is substantial. It can affect mechanical, thermal, and electrical properties, making it crucial for researchers to understand and control this phenomenon to tailor materials for specific applications. For instance, in magnetic alloys, phase segregation can influence magnetic properties, and in certain structural alloys, it may impact hardness or corrosion resistance. Engineers and researchers often manipulate alloy compositions and processing conditions to either promote or inhibit phase segregation based on the desired material characteristics. Optimizing these factors allows for the precise control of microstructures within alloys, enhancing the performance and reliability of materials in diverse applications. Thus, phase segregation is a crucial aspect of alloying processes that significantly influences the properties of the resulting material. Understanding and controlling phase segregation is essential for tailoring alloys with specific characteristics, ensuring their suitability for various industrial and technological applications.

Disorder-Order Transformations

Ball milling can induce disorder-order transformations in alloy systems. Disorder in the initial structure may be gradually replaced by a more ordered crystalline structure as milling progresses, influencing the material's properties. Disorder-order transformations [26] represent a critical aspect of material science, particularly during processes like ball milling, where the ordered crystalline structure of alloys changes greater organization or disorder. This transformation is driven by the intense mechanical forces and high-energy impacts that characterize ball milling, influencing the arrangement of atoms within the material. During disorder-order transformations, the initial crystalline structure of the alloy undergoes disruptions as a result of the mechanical deformation induced by ball milling. This disruption leads to a temporary increase in disorder within the material. However, as the milling process continues, the alloy's atoms reorganize, and the system transforms into a more ordered crystalline state.

The disorder-order transformations can manifest in several ways. In some cases, it involves the restoration of the original crystalline structure with improved order compared to the initial state. In other instances, the alloy may transition to a different, more ordered crystal structure, often driven by the rearrangement of atoms facilitated by the mechanical energy imparted during milling. The extent and nature of disorder-order transformations depend on various factors, including the alloy composition, milling parameters, and the duration of the ball milling process. Researchers carefully tailor these parameters to achieve specific disorder-order transformations that align with the desired material properties. Understanding and harnessing disorder-order transformations are crucial for materials engineering. The ability to control the level of disorder and subsequently induce ordered structures provides a powerful tool for tailoring materials with enhanced mechanical, thermal, and electronic properties. Disorder-order transformations contribute to advancements in the design of materials for applications in structural components, electronic devices, and other areas where precise control over material properties is essential.

Transformation to Equilibrium Phases

Prolonged ball milling tends to drive the alloy system toward thermodynamic equilibrium, resulting in the formation of stable phases and the elimination of metastable or amorphous structures. Transformation to equilibrium phases is a crucial aspect of materials science, particularly during processes like ball milling where alloyed materials change thermodynamically stable states. The concept is rooted in the drive of materials to reach a state of lower energy and greater stability under equilibrium conditions. During ball milling, the intense mechanical forces applied to alloyed powders induce a series of structural changes, leading to the creation of metastable phases or amorphous structures. However, as the milling process continues, the system tends to move towards thermodynamic equilibrium by transforming into more stable phases. The transformation to equilibrium phases involves the evolution of the material's microstructure, often characterized by the formation of well-defined crystalline structures that are energetically favourable. This process is influenced by factors such as composition, milling time, and temperature. Longer milling times or subsequent heat treatment steps are often employed to facilitate the completion of the transformation to equilibrium phases [27].

For example, in the synthesis of alloys, the initial phases formed during ball milling may be metastable or amorphous. However, through controlled processing and optimization of parameters, the material can progress to a state where the phases formed are in thermodynamic equilibrium. This results in materials with more predictable properties and stable structures. The ability to guide the transformation to equilibrium phases is crucial for tailoring material properties. It allows researchers to design alloys with specific characteristics such as improved mechanical strength, enhanced conductivity, or optimized magnetic properties. By understanding and manipulating these phase transformations, materials scientists can engineer advanced materials for a wide range of applications, from structural components to electronic devices, ensuring that the final products meet the desired performance criteria.

Mechanical Activation

The repeated mechanical impacts during ball milling can activate chemical reactions that may not occur under normal thermodynamic conditions, leading to the formation of unique phases. Mechanical activation is a fundamental process in materials science that involves the application of mechanical energy to induce structural changes in materials. This concept is particularly relevant in techniques like ball milling, where mechanical forces play a pivotal role in altering the properties and characteristics of powders or bulk materials. During mechanical activation, high-energy impacts, compression, and shear forces lead to deformation, fracturing, and rearrangement of atoms within the material. This mechanical energy input results in changes at the microstructural level, influencing properties such as particle size, crystallinity, and phase composition. In ball milling, a commonly used method for mechanical activation, metallic or non-metallic powders are subjected to repeated collisions with milling balls in a high-energy ball mill. The intense mechanical forces cause particles to undergo deformation, breakage, and welding, leading to the formation of finely dispersed powders with altered properties [28].

The mechanical activation process can have diverse outcomes, depending on factors such as milling time, milling speed, and the composition of the materials. For instance, in alloying processes, mechanical activation facilitates the creation of solid solutions, metastable phases, or nanocrystalline structures, enhancing mechanical and functional properties. One notable application of mechanical activation is in the synthesis of nanomaterials. By controlling the milling parameters, researchers can achieve a high degree of mechanical activation, resulting in the production of nanoparticles with unique size, shape, and surface characteristics. These nanoparticles find applications in areas like catalysis, sensors, and biomedical devices. Overall, mechanical activation is a versatile and powerful tool in materials processing. Its ability to induce structural changes at the atomic and microstructural levels provides a means for tailoring the properties of materials, making it a key technique in the development of advanced materials with enhanced performance for various technological applications.

These examples highlight the versatility of ball milling in inducing various phase transformations, offering researchers a powerful tool for tailoring material properties and achieving desired characteristics in alloys. Understanding and controlling these transformations are crucial for optimizing the performance of ball-milled materials in specific applications.

3.4.3 Applications of Mechanically Alloyed Powders

Mechanically alloyed powders find applications in various industries, including aerospace, automotive, electronics, and medical devices. Examples of applications include the development of high-strength lightweight materials, magnetic materials, and hydrogen storage materials.

Challenges

- Controlling the process parameters is crucial to achieving the desired alloy composition and properties.

- Care must be taken to avoid contamination from the milling media or the atmosphere.
- Mechanically alloyed powders play a significant role in advancing materials science by enabling the production of novel materials with tailored properties for specific applications. The process continues to be an active area of research for developing advanced materials with improved performance characteristics.

3.4.4 Advanced Characterization Techniques

Mechanical alloying (MA) is a powder processing technique used to produce homogeneous materials starting from powders with different compositions. The process involves repeated welding, fracturing, and rewelding of powder particles in a high-energy ball mill. Advanced characterization of mechanically alloyed powders is crucial to understanding the evolution of microstructure, phase transformations, and properties. Here are some advanced characterization techniques commonly employed:

X-Ray Diffraction (XRD)

Phases and Crystal Structure: XRD is a fundamental technique employed to identify phases and analyze the crystal structure of ball-milled alloys. This non-destructive method utilizes X-rays to probe the crystalline structure of materials, providing detailed information about the arrangement of atoms in the lattice. In the analysis of ball-milled alloys, XRD is instrumental in determining the phases formed during the milling process. It allows researchers to identify crystalline structures, observe phase transformations, and quantify the degree of crystallinity. The diffraction patterns obtained from ball-milled alloys provide unique signatures that can be matched with crystallographic databases, aiding in the precise identification of phases. Moreover, XRD is sensitive to changes in crystal structure, such as the formation of nanocrystalline or amorphous phases induced by ball milling. Researchers can analyze peak broadening and shifting in the diffraction patterns to assess grain size, lattice strain, and the presence of disordered structures.

Thus, XRD is a valuable tool for characterizing ball-milled alloys, offering insights into the phases present and the resulting crystallographic changes [29]. This information is essential for tailoring material properties and optimizing processing conditions in various industrial applications, including materials science, metallurgy, and nanotechnology.

Transmission Electron Microscopy (TEM)

Microstructural Analysis: TEM provides high-resolution images, allowing for the observation of microstructural features, such as grain size, defects, and dislocations. It helps identify the presence of nanocrystalline or amorphous phases. TEM is employed for even higher-resolution microstructural analysis, allowing direct visualization of individual grains, defects, and nanoparticles. TEM provides detailed information about crystallographic features and phase distribution within the alloy, offering a deeper understanding of the microstructure. It is a powerful technique extensively used for microstructural analysis of ball-milled alloys, providing detailed insights into the nanoscale features, crystallography, and defects within the material. The unique capabilities of TEM make it an invaluable tool for researchers studying the effects of ball milling on alloy microstructure.

High-Resolution Imaging: TEM offers extremely high-resolution imaging capabilities, allowing researchers to directly visualize and analyze the microstructure of ball-milled alloys at the atomic and nanoscale levels. This includes observing changes in grain size, the formation of nanocrystalline structures, and the presence of defects induced by ball milling.

Crystallographic Information: TEM provides crystallographic information about the phases present in the alloy. Researchers can use selected area diffraction (SAED) patterns obtained from specific regions of the sample to determine the crystal structures, orientations, and the presence of any preferred orientations resulting from the mechanical deformation during ball milling.

Analysis of Defects and Dislocations: TEM is highly effective in characterizing defects and dislocations within the alloy microstructure induced by ball milling. The detailed imaging capability allows researchers to quantify and analyze the density and distribution of defects, providing insights into the material's mechanical behaviour.

Nanoparticle Identification: TEM is particularly useful for identifying and characterizing nanoparticles formed during ball milling. It allows researchers to examine the size, shape, and distribution of nanoparticles on the alloy surfaces, contributing to a comprehensive understanding of the material's nanoscale features.

TEM enables in situ observations during dynamic processes, providing real-time insights into the evolution of microstructural changes during ball milling. This allows researchers to capture and understand transient states and phase transformations as they occur. TEM can be coupled with EDS to obtain compositional information at the nanoscale. This is especially valuable for studying the distribution of alloying elements within the microstructure and understanding the effects of alloy composition on the ball-milled material. Thus, TEM serves as a cornerstone for microstructural analysis of ball-milled alloys. Its high-resolution imaging, crystallographic capabilities, and ability to analyze defects and nanoparticles make it an indispensable tool for researchers seeking a detailed understanding of the structural changes induced by ball milling in nanoscale materials.

Scanning Electron Microscopy (SEM)

Surface Morphology: The surface morphology of ball-milled alloys undergoes significant transformations, shaping the material's properties and functionality. Ball milling, a high-energy mechanical alloying process, imparts intense mechanical forces, resulting in alterations at the microstructural and nanoscale levels, directly impacting the surface morphology of the alloy. Initially, the surfaces of the alloy particles experience cold welding and fracturing due to repeated collisions with milling balls. This leads to the creation of fresh surfaces and the formation of microcrystalline or nanocrystalline structures, contributing to increased surface area and reactivity. The refined surface morphology is characterized by smaller grain sizes, intricate nanostructures, and a higher density of defects and grain boundaries. The process of ball milling often results in the formation of a rough and irregular surface due to the mechanical deformation and fragmentation of alloy particles. The increased surface roughness is particularly prominent in the early stages of milling. As milling continues, the surfaces may undergo further refinement, leading to a more homogenized and controlled surface morphology. Surface alloying is another significant aspect of ball milling that influences surface morphology. The repeated impacts and grinding action enable the diffusion of alloying elements onto the surface layers, creating a modified surface composition. This surface alloying enhances the material's resistance to wear, corrosion, and other surface-related degradation mechanisms. Nanoparticles or nanoscale features may also be formed on the surface during ball milling, contributing to a unique surface morphology. These nanoparticles can have a profound impact on the material's optical, magnetic, and catalytic properties, making the alloy suitable for diverse applications such as sensing, catalysis, and biomedical devices. Thus, the surface morphology of ball-milled alloys reflects the intricate interplay of mechanical forces, deformation, and phase transformations. The resulting surfaces exhibit refined microstructures, increased roughness, and potentially the presence of nanoparticles, contributing to the tailored properties of the material. Understanding and controlling the surface morphology during ball milling is essential for optimizing the performance of alloys in various industrial applications.

SEM is a powerful tool extensively employed for studying the surface morphology of ball-milled alloys. SEM provides high-resolution imaging and detailed analysis, allowing researchers to visualize and characterize the microstructural features and surface topography of materials at a nanoscale level. In the context of ball-milled alloys, SEM is instrumental in several ways:

Visualization of Microstructural Changes: SEM allows researchers to observe the evolution of microstructure during ball milling. It provides detailed images of the alloy surfaces, revealing changes in grain size, morphology, and the development of nanocrystalline structures resulting from the mechanical deformation and fragmentation induced by ball milling.

Particle Size Analysis: SEM is invaluable for assessing the particle size distribution of ball-milled alloys. The high-resolution imaging capability enables precise measurements, helping researchers understand how milling parameters influence the size and distribution of particles on the alloy surfaces.

Surface Roughness and Texture Analysis: SEM provides insights into the surface roughness and texture of ball-milled alloys. The detailed images assist in quantifying surface irregularities and roughness, crucial for understanding the material's tribological and mechanical properties.

Identification of Nanoparticles and Surface Features: SEM is particularly effective in identifying nanoparticles and other nanoscale features formed during ball milling. Researchers can explore the presence and distribution of these features, providing information about the alloy's unique surface characteristics.

Surface Alloying Visualization: SEM enables the visualization of surface alloying effects on ball-milled alloys. By mapping the elemental composition using energy-dispersive X-ray spectroscopy (EDS) in conjunction with SEM, researchers can study how alloying elements are distributed across the alloy surfaces.

Thus, SEM is an indispensable tool for investigating the surface morphology of ball-milled alloys. Its capability to provide detailed imaging and analysis at the nanoscale makes it an essential technique for researchers aiming to understand the microstructural changes, particle size distribution, and surface characteristics resulting from ball milling processes.

Energy-Dispersive X-Ray Spectroscopy (EDS)

Chemical Composition: EDS [22] is a crucial analytical technique employed for determining the chemical composition of ball-milled alloys. Operating as an attachment to SEM or TEM, EDS provides valuable information about the elemental composition of materials at the nanoscale. During the analysis of ball-milled alloys, EDS detects characteristic X-rays emitted when the high-energy electron beam interacts with the sample. These X-rays are specific to each element, allowing for the identification and quantification of the elements present in the alloy.

In the context of ball-milled alloys, EDS is particularly useful for investigating elemental distribution, identifying alloying elements, and mapping the composition across the microstructure. It aids researchers in understanding how the alloy composition changes during ball milling and helps assess the homogeneity of the material. By providing real-time, quantitative data on the elemental makeup of the alloy, EDS contributes crucial insights for optimizing alloy design and processing parameters in various industrial applications.

High-Resolution Transmission Electron Microscopy (HRTEM)

Detailed Microstructural Analysis: HRTEM [22] provides even higher resolution than standard TEM, allowing for the observation of finer microstructural details, such as dislocations, stacking faults, and interfaces. HRTEM is a sophisticated technique that plays a pivotal role in the detailed analysis of ball-milled alloys at the atomic and nanoscale levels. Its exceptional imaging capabilities and ability to resolve fine structural details make HRTEM an invaluable tool for researchers studying the microstructural changes induced by ball milling.

Atomic-Level Imaging: HRTEM provides atomic-level resolution, allowing researchers to directly visualize individual atoms within the crystalline lattice of ball-milled alloys. This level of detail is crucial for understanding how the arrangement of atoms changes during the mechanical alloying process.

Characterization of Nanoscale Structures: HRTEM is particularly effective in characterizing nanoscale features, such as the formation of nanocrystalline structures and nanoparticles. Researchers can observe the size, shape, and distribution of these structures, gaining insights into the nanoscale transformations induced by ball milling.

Crystallographic Information: HRTEM enables the acquisition of crystallographic information, including lattice parameters, crystal defects, and orientation relationships between different phases. This information is essential for understanding the structural evolution of ball-milled alloys.

Observation of Dislocations and Defects: HRTEM allows for the visualization and analysis of dislocations, stacking faults, and other defects in the crystal lattice. The presence of these defects provides crucial insights into the material's mechanical behaviour and its response to the mechanical forces during ball milling.

HRTEM can be used for in situ observations during dynamic processes, offering real-time insights into the evolution of microstructural changes as they occur. This capability is particularly beneficial for capturing transient states and phase transformations induced by ball milling. The use of High-Resolution Transmission Electron Microscopy in ball-milled alloys provides unparalleled insights into the nanoscale structural changes, allowing researchers to observe, analyze, and understand the intricate transformations that occur during mechanical alloying. HRTEM is an essential tool for advancing our knowledge of materials science and optimizing the properties of ball-milled alloys for diverse applications.

Atomic Force Microscopy (AFM)

Atomic Force Microscopy (AFM) [30] is a powerful imaging technique employed for studying the surface topography and roughness of ball-milled alloys at the nanoscale. Its ability to provide high-resolution, three-dimensional images makes it an invaluable tool in characterizing the fine details of material surfaces.

Surface Topography: AFM excels in capturing surface topography with nanometer-scale resolution. For ball-milled alloys, AFM reveals detailed information about the morphology, grain boundaries, and features such as nanocrystalline structures that arise from the mechanical deformation during milling. The technique allows for the direct visualization of changes in surface topography induced by ball milling.

Measurement of Roughness Parameters: AFM enables the quantification of surface roughness parameters, including average roughness (Ra), root mean square (RMS) roughness, and peak-to-valley height. These parameters offer quantitative insights into the impact of ball milling on the smoothness or irregularities of the alloy surface.

Visualization of Nanoparticles and Defects: AFM is capable of detecting and characterizing nanoparticles formed during ball milling. By scanning the surface, AFM reveals the size, distribution, and arrangement of nanoparticles. Additionally, AFM is sensitive to surface defects, providing information about dislocations, grain boundaries, and other structural irregularities induced by the milling process.

In Situ Observations: AFM can be utilized for in situ observations, allowing researchers to monitor real-time changes in surface topography during dynamic processes. This is particularly advantageous for understanding the evolution of surface features as ball milling progresses.

Correlation with Other Techniques: AFM complements other analytical techniques, such as SEM and TEM, by offering additional insights into surface characteristics. By combining AFM with these techniques, researchers obtain a more comprehensive understanding of the microstructural changes induced by ball milling. AFM serves as a valuable tool for investigating the surface topography and roughness of ball-milled alloys at the nanoscale. Its high-resolution imaging capabilities, quantitative roughness analysis, and compatibility with in situ observations contribute to a comprehensive understanding of how ball milling influences the surface characteristics of alloys in materials science and metallurgy applications.

Differential Scanning Calorimetry (DSC)

Differential Scanning Calorimetry (DSC) [31] is an analytical technique used for the thermal analysis of ball-milled alloys, providing valuable insights into their thermodynamic and kinetic properties during heating or cooling processes.

Melting and Solidification Behavior: DSC measures the heat flow associated with phase transitions, such as melting and solidification. For ball-milled alloys, DSC reveals changes in melting temperatures, enthalpies of fusion, and solidification behaviour. These parameters offer crucial information about the alloy's thermal stability and phase transformations induced by ball milling.

Crystallization and Amorphization Studies: DSC is effective in detecting crystallization and amorphization events in ball-milled alloys. The intense mechanical forces during ball milling can lead to the formation of amorphous phases, and DSC helps identify these transitions by detecting changes in heat flow associated with crystallization or amorphization.

Evaluation of Thermal Stability: DSC aids in assessing the thermal stability of ball-milled alloys by monitoring changes in heat flow over a range of temperatures. This information is essential for understanding the effects of ball milling on the alloy's ability to retain its structural and phase integrity under thermal stress.

Determining Alloy Composition and Eutectic Reactions: DSC can assist in identifying alloy composition by detecting characteristic eutectic reactions and alloying events. The thermal analysis helps researchers understand the influence of ball milling on the alloy's composition and the occurrence of alloying reactions.

Quantification of Phase Fractions: DSC can be used to estimate phase fractions in multi-phase alloys by analyzing the areas under specific peaks associated with different phase transitions. This information is valuable for quantifying the extent of phase transformations induced by ball milling.

Kinetic Analysis: DSC allows for the determination of kinetic parameters, such as activation energy and reaction rates, associated with thermal events in ball-milled alloys. This aids in understanding the mechanisms and kinetics of phase transformations occurring during heating or cooling.

DSC is a versatile tool for thermal analysis, providing essential information about phase transitions, thermal stability, and kinetic properties of ball-milled alloys. The data obtained through DSC enhances the understanding of the alloy's behaviour under thermal conditions, guiding materials scientists and metallurgists in optimizing processing parameters for specific applications.

Raman Spectroscopy

Raman spectroscopy [32, 33] is a valuable analytical technique used for obtaining chemical and structural information about ball-milled alloys. This non-destructive method relies on the inelastic scattering of monochromatic light, providing insights into vibrational modes and molecular interactions in materials.

Chemical Composition Analysis [34]: Raman spectroscopy is effective in identifying the chemical composition of ball-milled alloys. Different elements and compounds exhibit characteristic Raman spectra, enabling researchers to discern the presence of specific phases, oxide layers, or other chemical species resulting from the milling process.

Phase Identification: Raman spectroscopy aids in identifying different phases within ball-milled alloys. It is sensitive to changes in crystal structure and provides distinctive spectral features corresponding to specific phases. This information is crucial for understanding how ball milling influences the alloy's phase composition.

Detection of Nanocrystalline Structures [33, 34]: Raman spectroscopy is particularly useful for detecting nanocrystalline structures formed during ball milling. The technique can reveal changes in crystallinity, grain size, and the development of fine-scale structures, offering valuable information about the nanoscale effects induced by mechanical deformation.

Stress and Strain Analysis: Raman spectroscopy allows for the investigation of stress and strain in ball-milled alloys. Changes in vibrational frequencies and peak shifts provide insights into the mechanical deformation and structural modifications induced by ball milling, offering a perspective on the alloy's mechanical behaviour.

Amorphization Detection: Raman spectroscopy [32] can detect the amorphization of materials during ball milling. The loss of well-defined crystalline peaks and the appearance of broad features in the Raman spectra indicate the transition from a crystalline to an amorphous state.

Complementary Technique: Raman spectroscopy [33] complements other characterization techniques such as X-ray diffraction and electron microscopy. Its ability to provide chemical and structural information at a micron-scale resolution adds a valuable dimension to the comprehensive analysis of ball-milled alloys.

Thus, Raman spectroscopy serves as a versatile tool for investigating the chemical composition, phase transitions, and structural changes in ball-milled alloys. Its non-destructive nature, high sensitivity to vibrational modes, and compatibility with nanoscale analysis make it an essential technique for materials scientists and metallurgists studying the effects of ball milling on alloy properties.

Magnetic Property Measurements

Magnetic property measurements of ball-milled alloys are crucial for understanding how mechanical alloying influences the magnetic behaviour of materials. Various techniques can be employed to assess the magnetic properties of ball-milled alloys:

Vibrating Sample Magnetometry (VSM) [35]: VSM is a widely used technique for measuring the magnetic properties of materials, including ball-milled alloys. It determines the magnetization of a sample as a function of an applied magnetic field. VSM provides information about saturation magnetization, coercivity, and magnetic susceptibility, offering insights into the magnetic behaviour induced by ball milling.

Superconducting Quantum Interference Device (SQUID) Magnetometry [36]: SQUID magnetometry is a highly sensitive method for measuring magnetic properties, especially in the case of superparamagnetic or weakly magnetic materials. It is suitable for studying ball-milled alloys with fine magnetic structures or nanoscale magnetic features.

Magnetic Hysteresis Measurements [37]: Hysteresis loops obtained through measurements of magnetic field strength vs. magnetization reveal information about the coercivity and remanence of ball-milled alloys. This type of measurement is useful for understanding the magnetic stability and response to external magnetic fields.

Mössbauer Spectroscopy [38]: While primarily a technique for studying hyperfine interactions, Mössbauer spectroscopy can also provide information about the magnetic properties of ball-milled alloys. It is particularly useful for investigating the magnetic behaviour of iron-containing alloys.

Alternating Gradient Magnetometry (AGM): AGM [39, 40] is suitable for measuring magnetic properties, especially in the case of fine powders or nanoscale materials. It can provide information on magnetic susceptibility and the influence of milling on magnetic domain behaviour.

The magnetic properties of ball-milled alloys [39, 41, 42] can be significantly influenced by factors such as grain size reduction, changes in crystal structure, and the formation of nanocrystalline or amorphous phases induced by the mechanical alloying process. Understanding these magnetic changes is essential for tailoring materials with specific magnetic characteristics for applications in electronics, magnetic storage, and other relevant fields. The chosen magnetic measurement technique depends on the specific characteristics of the ball-milled alloy and the desired information about its magnetic behaviour.

Positron Annihilation Spectroscopy (PAS)

Positron Annihilation Spectroscopy (PAS) [43, 44] is a valuable technique for the characterization of defects in ball-milled alloys. This non-destructive method involves the interaction of positrons (positively charged electrons) with the electrons in the material, providing insights into various types of defects, such as vacancies, vacancy clusters, and dislocations. In the context of ball-milled alloys, PAS offers several advantages for defect analysis [45–47].

Defect Sensitivity: Positrons are highly sensitive to defects in materials, particularly to open-volume defects such as vacancies. During the ball milling process, intense mechanical forces induce the formation of defects, and PAS can detect and quantify these defects at the atomic scale.

Defect Depth Profiling: PAS can provide depth profiling information, allowing researchers to determine the distribution of defects beneath the surface of ball-milled alloys. This depth sensitivity is crucial for understanding how defects are distributed in the material and how they evolve during milling.

Identification of Defect Types: Different types of defects exhibit distinct signatures in positron annihilation spectra. PAS enables the identification of specific defect types, aiding in the comprehensive characterization of the defect structure induced by ball milling.

Quantitative Defect Analysis: PAS provides quantitative information about defect concentrations and sizes. This is valuable for researchers seeking to correlate the mechanical alloying parameters with the resulting defect structure in ball-milled alloys.

Monitoring Defect Evolution: PAS can be employed to monitor changes in the defect structure during different stages of ball milling. This capability allows for dynamic analysis of defect evolution and helps establish the relationship between processing conditions and defect formation.

Complementary Technique: PAS is often used in conjunction with other analytical techniques, such as positron lifetime spectroscopy and positron Doppler broadening spectroscopy, to obtain a more comprehensive understanding of the defect structure and its influence on the material properties.

Positron Annihilation Spectroscopy [47] is a powerful tool for the characterization of defects in ball-milled alloys. Its sensitivity to open volume defects, ability to provide depth profiling, and capability to identify and quantify different defect types make it an invaluable technique for understanding the structural changes induced by ball milling at the atomic and nanoscale levels.

3.5 Conclusion

Combining these advanced characterization techniques provides a comprehensive understanding of the evolution of microstructure, phase transformations, and properties in mechanically alloyed powders. This information is essential for optimizing processing parameters and tailoring the properties of the final material for specific applications. Techniques such as X-ray Diffraction (XRD) provide crucial information on the crystalline phases formed, crystallographic structures, and the degree of disorder. High-Resolution Transmission Electron Microscopy (HRTEM) offers unparalleled insights into the nanoscale morphology, crystal defects, and the evolution of microstructure. Positron Annihilation Spectroscopy (PAS) allows for sensitive detection and quantification of defects, providing a unique perspective on the material's structural integrity. Magnetic property measurements, utilizing methods like Vibrating Sample Magnetometry (VSM) and SQUID magnetometry, shed light on the alterations in magnetic behaviour induced by mechanical alloying. Thermal analysis through Differential Scanning Calorimetry (DSC) unveils phase transitions, crystallization, and amorphization events, contributing to a comprehensive understanding of the alloy's thermal stability.

Moreover, Raman spectroscopy, Atomic Force Microscopy (AFM), and Energy-Dispersive X-ray Spectroscopy (EDS) complement the characterization toolkit, offering valuable insights into chemical composition, surface topography, and elemental distribution. This integrated approach facilitates not only a deep understanding of the material's behaviour during mechanical alloying but also guides the optimization of processing parameters to tailor materials for specific applications. The ability to control and manipulate the properties of mechanically alloyed powders through advanced characterization opens avenues for the development of innovative materials with enhanced performance in fields ranging from materials science and metallurgy to electronics and nanotechnology. The ongoing synergy between advanced characterization techniques and materials engineering continues to pave the way for advancements in the design and application of novel alloys and composite materials. In conclusion, advanced characterization techniques play a pivotal role in unravelling the intricate structural, chemical, and physical transformations that occur during the mechanical alloying of powders. The process of mechanical alloying, particularly ball milling, induces profound changes at the atomic and nanoscale, shaping the properties of the resulting materials. The amalgamation of various advanced analytical methods allows researchers to explore and understand these changes comprehensively.

References

1. Ding, J., Xu, H., Li, X., et al.: The similarity of elements in multi-principle element alloys based on a new criterion for phase constitution. Mater. Des. **207**, 109849 (2021). https://doi.org/10.1016/J.MATDES.2021.109849
2. Rambabu, P., Eswara Prasad, N., Kutumbarao, V.V., Wanhill, R.J.H.: Aluminium alloys for aerospace applications. 29–52 (2017). https://doi.org/10.1007/978-981-10-2134-3_2
3. Kumar, G.B.V., Rao, C.S.P., Selvaraj, N., Bhagyashekar, M.S.: Studies on Al6061-SiC and Al7075-Al_2O_3 metal matrix composites. J. Miner. Mater. Charact. Eng. **09**, 43–55 (2010). https://doi.org/10.4236/JMMCE.2010.91004
4. Metal alloys: principles, types, advantages and applications. https://www.iqsdirectory.com/articles/metal-alloys.html. Accessed 19 January 2024
5. Lu, L., Lai, M.O., Zhang, S.: Diffusion in mechanical alloying. J. Mater. Process. Technol. **67**, 100–104 (1997). https://doi.org/10.1016/S0924-0136(96)02826-9
6. Li, S.S., Yue, X., Li, Q.Y., et al.: Development and applications of aluminum alloys for aerospace industry. J. Market. Res. **27**, 944–983 (2023). https://doi.org/10.1016/J.JMRT.2023.09.274
7. Huang, W., Lei, L., Fang, G.: Optimizing heat treatment to improve the microstructures and mechanical properties of 5CrNiMoV steel. Metals (Basel) **13** (2023). https://doi.org/10.3390/met13071263
8. Suryanarayana, C.: Mechanical alloying: a critical review. Mater. Res. Lett. **10**, 619–647 (2022)
9. Luqman, M., Ali, Y., Zaghloul, M.M.Y., et al.: Grain refinement mechanism and its effect on mechanical properties and biodegradation behaviors of Zn alloys—a review. J. Market. Res. **24**, 7338–7365 (2023). https://doi.org/10.1016/J.JMRT.2023.04.219
10. Carrillo, A., Saurina, J., Escoda, L., Suñol, J.J.: Fe-x-b-cu (X = nb, nizr) alloys produced by mechanical alloying: influence of milling device. Metals (Basel) **11**, 1–14 (2021). https://doi.org/10.3390/MET11030379
11. Suñol, J.J.: Mechanical alloying: processing and materials. Metals **11**, 11–798 (2021). https://doi.org/10.3390/MET11050798
12. Kuziora, P., Wyszyńska, M., Polanski, M., Bystrzycki, J.: Why the ball to powder ratio (BPR) is insufficient for describing the mechanical ball milling process. Int. J. Hydrogen Energy **39**, 9883–9887 (2014). https://doi.org/10.1016/J.IJHYDENE.2014.03.009
13. Fecht, H.J., Hellstern, E., Fu, Z., Johnson, W.L.: Nanocrystalline metals prepared by high-energy ball milling. Metall. Trans. A **21**, 2333–2337 (1990). https://doi.org/10.1007/BF02646980
14. Suryanarayana, C.: Recent Developments in Mechanical Alloying (2008)
15. Suryanarayana, C., Al-Joubori, A.A., Wang, Z.: Nanostructured materials and nanocomposites by mechanical alloying: an overview. Met. Mater. Int. **28**, 41–53 (2022)
16. Suryanarayana, C.: Mechanical alloying: a novel technique to synthesize advanced materials. Research (2019). https://doi.org/10.34133/2019/4219812

17. Pochet, P., Bellon, P., Chaffron, L., Martin, G.: Phase transformations under ball milling: theory versus experiment. Mater. Sci. Forum **225–227**, 207–216 (1996). https://doi.org/10.4028/www.scientific.net/MSF.225-227.207
18. Phasha, M.J., Bolokang, A.S., Ngoepe, P.E.: Solid-state transformation in nanocrystalline Ti induced by ball milling. Mater. Lett. **64**, 1215–1218 (2010). https://doi.org/10.1016/j.matlet.2010.02.054
19. Bezzon, V.D.N., Pinto, R. da S., de Araújo, G.L.B., et al.: Describing the influence of ball-milling on the amorphization of flubendazole using the PDF and RMC methods with X-ray powder diffraction data. J. Pharm. Sci. **111**, 3054–3063 (2022). https://doi.org/10.1016/j.xphs.2022.06.018
20. Joy, J., Krishnamoorthy, A., Tanna, A., et al.: Recent developments on the synthesis of nanocomposite materials via ball milling approach for energy storage applications. Appl. Sci. **12**, 9312 (2022). https://doi.org/10.3390/app12189312
21. Shashanka, R., Chaira, D.: Optimization of milling parameters for the synthesis of nano-structured duplex and ferritic stainless steel powders by high energy planetary milling. Powder Technol. **278**, 35–45 (2015). https://doi.org/10.1016/j.powtec.2015.03.007
22. Pradeep, N.B., Hegde, M.M.R., Manjunath Patel, G.C., et al.: Synthesis and characterization of mechanically alloyed nanostructured ternary titanium based alloy for bio-medical applications. J. Market. Res. **16**, 88–101 (2022). https://doi.org/10.1016/j.jmrt.2021.11.101
23. Baricco, M., Palumbo, M., Baldissin, D., et al.: Metastable phases and phase diagrams. In: Metallurgia Fisica (2003)
24. Hara, K.O., Yamasue, E., Okumura, H., Ishihara, K.N.: Formation of metastable phases by high-energy ball milling in the Ti-O system. J. Phys. Conf. Ser. **144**, 012021 (2009). https://doi.org/10.1088/1742-6596/144/1/012021
25. Fogagnolo, J.B., Velasco, F., Robert, M.H., Torralba, J.M.: Effect of mechanical alloying on the morphology, microstructure and properties of aluminium matrix composite powders. Mater. Sci. Eng. A **342**, 131–143 (2003). https://doi.org/10.1016/S0921-5093(02)00246-0
26. Tawancy, H.M.: Correlation between disorder-order transformations in a Ni-based alloy and its mechanical properties. Mater. Sci. Eng. A **719**, 93–103 (2018). https://doi.org/10.1016/j.msea.2018.02.033
27. Stingaciu, M., Topole, M., McGuiness, P., Christensen, M.: Magnetic properties of ball-milled $SrFe_{12}O_{19}$ particles consolidated by spark-plasma sintering. Sci. Rep. **5**, 14112 (2015). https://doi.org/10.1038/srep14112
28. Amrute, A.P., De Bellis, J., Felderhoff, M., Schüth, F.: Mechanochemical synthesis of catalytic materials. Chem. Eur. J. **27**, 6819–6847 (2021). https://doi.org/10.1002/CHEM.202004583
29. Bunaciu, A.A., Udriştioiu, E.G., Aboul-Enein, H.Y.: X-Ray diffraction: instrumentation and applications. Crit. Rev. Anal. Chem. **45**, 289–299 (2015). https://doi.org/10.1080/10408347.2014.949616
30. Jakubowicz, J.: Application of atomic force microscopy in microstructure analysis of mechanically alloyed $Nd_2Fe_{14}B/\alpha$-Fe-type nanocomposites. J. Alloys Compd. **351**, 196–201 (2003). https://doi.org/10.1016/S0925-8388(02)01075-7
31. Elkordy, A.A.: Differential scanning calorimetry, isothermal titration calorimetry and microcalorimetry. In: Applications of Calorimetry in a Wide Context—Differential Scanning Calorimetry, Isothermal Titration Calorimetry and Microcalorimetry 484 (2013)
32. Shimodaira, N., Masui, A.: Raman spectroscopic investigations of activated carbon materials. J. Appl. Phys. **92**, 902–909 (2002). https://doi.org/10.1063/1.1487434
33. Kim, J.H., Hwang, I.S.: Development of an in situ Raman spectroscopic system for surface oxide films on metals and alloys in high temperature water. Nucl. Eng. Des. **235**, 1029–1040 (2005). https://doi.org/10.1016/J.NUCENGDES.2004.12.002
34. Chao, J., Cao, W., Su, S., et al.: Nanostructure-based surface-enhanced Raman scattering biosensors for nucleic acids and proteins. J. Mater. Chem. B **4**, 1757–1769 (2016). https://doi.org/10.1039/C5TB02135A
35. Dodrill, B., Lindemuth, J.R.: Vibrating sample magnetometry. In: Magnetic Measurement Techniques for Materials Characterization, pp. 15–37 (2021). https://doi.org/10.1007/978-3-030-70443-8_2/COVER

36. Schmelz, M., Stolz, R.: Superconducting quantum interference device (SQUID) magnetometers. Smart Sens. Meas. Instrum. **19**, 279–311 (2017). https://doi.org/10.1007/978-3-319-34070-8_10/COVER
37. Paterson, G.A., Zhao, X., Jackson, M., Heslop, D.: Measuring, processing, and analyzing hysteresis data. Geochem. Geophys. Geosyst. **19**, 1925–1945 (2018). https://doi.org/10.1029/2018GC007620
38. Goodman, B.A.: Mossbauer spectroscopy. In: Developments in Sedimentology, vol. 34, pp. 113–137 (1982). https://doi.org/10.1016/S0070-4571(08)70018-9
39. Dodrill, B., Reichard, H.S.: Alternating gradient magnetometry. In: Magnetic Measurement Techniques for Materials Characterization, pp. 139–149 (2021). https://doi.org/10.1007/978-3-030-70443-8_7/COVER
40. Harrell, J.W.: Effect of AC gradient field on magnetic measurements with an alternating gradient magnetometer. J. Magn. Magn. Mater. **205**, 121–129 (1999). https://doi.org/10.1016/S0304-8853(99)00487-4
41. O'Grady, K., Lewis, V.G., Dickson, D.P.E.: Alternating gradient force magnetometry: applications and extension to low temperatures (invited). J. Appl. Phys. **73**, 5608–5613 (1993). https://doi.org/10.1063/1.353613
42. Flanders, P.J.: An alternating-gradient magnetometer (invited). J. Appl. Phys. **63**, 3940–3945 (1988). https://doi.org/10.1063/1.340582
43. Hautojärvi, P.: Positrons in solids. Top. Current Phys. **12** (1979)
44. Sedov, V.L.: Annihilation of positrons in metals. Usp. Fiz. Nauk **94**, 3 (1968). https://doi.org/10.3367/ufnr.0094.196803b.0417
45. Dekhtyar, I.Y.: The use of positrons for the study of solids. Phys. Rep. **9**, 243 (1974). https://doi.org/10.1016/0370-1573(74)90021-0
46. Grafutin, V.I., Prokop'ev E.P.: Positron annihilation spectroscopy in materials structure studies. Phys. Usp. **45**, 59 (2002). https://doi.org/10.1070/PU2002V045N01ABEH000971
47. Siegel, R.W.: Positron Annihilation Spectroscopy. **10**, 393–425 (2003). https://doi.org/10.1146/ANNUREV.MS.10.080180.002141

Chapter 4
An Overview of Mechanical Alloying and Conventional Metallurgical Methods

Vishwanath F. Bhajantri, Gangadhar M. Kanaginahal, and Sudhakar Jambagi

Abstract Mechanical alloying (MA) is a powder processing technique that involves the cyclic process of cold welding, fracturing, and rewelding of powder particles within a high-energy ball mill. This method has the ability to create several forms of non-equilibrium phases, such as supersaturated solid solutions, nanocrystalline materials, metallic glasses, high-entropy alloys, and more. Materials with shape memory properties have been utilized in various applications. This article enlightens on MA and other conventional metallurgical methods. The article provides a description of the current and future applications of MA.

Keywords Mechanical alloying · Cold-welding · High-energy ball mill · High entropy alloys

4.1 Introduction

Materials scientists have always focused their scientific efforts on enhancing the characteristics and functionality of materials. Chemistry changes and standard thermal, mechanical, and thermomechanical processing procedures have led to substantial enhancements in the mechanical, chemical, and physical properties. Nevertheless, the growing need for materials that are more heat-resistant, stronger, more rigid, and lighter than conventional materials has prompted the creation and advancement of sophisticated resources. The technologically advanced sectors have given these efforts an extra boost.

V. F. Bhajantri (✉) · S. Jambagi
Surface Engineering Laboratory, Department of Mechanical Engineering, National Institute of Technology Karnataka Surathkal, Mangaluru, India
e-mail: vishwanath.177me023@nitk.edu.in

G. M. Kanaginahal
Department of Mechanical Engineering, Jain College of Engineering and Research, Belagavi, Karnataka 590008, India

S. Rajendrachari (ed.), *Mechanically Alloyed Novel Materials*, Advanced Structured Materials 220, https://doi.org/10.1007/978-981-97-6504-1_4

Innovative substances are distinguished by their emphasis on purposeful production and accurate regulation of material structure, with the goal of developing specialized attributes that are well-suited to challenging applications [1]. It is commonly accepted that processing advanced materials under unstable environments can more effectively modify their structure and content [2]. Several commercial procedures, such as fast solidification from a liquid state, mechanical alloying, plasma processing, and vapor deposition, have received significant interest from researchers [2–4]. All of these strategies address the fundamental principle of creating non-equilibrium materials by energizing and rapidly cooling them. Energization is the process of producing a highly unstable state in a material using external dynamic forces such as it melts, evaporation, exposure to radiation, pressure, or mechanical energy storage via deformation caused by plasticity [5, 6]. The process of energization often entails a potential transition from a solid state to a liquid or gaseous one. The material is rapidly chilled to a fixed configuration, which is able to serve as a starting point for achieving the ideal chemistry and/or morphology via subsequent thermal treatment or process. Research has shown that substances generated using this technology have superior physical as well as mechanical characteristics when compared to materials processed using typical ingot (solidification) processes.

To evaluate the ability of alternative processing algorithms to produce metastable structures, one can effectively measure the divergence from equilibrium by measuring or approximating the highest amount of energy that is able to be stored above that of the steady form. Various groups have conducted this using diverse non-equilibrium processing approaches. The energy is measured in kilojoules per mole (kJ/mol) [7–10]. The method of calculating the deviation varies in certain cases. Various calculations yield different results, therefore not corresponding perfectly in all circumstances. Nevertheless, it is clear that the deposition of vapor and implantation of ions techniques deviate significantly from a state of equilibrium resulting in large practical quenching rates. Furthermore, it is clear that metallic alloying is a procedure that allows for the processing of materials that are further from equilibrium than fast solidification. This approach has shown promising results in the production of non-equilibrium materials. Table 4.1 provides a summary of the departures that have been estimated for the various processing approaches.

This book chapter will look at recent advances in the production of stable and metastable amalgam phases employing a low-cost and simple process technique known as physical alloying/milling of metallic powders. The structure of the review will be as follows: Principles, Mechanisms, Applications, Advantages and challenges, Conventional metallurgical methods, Casting, Powder metallurgy, Heat treatment, Comparative analysis, and recent advances and future trends [2, 11].

Table 4.1 Characteristics of mechanical alloying [2]

Production of fine dispersion of second-phase (usually oxide) particles
Extension of solid solubility limits
Refinement of grain sizes down to the nanometer range
Synthesis of novel crystalline and quasicrystalline phases
Development of amorphous (glassy) phases
Disordering of ordered intermetallics
Possibility of alloying of difficult-to-alloy elements
Inducement of chemical (displacement) reactions at low temperatures
Scalable process

4.2 Mechanical Alloying

Mechanical alloying is a process that involves cold welding, breaking down, and rewelding of powdered materials in an extremely energetic ball mill. In 1966, John Benjamin at INCO Laboratories developed MA, or mechanical alloying, in answer to an industrial demand [12, 13]. The researchers were looking for a metal alloy that might be strengthened at ambient temperature using precipitate hardening and at higher temperatures using diffusion hardening. In the early 1960s, INCO successfully discovered a process for making graphitic aluminum alloys. This involved introducing nickel-coated graphite particles into a molten aluminum bath using argon sparring. An adaptation of the same process was used to introduce nickel-based alloys containing a suspension of thin resistant oxide particles covered with nickel [13–15]. The goal of applying a nickel coat was to improve the wetting capabilities of oxide particles, which are normally non-wettable, by employing a nickel–chromium alloy. The early efforts used zirconium oxide coated with metal, however this did not yield the desired results. A lengthy examination determined that the research failed because the seller provided powder that was layered with zirconia on top of nickel, rather than nickel on top of zirconia. Because of the extremely exothermic nature of the aluminum-nickel interaction, the thermal radiation generated efficiently purified the graphite surface and lowered its surface energy. Based on this assumption, it was believed that applying a layer of aluminum to the refractory oxide would be the optimal method to generate the exothermic process. This endeavor likewise failed to yield positive results. After other attempts proved unsuccessful, the ball milling method was considered as a last resort. This method was formerly used to cover strong phases like tungsten carbide with delicate phases like cobalt or nickel. It was additionally found that metallic powder fragments might be fractured by exposing them to extreme deformation due to plasticity. The use of particular compounds can be used to form smaller particles by inhibiting cold welding, implying that cold welding may occur as quickly as fracture at some points. The element's responsiveness has to be taken into account. After carefully considering all these considerations, the following methods were proposed by Benjamin to manufacture composite powder particles through a specific process [16].

- Using an energy-intensive mill can increase plastic deformation during cold welding, reducing procedure time.
- Combining basic and superior alloy powders reduces the element's activity. This is due to the fact that the chemical reaction in an alloy or mixture is frequently much lower than that in pure materials.
- Avoid using surface-active chemicals that can produce fine pyrophoric powder and pollute the powder.
- Creating a powder with a detailed inner structure, akin to extremely tiny particles, involves constant welding and fracturing. However, this powder has a relatively coarse overall particle size, which makes it stable.

4.2.1 Principles

Mechanical alloying is a dry and extremely energetic process of using ball mills that have been used to manufacture a variety of commercially lucrative and scientifically intriguing materials. Mechanical operations can cause the formation of an unstructured phase. The crushing of a Y Co intermetallic mixture in 1981 and the use of balls in the milling of integrated elemental powder mixtures in the Ni Nb system in 1983 revealed that MA is a potentially unstable processing technique. Since the mid-1980s, multiple investigations have been carried out to develop various stable and metastable phases, including over-saturated solid solutions, crystalline and quasicrystalline intermediate phases, and amorphous alloys. Furthermore, it was established that powdered mixtures can be mechanically activated to initiate chemical reactions, known as mechanochemical reactions, at ambient temperature or even significantly lower temperatures than those required for the production of pure metals, nanocomposites, and other commercially valuable materials. Modeling studies have been conducted since the early 1990s to gain a better understanding of the fundamental processes of MA. Due to its unique characteristics, this uncomplicated yet efficient processing method has been utilized on various materials such as metals, ceramics, polymers, and composite materials [2, 17].

4.2.2 Mechanisms

Extremely energetic milling causes granules of powder to flatten, cold weld, fracture, and reweld [18–20]. When two steel balls collide, a certain amount of powder becomes trapped between them. Each impact typically captures roughly 1000 particulates weighing roughly 0.2 mg. The collision puts a large force on the powder particles, causing plastic deformation, work hardening, and fracture. The freshly created interfaces enhance particle fusion, increasing their size. During the initial phases

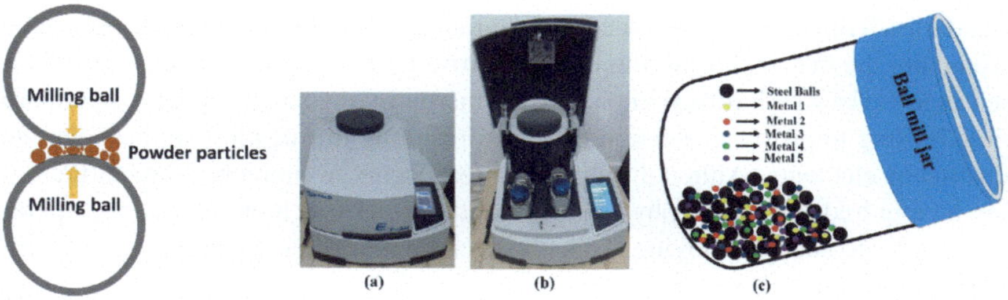

Fig. 4.1 Collision between balls containing a combination of powder particles during mechanical alloying [22]

of milling, when utilizing a mix of ductile-ductile or ductile–brittle materials, the particles are relatively soft. As a result, there is a significant likelihood that they may fuse together and create larger particles. A wide spectrum of particle sizes is formed, with certain particles being three times larger than the original ones. At this point, the composite elements have a distinct structure of layers that includes many combinations of the basic constituents. The particles harden as a result of continuous deformation and eventually shatter due to failure from fatigue or disintegration of fragile flakes. When there are no major constraints to cluster together, fragments created using this approach have the potential to shrink even further. At this time, the propensity for fracturing is more prevalent than cold welding. The grinding balls continue to improve the particle structure while keeping the size of the particles constant. As a result, the space between layers decreases while the total number of layers within a particle grows [20, 21] (Fig. 4.1).

Nonetheless, it is vital to note how the effectiveness of reducing the size of particles is surprisingly low, around 0.1%, when utilizing a standard rolling crusher. The effectiveness of extremely energetic ball milling methods may increase slightly, although it stays below 1%. The leftover energy is mostly wasted as heat, with a small percentage dedicated to the elasticity and plasticity of the particulates of powder. After a certain amount of milling time, a harmony is established between the pace of welding, which increases the average size of particles, and the rate of fracture, which decreases the normal composite size of the particles. Particles that are smaller may withstand strain without breaking and often merge together to form larger chunks. As a result, both extremely small and huge particles are pushed to intermediate sizes. At this point, every bit contains virtually all of the beginning components in the same quantities that they were combined. The particles become exceedingly hard due to the accumulation of strain energy. The existing particle size distribution is constrained by the simultaneous reduction in size of larger particles and the proliferation of smaller fragments caused by the aggregation of smaller particles [2, 23].

In light of the aforementioned information, it is evident that significant deformation occurs in the particles during MA. It is clear that various crystal defects exist, including fractures, vacant spaces, stacking faults, and a greater amount of grain boundaries. This fault pattern speeds up the rate at which liquid elements propagate throughout the matrix. Furthermore, the improved microscopic characteristics limit the ranges over which diffusion occurs. Furthermore, the small rise in temperature while milling improves dispersion behavior, resulting in true alloying between the constituent elements. Although alloying often occurs at ambient temperatures, it may be required to heat the physically alloyed powder to a more elevated temperature in order to achieve alloying. This is especially true when the purpose is to form intermetallic complexes.

The time required to build a specific structure in any system is determined by the beginning size and properties of the particles, the quality of the components, the particular machinery used for the MA process, and the equipment's operating parameters. However, in most cases, the rate at which the internal framework (such as particle size, crystallite size, lamellar spacing, etc.) refines is roughly exponential in relation to computation time. As a result, the particles' initial size is rather modest. Within a few minutes to an hour, the membrane-like gap is normally reduced, resulting in the improvement of the crystallized (or grain) size to nanoscale levels. The simplicity of synthesizing nanostructured materials is a key factor in the widespread use of MA for producing nanocrystalline materials.

As previously stated, it is feasible to do mechanical analysis on three distinct combinations of metals and alloys: (i) ductile-ductile, (ii) ductile–brittle, and (iii) brittle-brittle systems [2].

4.2.3 *Applications*

Mechanically alloyed materials are widely used in a variety of industrial sectors, including thermally processed glass interpreting, energy generation, aerospace, and others. These uses take advantage of the oxide-dispersion hardening effect achieved by mechanically alloying nickel, iron, and aluminum alloys (Fig. 4.2).

Mechanically alloyed materials have high strength at both normal and high temperatures. The increased thermal resistance of these materials is attributed to multiple mechanisms. The alloy's resistance to creep deformation is increased by the uniform dispersion of stable oxide particles (such as Y_2O_3, ThO_2, and La_2O_3) with a spacing

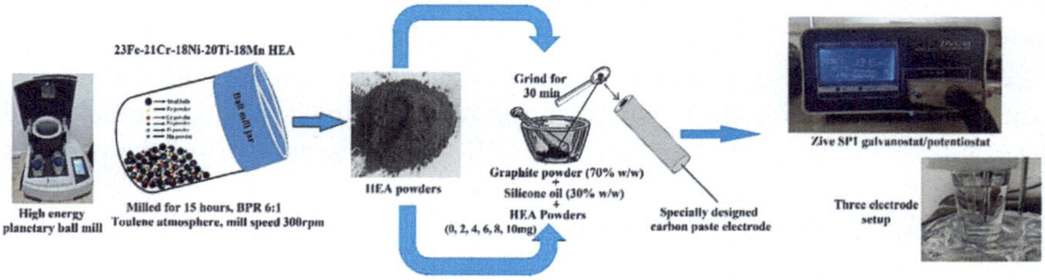

Fig. 4.2 Application of high entropy alloy in electrochemical sensor [11]

of approximately 100 nm and a size range of 5–50 nm [2, 15, 23]. The resulting dispersion significantly reduces the movement of dislocations in the metal solution. Dispersed particles also serve to impede the process of recovery and the recrystallization processes, resulting in the formation of a highly stable, large particle size. When deformed at high temperatures, these big grains successfully withstand granular rotation. Secondary recrystallization processes can also produce consistent grain sizes. Furthermore, the homogeneous dispersion of alloying elements during mechanical alloying (MA) improves resilience at elevated temperatures and overall properties for both solid-solution toughened and precipitation-hardened alloys. Mechanically alloyed materials are very resistant to oxidation and thermal corrosion. The increased resistance to oxidation-sulfidation attack can be attributed to the equal spread of the components of the alloy as well as the enhanced scale adhesion caused by the presence of the dispersion [24].

The benefit of increased solid solubility even in immiscible solutions is another benefit that MA offers in addition to nanoscale processing. The increased diffusion rates caused by the powder components' nano size prior to alloying can be the cause of it. Thus, in addition to enhancing configurational entropy, MA improves the stability of phases of solid solutions in High Entropy Alloys (HEAs). Because of the well-documented effects of various milling variables, including milling type, ball-to-powder ratio, speed, and time, the microstructure and properties of HEAs can be customized. Using equipment like planetary ball mills, attritor mills, and uniball mills, MA makes it possible to produce powder on a wide scale, in contrast to chemical methods and deposition procedures for nanomaterial synthesis [25]. This has previously demonstrated its value in the production of plasma-sprayed HEA coatings, and it would be appropriate for the industrial processing of nanocrystalline HEAs [26].

4.2.4 Advantages and Challenges

As previously stated, mechanical alloying has the ability to produce multiple metastable phases in various alloy systems, including over-saturated solid solutions and heterogeneous alloys (metallic glasses). Furthermore, a crucial aspect of MA is to enhance the precision and granularity of the particle. The diameters can be reduced to nanoscale levels if the powder remains in a crystalline state. Therefore, MA is a very suitable technique for synthesizing nanocrystalline HEAs. Furthermore, it was shown that the HEAs subjected to MA can undergo a transition to the amorphous state when the powders are continuously milled. As a result, it is possible to make both a nanocrystalline form and glassy HEAs, or combinations of them, by physically blending the powders for the appropriate time. This approach works entirely in the solid state. As a result, the state of the liquid and the diagram of phase limitations are irrelevant. In contrast, it is accurate in the case of HEAs because these alloys consist of a minimum of five components, each with potentially distinct melting temperatures. This scenario could present challenges in the processes of melting and casting, as well as potential issues of segregation. MA that has a positive heat of mixing, meaning they are not easily mixed together, can separate into different phases when in liquid and solid forms. This can cause difficulties in melting the alloy and achieving the desired properties. Uniformity of the molten substance. However, MA, as a fully solid-state processing technique, does not encounter any such issues as it does not entail any melting. MA has the ability to increase the maximum amount of one substance that can be dissolved in another in alloy systems, even systems where the substances do not normally mix. As a result, utilizing MA to handle HEAs greatly simplifies the process of establishing a stable solution phase. Occasionally, HEA alloys have been successfully produced in an amorphous state. However, the shift to a homogenous solid-state phase is not yet documented. The following raises the concern of whether the solid solution phase is the state of equilibrium phase in these complex alloys [24, 27].

4.3 Conventional Metallurgical Methods

4.3.1 Casting

The practice of casting has been extensively utilized for ages [28]. It is renowned for its capacity to manufacture components with exceptional surface polish, precise dimensions, and intricate shapes. It is especially useful for producing castings with complicated and nearly flawless form geometry, when grinding may be impracticable or inefficient. This is additionally regarded as the oldest method of producing art castings. Technological advancements have positioned it as the most contemporary and adaptable method among all metal casting techniques. The ideas can be dated back to 5000 BC, when early humans utilized this technique to create basic tools. A variety of products are made via the casting technique, such as electrical boxes, hip replacement implants, golf club heads, turbocharger wheels, and parts for general, aerospace, and defense applications. In terms of value, investment castings made of steel account for 33% of total output [26, 29]. Aluminum and its alloys have a diverse variety of applications among non-ferrous alloys. Investment castings do not have any metallurgical restrictions. The primary benefit of this method is its exceptional surface finish. It is possible to effectively cast splines, holes, bosses, writing, and even some threads. This method is capable of producing intricate details and slender parts. This procedure does not require complex or costly equipment. Nevertheless, the high physical effort involved in creating the wax-based patterns and ceramic slurry makes this process more expensive. The article gives a summary of the different uses for investment casting and comes to the conclusion that the casting's size and weight are usually the technique's limitations [28, 30].

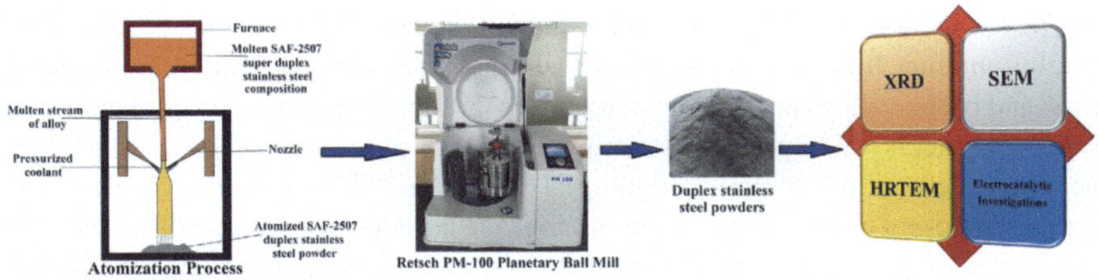

Fig. 4.3 Ball milling of duplex stainless steel powder [29]

4.3.2 Powder Metallurgy

Powder metallurgy is a manufacturing process that involves the production of metal parts and components using powdered metal materials [22]. It involves the reciprocal diffusion of reinforcements and the matrix material in a solid-state environment, facilitated by elevated temperature and pressure. The most preferred process for producing aluminum metal matrix composites with excellent mechanical characteristics is powder metallurgy (Fig. 4.3).

The incorporation of graphene and silicon carbide into the aluminum matrix is effectively addressed by powder metallurgy, overcoming the associated restrictions. Furthermore, this procedure is considered net-shape, meaning that minimal or no additional machining is needed after the composite is fabricated. As a result, material loss and expense are reduced. The process has two main stages: firstly, the combination of the basic matrix with reinforcement materials; secondly, the consolidation of the mixture into the desired shape using various techniques. The clustering of reinforcement inside the base matrix diminishes the strength of the composite. To have the best properties in the composite material, it is therefore desirable to have a homogeneous distribution of reinforcement. The first phase in powder metallurgy is the blending of basic powders comprising a base matrix and reinforcing materials. This can be achieved by employing ball milling and sonication in a liquid medium. It aids in achieving a consistent dispersion of reinforcement, ideal particle morphologies, and decreased powder particle size [22, 27].

4.3.3 Heat Treatment

Heat treating is a controlled process that involves heating and cooling metal in order to enhance its characteristics, performance, and durability. A material is subjected to regulated heating and cooling during the heat treatment (HT) process for the purpose to change its mechanical and physical characteristics [31, 32]. The purpose of controlling the microstructure of the metal is to enhance its processability, formability, and machinability, and to increase its qualities such as durability, corrosion resistance, wear resistance, strength, and fatigue resistance. HT can be applied to all types of metals and alloys. Aluminum alloys undergo the processes of annealing or solutionizing, followed by quenching and aging to achieve hardening. Copper, brass, and bronze alloys undergo heat treatment through annealing and age-hardening processes. Steels undergo heat treatment through processes such as annealing, quenching, tempering, or age hardening. This article will discuss the crucial heat-treating processes of annealing, solutionizing, age hardening, and controlled cooling, focusing on alloy systems and the need for process design and control. The debate is structured based on the characteristics of phase diagrams and the morphology and kinetics of phase transformations [33, 34].

4.4 Challenges and Future Trends

As mentioned earlier, MA is a versatile technique that has the potential to produce a variety of unstable (metastable) substances with interesting characteristics. Using mechanically alloyed materials raises serious concerns about milled powder contamination. While this is usually not desirable, there are a few situations in which it can be helpful. The presence of impurities, particularly carbon, oxygen, and nitrogen, in powdered materials, can significantly degrade their mechanical properties. These contaminants could have come from the milling media, the original powder, or the ambient. Consolidation may add impurities to the powder. Consequently, if the ground powder may be used as is, then the amount of impurities can be reduced. Conversely, when these chemicals are evenly distributed throughout the matrix, the material can exhibit increased strength and hardness, albeit with a higher likelihood of being brittle. Therefore, it is crucial to carefully select the appropriate material for a given application to ensure that these impurities do not negatively impact the material's qualities and performance. Therefore, if the application necessitates materials that cannot withstand contaminants, it is necessary to take specific measures to reduce these impurities to a level below the acceptable limit or eliminate them entirely. It is feasible and achievable, but the cost of the powder increases. However, certain applications such as paints, solders, coatings, and so on [2, 24].

4.5 Conclusion

Scholars are still sincerely drawn to mechanical alloying since it is a straightforward, sophisticated, and effective processing technique. The synthesis of several alloy phases, including solid solutions, quasicrystalline and crystalline intermetallic phases, and amorphous phases, has led to a large amount of study being done in recent years. The method has been extended to cover a greater variety of alloys, even though it was first designed to produce ODS superalloys. Compared to traditional methods, mechanical alloying presents a gray area for researchers.

References

1. Bloor, D., Brook, R.J., Flemings, M.C., Mahajan, S. (eds.): The Encyclopedia of Advanced Materials. Pergamon Press, Oxford (1994)
2. Suryanarayana, C. (ed.): Non-equilibrium Processing of Materials. Pergamon Press, Oxford (1999)
3. Anantharaman, T.R., Suryanarayana, C.: Rapidly Solidified Metals—A Technological Overview. Trans Tech Publications, Aedermannsdorf, Switzerland (1987)
4. Suryanarayana, C.: Bibliography on Mechanical Alloying and Milling. Cambridge International Science Publishing, Cambridge, UK (1995)
5. Suryanarayana, C.: Met. Mater. **2**, 195–209 (1996)
6. Murty, B.S., Ranganathan, S.: Internat. Mater. Rev. **43**, 101–141 (1998)
7. Shingu, P.H.: In: Henein, H., Oki, T. (eds.) Processing Materials for Properties, pp. 1275–1280. TMS, Warrendale, PA (1993)
8. Froes, F.H., Suryanarayana, C., Russell, K., Ward-Close, C.M. In: Singh, J., Copley, S.M. (eds.)
9. Singh, J., Steve, M.: Novel Techniques in Synthesis and Processing of Advanced Materials, pp. 1–21. TMS, Warrendale, PA (1994)
10. Froes, F.H., Suryanarayana, C., Russell, K., Li, C.-G.: Mater. Sci. Eng. **A19/193**, 612–623 (1995)
11. Benjamin, J.S.: Sci. Am. **234**(5), 40–48 (1976)
12. Rajendrachari, S., Adimule, V., Gulen, M., Khosravi, F., Somashekharappa, K.K.: Synthesis and characterization of high entropy alloy 23Fe-21Cr-18Ni-20Ti-18Mn for electrochemical sensor applications. Materials **15**, 1–12 (2022)
13. Benjamin, J.S.: In: Arzt, E., Schultz, L. (eds.) New Materials by Mechanical Alloying Techniques, pp. 3–18. DGM Informationgesellschaft, Oberursel, Germany (1989)
14. Shashanka, R.: Synthesis of nano-structured stainless steel powder by mechanical alloying-an overview. Int. J. Sci. Eng. Res. **8**, 1–7 (2017)
15. Benjamin, J.S.: Metal. Powder Rep. **45**, 122–127 (1990)
16. Pradeep, N.B., Rajath Hegde, M.M., Rajendrachari, S., Surendranathan, A.O.: Investigation of microstructure and mechanical properties of microwave consolidated TiMgSr alloy prepared by high energy ball milling. Powder Technol. **408**, 1–12 (2022)
17. Rajendrachari, S., Basavegowda, N., Vinaykumar, R., Narsimhachary, D., Somu, P., Lee, M.-J.: Electrocatalytic determination of methyl orange dye using mechanically alloyed novel metallic glass modified carbon paste electrode by cyclic voltammetry. **155**, 1–10 (2023)
18. Usef, A.P., Bhajantri, V., Kannoth, V., Jambagi, S.C.: Influence of carbon nanotube reinforcement on the heat transfer coefficient, microstructure, and mechanical properties of a die cast Al-7Si-0.35 Mg alloy. J. Alloy. Compd. **888**, 160844 (2021)
19. Shashanka, R., Chaira, D.: Optimization of milling parameters for the synthesis of nano-structured duplex and ferritic stainless steel powders by high energy planetary milling. Powder Technol. **278**, 35–45 (2015)
20. Chaira, D.: Development of nano-structured duplex and ferritic stainless steels by pulverisette planetary milling followed by pressureless sintering. Mater. Charact. **99**, 220–229 (2015)
21. Jayaprakash, G.K., Rajendrachari, S., Adimule, V.M.: Electrochemical oxidation of methylene blue dye in wastewater using mechanically alloyed high entropy alloy modified carbon paste electrode using cyclic voltammetry. Mater. Res. Express **10**, 1–12 (2023)
22. Kang, B., Lee, J., Ryu, H.J., Hong, S.H.: Ultra-high strength WNbMoTaV high-entropy alloys with fine grain structure fabricated by powder metallurgical process. Mater. Sci. Eng. A **712**, 616–624 (2018)
23. Basavarajappa, P.N., Hegde, M.M.R., Rajendrachari, S., Surendranathan, A.O.: Investigation of structural and mechanical properties of nanostructured TiMgSr alloy for biomedical applications. **13**, 1–10 (2023)
24. Suryanarayana, C.: Mechanical alloying: a critical review. Mater. Res. Lett. **10**(10), 619–647 (2022)

25. Vaidya, M., Muralikrishna, G.M., Murty, B.S.: High-entropy alloys by mechanical alloying: a review. J. Mater. Res. **34**(5), 664–686 (2019)
26. Bhajantri, V.F., Jambagi, S.C.: Factors influencing powders' flowability and favorable phases like crystalline (Mullite and quartz) and amorphous phases of plasma-sprayed fly ash coatings suitable for marine and offshore applications. Adv. Powder Technol. **34**(9), 104150 (2023)
27. Fu, Z., Chen, W., Jiang, Z., MacDonald, B.E., Lin, Y., Chen, F., Lavernia, E.J.: Influence of Cr removal on the microstructure and mechanical behaviour of a high-entropy Al0. 8Ti0. 2CoNiFeCr alloy fabricated by powder metallurgy. Powder Metall. **61**(2), 106–114 (2018)
28. Pattnaik, S., Karunakar, D.B., Jha, P.K.: Developments in investment casting process—a review. J. Mater. Process. Technol. **212**(11), 2332–2348 (2012)
29. Mahale, R.S., Vasanth, S., Chikkegouda, S.P., Rajendrachari, S., Narsimhachary, D., Basavegowda, N.: Electrochemical determination of ascorbic acid by mechanically alloyed super duplex stainless steel powders. **13**, 1–10 (2023)
30. Aleksandrov, V.M., Kulakov, B.: Increasing the strength and reducing the chemical activity of ceramic molds. Sov. Cast. Technol. **4**, 41–45 (1988)
31. Domanski, M., Webb, J.: A review of heat treatment research. Lithic Technol. **32**(2), 153–194 (2007)
32. Chauhan, K.P.S.: Influence of heat treatment on the mechanical properties of aluminium alloys (6xxx series): a literature review. Int. J. Eng. Res. **6**(03), 386–389 (2017)
33. Yumak, N., Aslantaş, K.: A review on heat treatment efficiency in metastable β titanium alloys: the role of treatment process and parameters. J. Market. Res. **9**(6), 15360–15380 (2020)
34. Mahale, R.S., Rajashekar, V., Vasanth, S., Chikkegowda, S.P., Rajendrachari, S., Mahesh, V.: Fabrication of mechanically alloyed super duplex stainless steel powder-modified carbon paste electrode for the determination of methylene blue by the cyclic voltammetry technique. ACS Omega 1–12 (2024)

Chapter 5
How to Achieve High-Density Materials

Ruchin Kacker, Sanjay Kumar Singh, and Sourabh Tandon

Abstract High-density materials serve as the bedrock of innovation across various industries, from aerospace to health care. This book chapter delves into the dynamic world of high-density materials, providing an extensive exploration of their properties, production techniques, applications, challenges, and future prospects. One of the core methodologies spotlighted in this chapter is mechanical alloying, a versatile technique for fabricating high-density materials. The chapter commences with a thorough introduction to high-density materials, elucidating their fundamental attributes and highlighting the paramount importance of achieving optimal density in diverse applications. Mechanical alloying is introduced as a pivotal process, offering precision, scalability, and versatility. It allows the blending and refinement of elemental or pre-alloyed powders through high-energy milling, unlocking an array of possibilities for material tailoring. This in-depth exploration encompasses factors affecting material density during mechanical alloying, including milling time, speed, temperature, starting materials, and ball-to-powder ratio. Real-world case studies exemplify successful high-density material production, shedding light on the materials involved, process parameters, and resulting properties. Yet, challenges and limitations are acknowledged in the pursuit of high-density materials, spanning concerns like temperature control, contamination, agglomeration, and equipment wear. Vital characterization techniques are presented, offering insight into material density measurement and confirmation. The chapter further delves into the multi-faceted applications of high-density materials across industries like aerospace, automotive, health care, electronics, defense, nuclear, and energy. Moreover, it explores emerging trends, such as nanoscale materials, sustainability, multi-functionality, and additive manufacturing, which signify the future of high-density materials in shaping innovative technologies. As we peer ahead into a future teeming with potential, high-density

R. Kacker (✉) · S. K. Singh
Department of Mechanical Engineering, Sagar Institute of Science & Technology, Bhopal, Madhya Pradesh, India
e-mail: ruchinkacker@sistec.ac.in; ruchinkacker@gmail.com

S. Tandon
Ph.D Scholar, Department of Mechanical Engineering, Indian Institute of Technology Bhubaneswar, Odisha, Bhopal, Madhya Pradesh, India

S. Rajendrachari (ed.), *Mechanically Alloyed Novel Materials*, Advanced Structured Materials 220, https://doi.org/10.1007/978-981-97-6504-1_5

materials crafted through mechanical alloying remain central to material science, propelling advancements in technology and opening new vistas of possibility.

Keywords High-density materials · Mechanical alloying · Material characterization · Nano materials · Density measurement

5.1 Introduction to High-Density Materials

High-density materials are a class of substances that exhibit exceptionally high mass or density relative to their volume. Density, in this context, refers to the amount of mass (usually measured in kilograms or grams) contained within a given volume (usually measured in cubic centimeters or cubic meters). In other words, high-density materials are characterized by having a substantial amount of mass tightly packed into a relatively small space (Table 5.1).

5.1.1 Importance of High-Density Materials

High-density materials play a pivotal role in various industrial, scientific, and technological applications due to their unique properties. Understanding their significance is crucial for researchers, engineers, and professionals in fields such as materials science, engineering, and manufacturing. Here are some key reasons why high-density materials are highly valued.

- Weight Efficiency: High-density materials are prized for their ability to provide significant mass in compact forms. This attribute is particularly beneficial in industries where weight efficiency is critical, such as using high-density materials that allows for the construction of lightweight yet robust components and structures, contributing to fuel efficiency and overall performance [8].
- Radiation Shielding: Another essential application of high-density materials is in radiation shielding. Materials with high density are effective at attenuating ionizing radiation, making them indispensable in nuclear power plants, medical facilities, and space exploration. Lead, for instance, is a commonly used high-density material in radiation shielding applications [9].
- Counterweights and Balancing: High-density materials are frequently employed as counterweights and balancing components. They help maintain stability in various systems, such as heavy machinery, elevators, and pendulum clocks, by counteracting the effects of uneven weight distribution.
- Medical Applications: In the medical field, high-density materials find use in diagnostic imaging equipment like X-ray machines and CT scanners. Their ability to absorb X-rays effectively enables the creation of detailed images of the human body [10].

Table 5.1 Characteristics of high-density materials

Material	Density/ (g/cm^3)	Applications	Key properties
Tungsten (W)	19.25	Aerospace (radiation shielding), electronics (filaments) [1]	High melting point, excellent radiation shielding
Osmium (Os)	22.59	Electrical contacts, fountain pen tips [2]	The densest naturally occurring element
Platinum (Pt)	21.45	Jewelry, electronics, catalytic converters	High resistance to corrosion, malleability [3]
Iridium (Ir)	22.56	Aerospace (rocket nozzles), electrical contacts	Exceptional corrosion resistance, high density [4]
Gold (Au)	19.32	Jewelry, electronics, dentistry	Excellent electrical conductivity, ductility
Uranium (U)	19.05	Nuclear reactors, munitions, radiation shielding	Radioactive, fissionable, dense
Lead (Pb)	11.34	Radiation shielding, batteries, roofing	Excellent radiation attenuation, malleability
Bismuth (Bi)	9.80	Pharmaceuticals, alloys, cosmetics	Low toxicity, high diamagnetism
Tantalum (Ta)	16.69	Aerospace (heat shields), electronics (capacitors)	High melting point, excellent corrosion resistance [5]
Thorium (Th)	11.72	Nuclear reactors, gas mantles	Radioactive, fertile, high density
Rhenium (Re)	21.04	Aerospace (jet engines), electrical contacts [6]	High melting point, excellent wear resistance
Copper (Cu)	8.96	Electronics, electrical wiring, plumbing	High electrical conductivity, ductility
Iron (Fe)	7.87	Construction (reinforcement), automotive (engine blocks)	Abundant, magnetic, versatile
Nickel (Ni)	8.91	Aerospace (alloys), electronics (batteries)	High corrosion resistance, magnetic
Cobalt (Co)	8.9	Aerospace (alloys), medical implants (alloys)	High magnetic properties, corrosion resistance

(continued)

Table 5.1 (continued)

Material	Density/ (g/cm^3)	Applications	Key properties
Molybdenum (Mo)	10.22	Electronics (filaments), aerospace (alloys)	High melting point, good electrical conductivity
Zinc (Zn)	7.14	Galvanization, batteries, die casting	Low melting point, corrosion resistance
Platinum Iridium (Pt-Ir)	21.45	Standard for international prototype of the kilogram (SI unit)	Extremely dense, stable, and corrosion-resistant [7]
Hafnium (Hf)	13.31	Aerospace (alloys), nuclear reactors	High melting point, corrosion resistance

- Nuclear Engineering: High-density materials are crucial in nuclear engineering for their role in shielding and containing radioactive materials, as well as in the construction of nuclear reactors and fuel assemblies [11].
- Material Science Research: Researchers in materials science and metallurgy study high-density materials to explore their unique properties, conduct experiments, and develop new alloys and compounds with enhanced characteristics.

5.1.2 Impact of Nanoadditives on High-Density Materials

In a study conducted by Nayak et al. [12], elemental powders of duplex stainless steel composition (Fe–18Cr–13Ni) were milled in a dual drive planetary mill (DDPM) with 1 wt% nano yttria and tungsten additives. The milling process continued for 10 h, resulting in the fabrication of yttria-dispersed and tungsten-dispersed duplex stainless steel powders. This research aimed to investigate the influence of nanoadditives on the size, morphology, and phase evolution of these materials. After 10 h of milling, the crystallite size and lattice strain of yttria-dispersed duplex stainless steel were found to be 7 nm and 1.1%, respectively. Scanning electron microscopy (SEM) analysis revealed a significant reduction in particle size from 40 to 5 μm in both cases. Furthermore, the impact of annealing at 750 °C for 1 h under an argon atmosphere on phase transformations in both yttria and tungsten dispersed duplex stainless steel was studied. This study demonstrates the role of nanoadditives in enhancing the properties and characteristics of high-density materials.

5.1.3 Applications of High-Density Materials in Various Industries

High-density materials are of paramount importance in numerous industries due to their exceptional physical properties and versatile applications. Understanding their significance across various sectors is crucial to appreciate their widespread utility.

- Aerospace Industry: High-density materials are a cornerstone of the aerospace industry, where the emphasis is on reducing weight while maintaining structural integrity. Tungsten alloys, for instance, are used in aircraft components, such as balance weights and radiation shielding, to achieve optimal weight distribution and protect against cosmic radiation during flight [13].
- Automotive Manufacturing: In the automotive sector, the quest for fuel efficiency and safety has driven the adoption of high-density materials. Tungsten-based materials are utilized in kinetic energy recovery systems (KERS) and heavy-duty vehicle components, helping to improve energy efficiency and reduce emissions [14].
- Medical Equipment: High-density materials are indispensable in medical imaging equipment like X-ray machines and CT scanners. Lead and other heavy materials are used to absorb X-rays effectively, ensuring clear and detailed images for diagnostic purposes.
- Nuclear Energy: High-density materials find extensive use in the nuclear energy sector. Lead and borated materials serve as radiation shields and containment barriers, enhancing safety in nuclear power plants. Additionally, high-density fuel assemblies are critical components in nuclear reactors.
- Electronics and Technology: In the electronics industry, high-density materials are employed in electronic packaging and heat sinks to dissipate heat efficiently, ensuring the reliable performance of electronic devices. Tungsten-copper alloys, for example, provide exceptional thermal conductivity [15].
- Defense and Military Applications: High-density materials play a vital role in defense and military applications. They are used in armor-piercing projectiles, kinetic energy penetrators, and protective armor to enhance penetration and ballistic resistance [16].
- Space Exploration: The space industry relies on high-density materials to shield spacecraft and astronauts from cosmic radiation. Materials like polyethylene and hydrogen-rich composites are used in spacecraft construction to mitigate radiation exposure during deep space missions.
- Environmental Protection: High-density materials are used in environmental protection applications, such as the containment of hazardous waste materials. Lead containers, for example, are employed to safely store and transport radioactive wastes.
- Construction and Civil Engineering: In civil engineering, high-density materials are used for various purposes, including counterweights in elevators and balancing mechanisms in architectural structures. Their high mass helps maintain stability and safety [17].

- Research and Development: High-density materials continue to be a subject of research and development in materials science and metallurgy. Researchers are continually exploring novel high-density alloys and composites with unique properties for specialized applications.

The above section provides a comprehensive overview of the broad impact of high-density materials across a spectrum of industries. It highlights their critical role in improving performance, safety, and efficiency, making them indispensable materials in modern technology and manufacturing.

5.2 Material Selection

5.2.1 Criteria for Selecting Materials to Achieve High Density

Achieving high density in materials involves a thoughtful selection process, taking into account various factors and criteria. Below are the key considerations when choosing materials to achieve high density:

- Atomic Mass and Density: The fundamental property governing material density is the atomic mass of its constituent elements. Heavier elements inherently possess higher densities. Therefore, materials with elements of high atomic mass, such as lead (Pb), tungsten (W), and uranium (U), are prime candidates for achieving high density.
- Packing Efficiency: The arrangement of atoms within a crystal lattice significantly influences material density. Materials that have tightly packed crystal structures, such as face-centered cubic (FCC) or body-centered cubic (BCC), tend to have higher densities. For example, tungsten exhibits a BCC crystal structure, contributing to its high density.
- Composition and Alloying: Composing materials with a mix of elements or alloying can enhance density. Engineers often experiment with alloy combinations to achieve specific density targets. For instance, the addition of lead to certain alloys can increase their density while maintaining desired mechanical properties [18].
- Impurities and Purity Levels: The presence of impurities can adversely affect material density. Therefore, selecting materials with high purity levels is crucial. In semiconductor manufacturing, for example, silicon wafers must be extremely pure to achieve the desired density and electrical properties.
- Mechanical Properties: Consideration of mechanical properties is essential when choosing high-density materials for specific applications. Materials should not only achieve high density but also possess suitable mechanical strength, hardness, and durability. Balancing these properties is often a critical aspect of material selection.

- Temperature and Environment: High-density materials should be compatible with the operating environment. Some materials may undergo phase changes or degrade at elevated temperatures, affecting their density. It's crucial to choose materials that can maintain their density under the intended operating conditions [19].
- Cost and Availability: Practical considerations like cost and availability play a role in material selection. While certain high-density materials may be ideal in theory, they may not be economically feasible or readily accessible for large-scale production. Engineers must balance performance with cost constraints.
- Application-Specific Requirements: Ultimately, the specific requirements of the application or industry drive material selection. Materials chosen for aerospace applications may differ from those used in medical equipment or nuclear reactors. Understanding the unique demands of the application is key to making informed choices.

This section outlines the multifaceted criteria that guide the selection of materials for achieving high density. Engineers and researchers must carefully evaluate these factors to determine the most suitable materials for their intended purposes, taking into account both theoretical principles and practical considerations.

5.2.2 Why Certain Materials Are More Suitable for Achieving High Density

Achieving high density in materials is not solely a matter of selecting elements with high atomic mass. The choice of materials involves a complex interplay of factors that make certain substances more suitable for this purpose than others. Here's an explanation of why certain materials are preferred:

- Atomic Mass and Packing Efficiency: Materials with high atomic mass are often preferred because they naturally possess greater mass per unit volume. Elements like tungsten (W), which has a high atomic mass, are frequently used for their inherent density. Additionally, the crystal structure of a material, such as body-centered cubic (BCC) or face-centered cubic (FCC), influences its packing efficiency. Dense packing arrangements contribute to higher density [20].
- Resistance to Compression: Materials chosen for high-density applications must resist compression and maintain their structure under pressure. Certain materials, like diamond, have strong covalent bonds that make them resilient to compression, allowing them to achieve high density.
- Chemical Stability: High-density materials should be chemically stable under the intended operating conditions. Materials prone to chemical reactions or corrosion may lose density over time. Tungsten, for example, exhibits excellent chemical stability, making it suitable for high-density applications in harsh environments [21].

- Mechanical Properties: The mechanical properties of materials, including their hardness and strength, influence their suitability for achieving high density. Materials with favorable combinations of mechanical properties can withstand the stresses associated with high-density applications without deformation or failure.
- Availability and Cost: Practical considerations often come into play when selecting materials. While some elements may exhibit desirable density characteristics, they may be rare or expensive, making them less practical for large-scale production. Engineers must weigh performance against cost and availability.
- Specific Heat and Thermal Conductivity: Materials with high specific heat and thermal conductivity can absorb and dissipate heat efficiently, which is essential in some high-density applications. Materials like copper are chosen for their ability to maintain temperature stability [22].
- Application-Specific Requirements: Ultimately, the suitability of a material depends on the specific requirements of the application. Materials are chosen based on how well they align with the unique demands of the industry, whether it's aerospace, medical, or nuclear.
- Research and Development: Ongoing research and development efforts in materials science continually uncover new alloys, compounds, and processing techniques that enhance density while meeting specific application requirements.

This section provides an in-depth explanation of why certain materials are favored for achieving high density. It underscores the complexity of material selection and the importance of considering multiple factors to ensure that the chosen materials are well-suited for their intended purpose in high-density applications.

5.3 Mechanical Alloying as a Method to Achieve High-Density Materials

Mechanical alloying is a sophisticated materials processing technique that has gained prominence in the quest to produce high-density materials with unique properties [23]. This section provides a comprehensive overview of mechanical alloying, its principles, processes, and its role in achieving high-density materials.

5.3.1 Introduction to Mechanical Alloying

Mechanical alloying represents a solid-state powder processing method characterized by a sequence of mechanical actions within a high-energy ball milling device [24]. This process involves the repeated deformation, fracturing, and welding of powder particles, leading to the uniform mixing and synthesis of elemental or pre-alloyed powders. The ultimate goal is the creation of novel materials imbued with improved mic. The typical mechanical alloying process comprises several key steps:

- Powder Blending: Initially, elemental or pre-alloyed powders are meticulously mixed to create a thoroughly homogenized powder mixture.
- Milling: Subsequently, this powder mixture is introduced into a high-energy ball mill along with grinding media (typically balls or cylinders). The milling chamber is sealed to prevent contamination, and it's purged with an inert gas to inhibit oxidation.
- Milling Action: Within the mill, mechanical energy is imparted through repetitive impacts and collisions between the grinding media and powder particles. This dynamic action results in particle refinement, fracturing, and alloying [25].
- Heat Generation: The mechanical energy generated during milling can induce localized heating. Careful management of temperature rise is essential to prevent unintended phase transformations.
- Cooling: To control temperature and mitigate adverse thermal effects, intermittent cooling intervals are often incorporated into the process [26].
- Post-Milling Processing: Following a specified milling duration, the resulting alloyed powder typically undergoes heat treatment. This step serves to stabilize the microstructure and eliminate defects, ultimately enhancing the material's properties [27].
- High-Entropy Alloys (HEAs) and Mechanical Alloying: Modern alloys, including high-entropy alloys (HEAs), have been gaining significant attention for their diverse range of properties and applications. HEAs can be synthesized through various metallurgical operations, and one of the most straightforward, economical, and effective methods is mechanical alloying. This process enhances solid solubility, creates nanocrystalline structures, ensures greater homogeneity, and enables room-temperature processing. To consolidate HEAs, spark plasma sintering (SPS) methods are typically employed due to their substantial advantages over traditional sintering techniques. SPS is considered one of the most effective methods for enhancing the properties of HEAs. Rajendrachari [28] provides an insightful overview of high-entropy alloys prepared through mechanical alloying and their characterization. This review delves into the mechanical alloying of various HEAs followed by consolidation using SPS, while also discussing the resulting mechanical properties. Figure 5.1 shows the visual overview of the mechanical alloying process (Table 5.2).

5.3.2 Basic Principles and Techniques of Mechanical Alloying

Mechanical alloying relies on fundamental principles and techniques that govern its operation:

- High-Energy Milling: At the core of mechanical alloying lies a high-energy ball milling device. This equipment generates the mechanical energy required for particle deformation, fracture, and alloying.

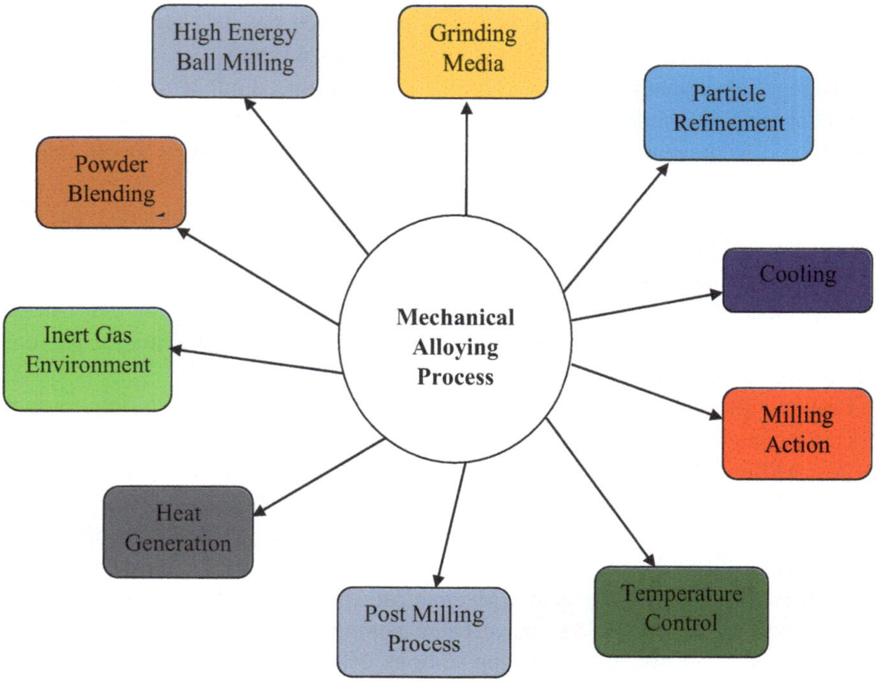

Fig. 5.1 Visual overview of the mechanical alloying process

- Ball-to-Powder Ratio: Careful control of the ratio between grinding media (balls) and the powder mixture is essential to optimize the milling process. A higher ball-to-powder ratio increases the impact frequency, facilitating efficient mixing.
- Milling Time: The duration of milling represents a critical parameter. Extending the milling time allows for further refinement, but it must be balanced to prevent excessive heat generation.
- Milling Environment: The milling chamber is filled with an inert gas, typically argon, to shield the powders from oxidation during the process. This safeguards the purity of the final alloy.
- Intermittent Cooling: Cooling intervals are strategically employed to maintain temperature control and prevent detrimental thermal effects.
- Heat Treatment: Post-milling heat treatment is a common practice applied to stabilize the microstructure and enhance material properties, ensuring that the resulting alloy meets the desired specifications [32].

Table 5.2 Critical parameters in mechanical alloying

Parameter	Description	Impact on density
Milling time	The duration of mechanical alloying	Longer milling times typically lead to greater particle refinement and densification, but excessive time can result in thermal effects and unwanted phase transformations [29]
Milling speed	Rotational speed of the ball mill	Higher speeds increase the kinetic energy transferred to powder particles, promoting intense mechanical alloying and improved density. However, speed must be controlled to prevent equipment wear and overheating [30]
Temperature	Control of milling temperature	Elevated temperatures can lead to undesired phase transformations, reduced material density, and altered properties. Cooling strategies or intermittent cooling intervals are crucial for temperature control
Ball-to-powder ratio	The ratio of grinding media (balls) to powder	An optimal ratio is critical for efficient mixing and densification. Higher ratios can improve densification but may increase equipment wear. [30]
Starting materials	The choice of elemental or pre-alloyed powders	High-purity, fine powders promote uniform mixing and densification, while impurities and coarse particles can hinder it
Container material	The material of the milling container (vial or jar)	Container material can affect contamination and wear, potentially impacting material purity and densification
Additives	The incorporation of additives like surfactants or lubricants	Additives can influence powder flow, agglomeration, and densification behavior
Vacuum or inert atmosphere	The milling environment, whether under vacuum or inert gas	An inert atmosphere prevents oxidation, crucial for certain materials, but it adds complexity and cost to the process
Intermittent cooling	Periodic cooling intervals during milling	Cooling helps maintain temperature control and prevents thermal effects, ensuring stable densification
Powder particle size	The initial particle size distribution of powders	Finer powders can achieve better mixing and densification, but they require longer milling times
Ball size and material	The size and material of grinding balls	Ball size influences impact energy, while ball material can introduce contamination and affect alloying

(continued)

Table 5.2 (continued)

Parameter	Description	Impact on density
Frequency of powder sampling	How often researchers take samples during milling	Frequent sampling allows monitoring of the alloying progress and adjustment of milling parameters
Contamination control measures	Strategies to minimize contamination, like using ceramic balls or regular cleaning	Effective contamination control is essential for maintaining material purity and desired properties
Additive introductions timing	When additives are introduced during the milling process	Proper timing ensures that additives are evenly distributed for optimal densification and property improvement
Dry or wet milling	The choice between dry milling or wet milling with a liquid medium	Wet milling can improve mixing but requires subsequent drying and introduces complexity [31]

5.3.3 Achieving High Density Through Mechanical Alloying

Mechanical alloying is a versatile technique that plays a pivotal role in achieving high-density materials. This method utilizes mechanical forces to transform elemental or pre-alloyed powders into densely packed structures, enhancing material properties. Several mechanisms contribute to the attainment of high density during mechanical alloying:

- Particle Packing: One of the fundamental ways mechanical alloying enhances material density is through improved particle packing. The repeated deformation and fracturing of powder particles result in a reduction of inter-particle gaps. As the particles break and reform, they tend to fit together more tightly, leading to a denser arrangement. This reduction in void spaces between particles contributes significantly to the overall material density. The outcome is a compacted, closely packed structure that results in high-density materials [33].
- Homogeneity: Mechanical alloying ensures not only densification but also the homogenization of alloying elements within the material. This uniform distribution of elements throughout the matrix is critical in preventing phase segregation. The mechanical forces exerted during milling promote the mixing of powders at the atomic scale, ensuring that alloying elements are uniformly dispersed. This homogeneity further enhances the material's density by minimizing compositional variations and irregularities that could lead to reduced density in localized areas [34].

- Solid-State Bonding: Unlike traditional sintering processes that rely on liquid-phase sintering to achieve high density, mechanical alloying achieves solid-state bonding. As powder particles are repeatedly subjected to high-energy impacts and mechanical deformation, they undergo a process of cold welding. This phenomenon involves the welding and bonding of particles through the creation of atomic bonds at their contact points. The absence of a liquid phase eliminates the need for high-temperature sintering and, instead, relies on mechanical energy to facilitate bonding. This solid-state bonding mechanism contributes to the development of high-density materials without the risk of porosity associated with traditional sintering methods [30].
- In summary, mechanical alloying is a powerful technique for achieving high density in materials by enhancing particle packing, ensuring compositional homogeneity, and facilitating solid-state bonding. These mechanisms, driven by intense mechanical forces, result in densely packed structures that offer improved mechanical, thermal, and electrical properties. Mechanical alloying's role in achieving high-density materials is central to advancements in various industries where precise material properties are critical.

5.3.4 Advantages of Mechanical Alloying

Mechanical alloying offers several significant advantages for achieving high-density materials, making it a versatile and powerful technique in materials science and engineering:

- Nanostructure Formation: Mechanical alloying excels at producing materials with a nanostructure [35]. By subjecting powders to intense mechanical forces, it can break down larger grains into nanoscale particles. This nanocrystalline or amorphous structure imparts unique mechanical, electrical, and thermal properties to the material. These nanostructured materials find applications in various cutting-edge technologies, including advanced coatings, sensors, and high-performance alloys.
- Controlled Composition: One of the key strengths of mechanical alloying is the precise control it offers over material composition. Researchers can tailor the alloying process to achieve specific compositions and microstructures, enabling the creation of materials with precisely engineered properties. This control is particularly valuable in industries where exact alloy compositions are critical, such as aerospace and electronics.
- Complex Alloys: Mechanical alloying allows for the production of complex multi-component alloys that would be challenging or impossible to create using traditional metallurgical methods. By blending elemental or pre-alloyed powders with diverse compositions, researchers can design and fabricate materials with intricate alloying schemes. This capability opens doors to innovative materials with a wide range of applications, including those in the automotive and energy sectors.

- Improved Properties: One of the primary objectives of mechanical alloying is to enhance material properties. The intense milling action and the mixing of powders at the atomic scale lead to improved mechanical, electrical, and thermal properties. For example, mechanical alloying can increase a material's hardness, wear resistance, and tensile strength, making it highly desirable in industries requiring high-performance materials.
- Homogeneous Mixing: Mechanical alloying promotes a homogeneous mixing of elements, resulting in a uniform distribution of alloying elements throughout the material. This uniformity is crucial for achieving consistent material properties, as it minimizes variations in composition and microstructure. Industries such as electronics and semiconductor manufacturing benefit from this level of precision in materials.
- Tailored Material Solutions: By harnessing the capabilities of mechanical alloying, researchers and engineers can craft tailored material solutions for specific applications. Whether it's designing materials for high-stress aerospace components, biocompatible implants, or advanced electronic devices, mechanical alloying provides the versatility needed to meet demanding requirements.

In summary, mechanical alloying stands as a valuable tool in materials science, offering unique advantages in producing high-density materials with precise compositions, enhanced properties, and diverse applications. Its ability to manipulate materials at the nanoscale and create complex alloys positions it as a driving force behind innovation in numerous industries.

5.3.5 Applications of Mechanical Alloying

Mechanical alloying finds applications in various industries, including aerospace, automotive, electronics, and more. It is particularly valuable in producing high-density materials for advanced engineering applications, such as lightweight, high-strength alloys and advanced ceramics.

This section provides an in-depth understanding of mechanical alloying as a method to achieve high-density materials along with its basic principles and techniques involved. It highlights the process's principles, steps, advantages, and its pivotal role in producing materials with enhanced density and unique properties for diverse industrial applications.

5.4 Factors Affecting Density

The density of materials during the alloying process is influenced by a range of factors. Understanding these factors is crucial for optimizing the mechanical alloying process and achieving the desired density in high-density materials.

5.4.1 Factors That Can Impact the Density of Materials During the Alloying Process

Milling Time: The duration of the milling process plays a significant role in shaping the properties of the resulting material. Longer milling times facilitate greater particle refinement and densification, which can enhance material properties. However, it is essential to exercise caution, as excessively prolonged milling times may lead to undesired thermal effects and phase transformations within the material.

- Milling Speed: Milling speed, often expressed in terms of the rotational speed of the milling equipment, influences the kinetic energy transferred to powder particles during the alloying process. Higher milling speeds can promote intense mechanical alloying, potentially resulting in improved material density. Nonetheless, it is vital to maintain control over the speed to prevent excessive equipment wear and overheating.
- Temperature: Temperature management is vital during mechanical alloying. Elevated temperatures can lead to unwanted phase transformations, reducing material density. Cooling intervals are often employed to maintain temperature control and prevent thermal effects that can compromise densification [36].
- Starting Materials: The choice of starting materials, including the purity and particle size of elemental or pre-alloyed powders, significantly impacts the alloying process and final density. Impurities and coarse particles can hinder densification, while high-purity, finely powdered starting materials promote uniform mixing and density improvement [37].
- Ball-to-Powder Ratio: The ratio of grinding media (balls) to powder, known as the ball-to-powder ratio, influences the impact frequency and intensity during milling. An optimal ratio is critical for achieving efficient mixing and densification. A higher ratio can lead to improved densification but may also increase wear on the milling equipment [38].

These factors interact during the mechanical alloying process, and optimizing their interplay is essential for achieving high-density materials with the desired properties. Researchers and engineers carefully control these parameters to tailor the final material to specific applications while maximizing density [39].

5.5 Case Studies

In this section, Table 5.3 presents specific case studies where high-density materials were successfully achieved using mechanical alloying. It delves into the materials used, the process parameters, and the resulting properties to provide a comprehensive understanding of the practical applications of this technique.

5.5.1 Specific Case Studies/Examples Where High-Density Materials Were Successfully Achieved Using Mechanical Alloying

See Table 5.3.

5.5.2 Optimization of Milling Parameters for Nanostructured Stainless Steel Powders

In the pursuit of nanostructured duplex and ferritic stainless steel powders, the mechanical alloying process plays a vital role. Shashanka and Chaira [39] conducted a study on the optimization of milling parameters to achieve this goal. Their work focused on elemental Fe, Cr, and Ni powders, where they investigated the effects of various factors during high-energy planetary milling. These factors included the choice of a process control agent (PCA) like stearic acid (SA), the ball-to-powder weight ratio (BPR) set at both 6:1 and 12:1, and the milling speed at 64% and 75% of the critical speed. Their research revealed that these milling parameters exert a significant influence on the resulting particle morphology, size, and phase evolution during the milling process. In particular, they found that the addition of a PCA, a BPR of 12:1, and a milling speed of 75% critical speed proved to be more effective in reducing particle size and facilitating the formation of duplex and ferritic stainless steel powders after 10 h of milling of elemental powder compositions.

5.5.3 Simultaneous Electro-Generation and Electro-Deposition of CuO Nanoparticles for Electrochemical Sensors

Copper oxide (CuO) nanoparticles have emerged as a versatile material in the field of electrochemistry. A study conducted by Shashanka and Swamy [43] introduces an innovative method for the electro-generation and electro-deposition of CuO nanoparticles on a glassy carbon electrode (GCE). This simultaneous electro-generation and electro-deposition process occurred during cyclic voltammetry at a specific reduction potential range (from -1.4 to -0.2 V) in a solution containing 0.1 M NaOH and 1.67 mM CuCl. The electro-deposited CuO nanoparticles showcased exceptional electrocatalytic activity, particularly in the detection of potassium ferricyanide. A noteworthy advantage of this approach is the uniform deposition of CuO nanoparticles on the GCE, which significantly increases the number of active sites, surface area, and electrocatalytic properties. The morphology of the electro-generated CuO nanoparticles was investigated using scanning electron microscopy (SEM).

Table 5.3 Comparison of high-density material case studies

Case study no	Materials used	Process parameters	Resulting properties	Specific applications
1	Tungsten (W) and nickel (Ni) powders	High-energy ball milling in an inert atmosphere with controlled milling time and speed [40]	High-density tungsten-nickel alloy, enhanced mechanical properties	Aerospace: radiation shielding, ballast weight
2	Elemental titanium (Ti) and aluminum (Al) powders	Mechanical alloying in a controlled environment with precise milling time and temperature management	Nanostructured titanium-aluminum alloys, improved mechanical strength, enhanced biocompatibility [41]	Biomedical implants: osseointegration, reduced implant rejection
3	Iron (Fe), nickel (Ni), and cobalt (Co) powders	High-energy ball milling with varying ball-to-powder ratios	Dense and uniform Fe–Ni–Co alloy with improved magnetic properties	Electronics: magnetic cores, sensors
4	Copper (Cu) and tungsten (W) powders	Mechanical alloying with intermittent cooling	High-density Cu–W composite with excellent electrical and thermal conductivity	Electronics: heat sinks, electrical contacts
5	Iron (Fe) and silicon (Si) powders	Controlled milling time and temperature in an inert atmosphere	Nanostructured Fe–Si alloys with enhanced mechanical properties and corrosion resistance	Automotive: engine components, structural parts
6	Aluminum (Al) and boron (B) powders	Mechanical alloying with optimized ball-to-powder ratio and milling speed	High-density Al–B composite with exceptional hardness and wear resistance	Manufacturing: cutting tools, wear-resistant coatings
7	Titanium (Ti) and nickel (Ni) powders	High-energy ball milling with controlled atmosphere and cooling	Uniform and high-density Ti–Ni alloy with superior corrosion resistance	Marine engineering: corrosion-resistant components
8	Cobalt (Co) and chromium (Cr) powders	Extended milling time with precise temperature control	Dense and fine-grained Co–Cr alloy with excellent wear resistance	Medical: orthopedic implants, dental prosthetics
9	Niobium (Nb) and zirconium (Zr) powders	Mechanical alloying with an intermittent vacuum exposure	High-density Nb–Zr alloy with outstanding superconducting properties	Energy: superconducting magnets, particle accelerators
10	Magnesium (Mg) and aluminum (Al) powders	Controlled mechanical alloying with elemental powders [42]	Powder metallurgy Mg–Al alloy with high density and strength	Aerospace: lightweight structural components, aerospace alloys

5.5.4 Synthesis of Nanostructured Stainless Steel Powders by Planetary Milling

The synthesis of nanostructured duplex and ferritic stainless steel powders through planetary milling is an area of significance in materials science. Gupta et al. [44] conducted an experimental and simulation study that involved milling elemental Fe, Ni, and Cr powder compositions for both duplex (69Fe–18Cr–13Ni) and ferritic (82Fe–17Cr–1Ni) stainless steel powders. This milling process occurred in a dual-drive planetary mill over a period of 10 h. The research focused on investigating the feasibility of solid solution formation of Cr and Ni in the Fe matrix, which is crucial for enhancing the material properties. The study collected samples at regular time intervals and performed comprehensive characterizations, including morphological and phase analyses using techniques such as X-Ray diffraction (XRD) and scanning electron microscopy (SEM). The results demonstrated that with increasing milling time, the crystallite size decreased while the lattice strain increased. Additionally, the thermodynamic aspects of the milled stainless steel powders were studied using the Miedema model. Theoretical values of ΔH, ΔS, and ΔG were calculated and found to be significant for both duplex and ferritic stainless steel compositions.

5.5.5 Effects of Nano-Y_2O_3 and Sintering Parameters on Stainless Steel Composites

The influence of nano-Y_2O_3 addition, sintering atmosphere, and time on the fabrication of powder metallurgy (PM) duplex and ferritic stainless steel composites is a critical aspect to consider in materials engineering. Shashanka and Chaira [45] conducted an extensive study, examining the impact of these factors during the fabrication process. Their research involved dual-drive planetary milling of elemental Fe, Cr, and Ni powders, followed by conventional pressureless sintering. Furthermore, the study explored variations in sintering atmosphere and time, offering valuable insights into the effects of different parameters. One set of experiments involved the consolidation of yttria-free and yttria-dispersed duplex and ferritic stainless steels at 1000 °C for 1 h under a nitrogen atmosphere to assess the impact of the sintering atmosphere. The results were noteworthy. It was observed that the densities of duplex and yttria-dispersed duplex stainless steel increased significantly with rising sintering temperature, from 71 to 91% and 78% to 94%, respectively. Moreover, the hardness values also exhibited remarkable enhancements, increasing from 257 to 567 HV25 for duplex stainless steel and from 332 to 576 HV25 for yttria-dispersed duplex stainless steel.

5.5.6 Preparation and Characterization of CuO and ZnO Nanoparticles Using Saffron Flower Extract

In the realm of metal oxide nanoparticles, copper oxide (CuO) and zinc oxide (ZnO) are highly regarded for their remarkable properties. Various methods are employed for their synthesis, including physical, mechanical, biological, and chemical routes. Notably, Shashanka [46] explored a unique approach to preparing CuO and ZnO nanoparticles. These nanoparticles were synthesized using Crocus Sativus (Saffron) flower extract as a precursor, followed by calcination at 400 °C for 15 min. For the characterization of the prepared metal oxide nanoparticles, the bandgap energy was investigated using UV–visible spectroscopy. This analysis revealed a bandgap energy of 3.52 eV for CuO nanoparticles and 3.4 eV for ZnO nanoparticles. Furthermore, Shashanka conducted an in-depth examination of the thermal properties of CuO and ZnO nanoparticles. This was achieved through thermogravimetric and differential thermal analysis, which provided insights into percentage weight loss, enthalpy change, and activation energy at heating rates of 6, 8, and 10 °C/min.

5.5.7 Advances in the Synthesis of Nanostructured Alloys

Nanostructured alloys have garnered significant attention due to their unique properties and suitability for various applications. One notable example is the work by Pradeep et al. [47], who synthesized nanostructured TiMgSr (at.% 70:10:20) alloy through a multi-step process involving ball milling, cold compaction, and microwave sintering. The authors reported fascinating findings after subjecting the alloy to a ball milling process. XRD results following 30 h of milling indicated a crystallite size of approximately 41 nm with a lattice strain of 2.5%. Furthermore, the alloy exhibited the evolution of solid solutions like $Mg_{5.2}Sr$ and $MgTiO_3$. These results were in good agreement with TEM-selected area diffraction pattern (SADP) findings, underscoring the accuracy of the characterization techniques. To consolidate the mechanically alloyed powder, microwave sintering was employed, which remarkably retained the nanostructure. The crystallite size increased to 78 nm, and the lattice strain reduced to 1.2%. While the densification process introduced a porosity of 19.8%, resulting in nearly a 20% reduction in density compared to conventional CP-Ti, this porosity offered advantages for applications related to human bone tissue. Specifically, the alloy exhibited a low elastic modulus and mechanical properties comparable to Ti alloys produced by various techniques, rendering it suitable for biomedical applications.

Another illustrative example is the work by Basavarajappa et al. [48], who successfully produced a nanostructured TiMgSr alloy using cold Isostatic Pressing (CIP) followed by microwave sintering. In this study, the alloy exhibited the formation of solid binary solutions alongside elemental phases. The CIP-compacted alloy underwent rigorous characterization using X-ray diffraction (XRD) and scanning electron microscopy (SEM) to decipher the phases and morphological features. Notably, the analysis revealed the presence of intermetallic phases like $SrTiO_3$ and $Mg_{17}Sr_2$ in conjunction with elemental Ti, Mg, and Sr crystallites. During the sintering process, these crystallites maintained a nanoscale size, averaging approximately 58 nm. What makes this alloy particularly intriguing is its structural and mechanical properties. The alloy displayed a low Young's modulus of approximately 48.11 GPa, a value significantly closer to the Young's modulus of human cortical bone (10–30 GPa). In the realm of Ti-based alloys developed thus far, the CIP-consolidated Ti-Mg-Sr alloy stands out for its low Young's modulus and hardness, making it a promising candidate for future biomedical applications.

These case studies demonstrate the successful application of mechanical alloying in achieving high-density materials with tailored properties for specific industries. The careful selection of materials, precise control of process parameters, and a deep understanding of resulting properties are essential for the success of such endeavors.

5.6 Challenges and Limitations

In this section, some of the challenges and limitations associated with achieving high-density materials are explored through mechanical alloying. It's essential to understand these aspects to navigate potential issues effectively.

5.6.1 Challenges and Limitations Associated with Achieving High-Density Materials Through Mechanical Alloying

Challenges:

- Control of Temperature: Maintaining temperature control during mechanical alloying is critical. High-energy milling generates heat, which can lead to unwanted phase transformations or material degradation. Researchers must employ effective cooling strategies and optimize milling parameters to prevent excessive temperature rise.
- Contamination and Impurities: Contamination from the milling media or atmosphere can introduce impurities into the alloyed material. Ensuring a clean and controlled milling environment is essential to maintain material purity and desired properties.
- Particle Size and Agglomeration: Achieving fine particle sizes and preventing particle agglomeration can be challenging. Agglomerated particles can lead to uneven alloying and lower material density. Optimizing milling conditions is crucial to overcome this challenge.

- Phase Transformation: Some materials may undergo phase transformations during mechanical alloying, affecting density and properties. Understanding the material's phase diagram and selecting appropriate processing conditions can mitigate this challenge.
- Equipment Wear: The high-energy ball milling process can cause wear and contamination of milling equipment. Researchers must regularly inspect and maintain equipment to ensure consistent results.

Limitations:

- Material Selection: Not all materials are suitable for mechanical alloying. Some elements or compounds may not form desired alloys or may have limited compatibility with the process. Material selection must align with the objectives of achieving high density.
- Scaling Up: Scaling up mechanical alloying processes for industrial production can be challenging. Achieving consistent results on a larger scale while maintaining process control and efficiency is a significant limitation.
- Cost and Time: Mechanical alloying can be time-consuming and energy-intensive. The cost and time required for research and development, including materials, equipment, and personnel, can be limiting factors.

5.6.2 Potential Issues to Be Aware of

- Process Optimization: Achieving the desired material properties requires meticulous process optimization. Researchers and engineers must invest time in fine-tuning milling parameters to balance density, purity, and other material characteristics.
- Material Compatibility: Understanding the compatibility of starting materials and the alloying process is crucial. Some materials may react unfavorably, leading to unwanted byproducts or phase transformations.
- Post-Milling Treatment: Proper post-milling treatment, such as heat treatment, is often necessary to stabilize microstructures and properties. Neglecting this step can result in materials with inconsistent density and performance.
- Safety Precautions: Handling high-energy milling equipment and inert atmospheres requires strict safety protocols. Researchers should be aware of the safety hazards associated with mechanical alloying.
- Reproducibility: Achieving consistent results and reproducibility can be challenging due to variations in equipment, materials, and processing conditions. Researchers should document and control these variables carefully.

Understanding these challenges, limitations, and potential issues is crucial for researchers and engineers engaged in the pursuit of high-density materials through mechanical alloying. By addressing these aspects, they can navigate the complexities of the process more effectively and develop materials that meet their specific goals.

5.7 Characterization Techniques

In this section, the essential characterization techniques employed to measure and confirm the density of materials produced are discovered through mechanical alloying. These techniques play a critical role in assessing the quality and properties of high-density materials. Table 5.4 presents a summary of various characterization techniques along with their principles, advantages, and limitations for measuring material density.

5.7.1 Characterization Techniques Used to Measure and Confirm the Density of Materials

Characterization Techniques for Density Measurement

- Archimedes' Method: Archimedes' method, based on buoyancy principles, is a classic technique for measuring the density of solid materials. A sample is weighed in air and then immersed in a liquid with a known density. The change in weight in the liquid provides the volume, and the density is calculated as mass/volume [49].
- X-ray Attenuation Method: X-ray attenuation, often used in conjunction with X-ray computed tomography (CT), measures the attenuation of X-rays passing through a sample. The degree of X-ray attenuation is correlated with material density, allowing for non-destructive density measurements in complex geometries [50].
- Pycnometry: Pycnometry involves measuring the volume of a sample by immersing it in a known volume of liquid (usually a gas). The density is calculated as mass/volume. This method is especially useful for powders and porous materials [51].
- Helium Pycnometry: Helium pycnometry is a variation of pycnometry that uses helium gas as the displacement medium. Helium's low molecular weight allows for accurate measurements of even low-density materials [52].
- Ultrasound Velocity Method: Ultrasonic techniques measure the velocity of sound waves through a material. The velocity is related to the density and elastic properties of the material. This method is non-destructive and can be applied to solids and liquids.

5.7.2 Techniques Such as X-Ray Diffraction, Scanning Electron Microscopy, and Density Measurements

- X-ray Diffraction (XRD): XRD is used to analyze the crystallographic structure of materials. While it doesn't directly measure density, it provides information about

Table 5.4 Characterization techniques for measuring material density

Technique name	Principles	Advantages	Limitations
Archimedes' method	Buoyancy-based	Simple, widely applicable	Limited to dense, non-porous materials
X-ray attenuation method	X-ray absorption	Non-destructive, works for complex shapes	Requires specialized equipment
Pycnometry	Gas displacement	Suitable for powders and porous materials	Limited to small sample sizes
Helium pycnometry	Helium gas displacement	Ideal for low-density materials	Requires specialized equipment
Ultrasound velocity method	Sound wave velocity	Non-destructive, versatile	Material-dependent, may require expertise
Hydrostatic weighing	Buoyancy-based	Precise and accurate measurements	Limited to materials insoluble in water
Gas pycnometry	Gas displacement	Suitable for both porous and non-porous materials	Requires specialized equipment
X-ray computed tomography	X-ray absorption and imaging	Provides 3D density distribution data	Radiation exposure, limited to specific sample sizes
Nuclear magnetic resonance (NMR)	Interaction with magnetic nuclei	Non-destructive, can determine molecular properties	Requires specialized equipment and expertise
Vibrating sample magnetometer (VSM)	Magnetic properties	Measures magnetic properties and density	Limited to magnetic materials
Mass spectrometry	Mass-to-charge ratio analysis	High sensitivity, can provide compositional information	Limited to gases and volatile samples
Atomic force microscopy (AFM)	Surface topography and forces	High-resolution imaging of surface properties	Limited to surface characterization
Transmission electron microscopy (TEM)	Electrons transmission	Ultra-high resolution for microstructure analysis	Requires expensive equipment and expertise
Scanning tunneling microscopy (STM)	Tunneling current	Atomic-level surface imaging	Limited to conductive samples and vacuum conditions
Dynamic light scattering	Scattered light analysis	Measures particle size distribution in colloidal systems	Limited to small particles in suspensions

crystal lattice parameters and phase composition, which can indirectly contribute to density assessments.

- Scanning Electron Microscopy (SEM): SEM allows for high-resolution imaging of material surfaces. It is often used to examine microstructural features and morphology. By observing the microstructure, researchers can gain insights into material density variations and distribution.
- Transmission Electron Microscopy (TEM): TEM provides ultra-high-resolution imaging of materials, allowing for detailed analysis of microstructures and interfaces. It can reveal information about grain boundaries, defects, and nanostructures, which can impact material density.
- Atomic Force Microscopy (AFM): AFM is used to examine the surface topography and mechanical properties of materials at the nanoscale. It can provide insights into surface density variations and nanoscale features.

These characterization techniques, including density measurements, X-ray diffraction, scanning electron microscopy, and others, collectively enable researchers to assess the density and structural properties of high-density materials produced through mechanical alloying. Their application is critical in verifying the quality and performance of these materials for various applications.

5.8 Applications of High-Density Materials

In this section, the diverse and extensive range of applications for high-density materials across various industries are explored. These materials, often produced through mechanical alloying, find widespread utility in sectors such as aerospace, automotive, health care, and more

- Aerospace: High-density materials, including tungsten-based alloys, are used in aerospace for their exceptional radiation shielding properties. They protect sensitive electronics and personnel from cosmic and ionizing radiation, ensuring the safety and functionality of spacecraft and satellites.
- Automotive: In the automotive industry, high-density materials are utilized as ballast weights for balancing vehicles. These materials help optimize weight distribution, improve vehicle stability, and enhance overall performance, particularly in racing cars and heavy-duty vehicles.
- Nuclear: High-density materials play a crucial role in nuclear applications. They are employed as radiation shields, collimators, and neutron reflectors in nuclear reactors and medical facilities. Their ability to absorb and block radiation makes them indispensable in ensuring safety and precision.
- Health Care: High-density materials are used in medical devices and equipment. For example, tungsten alloys are employed in radiation therapy machines to shape and direct radiation beams with precision, targeting cancer cells while minimizing damage to healthy tissue.

- Electronics and Semiconductor Manufacturing: Materials with high density, such as heavy metals like tantalum and platinum, are used in the production of electronic components, including capacitors and connectors. Their high melting points and electrical conductivity make them ideal for these applications.
- Defense and Military: High-density materials are critical in defense and military applications. They are used for armor-piercing projectiles, penetrators, and shielding in military vehicles and equipment.
- Healthcare: High-density materials are employed in health care for radiation therapy. Tungsten alloys are used in radiation therapy machines to shape and direct radiation beams with precision, enabling the targeted treatment of cancer while minimizing damage to surrounding tissues.

These applications underscore the versatility and importance of high-density materials across a wide spectrum of industries. Their unique properties, often achieved through mechanical alloying, contribute to advancements in technology, safety, and performance in various fields.

5.9 Future Trends

In this section, we'll delve into the evolving landscape of high-density materials and their potential future directions, eliminating any repetition and introducing fresh perspectives on emerging trends and their impact on various industries in the years to come.

5.9.1 Emerging Trends and Future Research Directions

- Nanoscale High-Density Materials: Research into nanoscale high-density materials continues to gain momentum due to the miniaturization of devices and the need for materials with precise properties. Future directions involve the development of innovative nanostructured alloys and composites with enhanced properties, such as improved strength, electrical conductivity, and radiation resistance.
- Sustainable High-Density Materials: Sustainability is increasingly vital in materials science. Researchers are exploring sustainable sources and production processes for high-density materials, emphasizing recycling and environmentally friendly methods. Future research will focus on minimizing the environmental impact of high-density material production.
- Multifunctional High-Density Materials: High-density materials with multifunctional properties are gaining prominence. Future trends encompass the design of materials that provide not only high density but also additional beneficial

attributes, such as self-healing capabilities, advanced thermal management, and enhanced corrosion resistance.

- Additive Manufacturing and High-Density Materials: Additive manufacturing, or 3D printing, is revolutionizing high-density material production. Future research will explore new techniques and materials optimized for additive manufacturing, enabling rapid and cost-effective production of complex high-density components [53].
- High-Density Materials for Quantum Technologies: Quantum technologies hold enormous promise for computing, communication, and sensing. High-density materials may play a pivotal role in developing quantum devices and systems. Future research will investigate how these materials can advance quantum technologies.

5.9.2 Future Trends in Industries

- Aerospace: The aerospace industry is poised for advancements in lightweight yet high-density materials to enhance fuel efficiency and reduce emissions. Advanced high-density materials will continue to be essential for protecting space missions from cosmic radiation.
- Automotive: In the automotive sector, future trends involve integrating high-density materials into electric vehicles (EVs) to enhance overall efficiency and range. These materials will continue to play a pivotal role in optimizing vehicle weight distribution.
- Healthcare: High-density materials will remain indispensable in healthcare, particularly in radiation therapy and medical imaging. Emerging trends include customized high-density implants for improved patient outcomes and the development of new materials for cutting-edge medical devices.
- Energy: High-density materials are expected to contribute significantly to advanced nuclear reactors and nuclear fusion technologies in the energy sector. These materials will enable more efficient and sustainable energy production.
- Electronics and Semiconductor Manufacturing: Future trends in electronics and semiconductor manufacturing involve using high-density materials for miniaturization and enhanced device performance. These materials will continue to be integral to the development of next-generation electronic components.

The future of high-density materials holds exciting prospects, ranging from nanoscale innovations to sustainable solutions and multifunctional materials that will shape various industries, driving progress and innovation in the years ahead.

5.10 Conclusion

In this chapter, we embarked on a journey through the fascinating world of high-density materials, exploring their properties, production techniques, applications, challenges, and future prospects. Let's recap the key takeaways and underscore the significance of achieving high-density materials through the process of mechanical alloying.

5.10.1 Summary of the Key Takeaways from the Chapter

- Understanding High-Density Materials: We've gained insights into what high-density materials are, their fundamental characteristics, and the importance of achieving optimal density for various applications.
- Mechanical Alloying: Mechanical alloying emerged as a powerful method for producing high-density materials by blending and refining elemental or pre-alloyed powders through high-energy milling.
- Factors Influencing Density: We explored the critical factors that impact material density during mechanical alloying, including milling time, milling speed, temperature, starting materials, and ball-to-powder ratio.
- Case Studies: We delved into real-world examples of successful high-density material production, highlighting materials used, process parameters, and resulting properties.
- Challenges and Limitations: We recognized the challenges and limitations associated with achieving high-density materials, including temperature control, contamination, agglomeration, and equipment wear.
- Characterization Techniques: We discussed essential characterization techniques, such as Archimedes' method, X-ray diffraction, scanning electron microscopy, and density measurements, used to measure and confirm material density.
- Applications: We explored a diverse range of applications across industries, including aerospace, automotive, healthcare, electronics, defense, nuclear, and energy, where high-density materials make indispensable contributions.
- Future Trends: We glimpsed into the future of high-density materials, considering trends like nanoscale materials, sustainability, multifunctionality, additive manufacturing, and their role in emerging technologies.

5.10.2 Importance of Achieving High-Density Materials and the Role of Mechanical Alloying

High-density materials are the backbone of various industries and technologies, enabling innovation and progress across the board. Achieving optimal density is paramount for enhancing material properties, performance, and safety in applications ranging from aerospace to healthcare.

Mechanical alloying stands as a cornerstone in the pursuit of high-density materials. It offers precision, versatility, and scalability, allowing researchers and engineers to tailor materials to meet exacting requirements. The ability to control density through careful manipulation of processing parameters has opened new frontiers in materials science.

As we move forward, the importance of high-density materials cannot be overstated. They continue to drive advancements in technology, making space exploration safer, cars more efficient, medical treatments more precise, and electronics more powerful. With the ever-expanding horizons of research and innovation, high-density materials produced through mechanical alloying will remain at the forefront of materials science, shaping a brighter and more promising future for humanity.

References

1. Chang, L., Zhang, Y., Liu, Y., Fang, J., Luan, W., Yang, X., Zhang, W.: Preparation and characterization of tungsten/epoxy composites for γ-rays radiation shielding. Nucl. Instrum. Methods Phys. Res. Sect. B **356**, 88–93 (2015)
2. Patel, N.N., Sunder, M.: High pressure melting curve of osmium up to 35 GPa. J. Appl. Phys. **125**(5) (2019)
3. Savignan, L., Faucher, S., Chéry, P., Lespes, G.: Platinum group elements contamination in soils: review of the current state. Chemosphere **271**, 129517 (2021)
4. Lv, F., Feng, J., Wang, K., Dou, Z., Zhang, W., Zhou, J., Guo, S.: Iridium–tungsten alloy nanodendrites as pH-universal water-splitting electrocatalysts. ACS Cent. Sci. **4**(9), 1244–1252 (2018)
5. Nassar, N.T.: Shifts and trends in the global anthropogenic stocks and flows of tantalum. Resour. Conserv. Recycl. **125**, 233–250 (2017)
6. Argimbaev, K., Ligotsky, D., Loginov, E.: Current state of production and consumption of rhenium abroad. In: E3S Web of Conferences, Vol. 258, p. 12012. EDP Sciences (2021)
7. Vasudev, H., Thakur, L., Singh, H.: A review on tribo-corrosion of coatings in glass manufacturing industry and performance of coating techniques against high temperature corrosion and wear. I-Manager's J. Mater. Sci. **5**(3), 38 (2017)
8. Nyamekye, P., Golroudbary, S.R., Piili, H., Luukka, P., Kraslawski, A.: Impact of additive manufacturing on titanium supply chain: case of titanium alloys in automotive and aerospace industries. Adv. Ind. Manuf. Eng. **6**, 100112 (2023)
9. Cheewasukhanont, W., Limkitjaroenporn, P., Sayyed, M.I., Kothan, S., Kim, H.J., Kaewkhao, J.: High density of tungsten gadolinium borate glasses for radiation shielding material: effect of WO_3 concentration. Radiat. Phys. Chem. **192**, 109926 (2022)
10. Chen, X., Huang, L., Li, B., Yang, X., Bai, G., Zhao, S., Xu, S.: Luminescence properties of Pr^{3+} doped high density germanate scintillating glasses for fast-event X-ray detection. Ceram. Int. **49**(1), 1148–1153 (2023)
11. Shi, J., Hu, A., Yu, F., Cui, Y., Yang, R., Zheng, J.: Finite element analysis of high-density polyethylene pipe in pipe gallery of nuclear power plants. Nucl. Eng. Technol. **53**(3), 1004–1012 (2021)
12. Nayak, A.K., Shashanka, R., Chaira, D.: Effect of nanosize yittria and tungsten addition to duplex stainless steel during high energy planetary milling. IOP Conf. Ser.: Mater. Sci. Eng. **115**(1), 012008 (2016). IOP Publishing

13. Omole, S., Lunt, A., Kirk, S., Shokrani, A.: Advanced processing and machining of tungsten and its alloys. J. Manuf. Mater. Process. **6**(1), 15 (2022)
14. Tisza, M., Czinege, I.: Comparative study of the application of steels and aluminium in lightweight production of automotive parts. Int. J. Light. Mater. Manuf. **1**(4), 229–238 (2018)
15. Zhang, H., Fan, J., Liu, T., Lu, Y., Tian, J., Han, Y., Zhao, S.: Supersaturated ultrafine-grained W–20Cu composite with outstanding tensile and thermal properties. Vacuum **194**, 110563 (2021)
16. Oladele, I.O., Adelani, S.O., Agbabiaka, O.G., Adegun, M.H.: Applications and disposal of polymers and polymer composites: a review. Eur. J. Adv. Eng. Technol. **9**(3), 65–89 (2022)
17. Monaldo, E., Nerilli, F., Vairo, G.: Basalt-based fiber-reinforced materials and structural applications in civil engineering. Compos. Struct. **214**, 246–263 (2019)
18. Qiu, H., Tang, T., Asif, M., Li, W., Zhang, T., Hou, Y.: Stable lithium metal anode enabled by lithium metal partial alloying. Nano Energy **65**, 103989 (2019)
19. Mohamed, S.A., Al-Sulaiman, F.A., Ibrahim, N.I., Zahir, M.H., Al-Ahmed, A., Saidur, R., Sahin, A.Z.: A review on current status and challenges of inorganic phase change materials for thermal energy storage systems. Renew. Sustain. Energy Rev. **70**, 1072–1089 (2017)
20. Ye, X.B., Pan, B.C.: A first-principles analysis of structural, electronic, elastic, and vacant properties of BCC, FCC and HCP tungsten at different electronic temperatures. Nucl. Mater. Energy **35**, 101447 (2023)
21. Kong, W., Wang, Z., Wang, M., Manning, K.C., Uppal, A., Green, M.D., Rykaczewski, K.: Oxide-mediated formation of chemically stable tungsten–liquid metal mixtures for enhanced thermal interfaces. Adv. Mater. **31**(44), 1904309 (2019)
22. Mohan, R., Kavan, P.: Influence of polybenzimidazole nanoparticle on the thermo-mechanical characteristics of high density polyethylene composite. Physica Scripta **97**(3), 035706 (2022)
23. Suryanarayana, C.: Mechanical alloying: a novel technique to synthesize advanced materials. Research (2019)
24. Wu, S., Cai, X., Cheng, Y., Zhou, L.: Evolution of intermetallic Cu_9Al_4 during the mechanical alloying of Cu–Al mixtures in high-energy ball milling. J. Electron. Mater. **50**(8), 4549–4556 (2021)
25. Wei, Z., Yan, Z., Chen, F., Khanlari, K., Wang, J., Huang, Z.: Refining of copper powder by a novel micro-abrasive milling method conducted in the air. Adv. Powder Technol. **33**(8), 103715 (2022)
26. Naik, D.N.S., Sharma, V.: Development of pulse-assisted cryo-micro lubrication architecture for machining of Ti-based alloys: A frugal innovative attempt for a hybrid lubri-cooling technique. J. Manuf. Process. **91**, 61–77 (2023)
27. Hosokawa, A., Suzuki, K., Yamaguchi, W., Takagi, K.: Mechanism of anomalous α-Fe formation from stoichiometric Sm_2Fe_{17} jet-milled powder during post-pulverization annealing. Acta Mater. **213**, 116981 (2021)
28. Rajendrachari, S.: An overview of high-entropy alloys prepared by mechanical alloying followed by the characterization of their microstructure and various properties. Alloys **1**(2), 116–132 (2022)
29. Toozandehjani, M., Matori, K.A., Ostovan, F., Abdul Aziz, S., Mamat, M.S.: Effect of milling time on the microstructure, physical and mechanical properties of Al-Al_2O_3 nanocomposite synthesized by ball milling and powder metallurgy. Materials **10**(11), 1232 (2017)
30. Shuai, C., He, C., Peng, S., Qi, F., Wang, G., Min, A., Wang, W.: Mechanical alloying of immiscible metallic systems: process, microstructure, and mechanism. Adv. Eng. Mater. **23**(4), 2001098 (2021)
31. Liu, X., Duan, Y., Guo, Y., Pang, H., Li, Z., Sun, X., Wang, T.: Microstructure design of high-entropy alloys through a multistage mechanical alloying strategy for temperature-stable megahertz electromagnetic absorption. Nano-Micro Lett. **14**(1), 142 (2022)
32. Olejarz, A., Huo, W.Y., Zieliński, M., Diduszko, R., Wyszkowska, E., Kosińska, A., Fang, F.: Microstructure and mechanical properties of mechanically-alloyed CoCrFeNi high-entropy alloys using low ball-to-powder ratio. J. Alloy. Compd. **938**, 168196 (2023)

33. Soler, R., Evirgen, A., Yao, M., Kirchlechner, C., Stein, F., Feuerbacher, M., Dehm, G.: Microstructural and mechanical characterization of an equiatomic YGdTbDyHo high entropy alloy with hexagonal close-packed structure. Acta Mater. **156**, 86–96 (2018)
34. Shen, X.F., Cheng, Z.Y., Wang, C.G., Wu, H.F., Yang, Q., Wang, G.W., Huang, S.K.: Effect of heat treatments on the microstructure and mechanical properties of Al–Mg–Sc–Zr alloy fabricated by selective laser melting. Opt. Laser Technol. **143**, 107312 (2021)
35. Chaira, D.: Development of nano-structured duplex and ferritic stainless steels by pulverisette planetary milling followed by pressureless sintering. Mater Charact **99**, 220–229 (2015)
36. Rajendrachari, S.: Effect of sintering temperature on the pitting corrosion of ball milled duplex stainless steel by using linear sweep voltammetry. Anal. Bioanal. Electrochem. **10**(3), 349–361 (2018)
37. Song, Y., Ma, G., Qiu, F., Stasiuk, O., Savvakin, D., Ivasishin, O., Cheng, T.: Nearly dense Ti–6Al–4V/TiB composites manufactured via hydrogen assisted BEPM. Ceram. Int. **48**(8), 10902–10910 (2022)
38. Biyik, S.E.R.K.A.N., Aydin, M.U.R.A.T.: Optimization of mechanical alloying parameters of $Cu_{25}W$ electrical contact material. Acta Phys. Pol. A **132**(3), 909–912 (2017)
39. Shashanka, R., Chaira, D.: Optimization of milling parameters for the synthesis of nano-structured duplex and ferritic stainless steel powders by high energy planetary milling. Powder Technol. **278**, 35–45 (2015)
40. Nokhrin, A.V., Malekhonova, N.V., Chuvil'deev, V.N., Melekhin, N.V., Bragov, A.M., Filippov, A.R., Sakharov, N.V., et al.: Effect of high-energy ball milling time on the density and mechanical properties of W-7% Ni-3% Fe alloy. Metals **13**(8), 1432 (2023)
41. Malikov, A., Vitoshkin, I., Orishich, A., Filippov, A., Karpov, E.: Microstructure and mechanical properties of laser welded joints of Al-Cu-Li and Ti-Al-V alloys. J. Manuf. Process. **53**, 201–212 (2020)
42. Seth, P.P., Parkash, O., Kumar, D.: Structure and mechanical behavior of in situ developed Mg_2Si phase in magnesium and aluminum alloys–a review. RSC Adv. **10**(61), 37327–37345 (2020)
43. Shashanka, R., Kumara Swamy, B.E.: Simultaneous electro-generation and electro-deposition of copper oxide nanoparticles on glassy carbon electrode and its sensor application. SN Appl. Sci. **2**, 1–10 (2020)
44. Gupta, S., Shashanka, R., Chaira, D.: Synthesis of nano-structured duplex and ferritic stainless steel powders by planetary milling: an experimental and simulation study. IOP Conf. Ser.: Mater. Sci. Eng. **75**(1), 012033 (2015). IOP Publishing
45. Shashanka, R., Chaira, D.: Effects of nano-Y_2O_3 and sintering parameters on the fabrication of PM duplex and ferritic stainless steels. Acta Metall. Sin. (Engl. Lett.) **29**, 58–71 (2016)
46. Shashanka, R.: Investigation of optical and thermal properties of CuO and ZnO nanoparticles prepared by Crocus Sativus (Saffron) flower extract. J. Iran. Chem. Soc. **18**(2), 415–427 (2021)
47. Pradeep, N.B., Hegde, M.R., Rajendrachari, S., Surendranathan, A.O.: Investigation of microstructure and mechanical properties of microwave consolidated TiMgSr alloy prepared by high energy ball milling. Powder Technol. **408**, 117715 (2022)
48. Basavarajappa, P.N., Hegde, M.M.R., Rajendrachari, S., Surendranathan, A.O.: Investigation of structural and mechanical properties of nanostructured TiMgSr alloy for biomedical applications (2023)
49. Arvieu, C., Galy, C., Le Guen, E., Lacoste, E.: Relative density of SLM-produced aluminum alloy parts: interpretation of results. J. Manuf. Mater. Process. **4**(3), 83 (2020)
50. Mathews, J.P., Campbell, Q.P., Xu, H., Halleck, P.: A review of the application of X-ray computed tomography to the study of coal. Fuel **209**, 10–24 (2017)

51. Carretier, V., Pucci, M.F., Lacoste, C., Regazzi, A., Lopez-Cuesta, J.M.: An efficient solution to determine surface energy of powders and porous media: application to untreated and treated lignin. Appl. Surf. Sci. **579**, 152159 (2022)
52. Columbu, S., Mulas, M., Mundula, F., Cioni, R.: Strategies for helium pycnometry density measurements of welded ignimbritic rocks. Measurement **173**, 108640 (2021)
53. Mahale, R.S., Shamanth, V., Hemanth, K., Nithin, S.K., Sharath, P.C., Shashanka, R., Shetty, D.: Processes and applications of metal additive manufacturing. Mater. Today: Proc. **54**, 228–233 (2022)

Chapter 6
Consolidation of Mechanically Alloyed Powders

Trinath Talapaneni and Vatsala Chaturvedi

Abstract The consolidation of mechanically alloyed powders is a critical process in the field of materials science and metallurgy. Mechanically alloyed powders are synthesized through high-energy ball milling, resulting in finely divided and intimately mixed powder particles. This chapter provides an overview of the consolidation techniques employed to transform these mechanically alloyed powders into functional and advanced materials. Various consolidation methods, such as hot isostatic pressing, spark plasma sintering, powder extrusion, powder forging, solid and liquid state sintering, reaction sintering, selective laser sintering, and microwave sintering are discussed. Also, the chapter explores the effects of consolidation parameters, such as temperature, pressure, and holding time, on the properties of the final product. Further, it includes various advantages, challenges and applications of several consolidation techniques.

Keywords Mechanical alloying · Consolidation techniques · Powder metallurgy · Sintering

6.1 Overview of Mechanical Alloying

Mechanical Alloying (MA) serves as a dry powder processing technique that has proven effective in synthesizing both stable and metastable phases of materials with commercial and scientific significance [1]. Originating in the mid-1960s at the Paul D. Merica Research Laboratory of the International Nickel Company (INCO), this method was pioneered by John S. Benjamin and colleagues. Their initial focus was on developing an alloy, combining oxide dispersion strengthening with γ' precipitation hardening in a nickel-based superalloy intended for gas turbine applications [2]. MA stands out as a straightforward and adaptable technique, offering economic viability in material processing. The process involves loading individual elemental powders

T. Talapaneni (✉) · V. Chaturvedi
Department of Metallurgical Engineering, OP Jindal University, Raigarh, Chhattisgarh 496109, India
e-mail: trinath.talapaneni@opju.ac.in

S. Rajendrachari (ed.), *Mechanically Alloyed Novel Materials*, Advanced Structured Materials 220, https://doi.org/10.1007/978-981-97-6504-1_6

or pre-alloyed powders, along with a grinding medium (typically hardened steel or tungsten carbide balls), into a high-energy ball mill. A ball-to-powder weight ratio of 10:1 or higher is typically maintained. The high-energy milling in this process results in the repeated flattening, fracturing, and rewelding of powder particles [3]. The versatility of MA is evident in its ability to be conducted in both small-capacity and high-energy mills, allowing for the production of powders ranging from 10 g per run to an impressive 5000 kg/h. MA stands out as a highly employed technique for plastic deformation, playing a crucial role in achieving an exceptionally refined structure [4]. Furthermore, the powder material produced through mechanical alloying effectively minimizes oxidation of its constituents, due to the brief processing duration [5]. MA superalloys find applications in aircraft, industrial gas turbines, power generation equipment, heat treatment furnace parts, etc. due to their high-temperature properties.

6.1.1 Objectives of Consolidation in the Powder Metallurgy Process

Consolidation is a crucial step in the powder metallurgy process, playing a vital role in determining the final properties and performance of the manufactured components some of the key objectives of consolidation in powder metallurgy are mentioned below.

(a) Densification and Strength: The primary objective of consolidation is to enhance the density of the powder compact [6]. Achieving a high level of density is essential for imparting strength and mechanical integrity to the final product. Proper consolidation ensures that the powder particles are closely packed, minimizing porosity and enhancing the structural integrity of the material.
(b) Improved Mechanical Properties: Consolidation contributes significantly to the mechanical properties of the final product. It helps in reducing voids and porosity, resulting in improved hardness, tensile strength, and fatigue resistance. Well-consolidated materials exhibit better overall mechanical performance, making them suitable for a wide range of applications.
(c) Controlled Microstructure: The consolidation process influences the microstructure of the material, which, in turn, affects its properties. Through careful control of consolidation parameters, such as temperature, pressure, and time, manufacturers can tailor the microstructure to achieve desired characteristics such as grain size, phase distribution, and texture.
(d) Enhanced Uniformity: Proper consolidation ensures uniformity in the distribution of alloying elements and phases within the material. This uniformity is essential for achieving consistent properties throughout the final product, reducing variations in performance, and enhancing predictability in the manufacturing process.

(e) Minimization of Defects: Consolidation helps in minimizing defects such as cracks, voids, and inhomogeneity. These defects can compromise the structural integrity and performance of the material. Effective consolidation techniques contribute to defect-free components with improved reliability.
(f) Dimensional Control: The consolidation process also plays a role in controlling the final dimensions of the manufactured components. By carefully managing the consolidation parameters, manufacturers can achieve precise and accurate shapes, meeting design specifications and tolerances.
(g) Enhanced Sintering: In powder metallurgy, consolidation is often accompanied by sintering, a process where the powder particles bond together through diffusion and solid-state reactions. Sintering, combined with consolidation, promotes atomic diffusion, leading to improved material properties and a more homogeneous structure.

6.2 Consolidation Techniques

6.2.1 Sintering

Sintering is the process of heating a loose or compacted assembly of metal powders below the melting point of the base metal. This transformation occurs with or without the application of external pressure, resulting in the creation of a denser material through interparticle bonding. Figure 6.1 illustrates the overall steps involved in the production of sintered parts.

Types of Sintering

Solid-State Sintering

Solid-state sintering and liquid-state sintering are called pressure-less sintering processes. Solid-state sintering is a manufacturing technique wherein a powdered material, typically a metal or ceramic, undergoes compaction and heating below its melting point [7]. The compaction is achieved using a die or mold. As the sintering

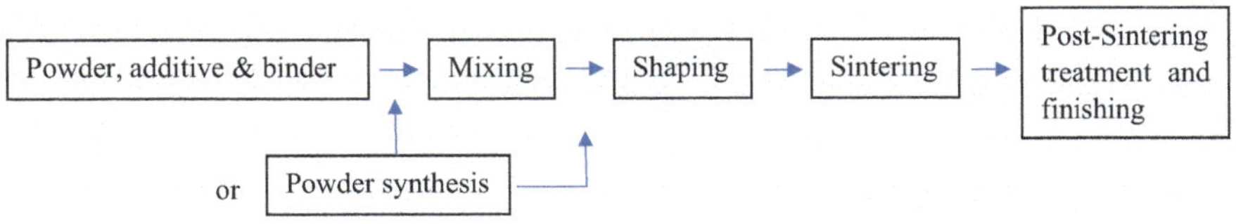

Fig. 6.1 General procedure for manufacturing sintered parts

process advances with rising temperatures, atoms along particle boundaries migrate, facilitating particle rearrangement and bonding. This transformative process leads to the creation of a denser and more cohesive solid mass. To mitigate surface oxidation of numerous metal particles, solid-state sintering is generally conducted in a controlled environment, often employing an atmosphere furnace or, occasionally, a vacuum furnace.

Advantages of Solid-State Sintering

Solid-state sintering offers several advantages in the manufacturing of materials, particularly metals and ceramics. Some key advantages include:

(a) Preservation of Material Purity

Solid-state sintering allows for the production of materials with a high degree of purity because there is no introduction of foreign substances during the process. This is particularly important in applications where the original material composition must be maintained.

(b) High-Density Products

The process results in high-density materials with improved mechanical properties. The compacting and bonding of particles during sintering lead to a more cohesive and closely packed structure, enhancing the strength and hardness of the final product.

(c) Fine Control over Microstructure

Solid-state sintering provides precise control over the microstructure of the material. This control is crucial in achieving specific mechanical, thermal, or electrical properties tailored to the requirements of the intended application.

(d) No Melting Required

As the name suggests, solid-state sintering occurs below the melting point of the material. This eliminates the need for energy-intensive melting and subsequent cooling processes, making it a more energy-efficient manufacturing method.

(e) Cost-Effective

The absence of the liquid phase and the reduced energy requirements contribute to the cost-effectiveness of solid-state sintering. It can be a more economical process compared to methods involving high-temperature melting and subsequent solidification.

(f) Improved Product Uniformity

Solid-state sintering allows for better control over the uniformity of the final product. The avoidance of phase changes associated with melting and cooling helps in maintaining consistent material properties throughout the manufactured part.

(g) Environmental Considerations

The process can be conducted in controlled atmospheres or vacuum furnaces to prevent or reduce surface oxidation of metal particles. This is important for maintaining the integrity and properties of the material, contributing to environmental sustainability.

(h) Versatility in Material Selection

Solid-state sintering can be applied to various materials, including metals, ceramics, and certain types of composites. This versatility makes it suitable for various industries and applications.

(i) Suitability for Complex Shapes

The process allows for the fabrication of parts with complex shapes and intricate details. The initial compaction of powders into molds or dies enables the creation of components with specific geometries.

Limitations of Solid State Sintering

While solid-state sintering has numerous advantages, it also comes with certain limitations and challenges. Here are some of the key limitations associated with solid-state sintering:

(a) Time-Consuming Process

Solid-state sintering can be a time-consuming process, particularly for materials that require extended exposure to high temperatures to achieve the desired densification. The slow diffusion kinetics can be a limiting factor in the production rate.

(b) Limited Densification in Some Materials

Certain materials may not achieve high levels of densification through solid-state sintering alone. This limitation is especially relevant for materials with complex compositions or those that inherently resist particle rearrangement.

(c) Energy Consumption

Although solid-state sintering is generally more energy-efficient than processes involving melting, it still requires elevated temperatures. The energy consumption can be a consideration, especially when compared to alternative manufacturing methods with lower energy requirements.

(d) Uniformity Challenges

Achieving uniform properties throughout the entire sintered product can be challenging. Variations in temperature, pressure, or powder distribution during the compaction stage may lead to non-uniform sintering and variations in the final product.

(e) Tooling and Equipment Costs

The need for specialized tooling and equipment, such as dies and molds for powder compaction, can contribute to the overall cost of the manufacturing process. Initial setup costs may be a consideration, especially for small-scale production.

(f) Limited Shape Complexity

While solid-state sintering enables the formation of relatively complex shapes, it may have limitations in producing intricate designs or components with extremely fine features compared to other manufacturing processes.

(g) Scale-Up Challenges

Scaling up the solid-state sintering process for large-volume production can pose challenges. Uniformly applying heat and pressure to large quantities of material may require sophisticated and expensive equipment.

Application of Solid-State Sintering Process

The process of Solid-State Sintering is esteemed for its capability to create components with elevated density, strength, and precision. It finds extensive use in manufacturing diverse products, ranging from metal and ceramic components to specific electronic components and cutting tools. This method is particularly advantageous in applications where preserving the original material composition is paramount. Below are some typical instances of its applications:

(a) Metal Components in Aerospace

Solid-state sintering is often used to produce high-strength and lightweight metal components for aerospace applications [8], where the specific mechanical properties of materials are crucial.

(b) Ceramic Parts for Electronics

In the electronics industry, solid-state sintering is employed to manufacture ceramic components, such as insulators [9], substrates, and electronic packaging materials with tailored electrical and thermal properties.

(c) Cutting Tools and Hard Metals

Solid-state sintering is utilized to manufacture cutting tools, hard metal inserts, and wear-resistant components in the tooling industry [10]. The process helps achieve high hardness and wear resistance.

(d) Powder Metallurgy Products

Solid-state sintering is a key technique in powder metallurgy, where it is used to produce a variety of components such as gears, bearings, and bushings [11].

(e) Medical Implants

In the medical field, solid-state sintering is applied to produce biocompatible metal and ceramic implants with specific shapes and properties for orthopedic and dental applications [12, 13].

(f) Catalysts and Chemical Industry

Solid-state sintering is employed to fabricate catalysts and catalytic supports used in the chemical industry. The controlled microstructure and composition are critical for catalytic activity.

(g) Fuel Cells and Batteries

Solid-state sintering is utilized in the production of components for fuel cells and batteries, where it helps create dense and conductive materials for efficient energy storage and conversion [14].

(h) Automotive Components

Various automotive components, including engine parts, sensors, and brake components, are manufactured using solid-state sintering to achieve the required strength and durability.

(i) Magnetic Materials

Magnetic materials, such as permanent magnets, are produced through solid-state sintering [15]. The process allows for the fabrication of materials with tailored magnetic properties.

(j) High-Performance Ceramics

Solid-state sintering is applied to produce high-performance ceramics used in applications like cutting tools, industrial wear parts, and advanced structural components.

(k) Electronic Substrates and Components

The electronics industry utilizes solid-state sintering to manufacture substrates, heat sinks, and various electronic components with specific thermal and electrical properties.

(l) Precision Engineering Components

Solid-state sintering is employed in the production of precision engineering components where the controlled microstructure and high density are critical for meeting strict tolerances.

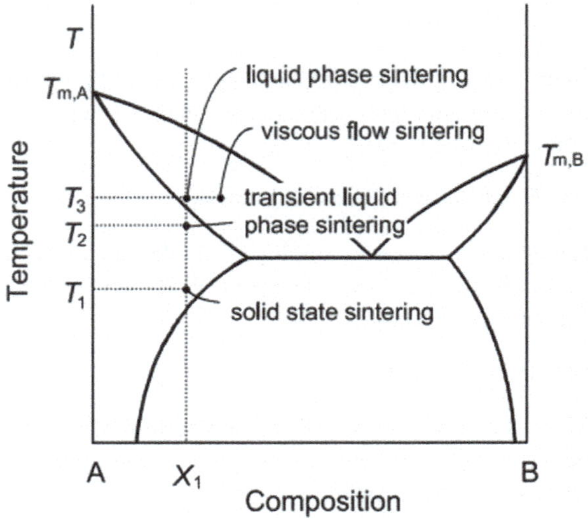

Fig. 6.2 Representing various types of sintering in phase diagram [17]

(m) Solar Cell Manufacturing

Solid-state sintering is used in the production of certain types of solar cells, helping create materials with desirable electrical and structural properties.

(n) Heat Exchangers

Components for heat exchangers, where thermal conductivity and durability are important, can be produced using solid-state sintering.

Liquid-State Sintering

The liquid-state sintering process is conducted at a temperature where a liquid phase is present [16]. This temperature can surpass the melting point of either one of the alloy constituents or extend beyond the melting point of an alloy created during the sintering process. Figure 6.2 illustrates the liquid state sintering in a phase diagram at T_3 temperature in an A–B powder compact at X_1 composition.

Advantages of Liquid-State Sintering

Liquid-state sintering offers several advantages in the manufacturing of materials, particularly in comparison to traditional solid-state sintering. Some of the key advantages of liquid-state sintering are mentioned below:

(a) Enhanced Densification

The presence of a liquid phase promotes greater particle mobility and rearrangement, leading to enhanced densification of the material. This results in a higher final density and improved mechanical properties.

(b) Reduced Sintering Temperature

Liquid-state sintering often allows for sintering at lower temperatures compared to solid-state sintering. This can be beneficial for materials that are sensitive to high temperatures or for reducing energy consumption during the manufacturing process.

(c) Improved Homogeneity

The liquid phase facilitates the homogenization of the material, leading to a more uniform distribution of particles. This can result in improved microstructural homogeneity and better control over the final material's properties.

(d) Facilitates Alloy Formation

Liquid-state sintering enables the formation of alloys by allowing the melting and mixing of different components. This can lead to the development of materials with unique and tailored properties.

(e) Enhanced Grain Growth Control

The presence of a liquid phase can help control grain growth during sintering, leading to a fine and controlled microstructure. This is important for achieving specific material characteristics and properties.

(f) Reduced Pore Formation

The liquid phase aids in reducing or eliminating pore formation within the material. This is crucial for achieving higher densities and improving the overall strength and integrity of the final product.

(g) Faster Processing Time

Liquid-state sintering often results in faster processing times compared to solid-state sintering. This can be advantageous for increasing production efficiency and reducing manufacturing cycle times.

(h) Complex Geometries

The fluid nature of the liquid phase allows for the fabrication of components with complex shapes and intricate details. This is particularly useful in industries where precise geometries are essential.

(i) Improved Thermal Conductivity

Certain materials processed through liquid-state sintering may exhibit improved thermal conductivity due to enhanced particle bonding and reduced porosity.

(j) Tailored Material Properties

Liquid-state sintering provides flexibility in tailoring material properties by adjusting the composition, temperature, and duration of the sintering process. This adaptability is valuable for meeting specific requirements in various applications.

(k) Applicable to Various Materials

Liquid state sintering can be applied to various materials, including metals, ceramics, and composites, making it a versatile manufacturing method.

Limitations of Liquid-State Sintering

While liquid state sintering offers several advantages, it also comes with certain limitations and challenges. Some of the limitations associated with liquid-state sintering are mentioned below:

(a) Controlled Atmosphere Requirements

Liquid-state sintering often requires a controlled atmosphere or a vacuum to prevent undesirable reactions, such as oxidation, with the environment. This may add complexity and cost to the manufacturing process.

(b) Higher Equipment Costs

The equipment used for liquid-state sintering, including furnaces capable of maintaining specific temperatures and atmospheres, can be more expensive than equipment used in solid-state sintering.

(c) Material Compatibility Challenges

Not all materials are suitable for liquid-state sintering. Some materials may react unfavorably with the liquid phase, limiting the range of materials that can be processed using this method.

(d) Difficulties in Handling Reactive Materials

For reactive materials, maintaining the correct balance of temperature and atmosphere during liquid-state sintering can be challenging. Reactive materials may undergo undesirable reactions with the liquid phase.

(e) Uniformity Issues

Achieving uniform properties throughout the entire sintered product can be challenging in liquid-state sintering. Variations in temperature, pressure, or powder distribution during the process may lead to non-uniform sintering.

(f) Porosity in Certain Materials

In some cases, liquid-state sintering may result in the formation of pores within the material, especially if the process conditions are not carefully controlled. This can impact the final product's mechanical properties.

(g) Limited Applicability to High-Melting-Point Materials

Liquid state sintering may not be suitable for materials with very high melting points. The required temperatures for achieving a liquid phase may be impractical or result in excessive energy consumption.

(h) Complex Process Optimization

Achieving the desired material properties through liquid-state sintering often requires careful optimization of process parameters, including temperature, time, and atmosphere. This complexity can make process development and control challenging.

Application of Liquid-State Sintering

Liquid state sintering is a versatile manufacturing process with applications in various industries. Some of the common applications of the liquid-state sintering process are mentioned below:

(a) Metal Alloy Production

Liquid-state sintering is commonly used in the production of metal alloys. The process allows for the controlled melting and mixing of different metal powders to create alloys with specific properties.

(b) Ceramic Manufacturing

In the ceramics industry, liquid-state sintering is applied to produce dense and high-strength ceramic components. This includes applications in the production of advanced ceramics for electronic devices, cutting tools, and wear-resistant components.

(c) Composite Materials

Liquid-state sintering is used to manufacture composite materials by combining different powder components with a liquid phase. This process enhances the bonding between particles, resulting in materials with improved mechanical and thermal properties.

(d) Tooling and Cutting Tools

The process is employed in the production of cutting tools, such as carbide tools used in machining and drilling. Liquid-state sintering enables the creation of hard and wear-resistant tools with complex geometries.

(e) Powder Metallurgy

Liquid-state sintering plays a significant role in powder metallurgy applications. It is used to produce a wide range of components, including gears, bearings, and other parts with specific mechanical properties.

(f) Medical Implants

Liquid-state sintering is utilized in the production of medical implants, such as hip and knee implants. The process allows for the fabrication of biocompatible materials with tailored properties for orthopedic applications.

(g) Electronic Components

The electronics industry uses liquid-state sintering to manufacture components such as substrates, connectors, and heat sinks. The process helps achieve precise shapes and controlled properties for electronic applications.

(h) Aerospace Components

Liquid state sintering is applied in the aerospace industry to produce high-performance components, including turbine blades and structural parts. The process allows for the creation of materials with desirable strength-to-weight ratios.

(i) Magnetic Materials

The process is used to manufacture permanent magnets with specific magnetic properties. Liquid-state sintering enables the precise control of the microstructure, contributing to the performance of magnetic materials.

(j) Fuel Cell Components

Liquid-state sintering is employed in the production of components for fuel cells. It helps create materials with high density, good electrical conductivity, and resistance to corrosion for efficient energy conversion.

(k) Automotive Parts

Liquid-state sintering is used to produce various automotive components, such as engine parts, gears, and brake components. The process contributes to the manufacturing of durable and high-performance materials.

(l) High-Performance Alloys

The production of high-performance metal alloys, such as those used in the aerospace and automotive industries, often involves liquid-state sintering. This allows for the formation of alloys with specific strength and thermal properties.

(m) Advanced Structural Materials

Liquid-state sintering is employed to manufacture advanced structural materials used in applications requiring high strength, toughness, and wear resistance.

(n) Energy Storage Devices

The process is utilized in the production of components for energy storage devices, including batteries. Liquid state sintering contributes to the development of materials with improved conductivity and durability.

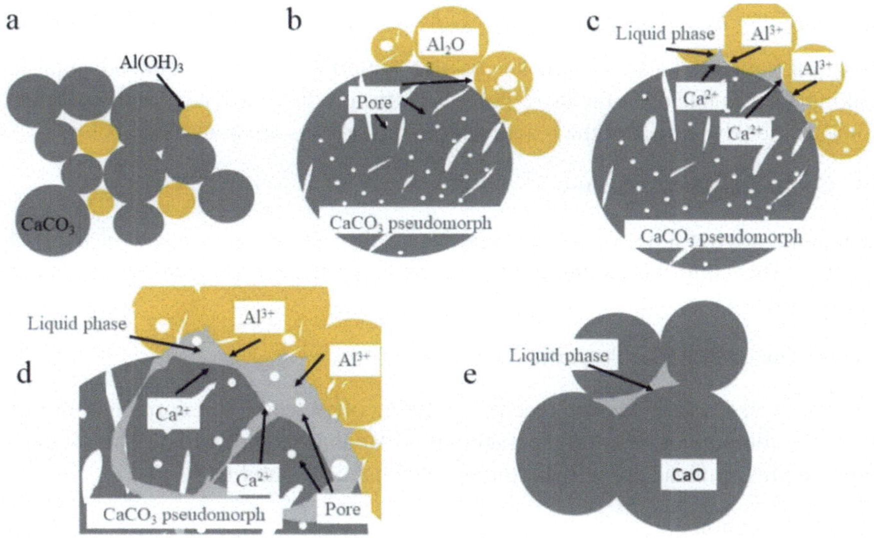

Fig. 6.3 Schematic representation of the reaction sintering mechanism of CaO granules [19]. **a** packing of the particles; **b** decomposition of the $CaCO_3$ and $Al(OH)_3$; **c** liquid formation; **d** penetration of the liquid and elimination of pores; **e** growth of the CaO grain

Reaction Sintering

Reaction sintering, also known as reactive sintering or reaction-bonded sintering, is a manufacturing process used to produce ceramics, composite materials, and certain alloys by inducing chemical reactions between the powder constituents during the sintering process [18]. This technique involves the simultaneous application of heat and pressure to promote chemical reactions that result in the formation of a dense, consolidated material. Figure 6.3 shows an example of the reaction sintering mechanism of CaO granules. In the initial phase, the contacts between.

$CaCO_3$ and $Al(OH)_3$ were characterized by point contacts within the $CaCO_3$ granules (Fig. 6.3a). As the heating temperature rose to the range of 300–800 °C, both $CaCO_3$ and $Al(OH)_3$ underwent decomposition, resulting in the formation of CaO and Al_2O_3, each exhibiting micro-sized pores (Fig. 6.3b). A local liquid phase emerged through reactions involving impurities, CaO, and Al_2O_3 as the heating temperature reached 1300 °C. Subsequently, the $CaCO_3$ pseudomorphs and Al_2O_3 both dissolved into this liquid phase (Fig. 6.3c), creating favorable conditions for the wetting and spreading behaviors of the liquid phase. With the continued increase in heating temperature, the quantity of the liquid phase grew, subsequently enhancing the dissolution of CaO and Al_2O_3 into the liquid phase (Fig. 6.3d). As sintering progressed, there was a notable increase in CaO grain size, accompanied by a simultaneous increase in densification.

Advantages of Reaction Sintering [20–27]

Reaction sintering, also known as reactive sintering, offers several advantages, making it a valuable manufacturing process for the production of advanced materials. Some of the advantages of the reaction sintering process are mentioned below:

(a) Controlled Microstructure

Reaction sintering allows for the precise control of the microstructure of the final material. This control is essential for tailoring properties such as hardness, strength, and wear resistance.

(b) Enhanced Material Properties

The chemical reactions during reaction sintering often lead to the formation of new phases or compounds, resulting in materials with enhanced properties. This includes improved mechanical strength, hardness, and chemical stability.

(c) Densification

The process promotes the densification of the material, resulting in high-density products. This is crucial for achieving optimal mechanical properties and durability.

(d) Formation of Complex Structures

Reaction sintering enables the formation of complex structures and compositions that may be challenging to achieve through conventional sintering methods.

(e) Tailoring Thermal Properties

The ability to control chemical reactions allows for the tailoring of thermal properties, making reaction-sintered materials suitable for specific applications where heat resistance or thermal conductivity is important.

(f) Production of Advanced Ceramics

Reaction sintering is widely used in the production of advanced ceramics, including silicon carbide (SiC) and silicon nitride (Si_3N_4), which exhibit superior properties compared to traditional ceramics.

(g) Versatility Across Materials

The process applies to various materials, including ceramics, metals, and composites. This versatility makes it suitable for a wide range of industrial applications.

(h) Customization for Specific Applications

The selection of raw materials and reaction conditions can be customized to achieve specific material properties tailored for particular applications.

(i) Formation of Hard and Wear-Resistant Materials

Reaction sintering is often employed to produce hard and wear-resistant materials, such as carbides and nitrides, which find applications in cutting tools and wear-resistant components.

(j) Cost Efficiency

In some cases, reaction sintering can lead to cost savings by reducing the need for high-temperature processing or expensive starting materials.

(k) Reduced Processing Time

Compared to traditional sintering methods, reaction sintering can result in shorter processing times due to the accelerated chemical reactions involved.

(l) Potential for Complex Geometries

The process allows for the fabrication of components with intricate shapes and complex geometries, providing design flexibility in various applications.

(m) Improved Bonding between Phases

The chemical reactions often lead to improved bonding between different phases in the material, contributing to better overall performance.

(n) Eco-Friendly Approaches

In some instances, reaction sintering can be designed to use environmentally friendly or sustainable materials, contributing to eco-friendly manufacturing practices.

(o) Controlled Atmosphere Options

Depending on the materials and reactions involved, reaction sintering may be performed in a controlled atmosphere, providing flexibility in processing materials under specific conditions.

Limitations of Reaction Sintering

While reaction sintering offers several advantages, it also comes with certain limitations and challenges. Some of the limitations of the reaction sintering process are mentioned below:

(a) Complex Process Optimization

Achieving the desired material properties through reaction sintering often requires careful optimization of process parameters, including temperature, time, pressure, and composition. This complexity can make process development and control challenging.

(b) Material Compatibility Challenges

Not all materials are suitable for reaction sintering, and compatibility issues can arise when attempting to induce specific chemical reactions. The selection of appropriate raw materials is critical.

(c) Controlled Atmosphere Requirements

Reaction sintering may require a controlled atmosphere or a vacuum to prevent undesirable reactions with the environment. This adds complexity to the process and may increase operational costs.

(d) Energy Consumption

In some cases, the reaction sintering process may require higher energy consumption, particularly when high temperatures and extended processing times are necessary to induce desired reactions.

(e) Limited Applicability to High-Melting-Point Materials

The process may not be suitable for materials with very high melting points, as achieving the required temperatures for specific reactions can be challenging or energy-intensive.

(f) Formation of Unwanted Phases

In certain cases, unintended reactions may occur, leading to the formation of unwanted phases or compounds. Controlling the reaction pathway to achieve the desired product can be challenging.

(g) Uniformity Issues

Achieving uniform properties throughout the entire sintered product can be challenging in reaction sintering. Variations in temperature, pressure, or powder distribution during the process may lead to non-uniform sintering.

(h) Difficulty in Scaling Up

Scaling up reaction sintering processes for large-scale production can be difficult due to challenges in maintaining uniform conditions and achieving consistent results.

(i) Reactivity with the Environment

Some reactive materials used in reaction sintering may be sensitive to environmental conditions, such as moisture or oxygen. Special precautions may be required to prevent unwanted reactions during processing.

(j) Cost Considerations

The raw materials used in reaction sintering can be expensive, and the complexity of the process may lead to higher overall production costs compared to simpler sintering methods.

Applications of Reaction Sintering

Some common applications of the reaction sintering process are mentioned below:

(a) Advanced Ceramics

Reaction sintering is widely used in the production of advanced ceramics, including silicon carbide (SiC), silicon nitride (Si_3N_4), and alumina-toughened zirconia (ATZ). These ceramics exhibit enhanced hardness, wear resistance, and high-temperature stability.

(b) Metal Matrix Composites (MMCs)

The process is employed in the fabrication of metal matrix composites, where a metal matrix is reinforced with ceramic particles. This results in materials with improved mechanical properties suitable for applications in aerospace, automotive, and structural components.

(c) Ceramic Matrix Composites (CMCs)

Reaction sintering is utilized in the manufacturing of ceramic matrix composites, which offer a combination of high-temperature resistance, strength, and toughness. These materials find applications in aerospace, gas turbines, and other high-performance engineering components [28].

(d) Cutting Tools and Wear-Resistant Components

Hard and wear-resistant materials, such as tungsten carbide (WC) and titanium carbide (TiC), can be produced using reaction sintering for applications in cutting tools, drills, and wear-resistant components.

(e) Automotive Components

Reaction sintering is applied in the production of various automotive components, including brake discs, engine parts, and components that require high strength, wear resistance, and thermal stability [29].

(f) Cemented Carbides

Cemented carbides, which are widely used in the production of cutting tools, mining tools, and wear-resistant components, are often manufactured using reaction sintering processes.

(g) Electronic and Electrical Components

Reaction sintering is used to produce materials for electronic and electrical applications, including substrates, insulators, and components with specific electrical and thermal conductivity properties.

(h) Bioceramics and Medical Implants

Certain bioceramics and materials for medical implants are manufactured using reaction sintering. These materials need to exhibit biocompatibility, high strength, and corrosion resistance.

(i) Refractory Materials

High-performance refractory materials, which are resistant to extreme temperatures and harsh environments, are produced using reaction sintering. These materials find applications in kiln linings, furnaces, and other high-temperature processes.

(j) Metal Nitrides and Carbides

Reaction sintering is employed to produce metal nitrides (e.g., titanium nitride) and metal carbides (e.g., tungsten carbide), which are used in various industrial applications due to their hardness and wear resistance.

(k) Aerospace Components

Components for aerospace applications, including turbine blades and structural parts, benefit from the high-temperature resistance and mechanical properties achieved through reaction sintering.

(l) Fuel Cell Components

Reaction sintering is utilized in the production of components for fuel cells, such as electrodes and electrolytes. The process helps create materials with high density, good electrical conductivity, and corrosion resistance.

(m) Powder Metallurgy

The process is applied in powder metallurgy for the production of components with specific properties, such as increased hardness, wear resistance, and improved structural integrity.

(n) Magnetic Materials

Certain magnetic materials, including permanent magnets, can be produced using reaction sintering, allowing for the precise control of microstructure and magnetic properties.

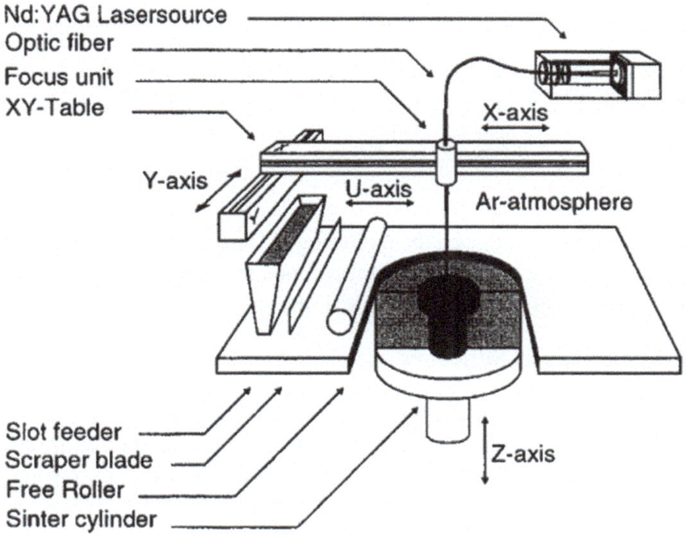

Fig. 6.4 The SLS machine set up at the Catholic University of Leuven, Belgium [31]

Selective Laser Sintering

Selective Laser Sintering (SLS) is an additive manufacturing (AM) or 3D printing process that utilizes a laser to selectively fuse or sinter powdered material, typically polymers or metal powders, layer by layer, to create three-dimensional objects [30]. Figure 6.4 represents the SLS machine developed at the Catholic University of Leuven, Belgium. This setup comprises a Lumonics MW500 Nd: YAG (neodymium: yttrium aluminum garnet) laser and a controlled process chamber, allowing for the regulation of both pressure and atmosphere conditions. The Nd: YAG laser is capable of delivering a maximum continuous power of 500 W, adjustable in increments of 10 W. Operating at a wavelength of 1.064 μm, the laser beam can be focused to a diameter of 0.8 mm. The laser light is directed through an optical fiber and traverses the powder surface either based on the product's cross-section or by following a predefined 2D pattern. This movement is facilitated by an X, Y-table holding the focusing lens. The machine's working speed is constrained by the table's maximum acceleration of 500 mm s^{-2}, limiting the machine to a maximum speed of 30 mm s^{-1}. Consequently, the system is suitable only for operations within the low-power range (up to 60 W effective). The powder deposition system includes a sintering container with a diameter of 100 mm, a slot feeder, a free-rolling roller, and a scraper. The powder is situated within a vacuum chamber initially evacuated to about 100 Pa and then filled with argon to a pressure of 0.5 bar, effectively reducing the initial oxygen concentration to 0.04%.

Advantages of Selective Laser Sintering

Selective Laser Sintering (SLS) offers several advantages, making it a popular additive manufacturing (3D printing) process for various applications. Here are some key advantages of Selective Laser Sintering:

(a) Material Versatility

SLS can work with various materials, including polymers (nylon, polyamide), metals (aluminum, stainless steel), ceramics, and composites. This versatility allows for the production of parts with diverse mechanical and thermal properties.

(b) No Need for Support Structures

Unlike some other 3D printing processes, SLS doesn't require support structures for overhangs or complex geometries. The surrounding unsintered powder acts as natural support during the printing process, simplifying post-processing and reducing material waste.

(c) Complex Geometries

SLS excels in producing intricate and complex designs, including internal features and cavities. This capability makes it suitable for creating parts with complex shapes that would be challenging or impossible to achieve through traditional manufacturing methods [32].

(d) High Resolution and Accuracy

SLS offers high resolution and accuracy, enabling the production of detailed and precise parts. This makes it suitable for applications where tight tolerances and fine details are crucial.

(e) Rapid Prototyping

SLS is widely used for rapid prototyping due to its ability to quickly produce functional prototypes. Designers and engineers can iterate on designs rapidly, accelerating the product development cycle.

(f) Batch Production

SLS is suitable for small to medium-scale batch production. It allows for the cost-effective production of multiple parts in a single build, making it an attractive option for customized or low-volume manufacturing.

(g) Reduced Material Waste

The unsintered powder that surrounds the printed object can be recycled and reused in subsequent builds, minimizing material waste and contributing to a more sustainable manufacturing process.

(h) Wide Range of Applications

SLS is used across various industries, including aerospace, automotive, healthcare, consumer goods, and more. It can produce functional prototypes, end-use parts, and components for various applications.

(i) No Tooling Required

Unlike traditional manufacturing methods, SLS does not require the creation of molds or tooling. This eliminates the need for costly and time-consuming tooling processes, reducing lead times and costs.

(j) Design Freedom

The layer-by-layer additive nature of SLS allows for design freedom, enabling the creation of complex and innovative shapes. This is particularly beneficial for designers seeking unconventional and optimized designs [33].

(k) Scalability

SLS is scalable for both prototyping and production. It can be used for producing a single prototype or scaled up for the manufacturing of multiple parts, making it adaptable to different production requirements.

(l) Reduced Time-to-Market

The rapid prototyping capabilities, design flexibility, and quick turnaround times of SLS contribute to reducing the time-to-market for new products.

(m) Tool-Free Iterations

Design changes and iterations can be implemented without the need for modifying or creating new molds or tools, allowing for more agile and cost-effective product development.

Limitations of Selective Laser Sintering

Some of the limitations of Selective Laser Sintering (SLS) are mentioned below:

(a) Material Limitations

The range of materials suitable for SLS is not as extensive as some other 3D printing technologies. While various materials are available, the selection might be limited compared to processes like stereolithography (SLA) or fused deposition modeling (FDM).

(b) Surface Finish

SLS parts may have a rougher surface finish compared to other additive manufacturing technologies like stereolithography. Post-processing steps, such as sanding or polishing, may be required to achieve a smoother surface.

(c) Resolution and Accuracy

SLS may not offer the same level of resolution and accuracy as some other 3D printing methods, such as stereolithography. This can affect the level of detail that can be achieved in the printed parts.

(d) Support Structures

SLS typically requires the use of support structures to prevent the collapse of overhanging features during the printing process. Removing these support structures can be labor-intensive and may require additional post-processing steps.

(e) Part Distortion

During the SLS process, heat is applied to fuse the powdered material. This can lead to thermal stress and part distortion, particularly in larger or more complex geometries. Cooling strategies and part orientation can help mitigate this issue, but it remains a consideration.

(f) Equipment Cost and Size

SLS machines can be expensive to purchase and maintain. Additionally, the size of the equipment may limit the types of parts that can be produced in terms of build volume.

(g) Powder Handling

Handling and recycling of powder materials can be challenging. It requires careful management to ensure that the powder remains consistent and uncontaminated between builds.

(h) Limited Color Options

While SLS is capable of producing parts in various materials, the color options are often limited compared to some other 3D printing technologies. This may be a consideration for applications where color is essential.

Applications of Selective Laser Sintering

Selective Laser Sintering (SLS) is a versatile additive manufacturing technology that finds applications in various industries due to its ability to produce complex and functional parts. Some of the notable applications of Selective Laser Sintering are mentioned below:

(a) Prototyping

SLS is widely used for rapid prototyping, allowing engineers and designers to quickly iterate and test designs before moving to full-scale production. Its ability to create complex geometries without the need for support structures makes it particularly useful for prototyping intricate parts.

(b) Customized and Low-Volume Production

SLS is suitable for producing low volumes of customized or small-batch production parts. This is valuable in industries where personalized products or low-production runs are required, such as medical devices, automotive components, and aerospace components.

(c) Aerospace Components

In the aerospace industry, SLS is utilized to manufacture lightweight and durable components, including complex geometries such as brackets, housings, and interior components. The technology's ability to work with a range of materials, including high-performance polymers and metals, makes it well-suited for aerospace applications.

(d) Automotive Parts

SLS is employed in the automotive sector for the production of prototypes, functional parts, and components with complex geometries [34]. It allows for the creation of lightweight and structurally sound parts, contributing to improved fuel efficiency and performance.

(e) Medical Devices

SLS is used to manufacture prototypes and end-use parts for medical devices. The technology enables the production of patient-specific implants, prosthetics, surgical guides, and other customized medical components with high precision [35].

(f) Tooling and Molds

SLS is employed in the production of tooling and molds for various manufacturing processes. The ability to quickly produce complex tooling without the need for traditional machining methods is advantageous in industries such as injection molding and die casting.

(g) Consumer Goods

The consumer goods industry utilizes SLS for creating prototypes and small batches of customized or intricate consumer products. This includes items such as consumer electronics, sporting goods, and household appliances.

(h) Dental Applications

SLS is applied in the dental field for producing dental models, crowns, bridges, and other dental prosthetics. The technology's precision and ability to work with biocompatible materials make it suitable for dental applications.

(i) Educational and Research Applications

SLS is used in educational institutions and research settings for teaching and research purposes. It allows students and researchers to explore and experiment with additive manufacturing concepts and produce functional prototypes.

(j) Architectural Models

SLS is utilized in architecture for creating detailed and complex architectural models. It enables architects to visually represent their designs in a physical form with a high level of detail.

Microwave Sintering

Microwave sintering represents a fundamentally distinct approach from conventional sintering methods. Traditional sintering relies on radiant, resistance, and/or convection heating, with subsequent transfer of thermal energy occurring through conduction. This process is comparatively slow, requiring significant time to reach thermal equilibrium and achieve material consolidation, regardless of the material's nature. In contrast, microwave sintering operates by absorbing or coupling with microwave fields, leading to volumetric heating of the entire material. Unlike conventional methods, there is no involvement of thermal conductivity mechanisms. Microwave sintering is characterized by instantaneous and rapid heating, determined by the specific properties of the material [36]. The heat is internally generated within the material, originating from electromagnetic energy conversion to thermal energy, and radiates outward. This results in an inverse heating profile, from inside-out, as opposed to the outside-in heating of conventional methods. The microwave heating process is notably swift due to energy conversion, distinct from the energy transfer mechanism seen in conventional methods. Figure 6.5 provides a visual representation of the distinguishing features between conventional and microwave heating. Microwaves, falling within the electromagnetic spectrum, have wavelengths ranging from 1 mm to 1 m in free space, and frequencies between 300 GHz and 300 MHz.

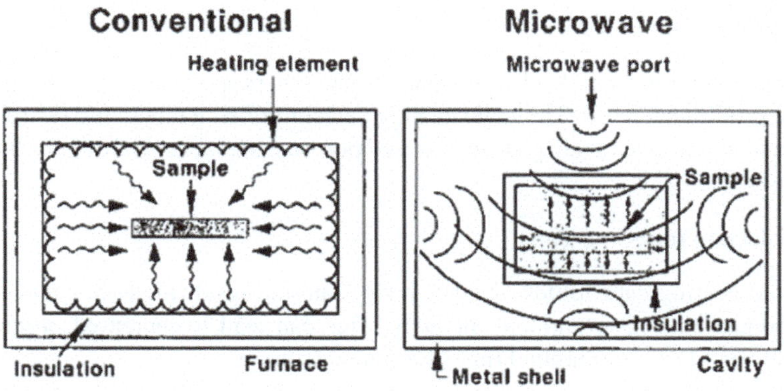

Fig. 6.5 Heating patterns in conventional and microwave furnaces [37]

Advantages of Microwave Sintering

Microwave sintering offers several advantages over conventional sintering methods, making it an attractive option for certain applications. Some of the advantages of microwave sintering are mentioned below:

(a) Rapid Heating

Microwave sintering enables rapid and uniform heating of materials. The direct absorption of microwave energy by the material leads to volumetric heating, allowing for quicker temperature increases compared to conventional methods.

(b) Energy Efficiency

The selective absorption of microwaves by the material results in efficient energy utilization. Microwave sintering is often more energy-efficient than traditional sintering processes, as there are fewer heat losses during the heating process.

(c) Uniform Temperature Distribution

Microwave sintering tends to produce a more uniform temperature distribution within the material. This uniformity contributes to improved microstructural homogeneity and consistency in the properties of the sintered product.

(d) Reduced Thermal Residual Stresses

The rapid and uniform heating characteristic of microwave sintering can minimize the development of thermal residual stresses within the material. This is beneficial for maintaining the integrity and dimensional stability of the sintered components.

(e) Selective Heating

Microwaves can selectively heat certain components or phases within a material based on their dielectric properties. This selectivity can be advantageous for processing complex mixtures or composite materials with multiple phases.

(f) Improved Properties

Microwave sintering can lead to enhanced material properties due to the unique thermal profiles achieved. The rapid heating and controlled temperature distribution contribute to improved grain growth, reduced porosity, and better overall mechanical properties.

(g) Reduced Processing Time

The rapid heating capabilities of microwave sintering result in shorter processing times compared to conventional methods. This can lead to increased production efficiency and faster turnaround times for manufacturing.

(h) Environmental Benefits

The energy efficiency of microwave sintering can contribute to reduced energy consumption, making it a more environmentally friendly option. Shorter processing times also contribute to a smaller environmental footprint.

(i) Applicability to Various Materials

Microwave sintering applies to various materials, including ceramics, metals, and composites. Its versatility makes it suitable for diverse industrial applications across different sectors.

(j) Controlled Heating Profiles

Microwave sintering allows for precise control over the heating profiles, enabling customization based on the material's specific requirements. This control is crucial for achieving desired material properties and avoiding overheating or thermal degradation.

Limitations of Microwave Sintering

Some of the limitations associated with microwave sintering are mentioned below:

(a) Material Compatibility

Not all materials are suitable for microwave sintering. The effectiveness of the process depends on the dielectric properties of the material. Materials with low dielectric loss may not absorb microwaves efficiently, limiting their suitability for this sintering method.

(b) Size and Shape Constraints

Microwave sintering equipment may have limitations on the size and shape of the materials that can be processed. Larger or irregularly shaped objects may experience uneven heating, potentially leading to non-uniform sintering.

(c) Uneven Heating in Complex Geometries

Complex geometries or components with varying thicknesses may experience uneven heating during microwave sintering. Achieving uniform temperature distribution in intricate structures can be challenging.

(d) Cost of Equipment

Microwave sintering equipment can be expensive to purchase and maintain. The initial investment may be higher compared to conventional sintering equipment, which could be a consideration for smaller-scale operations.

(e) Limited Penetration Depth

Microwaves have limited penetration depth into materials, especially for certain frequencies. This limitation can affect the uniformity of heating in thicker samples, as the microwaves may not penetrate deep enough.

(f) Temperature Control Challenges

Achieving precise temperature control in microwave sintering can be challenging. Rapid and uneven heating may lead to temperature spikes, requiring careful process optimization to avoid overheating and ensure the desired material properties.

(g) Potential for Thermal Runaway

In some cases, microwave sintering may be prone to thermal runaway, where the temperature increases uncontrollably. This can lead to undesired effects such as material degradation or changes in microstructure.

(h) Safety Concerns

Microwave sintering involves the use of electromagnetic radiation, and safety precautions are necessary to protect operators from potential exposure. Adequate shielding and safety measures must be implemented to ensure a safe working environment.

(i) Limited Applicability for Some Materials

While microwave sintering is versatile, it may not be the optimal choice for certain materials or applications. Some materials may be better suited to traditional sintering methods.

(j) Scale-Up Challenges

Scaling up microwave sintering processes for large-scale production may pose challenges. Ensuring consistent and uniform heating across larger volumes can be complex, and the equipment may need to be designed accordingly.

Applications of Microwave Sintering

Microwave sintering has found applications in various industries due to its advantages in terms of rapid heating, energy efficiency, and improved material properties. Some of the notable applications of the microwave sintering are mentioned below:

(a) Ceramic Industry

Microwave sintering is widely used in the ceramic industry for the production of various ceramic components, including tiles, cutting tools, and advanced ceramics [36]. The rapid and uniform heating enables improved densification and enhanced material properties.

(b) Metallurgy and Metal Processing

In metallurgy, microwave sintering is employed for the processing of metal powders, including the sintering of metal parts for aerospace and automotive applications. The technology can contribute to improved microstructural characteristics and mechanical properties of the sintered metals.

(c) Electronics and Semiconductor Manufacturing

Microwave sintering is applied in the electronics industry for the sintering of conductive inks, metallization pastes, and other electronic components. The rapid heating allows for the efficient processing of delicate electronic materials.

(d) Powder Metallurgy

Microwave sintering is used in powder metallurgy applications, such as the production of sintered metal parts, including gears, bearings, and filters. The technology provides advantages in terms of energy efficiency and reduced processing times.

(e) Medical Devices

Microwave sintering is utilized in the production of medical devices, including implants, prosthetics, and dental components. The ability to achieve rapid and uniform heating is advantageous for manufacturing precise and high-quality medical components.

(f) Catalyst Production

The catalysis industry employs microwave sintering for the production of catalysts. The technology enables the synthesis of catalyst materials with enhanced properties, contributing to improved catalytic performance.

(g) Composite Materials

Microwave sintering is used in the production of composite materials, such as ceramic-matrix composites and metal-matrix composites. The technology facilitates the bonding of dissimilar materials, creating high-performance composite structures.

(h) Glass Industry

In the glass industry, microwave sintering is applied for the production of glass–ceramic materials. The technology allows for controlled heating and improved crystallization of glass compositions.

(i) Research and Development

Microwave sintering is widely used in research and development settings for studying and developing new materials. Researchers leverage the advantages of rapid heating and controlled temperature profiles for investigating novel material compositions.

(j) Additive Manufacturing

Microwave sintering is explored in the field of additive manufacturing for post-processing steps, such as the sintering of metal or ceramic components produced through techniques like 3D printing. It offers a rapid and energy-efficient method for achieving the desired material properties.

(k) Environmental Applications

Microwave sintering has been investigated for applications related to environmental protection, such as the synthesis of photocatalytic materials for pollutant degradation or the production of absorbent materials for water purification.

Spark Plasma Sintering

Spark plasma sintering (SPS), also referred to as pulsed electric current sintering (PECS), is an innovative sintering technology that has emerged in recent years. This novel processing technique for material development employs a direct electrical current (DEC) to rapidly consolidate powders within a short timeframe, utilizing comparatively high sintering pressures. In the SPS process, electrical energy is charged into the gap between powder particles, and a momentary application of high-temperature spark plasma facilitates sintering at lower temperatures compared to conventional methods, achieving nearly complete densification [38]. While temperatures during SPS can briefly spike from 5000 to 10,000 °C, they are momentary, preventing any melting. In contrast to conventional sintering processes that may take hours, SPS accomplishes the same or even superior consolidation in just a few minutes, seamlessly integrating compaction and sintering stages into a single operation. This process stands out for its exceptional energy efficiency and ease of operation. SPS is versatile and capable of processing both conductive and non-conductive materials to achieve exceptionally high levels of densification. Notably, SPS ensures uniform sintering and minimizes grain growth, making it particularly well-suited for processing nanostructured materials. Figure 6.6 illustrates the fundamental stages of the SPS process.

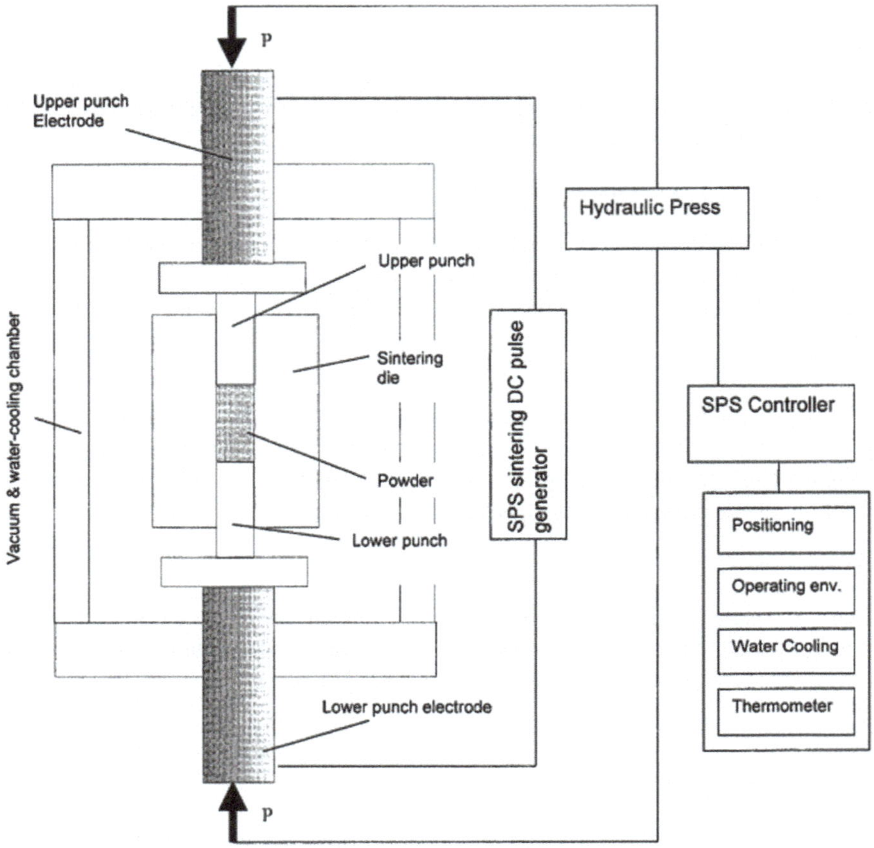

Fig. 6.6 Basic stages of the SPS process [39]

Advantages of Spark Plasma Sintering

Spark Plasma Sintering (SPS), also known as pulsed electric current sintering (PECS), offers several advantages that make it a preferred choice for material consolidation in various industries. Some of the advantages of Spark Plasma Sintering are mentioned below:

(a) Rapid Consolidation

SPS enables rapid consolidation of powder materials within a short period, typically in minutes. This is in contrast to conventional sintering processes that may take hours.

(b) Low Sintering Temperatures

SPS allows for sintering at relatively low temperatures compared to conventional methods. While temperatures during the process can momentarily reach high levels, the overall sintering temperature is lower, preventing excessive thermal exposure and allowing for the processing of temperature-sensitive materials.

(c) Almost Full Densification

SPS achieves nearly complete densification of materials, resulting in high-density and well-consolidated products. This is beneficial for producing materials with improved mechanical properties.

(d) Uniform Sintering

The SPS process provides uniform heating throughout the material, leading to uniform sintering. This ensures consistent material properties and avoids the uneven sintering that may occur in some conventional processes.

(e) Single-Step Operation

SPS combines the compaction and sintering stages into a single operation, streamlining the manufacturing process [40]. This simplification reduces production time and enhances overall efficiency.

(f) Energy Efficiency

SPS is known for its energy efficiency. The rapid consolidation and lower sintering temperatures contribute to reduced energy consumption compared to conventional sintering methods.

(g) Suitability for Various Materials

SPS can be applied to various materials, including metals, ceramics, and composites. It is versatile and can process both conductive and non-conductive materials effectively.

(h) Minimal Grain Growth

Spark Plasma Sintering minimizes grain growth during the consolidation process [41]. This is particularly advantageous for processing nanostructured materials, where maintaining fine-grained structures is crucial.

(i) High Densification Levels

SPS can achieve very high levels of densification, leading to materials with superior mechanical, thermal, and electrical properties. This is essential for applications requiring high-performance materials.

(j) Easy Operation

The SPS process is relatively easy to operate. It doesn't require complex setups or extensive post-processing steps, contributing to its practicality and suitability for various applications.

(k) Improved Mechanical Properties

Materials sintered using SPS often exhibit improved mechanical properties, including higher hardness, strength, and fracture toughness. This is desirable for applications in industries such as aerospace, automotive, and tool manufacturing.

Limitations of Spark Plasma Sintering

Some of the limitations of Spark Plasma Sintering are mentioned below:

(a) Size and Shape Constraints

The size and shape of the materials that can be processed using SPS may be limited by the specific design of the equipment [42]. Larger or irregularly shaped samples may experience challenges in achieving uniform heating.

(b) Limited Scalability

While SPS is highly effective for small to medium-sized samples, scaling up the process for large-scale industrial production may pose challenges. Ensuring consistent and uniform heating across larger volumes can be complex.

(c) Material Compatibility

The effectiveness of SPS can vary depending on the type of material being processed. Some materials may not respond well to the rapid heating and high pressures involved in the SPS process.

(d) Electric Field Non-Uniformity

Achieving a uniform electric field distribution, especially in larger samples, can be challenging. Non-uniform electric fields may result in uneven heating and affect the overall quality of the sintered material.

(e) Sample Contamination

In certain cases, there may be concerns about sample contamination during SPS, particularly when reactive materials or materials with low melting points are involved. Precautions must be taken to minimize potential contamination.

(f) Limited Control over Heating Rate

While SPS offers rapid consolidation, there may be limited control over the heating rate, especially in certain commercial systems. Fine-tuning the heating profile for specific materials may require additional modifications to the equipment.

(g) Maintenance Requirements

SPS equipment may require regular maintenance to ensure optimal performance. This includes maintaining the integrity of the graphite die, electrodes, and other components subjected to high temperatures and pressures.

(h) Die Wear and Degradation

The repeated use of the graphite die in the SPS process can lead to wear and degradation over time. This may necessitate periodic replacement of the die, adding to operational costs.

(i) Lack of In-Process Monitoring

In some SPS systems, there may be a lack of real-time monitoring capabilities during the sintering process. The ability to monitor and control parameters in real time could enhance process control and repeatability.

Applications of Spark Plasma Sintering

Spark Plasma Sintering (SPS) is a versatile consolidation technique that finds applications in various industries due to its ability to produce high-density and well-consolidated materials with unique properties. Some of the notable applications of Spark Plasma Sintering are mentioned below:

(a) Ceramic Industry

SPS is widely used in the ceramic industry for the fabrication of advanced ceramics, such as cutting tools, abrasives, and structural components. It allows for rapid consolidation at lower temperatures, preserving the fine-grained structure of ceramics.

(b) Metal Processing and Powder Metallurgy

In metal processing and powder metallurgy, SPS is employed for the production of sintered metal parts, including high-performance alloys, tool materials, and magnetic materials. The rapid consolidation and low sintering temperatures contribute to improved material properties.

(c) Research and Development

SPS is extensively used in research and development laboratories for the synthesis and investigation of novel materials, including nanostructured materials, intermetallic compounds, and composites. Its ability to achieve almost full densification in a short time makes it valuable for exploring new material possibilities.

(d) Aerospace Industry

The aerospace industry utilizes SPS for the production of components requiring high strength, wear resistance, and thermal stability. SPS is applied in the fabrication of aerospace alloys, composite materials, and components for aircraft and spacecraft.

(e) Medical Implants and Devices

SPS is employed in the manufacturing of medical implants and devices, such as dental implants, orthopedic implants, and biomedical components. The technique allows for precise control over material properties, ensuring compatibility with biological tissues.

(f) Electronic and Electrical Components

SPS is used in the production of electronic and electrical components, including conductive materials, semiconductors, and ceramic substrates. The uniform sintering achieved with SPS is beneficial for maintaining the desired electrical and thermal conductivity.

(g) Fuel Cells and Batteries

In the field of energy storage and conversion, SPS is applied for the fabrication of components in fuel cells and batteries. It enables the production of materials with enhanced electrochemical properties, contributing to improved performance in energy storage devices.

(h) Tool and Die Manufacturing

SPS is employed in tool and die manufacturing for the production of wear-resistant and high-strength cutting tools, dies, and molds. The ability to achieve high densification levels and controlled microstructures is advantageous for tooling applications.

(i) Nuclear Industry

SPS is used in the nuclear industry for the fabrication of materials with specific properties required for nuclear applications. This includes the production of nuclear fuels, control rods, and other components with tailored characteristics.

(j) Magnets and Magnetic Materials

SPS is applied in the production of permanent magnets and magnetic materials with improved magnetic properties. The rapid consolidation and controlled sintering conditions contribute to achieving the desired magnetic characteristics.

(k) Defense Applications

SPS is utilized in the defense industry for manufacturing components with superior mechanical and thermal properties. This includes the production of armor materials, aerospace components, and high-performance structural materials.

Sintering Variables [43]

The key factors influencing the sinterability and resultant microstructure of a powder compact can be categorized into two main groups: (i) material variables and (ii) process variables. An elaboration on the two categories is mentioned below.

(i) **Material Variables**

(a) Particle Size and Distribution

The initial particle size and distribution of the powder feedstock can affect sintering. Fine powders tend to sinter more rapidly, but controlling particle size distribution is crucial for achieving uniform densification.

(b) Particle Morphology

The shape of the powder particles, whether spherical, irregular, or elongated, can affect the packing density and, consequently, the sintering behavior.

(c) Powder Composition

The chemical composition of the powder, including impurities and additives, can affect sintering. Certain elements or compounds may aid in densification, control grain growth, or modify the final properties of the sintered material.

(d) Sintering Aids or Additives

Various additives, such as sintering aids or binders, may be introduced to improve densification, control grain growth, or modify the sintering kinetics. These additives can play a crucial role in optimizing the sintering process.

(ii) **Process Variables**

(a) Temperature

The sintering temperature is a critical parameter that influences the rate of diffusion, chemical reactions, and densification of the powder particles. Different materials and applications may require specific temperature profiles for optimal sintering.

(b) Time

Sintering time refers to the duration for which the material is exposed to an elevated temperature. It affects the degree of particle bonding, grain growth, and overall densification. Longer sintering times may lead to enhanced material properties but could also result in undesirable effects like over-sintering.

(c) Pressure

Pressure, especially in processes like hot isostatic pressing (HIP), can significantly impact the sintering process. It helps reduce porosity, improve density, and enhance mechanical properties.

(d) Atmosphere or Environment

The surrounding atmosphere during sintering can influence the material properties. Common atmospheres include vacuum, inert gases (e.g., argon), or controlled atmospheres (e.g., reducing or oxidizing environments), depending on the material and the desired outcome.

(e) Heating Rate

The rate at which the temperature is increased during the initial stages of sintering can impact the microstructure and properties of the final product. Controlled heating rates help prevent thermal shock and uneven temperature distribution.

(f) Cooling Rate

The rate at which the material is cooled after sintering can influence the final microstructure and properties. Slow cooling rates may reduce the risk of thermal stresses and enhance certain material characteristics.

Sintering Atmosphere

The sintering atmosphere refers to the gaseous environment or conditions in which the sintering process takes place. It plays a crucial role in influencing the final properties and characteristics of the sintered material. The choice of the sintering atmosphere depends on factors such as the material composition, sensitivity to oxidation, desired microstructure, and specific applications. Selecting the appropriate atmosphere is crucial for achieving the desired material properties and preventing undesired reactions or defects during the sintering process. A few common types of sintering atmospheres are mentioned below:

(a) Vacuum

Vacuum sintering involves conducting the sintering process in a low-pressure environment, typically below atmospheric pressure. This minimizes the presence of reactive gases and prevents oxidation of certain materials. Vacuum atmospheres are often used for processing materials sensitive to oxygen or for achieving a high-purity final product. Refractory metal carbides, stainless steels, beryllium, titanium, zirconium, tantalum, niobium, vanadium, thorium, uranium, and cermet are the best examples of vacuum sintering.

(b) Inert Gas (e.g., Argon or Nitrogen)

Inert gas atmospheres, such as argon or nitrogen, are commonly employed to prevent oxidation of metal powders during sintering. These gases create a non-reactive environment, reducing the likelihood of chemical reactions with the sintered material. Inert gas atmospheres are particularly suitable for materials that are prone to oxidation at elevated temperatures. Nitrogen-enriched atmospheres find application in the sintering and infiltration processes of iron, carbon steel, and a range of ferrous and nonferrous alloys. Stainless steel and refractory metals can undergo sintering within nitrogen-based atmospheres, particularly in situations where nitriding is not a crucial requirement.

(c) Reducing Atmosphere

A reducing atmosphere involves the introduction of gases that have a reducing effect, such as hydrogen or dissociated ammonia. This atmosphere is used to prevent the oxidation of certain metals and can contribute to the removal of oxides from the material during sintering. Typical applications of dissociated ammonia include sintering of brass, copper, iron–copper, tungsten and tungsten alloys, aluminum and its alloys, and stainless steels.

(d) Oxidizing Atmosphere

In contrast to reducing atmospheres, an oxidizing atmosphere involves the presence of oxygen or other oxidizing gases. This atmosphere may be used intentionally to promote the formation of oxides in the sintered material or induce specific chemical reactions. Oxidizing atmospheres are often utilized for materials that benefit from controlled oxidation.

6.2.2 Hot Isostatic Pressing

The Hot Isostatic Pressing (HIP) process is a specialized manufacturing technique used to consolidate and densify materials through the simultaneous application of high temperature and pressure. This process is particularly effective in achieving uniform and high-density components, reducing porosity, and improving material properties. Initially, this method found its application in compacting components for nuclear reactors and sintering carbide tools. It is a process chosen for consolidating powders of superalloys based on Ni and Co, tool steels, maraging steels, Ti alloys, refractory metal powders, ceramic powders, and cermet. HIP is also adaptable for manufacturing components from alloys based on Al, Mg, Cu, cemented tungsten carbides, magnetic materials, and metal matrix composites. Figure 6.7 shows the schematic view of the HIP unit. The major components of a typical HIP unit include:

(a) Pressure Vessel

The pressure vessel is a robust and thick-walled chamber made of low alloy steel designed to withstand high pressures. It houses the material to be processed, typically in the form of a powder or pre-formed compact. The vessel is sealed tightly to prevent gas leakage during the HIP process.

(b) Heating System

The heating system is responsible for raising the temperature inside the pressure vessel to the desired level. This is crucial for achieving the necessary sintering or

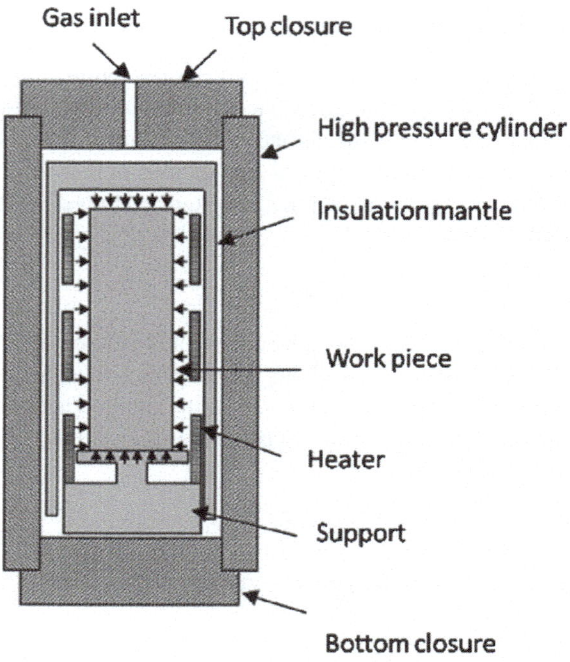

Fig. 6.7 Schematic view of HIP setup [44]

consolidation of the material. Resistance heating elements or induction heating coils are commonly used in HIP units.

(c) Pressure Generation System

The pressure generation system creates and controls the high pressure needed for the HIP process. It often involves hydraulic systems or gas pressure intensifiers. The pressure is applied uniformly in all directions, creating an isostatic condition for the material being processed.

(d) Cooling System

After the material has been subjected to high temperature and pressure, a cooling system is employed to bring it back to ambient temperature. Controlled cooling is essential to avoid thermal shock and ensure the material's stability and uniform properties.

(e) Control and Monitoring System

The control and monitoring system is a critical component that oversees and regulates the entire HIP process. It includes sensors to measure temperature and pressure, as well as a control unit that adjusts parameters based on the desired sintering conditions.

Table 6.1 Temperature and pressure used in hot isostatic pressing [45]

S. no	Materials	Temperature (K)	Pressure (MPa)
1	Al alloy	725–805	100
2	Ti alloy	1123–1223	100
3	Ni alloy	1373–1473	100–150
4	Cu alloy	773–1173	100
5	Al_2O_3	1623–1723	100
6	Fe	1273	175

(f) Insulation

The pressure vessel is typically surrounded by insulation to minimize heat loss and maintain uniform temperature distribution. This ensures efficient energy utilization and helps achieve consistent sintering results.

(g) Loading and Unloading System

A mechanism for loading and unloading the material into and from the pressure vessel is integrated into the HIP unit. This is often a part of the overall automation system, allowing for efficient and safe handling of materials.

(h) Gas Management System

The gas management system controls the atmosphere within the pressure vessel. Inert gases such as argon or nitrogen are commonly used to prevent oxidation during the high-temperature sintering process. This system may include gas supply, purification, and exhaust components.

(i) Safety Features

HIP units are equipped with safety features to protect personnel and prevent accidents. These may include pressure relief systems, emergency shutdown mechanisms, and interlocks to ensure that the unit operates within safe parameters.

(j) Automation and Control Interface

Many modern HIP units are equipped with advanced automation features and a user-friendly control interface. Automation allows for precise control over the process parameters, making it easier to achieve specific material properties.

Typical temperatures and pressure used in processing different materials using the HIP process are mentioned in Table 6.1.

Advantages of HIP Process

The Hot Isostatic Pressing (HIP) process offers several advantages, making it a valuable manufacturing technique for a variety of materials. Some of the key advantages of the HIP process include:

(a) Uniform Densification

HIP allows for the uniform densification of materials, reducing porosity and enhancing overall density. This results in improved mechanical properties, such as increased strength, hardness, and wear resistance.

(b) Isostatic Pressing

The isostatic nature of the pressing in the HIP process ensures that pressure is applied uniformly in all directions. This results in a more even distribution of material, minimizing defects and promoting consistency throughout the processed component.

(c) Reduced Porosity

HIP effectively reduces the porosity in materials, leading to enhanced structural integrity and improved performance. The elimination of voids and gaps contributes to better mechanical and thermal properties.

(d) Improved Material Properties

The combination of high temperature and pressure during HIP induces desirable changes in the microstructure of materials. This can lead to improved material properties, such as increased hardness, toughness, and resistance to wear and corrosion.

(e) Versatility in Materials

HIP can be applied to various materials, including metals, ceramics, composites, and alloys. This versatility makes it suitable for various industries, including aerospace, automotive, medical, and energy.

(f) Complex Geometries

The HIP process allows for the fabrication of components with complex geometries. This is particularly valuable in industries where intricate and customized shapes are required, such as aerospace and medical device manufacturing.

(g) Enhanced Bonding of Powders

HIP promotes strong bonding between powder particles during sintering, resulting in a more homogeneous and well-bonded material. This is especially beneficial for materials that undergo powder metallurgy processes.

(h) Consolidation of Difficult-to-Process Materials

Certain materials, such as refractory metals, ceramics, and metal matrix composites, can be challenging to process using conventional methods. HIP overcomes these challenges, enabling the consolidation of difficult-to-process materials.

(i) High-Temperature Processing

HIP involves high-temperature processing, allowing for sintering and densification at temperatures that may not be achievable through other methods. This is advantageous for materials that require elevated temperatures for optimal performance.

(j) Increased Material Homogeneity

The HIP process contributes to the homogeneity of the material, resulting in a more consistent microstructure. This is important for achieving reliable and predictable material properties.

(k) Production of Near-Net Shape Components

HIP can be used to produce near-net shape components, reducing the need for extensive post-processing and machining. This leads to cost savings and improved efficiency in manufacturing.

Limitations of the HIP Process

Some of the limitations and challenges associated with the HIP process are mentioned below:

(a) High Equipment Cost

The initial investment and maintenance costs for HIP equipment are relatively high. This can be a barrier for smaller manufacturers or industries with budget constraints.

(b) Limited Size of Components

The size of components that can be processed using HIP is restricted by the dimensions of the pressure vessel. Large and bulky components may pose challenges in terms of uniform pressure distribution and heating.

(c) Cycle Time

The HIP process typically has longer cycle times compared to some alternative manufacturing methods. The extended processing time may impact overall production efficiency.

(d) Energy Consumption

The process involves high temperatures and pressures, leading to elevated energy consumption. This can contribute to higher operational costs and environmental considerations.

(e) Limited Material Compatibility

Not all materials are suitable for the HIP process. Some materials may not respond well to the required combination of high temperature and pressure, limiting the range of applicable materials.

(f) Surface Finish

Achieving a smooth surface finish on components processed through HIP can be challenging. Post-processing steps may be necessary to meet specific surface finish requirements.

(g) Powder Handling Challenges

Handling and consolidating certain types of powders can be challenging in the HIP process. Factors such as powder particle size, morphology, and flowability can impact the effectiveness of the consolidation.

(h) Risk of Contamination

Despite efforts to maintain a clean environment, there is a risk of contamination during the loading and unloading phases of the HIP process. This can affect the purity of the final product, especially in applications where material purity is critical.

(i) Complexity of Operation

Operating a HIP system requires specialized knowledge and expertise. The complexity of the process may necessitate skilled personnel to ensure the equipment is used effectively and safely.

(j) Limited Precision in Cooling

The cooling phase of the HIP process may have limitations in terms of precision. Rapid or uneven cooling can potentially lead to thermal stresses and affect the final material properties.

(k) Material and Design Constraints

Some materials may not be suitable for HIP due to their specific properties, and certain component designs may be challenging to process effectively using this method.

Applications of HIP Process

The Hot Isostatic Pressing (HIP) process finds applications in various industries due to its ability to enhance material properties, reduce porosity, and achieve uniform densification. Some notable applications of the HIP process include:

(a) Aerospace Components

HIP is widely used in the aerospace industry for producing high-performance components, such as turbine blades, structural parts, and components for jet engines [46]. The process enhances material properties and ensures the integrity of critical parts.

(b) Medical Implants

Components for medical implants, such as hip and knee joint replacements, are often manufactured using the HIP process. This ensures the production of dense and biocompatible materials with improved mechanical properties.

(c) Tool and Die Manufacturing

HIP is employed in the production of cutting tools, dies, and molds used in various industries. The process enhances the hardness and wear resistance of tool materials, resulting in longer tool life and improved performance.

(d) Nuclear Industry

Components for nuclear reactors, including fuel elements and structural parts, are processed using HIP to achieve high density, reduce porosity, and enhance material properties.

(e) Automotive Components

HIP is used in the automotive industry to manufacture components such as gears, pistons, and engine parts. The process improves the strength and durability of these components, contributing to enhanced performance and reliability.

(f) Oil and Gas Industry

Components for the oil and gas sector, including valves, pumps, and drill bits, can benefit from the HIP process. It ensures the production of materials with improved corrosion resistance and mechanical properties.

(g) Metal Injection Molding (MIM)

HIP is used in conjunction with metal injection molding to produce high-density metal parts with complex shapes. This is particularly valuable in the production of small, intricate components used in various industries.

(h) Additive Manufacturing (3D Printing)

HIP is employed as a post-processing step for components produced through additive manufacturing techniques. It helps improve the material properties and integrity of 3D-printed parts.

(i) Ceramic and Composite Materials

The HIP is utilized in the production of advanced ceramic materials and metal matrix composites. It helps achieve uniform densification and enhances the mechanical and thermal properties of these materials.

6.2.3 Powder Extrusion Process

During the extrusion process, significant hydrostatic compression forces are applied, accompanied by a unidirectional force component that guides the compact through the die (refer to Fig. 6.8). Frictional forces generate a shear component, resulting in a distinct shear pattern in the extruded metal, as illustrated in Fig. 6.8. The three fundamental methods of hot extrusion for metal powders are depicted in the figure.

In the first approach, loose powder is introduced into a heated extrusion container and directly extruded through the die. This method is specifically designed for certain magnesium alloy powders. The second method involves cold compacting the powder, followed by hot pressing. The resulting hot-pressed compact is then extruded using conventional methods, a technique commonly employed for extruding billets of aluminum alloy powders. The third method involves placing metal powders into a metallic capsule or 'can,' heating them, and extruding the material along with the can. The green metal powder compact can either be canned, or the powder may be cold pressed into a metal can under moderate pressure. The can is evacuated at room or elevated temperature, sealed off, and then heated for extrusion. To ensure smooth extrusion without turbulent flow, the end of the can is conical, fitting into an extrusion die with a corresponding conical opening. To prevent folding, especially when the powder in the can is loosely packed, a penetrator ram, as depicted in Fig. 6.9, may be employed. Copper and low-carbon steel are commonly used materials for the can.

Initially developed for consolidating beryllium powder and powders containing fissile material dispersed in a zirconium and stainless steel matrix, the hot extrusion

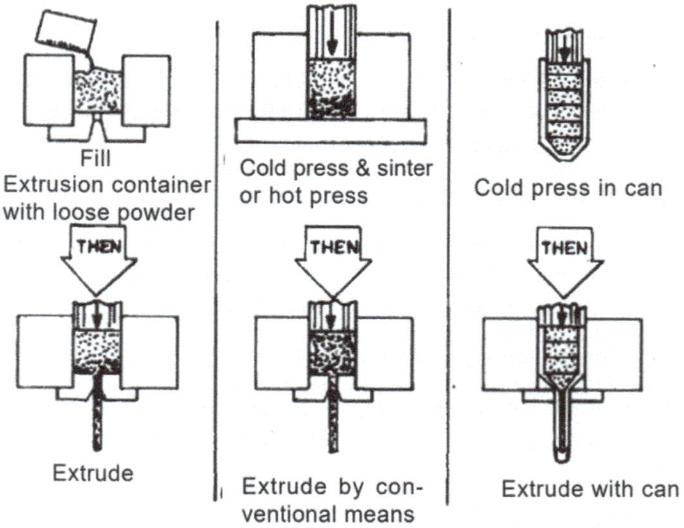

Fig. 6.8 Three hot extrusion methods [47]

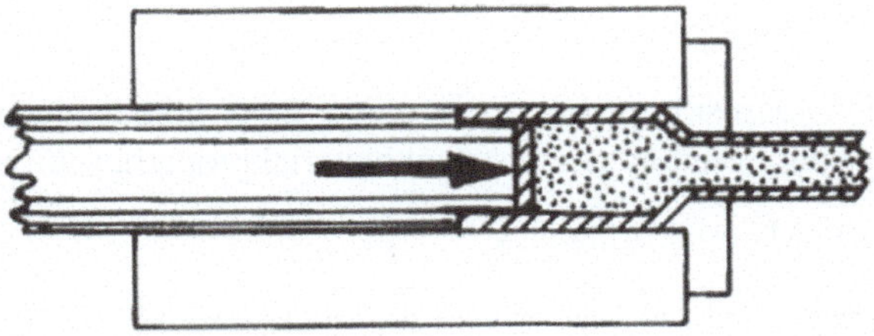

Fig. 6.9 Penetrator technique in powder extrusion to avoid folding [47]

process encapsulated in cans has become a prevalent method for copper- and nickel-based dispersion-strengthened alloys. Notably, this technique was first utilized in Sweden to produce seamless tubing from stainless steel powder.

Advantages of the Powder Extrusion Process

The powder extrusion process offers several advantages, making it a valuable manufacturing method for various industries. Some of the key advantages include:

(a) Versatility in Materials

Powder extrusion applies to various materials, including metals, ceramics, polymers, and composites. This versatility allows for the production of diverse components with tailored properties.

(b) Complex Geometries

The process enables the fabrication of components with intricate and complex geometries. This is particularly advantageous in industries where customized and intricate shapes are required.

(c) Near-Net Shape Manufacturing

Powder extrusion allows for near-net shape manufacturing, reducing the need for extensive post-processing and machining. This leads to material and cost savings.

(d) Enhanced Mechanical Properties

Components produced through powder extrusion often exhibit improved mechanical properties, including higher strength, hardness, and wear resistance, due to the densification and deformation during the extrusion process.

(e) High Production Rates

Powder extrusion is a continuous process that can be adapted for high production rates. This makes it suitable for mass production of components, contributing to cost-effectiveness.

(f) Homogeneous Microstructure

The extrusion process promotes a more homogeneous microstructure in the final product, leading to consistent material properties throughout the component.

(g) Material Efficiency

Powder extrusion allows for the efficient use of materials, minimizing waste. This is particularly beneficial in terms of resource utilization and cost reduction.

(h) Enhanced Surface Finish

Components produced through powder extrusion often have a smoother surface finish compared to some alternative manufacturing methods. This can reduce the need for additional finishing processes.

(i) Tailored Material Compositions

The process allows for the incorporation of multiple materials or the creation of composite structures, enabling the production of components with tailored material compositions and properties.

(j) Applicability to Sensitive Materials

Powder extrusion is suitable for processing materials that may be sensitive to high temperatures, as the process can be performed at lower temperatures compared to traditional melting and casting methods.

(k) Reduced Porosity

The extrusion process helps in reducing porosity in the final product, contributing to improved structural integrity and material properties.

(l) Cost-Effective Tooling

Tooling costs in powder extrusion are often lower compared to some traditional manufacturing methods, making it a cost-effective solution for certain applications.

Limitations of the Powder Extrusion Process

While the powder extrusion process offers various advantages, it is important to consider its limitations and challenges. Some of the limitations include:

(a) Limited Material Selection

Certain materials may not be well-suited for powder extrusion, particularly those with poor flow characteristics or materials that are highly sensitive to temperature.

(b) Tool Wear and Maintenance

The tooling used in powder extrusion can experience wear over time, especially when processing abrasive materials. Regular maintenance and tool replacement may be required, impacting production costs.

(c) Complexity of Process Optimization

Optimizing the powder extrusion process for specific materials and applications can be complex. Fine-tuning parameters such as temperature, pressure, and feedstock composition may require extensive testing and experimentation.

(d) Tooling Costs for Complex Shapes

Tooling costs may increase significantly for components with intricate and complex shapes. The design and manufacturing of specialized dies and molds can contribute to higher initial expenses.

(e) Risk of Oxidation

Certain materials are prone to oxidation during high-temperature processing, which can affect the properties of the final product. Protective atmospheres or encapsulation methods may be needed to mitigate oxidation.

(f) Size Limitations

The size of components that can be produced through powder extrusion is limited by the dimensions of the extrusion equipment. Large and bulky components may pose challenges in terms of uniformity and production efficiency.

(g) Powder Homogeneity

Achieving uniform powder homogeneity is crucial for consistent material properties. Inhomogeneous powder distribution can lead to variations in the final product's properties.

(h) Energy Consumption

Powder extrusion may require elevated temperatures, contributing to higher energy consumption. The environmental impact and operational costs associated with energy use should be considered.

(i) Risk of Contamination

There is a potential risk of contamination during the powder handling and extrusion process. Careful handling and maintenance of equipment are necessary to minimize contamination issues.

Applications of the Powder Extrusion Process

The powder extrusion process finds application in various industries due to its versatility in producing components with specific properties and complex shapes. Some notable applications of the powder extrusion process include:

(a) Automotive Parts

The automotive industry utilizes powder extrusion to produce components like gears, pistons, and structural parts. The process contributes to improved material properties, including enhanced strength and wear resistance.

(b) Tool and Die Manufacturing

Powder extrusion is employed in the production of cutting tools, dies, and molds used in various industries. The resulting components exhibit improved hardness, wear resistance, and durability.

(c) Electronics and Semiconductors

Components for the electronics industry, including heat sinks and specialized connectors, can be produced through powder extrusion. The process allows for the incorporation of materials with enhanced thermal conductivity.

(d) Metal Matrix Composites

Powder extrusion is employed in the production of metal matrix composites, where powders of different materials are combined to achieve specific material properties. This is particularly useful in industries requiring lightweight and high-strength materials.

(e) Dispersion-Strengthened Alloys

Powder extrusion is commonly used for the production of dispersion-strengthened alloys, including copper- and nickel-based alloys. These alloys exhibit enhanced strength and mechanical properties.

(f) Ceramic Components

The process is utilized in the production of ceramic components with complex shapes, such as insulators, electronic substrates, and wear-resistant parts.

(g) Defense and Military Applications

Powder extrusion is utilized in the defense and military industries for the production of components with specific material properties, such as armor materials and structural elements.

6.2.4 Powder Forging

Powder forging can be broadly categorized into two main classes: (i) conventional powder forging and (ii) preform powder forging. In conventional powder forging, the loose powder is initially enclosed in a metal container, a process referred to as "canning." The enclosed powder is then evacuated, and the assembly is subjected to traditional forging in a forging press. After forging, the can material may be removed through chemical or mechanical means. The resulting compact exhibits high density and may not require sintering if the powder is heated before the forging step. Powder preform forging represents a synergistic approach combining powder metallurgy and forging.

The powder used in performing forging aligns with that employed in conventional powder metallurgy practices. The processes in powder preform forging have evolved along two distinct lines:

1. Hot Repressing

This involves compressing a preform with a cross-section closely resembling the final part (refer to Fig. 6.10). This method, known as hot repressing, achieves densification with minimal lateral flow.

2. True Forging

In this approach, a preform of a relatively simple shape is compressed into a part with a significantly different shape (refer to Fig. 6.10b). This process, similar to

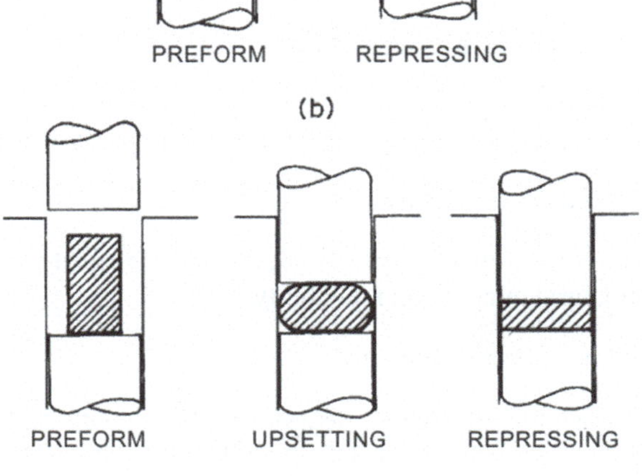

Fig. 6.10 **a** Schematic diagram of hot repressing. **b** Forging in a confined die shows upsetting and repressing modes [47]

true forging, achieves both densification and shape change simultaneously through substantial plastic deformation and lateral flow.

In true forging, the associated lateral flow subjects the pores to a combination of normal pressure and shear. This shearing action assists the normal pressure in closing pores, requiring lower pressures for material densification. Moreover, the shearing action enhances bond strength between opposite sides of collapsed pores, contributing to the overall soundness of the metallurgical structure.

Advantages of Powder Forging

Powder forging, encompassing both conventional powder forging and preform powder forging, offers several advantages that make it a valuable manufacturing process. Some of the key advantages include:

(a) High-Density Components

Powder forging can produce components with very high density, leading to improved mechanical properties such as strength, hardness, and wear resistance.

(b) Enhanced Strength-to-Weight Ratio

The process allows for the production of lightweight components with enhanced strength, contributing to a favorable strength-to-weight ratio.

(c) Near-Net Shape Manufacturing

Powder forging facilitates near-net shape manufacturing, reducing the need for extensive post-processing and machining. This leads to material and cost savings.

(d) Tailored Material Properties

Powder forging allows for precise control over material properties, making it possible to tailor the composition for specific applications. This is particularly valuable in industries where customized material characteristics are crucial.

(e) Improved Mechanical Properties

Components produced through powder forging exhibit improved mechanical properties, including higher tensile strength, impact resistance, and fatigue life.

(f) Reduced Porosity

The process effectively reduces porosity in the final product, contributing to improved structural integrity and reliability.

(g) Complex Geometries

Powder forging enables the production of components with complex and intricate geometries. This is valuable in industries where detailed shapes are essential.

(h) Cost-Efficient Tooling

Tooling costs in powder forging are often more economical compared to some traditional manufacturing methods, making it a cost-effective solution for certain applications.

(i) Energy Efficiency

The process can be performed at lower temperatures compared to traditional forging methods, contributing to energy efficiency and reducing environmental impact.

(j) Suitability for Various Materials

Powder forging applies to various materials, including metals, ceramics, and composites. This versatility allows for diverse applications across different industries.

(k) Combination of Powder Metallurgy and Forging

Powder preform forging combines the benefits of powder metallurgy and forging, allowing for simultaneous densification and shape change, leading to improved material properties and complex geometries.

(l) Sound Metallurgical Structure

True forging in powder forging results in a sound metallurgical structure, with enhanced bond strength between opposite sides of collapsed pores, contributing to the overall quality of the final product.

Limitations of Powder Forging

Some limitations and challenges associated with the powder forging include:

(a) High Tooling Costs for Complex Shapes

Tooling costs can be significant, especially for components with intricate and complex shapes. The design and manufacturing of specialized dies and molds may increase initial expenses.

(b) Material Selection Challenges

Certain materials may pose challenges in terms of flow characteristics and suitability for powder forging. The process may not be suitable for all types of materials, limiting the range of applicable alloys.

(c) Surface Finish

Achieving a smooth surface finish can be challenging in powder forging, especially for materials prone to surface irregularities. Additional post-processing may be required to meet specific surface finish requirements.

(d) Tool Wear and Maintenance

Tooling used in powder forging may experience wear over time, especially when processing abrasive materials. Regular maintenance and tool replacement may be necessary, impacting production costs.

(e) Limited Size of Components

The size of components that can be produced through powder forging is limited by the dimensions of the equipment. Large and bulky components may pose challenges in terms of uniformity and production efficiency.

(f) Energy Consumption

While powder forging can be performed at lower temperatures compared to traditional forging methods, it still involves energy consumption. The environmental impact and operational costs associated with energy use should be considered.

(g) Powder Homogeneity

Achieving uniform powder homogeneity is crucial for consistent material properties. Inhomogeneous powder distribution can lead to variations in the final product's properties.

(h) Complex Process Optimization

Optimizing the powder forging process for specific materials and applications can be complex. Fine-tuning parameters such as temperature, pressure, and feedstock composition may require extensive testing and experimentation.

(i) Risk of Material Contamination

There is a potential risk of material contamination during the powder handling and forging process. Careful handling and maintenance of equipment are necessary to minimize contamination issues.

(j) Limited Precision in Cooling

The cooling phase of the powder forging process may have limitations in terms of precision. Rapid or uneven cooling can potentially lead to thermal stresses and affect the final material properties.

(k) Size Limitations for Hot Repressing

In the hot repressing method of powder preform forging, the size of the preform is limited by the desired cross-section of the final part. This can pose constraints on the range of part sizes that can be effectively produced.

(l) Process Complexity

The combination of powder metallurgy and forging in preform forging adds complexity to the process. This complexity may require specialized knowledge and expertise for successful implementation.

Applications of Powder Forging

Powder forging, including both conventional powder forging and preform powder forging, finds applications in various industries where the advantages of the process are beneficial. Some notable applications include:

(a) Automotive Components

Powder forging is extensively used in the automotive industry to manufacture critical components such as connecting rods, gears, and other engine parts. The process enables the production of lightweight, high-strength components with improved performance.

(b) Aerospace Parts

The aerospace industry utilizes powder forging for the production of components like turbine blades, structural parts, and landing gear components. The process contributes to the creation of high-performance components with complex geometries.

(c) Hand Tools and Cutlery

Components of hand tools and cutlery, such as blades and handles, can be manufactured using powder forging. This ensures high strength, durability, and precision in the final products.

(d) Industrial Machinery Parts

Various parts for industrial machinery, such as shafts, gears, and bushings, can be produced using powder forging. The process enhances material properties and ensures high wear resistance.

(e) Sporting Goods

Powder forging is applied in the production of sporting goods such as golf club heads, bicycle components, and other high-performance equipment. The process contributes to lightweight and durable products.

(f) Firearms Components

Certain components of firearms, including triggers, hammers, and other precision parts, can be manufactured using powder forging. The process ensures high strength and dimensional accuracy.

(g) Electrical Connectors

Powder forging is employed in the production of electrical connectors and components that require high conductivity. The process allows for the creation of components with specific electrical and mechanical properties.

(h) Consumer Electronics

Certain components of consumer electronics, including connectors, housings, and structural parts, can be manufactured using powder forging. The process contributes to the production of lightweight and durable electronic devices.

6.2.5 *Powder Rolling*

Traditional powder metallurgy consolidation techniques are unsuitable for producing certain geometries, such as sheets, strips, and plates. To overcome this limitation, the powder rolling process has been developed as an alternative method [48].

Applications of Powder Rolling

The powder rolling process finds applications in various industries where the unique advantages of this manufacturing technique are beneficial. Some notable applications include:

(a) Electronics

Powder rolling is utilized in the production of thin metal sheets and strips used in electronic components. The process allows for the creation of materials with precise thickness and improved electrical conductivity, making them suitable for applications in semiconductors, connectors, and electronic packaging.

(b) Aerospace

Aerospace applications often require materials with enhanced strength-to-weight ratios and specific geometries. Powder rolling is employed to produce lightweight yet strong components such as aircraft skins, structural panels, and other parts critical for aerospace engineering.

(c) Automotive Industry

In the automotive sector, powder rolling is used to manufacture components like thin metal sheets for body panels, structural elements, and parts that require a balance between strength and weight. The process helps achieve precise dimensions and mechanical properties for these automotive components.

(d) Energy Sector

Powder rolling is applied in the production of components for energy-related applications. This includes the manufacturing of current collectors and electrodes for batteries, fuel cells, and other energy storage systems where controlled thickness and material properties are crucial.

(e) Medical Devices

The medical industry benefits from the powder rolling process in the production of components for medical devices. This includes thin metal sheets used in diagnostic equipment, implants, and other medical instruments that require precise dimensions and tailored material properties.

(f) Microelectronics

Powder rolling is suitable for producing fine, thin metal foils used in microelectronic applications. These foils may be used in flexible electronics, sensors, and other miniaturized electronic components.

(g) Defense and Aerospace

Applications in defense industries often involve the production of specialized materials for components such as armor plates, sensor housings, and aerospace structural elements. Powder rolling can provide the necessary material properties and geometries for these applications.

(h) Renewable Energy

Components in renewable energy systems, such as solar panels and wind turbines, can benefit from powder rolling. The process allows for the production of lightweight and durable materials with controlled thickness, contributing to the efficiency and reliability of renewable energy technologies.

(i) High-Performance Alloys

Powder rolling is used to manufacture high-performance alloys with tailored properties for demanding applications, such as aerospace engine components, where a combination of strength, temperature resistance, and weight savings is crucial.

6.3 Conclusion

1. Sintering is a metal powder manufacturing process involving sub-melting point heating for dense materials. Solid-state sintering offers purity and microstructure control but is time-consuming with uniformity challenges, applied in aerospace and medical implants. Liquid-state sintering, with a liquid phase, enhances densification, allows alloy formation, and offers faster processing at lower temperatures. Challenges include controlled atmospheres and uniformity, with applications in metal alloy production, advanced structural materials, and energy storage devices.
2. Reaction sintering is a versatile method known for its ability to control microstructure, enhance material properties, and shape intricate structures. Despite challenges like complex optimization and material compatibility issues, it plays a crucial role in various industries such as advanced ceramics, metal matrix composites, cutting tools, and automotive components. Its wide-ranging applications in aerospace, automotive, bioceramics, and electronics emphasize its substantial impact on modern manufacturing, enabling the production of tailored materials with improved functionalities.

3. Selective Laser Sintering (SLS) is a versatile additive manufacturing process known for its capacity to produce intricate designs without support structures, minimizing material waste. Recognized for high resolution, accuracy, and rapid prototyping capabilities, SLS contributes to accelerated product development. While facing challenges such as material restrictions and surface finish considerations, SLS is widely employed in diverse industries like aerospace, automotive, healthcare, and consumer goods. Key considerations include resolution limitations, part distortion, equipment costs, and powder handling complexities. Despite these challenges, SLS proves valuable in manufacturing various components, underscoring its broad applicability.
4. Microwave sintering marks a paradigm shift in material processing, employing rapid microwave energy absorption for volumetric heating and introducing an inside-out heating approach. Despite challenges like material compatibility and size limitations, it offers swift, uniform heating, energy efficiency, and enhanced material properties. Its versatile applications span ceramics, metallurgy, electronics, and medical devices, hinting at a promising future. Ongoing advancements also point towards potential uses in additive manufacturing and environmental protection.
5. Spark Plasma Sintering (SPS) is a cutting-edge material consolidation technique utilizing direct electrical current, enabling rapid powder consolidation at lower temperatures for nearly complete densification. Recognized for its quick and efficient process, low sintering temperatures, and uniform heating, SPS is versatile, applicable to various materials, and streamlines operations into a single step. Widely used in ceramics, metallurgy, aerospace, medical implants, electronics, and energy storage, SPS, despite facing challenges like size constraints and maintenance needs, remains indispensable due to its capability to produce high-density materials with superior properties across diverse industries.
6. The Hot Isostatic Pressing (HIP) process, employing high temperature and pressure, excels in consolidating materials like superalloys and ceramics. Initially applied in nuclear reactors and tool sintering, HIP offers advantages such as uniform densification and improved properties. However, challenges like equipment costs, size constraints, and operational complexities persist. HIP finds broad applications across aerospace, medical implants, tool manufacturing, nuclear components, automotive parts, and sectors like oil and gas, metal injection molding, additive manufacturing, and advanced ceramics production.
7. The powder extrusion process utilizes hydrostatic compression and unidirectional forces, offering advantages like material versatility and enhanced mechanical properties. Employing three hot extrusion methods, it was initially designed for beryllium powder but is now widely used in dispersion-strengthened alloys, particularly copper- and nickel-based alloys. Despite considerations such as limited material selection and tool wear, its applications range from automotive parts to defense and military sectors.
8. Powder forging, divided into conventional and preform types, uses canning for traditional forging, producing high-density components. Preform powder forging combines metallurgy and forging, achieving densification and shape change,

offering benefits like high-density components and tailored properties. Despite challenges such as tooling costs, it finds applications in automotive, aerospace, hand tools, industrial machinery, sporting goods, firearms, electrical connectors, and consumer electronics.

9. Powder rolling emerged to address traditional powder metallurgy limitations in shaping various components, finding widespread application across industries. This versatile process creates precise sheets for electronics, lightweight aerospace components, and specialized materials for defense, while also ensuring controlled thickness in energy applications and tailoring high-performance alloys for demanding industrial needs.

References

1. Suryanarayana, C., Ivanov, E., Boldyrev, V.V.: The science and technology of mechanical alloying. Mater. Sci. Eng. A **304**, 151–158 (2001)
2. Suryanarayana, C.: Mechanical alloying and milling. Progr. Mater. Sci. **46**(1–2), 1–184 (2001)
3. Suryanarayana, C.: Mechanical alloying: a novel technique to synthesize advanced materials. Research 1–15 (2019)
4. Chaira, D.: Development of nano-structured duplex and ferritic stainless steels by pulverisette planetary milling followed by pressure less sintering. Mater Charact **99**, 220–229 (2015)
5. Stainless Steels - Introduction to Grades, Properties and Applications, Supplier Data by Aalco (2005)
6. Majzoobi, G.H., Atrian, A., Pipelzadeh, M.K.: Effect of densification rate on consolidation and properties of Al7075–B4C composite powder. Powder Metall. **58**(4), 281–288 (2015)
7. Choudhuri, D., Blake, L.: Particle curvature effects on microstructural evolution during solid-state sintering: phenomenological insights from phase-field simulations. J. Mater. Sci. **56**(12), 7474–7493 (2021)
8. Seth, P.P., Singh, N., Singh, M., Prakash, O., Kumar, D.: Formation of fine Mg_2Si phase in Mg–Si alloy via solid-state sintering using high energy ball milling. J. Alloys Compd. **821**, 153205 (2020).
9. Long, K., Zhong, Y., Li, M., Shu, C., Zhang, C., Shi, L., He, X.: Three-dimensional sintering morphology simulation of silica fibrous insulator and its quantitative description. Ceram. Int. **48**(14), 19557–19566 (2022)
10. Gille, G., Szesny, B., Dreyer, K., Van Den Berg, H., Schmidt, J., Gestrich, T., Leitner, G.: Submicron and ultrafine grained hardmetals for microdrills and metal cutting inserts. Int. J. Refract. Metals Hard Mater. **20**(1), 3–22 (2002)
11. Onyishi, H.O., Oluah, C.K.: An overview of the state of the art and applications of sintered metals. IOSR J. Mech. Civ. Eng **17**, 1–10 (2020)
12. Olmos, L., Cabezas-Villa, J.L., Bouvard, D., Lemus-Ruiz, J., Jiménez, O., Falcón-Franco, L.A.: Synthesis and characterisation of Ti_6Al_4V/xTa alloy processed by solid state sintering. Powder Metall. **63**(1), 64–74 (2020)
13. Pramanik, S., Agarwal, A.K., Rai, K.N., Garg, A.: Development of high strength hydroxyapatite by solid-state-sintering process. Ceram. Int. **33**(3), 419–426 (2007)
14. Meng, X., Hao, J., Cao, H., Lin, X., Ning, P., Zheng, X., Chang, J., Zhang, X., Wang, B., Sun, Z.: Recycling of $LiNi_{1/3}Co_{1/3}Mn_{1/3}O_2$ cathode materials from spent lithium-ion batteries using mechanochemical activation and solid-state sintering. Waste Manag. **84**, 54–63 (2019)
15. Lalana, E.H.: Permanent magnets and its production by powder metallurgy. Rev. Metal. **54**(2), 1–10 (2018)

16. de Oro Calderon, R., Gierl-Mayer, C., Danninger, H.: Fundamentals of sintering: liquid phase sintering. Encycl. Mater. Metals Alloys **3**, 481–492 (2022)
17. Kang, S.-J.L.: Sintering: Densification, Grain Growth and Microstructure. Elsevier (2004).
18. Hausner, H.: Reaction sintering. In: Concise Encyclopedia of Advanced Ceramic Materials, pp. 389–391 (1991)
19. Li, B., Wei, Y., Li, N., Zhang, T., Wang, J.: Effect of $Al(OH)_3$ and La_2O_3 on the sintering behavior of CaO granules via $CaCo_3$ decomposition. Sci. Sinter. **52**(2), 195–206 (2020)
20. Zou, J., Ma, H.-B., Liu, J.-J., Wang, W.-M., Zhang, G.-J., Fu, Z.-Y.: Nanoceramic composites with duplex microstructure break the strength-toughness tradeoff. J. Mater. Sci. Technol. **58**, 1–9 (2020)
21. Zou, J., Liu, J.J., Zhang, G.J., Fu, Z.Y.: Tuning the combustion process during reactive sintering of high-performance ceramics by employing solid solutions as reactants. J. Eur. Ceram. Soc. **41**, 101–113 (2021)
22. Chamberlain, A.L., Fahrenholtz, W.G., Hilmas, G.E.: Low-temperature densification of zirconium diboride ceramics by reactive hot pressing. J. Am. Ceram. Soc. **89**, 3638–3645 (2006)
23. Zou, J., Zhang, G.-J., Fu, Z.-Y.: In-situ ZrB_2-hBN ceramics with high strength and low elasticity. J. Mater. Sci. Technol. **48**, 186–193 (2020)
24. Zou, J., Zhang, G.J., Shen, Z.J., Binner, J.: Ultra-low temperature reactive spark plasma sintering of ZrB_2-hBN ceramics. J. Eur. Ceram. Soc. **36**, 3637–3645 (2016)
25. Zou, J., Bin Ma, H., Chen, L., Wang, Y.J., Zhang, G.J.: Key issues on the reactive sintering of ZrB_2 ceramics from elementary raw materials. Scr. Mater. **164**, 105–109 (2019)
26. Gu, J., Zou, J., Ma, P., Wang, H., Zhang, J., Wang, W., Fu, Z.: Reactive sintering of B_4CTaB_2 ceramics via carbide boronizing: reaction process, microstructure and mechanical properties. J. Mater. Sci. Technol. **35**, 2840–2850 (2019)
27. Feng, L., Lee, S.-H., Kim, H.-N.: Effects of high-energy ball milling and reactive spark plasma sintering on the densification of HfC-SiC composites. J. Eur. Ceram. Soc. **37**, 1891–1898 (2017)
28. Wu, H.-Y., Zou, Ji., Eriksson, M., Liu, J.-J., Wang, W.-M., Fu, Z.-Y.: Reactive sintering of 2.5D Cf/ZrC-SiC ceramic matrix composite. J. Eur. Cer. Soc. **41**(13), 6189–6195 (2021)
29. Luyten, J., Cooymans, J.F.C., Snijkers, F.M.M.: Reaction-based processing methods. Key Eng. Mater. **264**, 707–712 (2004)
30. Akhtar, S., Saad, M., Misbah, M.R., Sati, M.C.: Recent advancements in powder metallurgy: a review. Mater. Today: Proc. **5**(9), 18649–18655 (2018)
31. Wang, X.C., Laoui, T., Bonse, J., Kruth, J.-P., Lauwers, B., Froyen, L.: Direct selective laser sintering of hard metal powders: experimental study and simulation. Int. J. Adv. Manufa. Technol. **19**, 351–357 (2002)
32. Natarajan, J. (ed.): Advances in Additive Manufacturing Processes. Bentham Science Publishers (2021)
33. Ronca, A., Rollo, G., Cerruti, P., Fei, G., Gan, X., Buonocore, G.G., Lavorgna, M., Xia, H., Silvestre, C., Ambrosio, L.: Selective laser sintering fabricated thermoplastic polyurethane/graphene cellular structures with tailorable properties and high strain sensitivity. Appl. Sci. **9**(5), 864 (2019)
34. Tiwari, S.K., Pande, S., Agrawal, S., Bobade, S.M.: Selection of selective laser sintering materials for different applications. Rapid Prototyp. J. **21**(6), 630–648 (2015)
35. Yang, Y., Ragnvaldsen, O., Bai, Y., et al.: 3D non-isothermal phase-field simulation of microstructure evolution during selective laser sintering. NPJ Comput. Mater. **5**, 81 (2019)
36. Hong, D., Yuan, J., Yin, Z., Peng, H., Zhu, Z.: Ultrasonic-assisted preparation of complex-shaped ceramic cutting tools by microwave sintering. Ceram. Int. **46**(12), 20183–20190 (2020)
37. Rahaman, M.N.: Ceramic Processing and Sintering, vol. 1. CRC Press (2017)
38. Sharma, N., Alam, S.N., Ray, B.C.: Fundamentals of spark plasma sintering (SPS): an ideal processing technique for fabrication of metal matrix nanocomposites. In: Spark Plasma Sintering of Materials: Advances in Processing and Applications, pp. 21–59 (2019)
39. Mamedov, V.: Spark plasma sintering as advanced PM sintering method. Powder Metall. **45**(4), 322–328 (2002)

40. Dudina, D.V., Mukherjee, A.K.: Reactive spark plasma sintering: successes and challenges of nanomaterial synthesis. J. Nanomater. **2013**, 5–5 (2013)
41. Shashanka, R., Chaira, D., Chakravarty, D.: Fabrication of nano-yttria dispersed duplex and ferritic stainless steels by planetary milling followed by spark plasma sintering and non-lubricated sliding wear behaviour study. J. Mater. Sci. Eng. B **6**, 111–125 (2016)
42. Cavaliere, P., Sadeghi, B., Shabani, A.: Spark plasma sintering: process fundamentals. In: Spark Plasma Sintering of Materials: Advances in Processing and Applications, pp. 3–20 (2019)
43. Angelo, P.C., Subramanian, R., Ravisankar, B.: Powder Metallurgy: Science, Technology and Applications. PHI Learning Pvt. Ltd. (2022)
44. Ageev, S.V., Girshov, V.L.: Hot isostatic pressing of metal powders. Metallurgist **59**(7–8), 647–652 (2015)
45. Lü, L., Lai, M.O.: Densification. In: Mechanical Alloying, pp. 173–187. Springer US, Boston, MA (1998)
46. Zimmerman, F.X., Toops, J.: Hot Isostatic Pressing: Today and Tomorrow. Avure Technologies. www.avure.com (2008)
47. Upadhyaya, G.S.: Powder Metallurgy Technology. Cambridge International Science Publishing (1997)
48. Fang, Z.Z., Paramore, J.D., Sun, P., Ravi Chandran, K.S., Zhang, Y., Xia, Y., Cao, F., Koopman, M., Free, M.: Powder metallurgy of titanium–past, present, and future. Int. Mater. Rev. **63**(7), 407–459 (2018)

Chapter 7
Importance of Secondary Processing Techniques in Sintered Compacts

J. Lario and V. Amigo

Abstract Advanced alloy production with high mechanical properties and corrosion resistance is a challenging task, sometimes technical and non-viable by conventional metallurgy process due to the need for special temperatures, inert atmospheres or requirements related to the high plastic deformation process. These alloys, crucial for aerospace, energy production and biomedical industries, demand continuous innovation in metallurgical processes and manufacturing technologies. Traditional casting methods face issues like phase segregation and high costs due to the required high temperatures for fusing refractory elements. Powder metallurgy, which presents advanced solid-state production, is an alternative route to achieve advanced alloys. Advanced secondary powder metallurgy processes can eliminate residual porosity and achieve submicron or nanoscale grain sizes, enhancing mechanical alloy properties. Secondary treatments, like hot isostatic pressing, plastic deformation and heat treatments, are necessary to optimize these alloys for high-performance applications. These processes help match or surpass the properties of traditionally manufactured alloys. The development of advanced manufacturing processes is driven by the need for materials with tailored properties for specific applications. These processes can reduce production costs and improve the properties of casting alloys, allowing for industrial-scale production that meets the severe demands of various industrial sectors.

Keywords Advanced alloys · Powder metallurgy · Secondary treatments · Powder alloying · Plastic deformations · Thermomechanical treatment · Metallurgy subprocessing

J. Lario (✉)
Departamento de Organización de Empresas, Universitat Politècnica de València (UPV), 46022 Valencia, Spain
e-mail: jlario@cigip.upv.es

Research Centre on Production Management and Engineering (CIGIP), UPV, 03801 Alicante, Spain

V. Amigo
Instituto de Tecnología de Materiales, UPV, 46022 Valencia, Spain

S. Rajendrachari (ed.), *Mechanically Alloyed Novel Materials*, Advanced Structured Materials 220, https://doi.org/10.1007/978-981-97-6504-1_7

7.1 Introduction

Advanced alloys with high mechanical properties and corrosion resistance are complex or non-viable to produce with conventional casting methods. The non-equilibrium phases or the high content of refractory elements usually require particular working temperatures or an inert atmosphere, which increases manufacturing costs.

In order to manufacture advanced metals (steels, hard metals, magnesium or titanium alloys) several metallurgical processes must be incorporated into the consolidation and transformation metals supply chain. The high-value alloys or materials require continuous research and implementation of new manufacturing technologies and material sciences to obtain the necessary properties to meet the requirements for advanced industries (such as aerospace, biomedical or energy production).

Traditional manufacturing processes of casting present the associated problems of phase segregation. Together, the high cost due to high temperatures required to fuse the refractory alloying elements, as well as the longer delivery times caused by the batch size to be produced. To overcome the shortcomings of conventional production methods, powder metallurgy and its advanced solid-state production processes have attracted great interest in advanced alloy manufacturing. Furthermore, future lines to consolidate and improve the mechanical properties of powder metallurgy alloys are the implementation of manufacturing subprocesses that can eliminate residual porosity and provide the alloys with a grain size submicron range, crucial factors in decision-making for the advanced materials.

Advanced alloys manufactured through powder metallurgy technology must present elevated mechanical and physical properties for high-demand applications (such as aerospace, energy production or biomaterials). Among the main factors to consider in the selection of the material are the maximum resistance to breakage, the elastic modulus, the fatigue resistance limit and the resistance to wear and corrosion [1–3]. Advanced alloys not only need to have elevated mechanical properties, which are achieved through a combination of high resistance to breakage and tenacity, but they must also have high fatigue resistance.

Powder metallurgy alloys require secondary treatments when the materials are used in high-performance applications, where the fatigue life and toughness of the material play a critical role in the part during its service life. The Hot Isostatic Pressing (HIP) process is one of the manufacturing methods used to obtain these advanced materials, as well as lamination, extrusion or angular channel pressing.

Increasing demand for materials with tailored properties for orthopaedic, cutting tools, aerospace or energy production applications has encouraged the development of advanced alloys. The current costs of forming ingots and high vacuum casting, as well as machining costs, limit the introduction of advanced alloys in several industrial sectors. Advanced manufacturing processes such as mechanical powder alloying, powder metallurgy and hot pressing have the potential to reduce the forming and finishing costs of advanced alloys.

With the introduction of hot isostatic compaction processes, heat treatments and plastic deformations, it will be possible to match or surpass the mechanical properties of powder metallurgical alloys with those obtained by conventional casting and forging manufacturing route. The improvements in properties introduced by advanced manufacturing processes, such as mechanical alloying of powders, hot isostatic compaction or high plastic deformation processes, will allow the leap from small-scale laboratory production to industrial production level, which can supply the specific demand of the metallurgy industry.

It is crucial to include advanced manufacturing processes that eliminate porosity, and achieve chemical homogeneity and microstructure on a nanometric scale, all of which will allow for predicting the behaviour of advanced alloys for their design and use in high-demand devices. Similar processing routes currently employed in casting alloys subject to high requirements, powder metallurgical materials require secondary processes to improve their properties. Thermomechanical deformation processes (hot forging, cold/hot working, cold/hot rolling, ECAP) and their consequent thermal treatments are necessary to obtain homogeneous alloys (steel, aluminium or titanium alloys) with a specific microstructure [4]. To avoid problems related to fracture during the service life of powder metallurgical alloys, the behaviour of the deformations introduced during advanced manufacturing processes (HIP, ECAP, Hot/Cold Rolling and Hot/Cold drawing) must be understood and how these influence the microstructure, present phases and mechanical properties of the material [5].

The mechanical properties of advanced powder metallurgical alloys must meet or exceed the properties of conventional casting parts used today in the industry (Tables 7.1). The availability, as well as the competitive improvement of mechanical properties, chemical and corrosion resistance of advanced powder metallurgy alloys, will increase the competitive position compared to conventional casting and forging alloys. The deployment for industrial use of powder metallurgical parts will provide an alternative supply for growing customer demands and reducing delivery times [7].

Depending on the thermomechanical treatment applied to powder metallurgic alloys, specific properties (mechanical, corrosion or chemical) can be achieved in the material (Fig. 7.1). Depending on the specific properties required, according on the application to which it is going to be used, there are different microstructures that can be obtained through these thermomechanical treatments (solid solution hardening, dispersion hardening, texturing hardening, annealing, stress relieve, etc.) and that modify properties such as elongation, toughness, elastic modulus, hardness, resistance to creep and corrosion [6].

Table 7.1 Mechanical properties of $\alpha + \beta$ titanium alloys [6]

Alloy designation (wt%)	Tensile strength (MPa)	Yeild strength (MPa)	Elongation (%)	Elastic modulus (GPa)
Ti–6AI–4V ELI (annealed)	860–965	795–875	10–15	101–110
Ti–6AI–4V (annealed)	895–930	825–869	6–10	110–114
Ti–6AI–7Nb (wrought)	900–1050	880–950	8.1–15	114
Ti–5AI–2.5Fe (cast)	1020	895	15	112
Ti–5AI–2Sn–2Zr–4Mo–4Cr (STA)	1105–1240	1035–1075	8–15	114
Ti–6AI–2Sn–4Zr–6Mo (STA)	1170	1100	6	114
Ti–6AI–6V–2Sn (air-cooled annealed)	993–1013	903–937	8	110
Ti–7AI–4Mo (annealed)	1000	930	10	111
Ti–8Mn (annealed)	1000	930	15	113

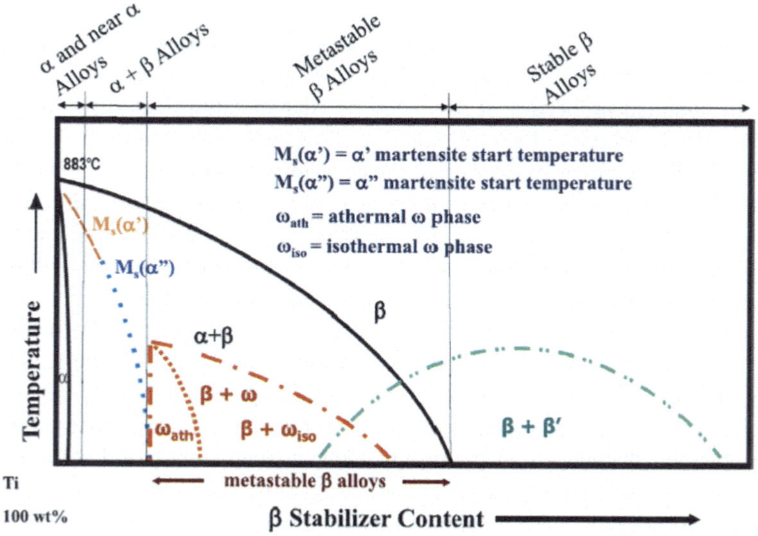

Fig. 7.1 Phase diagram for the classification of titanium alloys and the formation of non-equilibrium phases in metastable titanium alloys [6]

7.2 Mechanical Alloying

Mechanical Alloying (MA) is a powder metallurgy process that produces advanced materials through solid-state diffusion. This process involves repeatedly deforming the powder plastically at low temperatures to create new alloys or microstructures. Mechanical Alloying is a powder processing technique that allows obtaining advanced materials with equilibrium and non-equilibrium phases [8].

The application of this process in titanium and magnesium alloys is relatively recent. It consists of the following stages, initially, the powder mixture is loaded into the jars in a controlled medium (argon atmosphere) to avoid contamination and ensure an inert environment that can generate future oxidation phenomena. In addition to the alloying powders, a small amount of control elements (Process Control Agent, PCA) is added, the objective of which is to avoid excessive cold welding of the powders during mechanical alloying. Finally, a ball ratio (made of a hard material such as zirconium or high-speed steel) is added to the jars and sealed using an o-ring and thread system, before removing it from the argon chamber. Subsequently, the jug is installed in the equipment where the process parameters (revolutions, times and cooling stops) will be selected (Fig. 7.2).

The distribution of the allowing elements depends on their solubility, the rate of diffusion, as well as the time necessary for diffusion to occur. The mechanical alloying metallurgy processing step can enhance this process. The mechanical alloying process allows us to obtain pre-alloyed powders in a solid state, with high contents of refractory elements that could not be obtained by conventional techniques such as water atomization. From the point of view of the microstructure during the mechanical alloying process, each of the powder particles undergoes two states during the MA process: cold welding and breaking. During the grinding process, the powder particles are cold-welded and subsequently fractured due to the kinetic energy supplied by the impact of the balls. Breakage occurs when long particles break into smaller powder particles, due to the effect of the balls inside the jars. This process improves the chemical homogeneity of powder alloys with high contents of

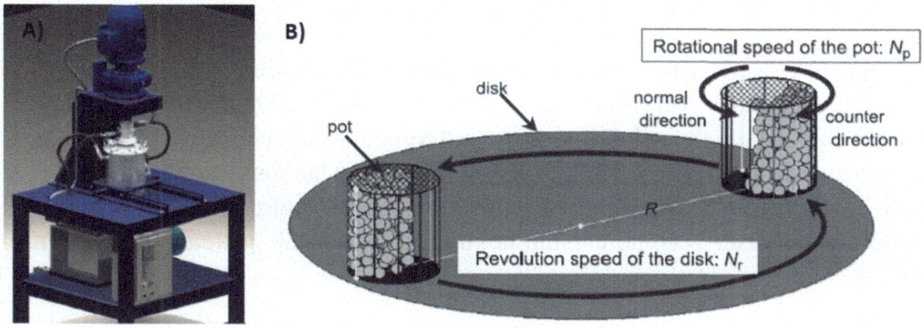

Fig. 7.2 Schematization of the mechanical alloying process. **a** Planetary ball mill. **b** Schematic description of the elements that make up the grinding jar, movements, and rotations during the grinding process. *Source* Adapted from Suárez et al. [9], Ramasamy et al. [10]

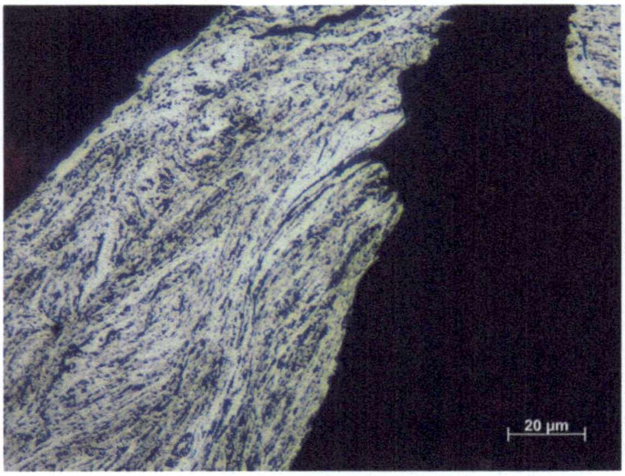

Fig. 7.3 Microstructure of titanium alloy powder with a percentage of 0.250% by weight of calcium, ground for 80 min in vacuum and etched with Kroll reagent. *Source* Zadra [11], Mechanical alloying of titanium

refractory elements, which is difficult to achieve during the sintering stage due to the difference in densities and fusion temperatures to the metal matrix (Fig. 7.3).

Among the factors influencing the mechanical alloying process, the following factors present high importance: the balls/powder ratio, the nature of the powders used, the size of the starting powder, the process temperature, the alloying energy/revolutions used, and the process control agents employed.

The process control agents are a source of impurities, interstitial elements, precipitates or oxides that reduce the mechanical properties of the consolidated materials. The PCA can introduce new, potentially undesirable elements into the powder mixture and, therefore into the final alloy composition. These elements can be included in the alloy matrix as carbides, oxides and/or precipitates during the consolidation, plastic deformation and sintering process. Typically, PCA is partially removed by "hot vacuum degassing" to avoid the formation of porosity/blistering during the sintering process or heat treatments.

Long milling times can reduce the powder's compatibility, reducing the mechanical properties of the sintered alloys. The hardening of the powder particles by "work-hardening" introduced during the high milling times causes a reduction in the compatibility of the powders [12].

In recent years, research into the properties of nanocrystalline materials compared to coarse grains (10–100 μm) has attracted great interest. The mechanical alloying process can change the solubility of the different alloying elements, as it is a non-equilibrium process. The XRD results obtained by Mahboubi Soufiani et al. [13] show how after 90 h of milling the powder's microstructure becomes amorphous. These powders can become crystalline after short annealing treatments (30 min) without excessive grain coarsening.

7.3 Hot Isostatic Pressing

The pressing and sintering powder metallurgical manufacturing processes have the most economical costs, but have the associated problem of residual porosity after sintering, which causes low fatigue resistance. Residual porosity can be eliminated using processes after sintering, such as the use of Hot Isostatic Pressing (HIP) (Fig. 7.4). Over the past three decades, the advantages of the powder metallurgical processing consolidated by HIP have been employed to manufacture tool steel for high performance applications [14].

For high-performance applications, thermomechanical treatments are necessary to homogenize the microstructure, eliminate porosity, and increase fatigue resistance. The rate of mass transfer can be accelerated by improving interparticle contacts. Any manufacturing factor or subprocess that improves mass transfer during sintering is of great help to improve the strength of powder metallurgical alloys.

HIP technology has been employed over the past 20 years by Synertech PM, which has produced titanium parts for a variety of rocket engines, gas compressors, and test parts for jet engines and airframes, and has opened up the possibilities of significantly reducing costs and introduction of powder metallurgical titanium for various profitable applications in oil and gas exploration, and power generation [16].

Historically, to manufacture high-performance parts, the sintering process has been used to achieve the minimum density (>90%) in which the material does not present interconnected porosity. This density limit is where the powder metallurgical material can be subjected to the HIP process without the need to be encapsulated, making the process economically viable [17]. New materials will be needed to manufacture the fusion power reactors for future needs of clean energy. These new materials will be subjected to high demanding conditions, such as high heat inputs and

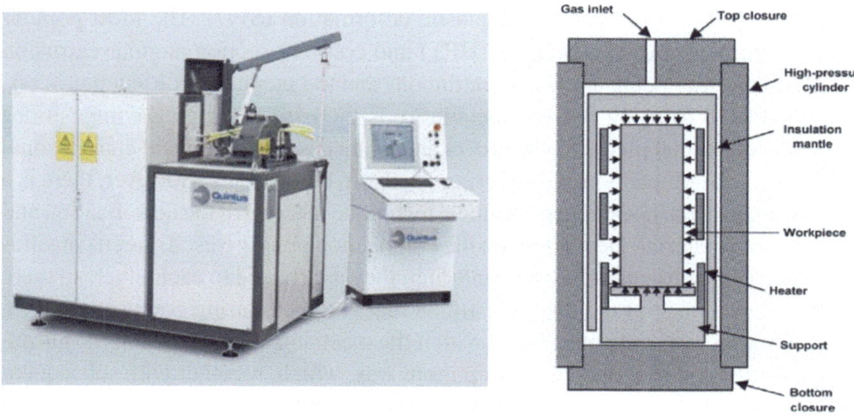

Fig. 7.4 Quintus hot isostatic pressing equipment, model QIH21. *Source* Adapted from www.quintus.com and Muley et al. [15]

neutron loads during reactor operation. Tungsten-vanadium-yttria alloys are considered candidate materials due to their high melting point, high temperature resistance, low thermal expansion and high resistance under neutron irradiation [18]. Palacios et al. [19] demonstrated that through the addition of vanadium (2–4 wt%) and yttria (0.5 wt%) to pure tungsten consolidated through the HIP route drastically reduces porosity, increasing flexural strength and stabilizing high-temperature degradation [19].

HIP equipment with high cooling rates allows HIP cycle times to be reduced. These high cooling rates (4000 °C/min) allow the HIP cycle to be combined with conventional thermal treatments where high cooling rates are needed to obtain a microstructure with specific mechanical properties. The Ti–6Al–4V additive manufacturing samples where HIP treatment was applied at 900 °C and 100 MPa for 2 h, followed by annealing at 650–700 °C for 1 h, significantly improved ductility and fatigue resistance by reducing porosity and therefore resulting in axial fatigue resistance comparable to wrought material produced by casting and forging [20].

In the study carried out by Lario et al. [7], the influence of hot isostatic post-processing on Ti–35Nb–2Sn alloys was observed (Fig. 67). It was found that the HIP treatment allows the relative density to be increased from 95%, obtained by conventional powder metallurgy, to 100% after being processed by HIP. The same study showed that high cooling rates (500 °C/min) after the HIP cycle allow a higher amount of β phase to be retained.

7.4 Equal Channel Angular Pressing

Powder metallurgy continues to present the problem of a lack of microstructural control. One way to control the microstructure and properties is the application of plastic deformation and even severe plastic deformation (SPD). The most popular processes are high-pressure twisting (HPT) and constant section angular extrusion (ECAP), the former being a process continuous and the second a multi-step process.

The objective of using powder metallurgy is the ease of varying compositions through the elemental mixture of powders and using a very economical conventional system, uniaxial compaction and sintering, to obtain the samples. However, there is a problem linked to the powder metallurgical technique in terms of the densification and porosity of the material. Depending on the compaction pressure used, green densities are reached and, subsequently, after sintering, they are related to each other, so there is a clear dependence of the final density on that obtained during compaction. As a means of microstructural control, and with it the mechanical properties of the alloys, plastic deformation is applied in an important way, which, together with subsequent thermal treatments, allows this microstructural control to be carried out.

Cold plastic deformation can induce the precipitation of interstitial phases, decrease the material's elastic modulus and control the grain size by inducing recrystallization when carried out Thermal treatments. On the other hand, applying severe plastic deformation is interesting for obtaining nanostructured materials. Among the most popular processes of severe plastic deformation are constant section angular extrusion (ECAP, Equal-Channel Angular Pressing) and high-pressure torsion (HPT, High-Pressure Torsion).

This section summary highlights the incorporation of metallurgical subprocesses in the manufacture of advanced powder alloys for producing nanometer-scale grain-size materials to be employed in high-demanding applications. The Equal Channel Angular Pressing (ECAP) is a technique that can be used to substantially improve the mechanical strength of powder metallurgy alloys (steels, aluminum, magnesium, titanium, or others), thanks to the formation of grain at a submicrometer or nanometer scale. Current metallurgy subprocessing techniques may be used to introduce an ultrafine grain size into polycrystalline materials [21]. The repetitive pressing conducted during ECAP does not change the cross-sectional dimensions and applies an intense plastic strain to the material, which makes it possible to add severe plastic deformation and reduce grain structure to sub-micrometer sizes [22].

In the past three decades, several technological advances have been implemented in industrial processes, allowing the fabrication of nanostructured materials [24]. These advances in manufacturing technologies have made it possible to obtain hard nanostructured materials. An example of its industrial application is the WC/Co cemented carbide nanocomposites and TiC/Fe, employed in the current cutting tools manufacture. The grain refinement of these nanocomposites has increased the values of hardness, toughness and wear resistance of the tools. It allows to manufacture machining tools that can work at higher speeds and feeding rates, with a longer tool life than non-nanostructured materials.

The high degree of grain refinement produced after applying the equal channel angular pressing process results from the high plastic deformations introduced into the material by forcing it to pass through a cavity/channel or matrix, which presents an abrupt change in angle through the application of hydrostatic pressure (Fig. 7.5). Valiev et al. [25] obtained an equation that relates the degree of plastic deformation introduced to a material as a function of the number of passes and the angle of the passage channel (Eq. 7.1) (M. Furukawa). The application of high plastic deformation during repeated passes through the ECAP matrix produces a reduction in the grain size of the alloy, and improves its mechanical properties.

$$\epsilon = \left(N\sqrt{3}\right)\left[2\cot\left\{\left(\frac{\Phi}{2}\right)+\left(\frac{\Psi}{2}\right)\right\} + \Psi\text{cosec}\left\{\left(\frac{\Phi}{2}\right)+\left(\frac{\Psi}{2}\right)\right\}\right] \tag{7.1}$$

Source Gao et al. [26].

Through angular channel extrusion, it is possible to achieve severe plastic deformation that allows extreme grain refinement in titanium alloys [27]. Processing powder metallurgic alloys using ECAP offers two important benefits. Firstly, it allows for avoiding alloying elements for grain refinement, with high cost. Instead, grain refinement occurs by solid solution and precipitation hardening. Powder metallurgic alloys with submicrometer size, obtained by HIP + ECAP processes, together with surface treatments, are considered the two ways to improve the chemical and corrosion resistance of advanced materials for the coming decades.

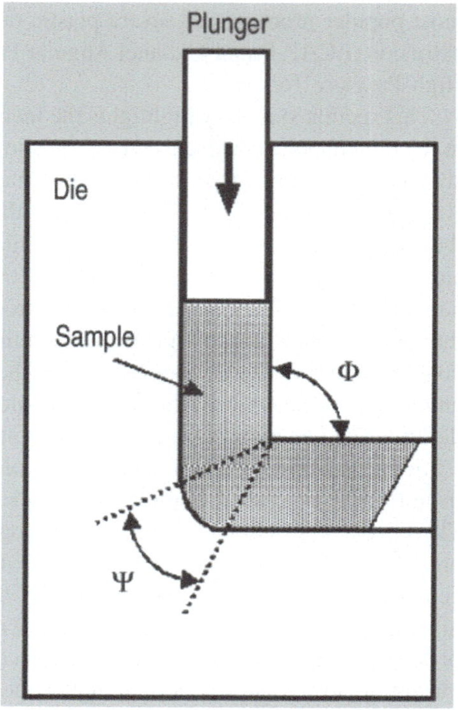

Fig. 7.5 Schematic illustration of ECAP showing the three orthogonal planes x, y and z. *Source* Semenova et al. [23]. Strength and fatigue properties enhancement in ultrafine-grained Ti produced by severe plastic deformation

Three possible processing routes are based on the effect of sample rotation between repetitive pressings. At route A samples no rotation is performed between repetitive pressings. The route B samples are rotated by 90° between pressing steps, and at route C the 180° rotation is performed each time. A schematic representation for each ECAP pressing route is represented in Fig. 7.6. The main difference between the processing routes is on which plane the grains' elongation takes place.

In this section, the state of the art on new mechanical properties of nanocrystalline materials is reviewed. There is a large amount of evidence that these new advanced materials have a higher degree of mechanical properties and chemical resistance than current alloys obtained by casting and forging [28, 29]. This class of Powder metallurgy superalloys is expected to form the technological base for the second half of the twenty-first century. As manufacturing techniques advance, it is becoming possible to obtain dense alloys with non-equilibrium phases in a nanosize range, allowing reliable results. The well-accepted intrinsic mechanical results for nanosized titanium alloys are summarized as:

- The elastic modulus of nanomaterials is approximately the same as materials with conventional grain size (10–100 μm), until the grain size becomes very small (<5 nm).
- The hardness and resistance to breakage of nanomaterials increase by reducing the grain size.

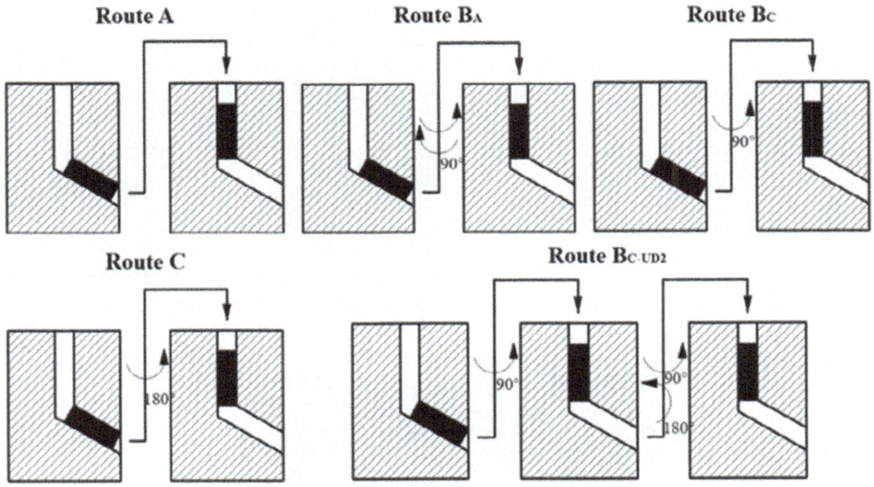

Fig. 7.6 Processing routes which may be used for repetitive pressings. *Source* Gao et al. [26]. Effects of different deformation routes of ECAP on AA6063 mechanical properties and microstructure

Semenova et al. [23] demonstrated that using the ECAP technique, it is possible to increase the tensile strength of CP Grade 4 Ti alloys from 700 to 1240 MPa, reaching values higher than 940 MPa for the Ti6Al4V alloy (Table 7.2), by reducing the grain size from 25 μm to 150 nm (Fig. 7.7).

The grain size reduction also increases the fatigue resistance of the alloy. Therefore, the ECAP process has two main advantages. The first of them is that it is possible to increase the mechanical properties without adding alloying elements. This increase

Table 7.2 Mechanical properties of Ti Cp alloys as a function of their processing condition [23]

No	Processing regimes	UTS (MPa)	Yeild stress (MPa)	Elongation (%)	Uniform elongation (%)
1	As-received (according to the certificate)	700	530	25	–
2	UGF (ECAP + TMT)	1240 ± 10	1150 ± 20	11 ± 1	1.6
3	UGF (ECAP + TMT + annealing 350 °C, 1 h)	1230 ± 10	1170 ± 30	12 ± 1	4.5
4	UGF (ECAP + TMT + annealing 350 °C, 6 h)	1250 ± 10	1100 ± 10	13 ± 1	5.1

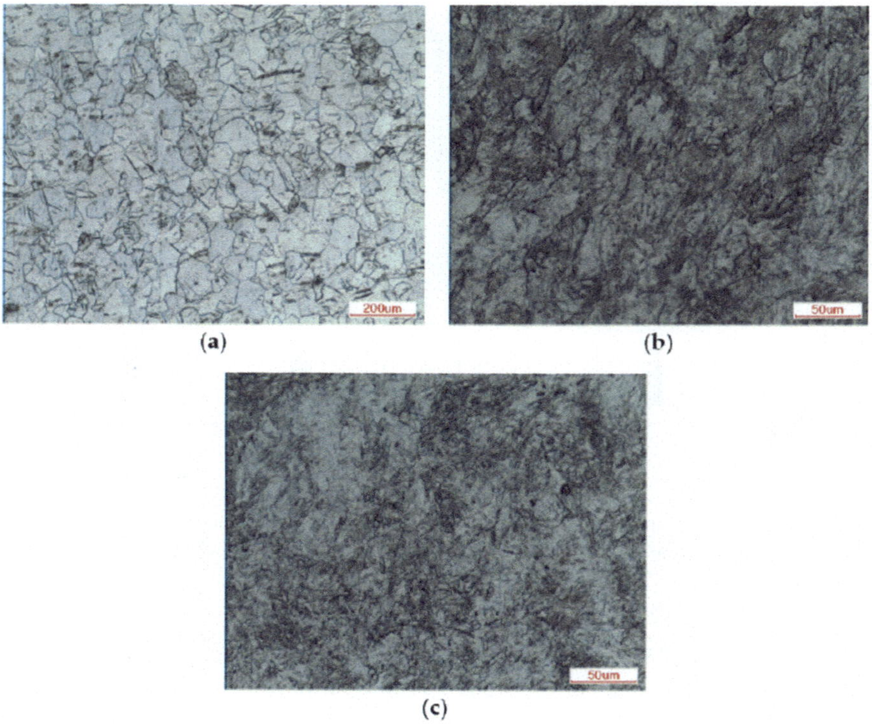

Fig. 7.7 Microstructure of CP grade 4 Titanium depending on its processing condition **a** equiaxial grain. **b** 4 passes through ECAP. **c** 8 passes through ECAP [30]

in properties is produced by a grain size refinement rather than solid solution solidification or precipitation hardening. Secondly, the improvement of mechanical properties allows a reduction in the size of the final part, consuming less material or with lower weight, which presents their advantages in aerospace, energy and biomedical sectors.

The ECAP processing route introduces an accumulation of crystalline defects in the structure of the alloy through dislocations, which hinders or reduces the degree of deformation. The materials processed by ECAP, with multiple passes, require subsequent thermal treatments to improve the degree of deformation before breaking. The application of annealing treatment at low temperature, allows increasing the ductility and fatigue resistance of alloys that have been subjected to high plastic deformation processes. These treatments reduce the density of the dislocations, allowing the degree of deformation before fracture to increase.

7.5 Heat Treatments

The mechanical properties of powder metallurgy alloys depend on the size, morphology, distribution and amount of different phases, which in turn depend directly on the parameters of the thermal treatments, such as temperature, duration, cooling rate and heating [20].

Heat treatments make it possible to relieve residual stresses and increase the toughness and fatigue resistance of the alloy [31]. As already mentioned, the microstructure (phases, grain sizes, precipitates, inclusions, porosity, etc.) determines the properties of titanium alloys.

Conventionally, the microstructure of powder metallurgic alloys is usually described by the organization of their main phases. The resulting microstructure depends on the thermomechanical history to which it has been subjected for its manufacture. These manufacturing stages are complex; among them, plastic deformations, ageing, annealing or stress relief treatments are usually carried out.

Likewise, chemical equilibrium temperatures depend on the chemical composition and homogeneity of the elements that make up the alloy. Below this temperature, a specific phase of the alloy stops transforming and remains in a metastable state. Therefore, depending on the temperature, time, and cooling rate during the treatment, different microstructures will result in different proportions of phases and different grain sizes.

Solution Heat Treatments

After the solution treatment, the beta alloys are aged at temperatures between 450 and 650 °C, depending on the alloy treated, and below the martensitic transformation temperature. Solution thermal treatments carried out at temperatures several degrees below the β-transus allow bimodal microstructures to be obtained. Bimodal microstructures combine the advantages of lamellar and equiaxed structures, presenting a well-balanced property profile. Solution treatments consist of heating the material above the beta transformation temperature, followed by rapid cooling, and in some cases may include an ageing step. Through this type of treatment, it is possible to increase the resistance and ductility values.

7.6 Influence of Metallurgical Processes on Fatigue Resistance

Today, the mechanical properties of powder metallurgical parts meet the requirements of the automotive sector, where the cost of parts plays a priority role. However, when this manufacturing method is considered for aerospace, energy or biomedical applications, the risk of premature failure is an issue of concern mainly due to residual porosity and grain size. As in casting alloys, which are subject to high requirements, powder metallurgical materials require secondary processes. Thermomechanical deformation processes (hot isostatic pressing, forging, hot/cold working, ECAP) and their consequent thermal treatments are necessary to obtain a homogeneous alloy with a specific microstructure.

Powder metallurgic alloys employed in advanced applications are subjected to cyclic loads during their service life, which is why their fatigue performance is one of the main criteria that must be considered when designing a manufacturing process to obtain advanced materials.

Material failure occurs when the crack is large enough that it grows unstable and causes the part to break. Fatigue failure occurs at stresses much lower than the maximum breaking strength of the materials, due to the cyclic nature of the application of stresses to parts in service. This failure mechanism is a common failure mode in service alloys, representing more than 50% of all failure mechanisms. Materials with high resistance lead to longer times for crack initiation to occur [32].

The number of cycles to failure will depend on the applied load, the geometry of the part, the surface finish and the composition and microstructure of the alloy. Residual porosity in powder metallurgical parts is difficult or virtually impossible to eliminate during conventional manufacturing processes. Therefore, additional operations are required to eliminate this effect and improve the properties of the material that allow for increasing service life [32].

PM alloys require secondary treatments when the materials are used in high-performance applications, where the fatigue life and toughness of the material play a critical role in the part during its service life. The pores present in powder metallurgical alloys act as stress concentrators, and when they are more elongated, they more easily cause the failure of the material at low stresses. To improve the mechanical behaviour of powder metallurgy alloys, they must be processed by hot isostatic pressing and severe plastic deformation to eliminate porosity and obtain the desired microstructure.

The HIP process is an effective thermomechanical treatment to improve the fatigue performance of beta titanium powder metallurgical alloys. [33] showed that Ti–6Al–4V additive manufacturing samples where HIP treatment was applied at 900 °C and 100 MPa for 2 h, followed by annealing at 650–700 °C for 1 h, significantly improved ductility and fatigue resistance by reducing porosity, resulting in axial fatigue resistance comparable to wrought material produced by casting and forging.

Therefore, metal parts obtained by powder metallurgy techniques must use advanced manufacturing processes to have high mechanical properties and thus withstand fatigue damage during service life. They cause considerable changes in breaking strength, flexural strength, fatigue, crack propagation, hardness, wear, corrosion and elastic modulus. These last processes can introduce grain refinement, at a nanometric scale, to improve the fatigue life and chemical resistance of PM alloys. The improvement is a consequence of the diffusion of alloying elements and therefore of the chemical/phase homogeneity and the reduction in grain size introduced by the manufacturing subprocesses.

References

1. Long, M., Rack, H.J.: Titanium alloys in total joint replacement—a materials science perspective. Biomaterials **19**, 1621–1639 (1998). https://doi.org/10.1016/S0142-9612(97)00146-4
2. Niinomi, M.: Mechanical biocompatibilities of titanium alloys for biomedical applications. J. Mech. Behav. Biomed. Mater. **1**(1), 30–42 (2008). https://doi.org/10.1016/j.jmbbm.2007.07.001
3. Okazaki, Y., Gotoh, E.: Comparison of metal release from various metallic biomaterials in vitro. Biomaterials **26**(1), 11–21 (2005). https://doi.org/10.1016/j.biomaterials.2004.02.005
4. Liu, Y., Chen, L.F., Tang, H.P., Liu, C.T., Liu, B., Huang, B.Y.: Design of powder metallurgy titanium alloys and composites. Mater. Sci. Eng. A **418**(1–2), 25–35 (2006). https://doi.org/10.1016/j.msea.2005.10.057
5. Lee, K.O., Lee, S.B.: Modeling of materials behavior at various temperatures of hot isostatically pressed superalloys. Mater. Sci. Eng. A **541**, 81–87 (2012). https://doi.org/10.1016/j.msea.2012.02.005
6. Amigó-Borrás, V., Lario-Femenía, J., Amigó-Mata, A., Vicente-Escuder, Á.: Titanium, titanium alloys and composites. In: Encyclopedia of Materials: Metals and Alloys, vol. 1, pp. 179–199 (2021). https://doi.org/10.1016/B978-0-12-819726-4.00044-2
7. Lario, J., Vicente, Á., Amigó, V.: Evolution of the microstructure and mechanical properties of a $Ti_{35}Nb_2Sn$ alloy post-processed by hot isostatic pressing for biomedical applications. Metals **11**(7) (2021). https://doi.org/10.3390/met11071027
8. Neikov, O.D., Gopienko, V.G.: Production of titanium and titanium alloy powders. In: Handbook of Non-Ferrous Metal Powders, 2nd ed. Elsevier Ltd. (2019). https://doi.org/10.1016/b978-0-08-100543-9.00018-x
9. Suárez, H.E.J., de Sanchez, N.A., Diaz, J.A.A.: Titanium carbide (TiC) production by mechanical alloying. Powder Technol. (2018). https://doi.org/10.5772/intechopen.76690
10. Ramasamy, D., Mohanesen, K., Maria, R., Samykano, M., Kadirgama, K., Rahman, M.M.: Analysis of alumina particles size and shape formation from developed planetary ball mill. IOP Conf. Ser.: Mater. Sci. Eng. **736**(5) (2020). https://doi.org/10.1088/1757-899X/736/5/052032
11. Zadra, M.: Mechanical alloying of titanium. Mater. Sci. Eng. A **583**, 105–113 (2013). https://doi.org/10.1016/j.msea.2013.06.064
12. Nazari, K.A., Nouri, A., Hilditch, T.: Effects of milling time on powder packing characteristics and compressive mechanical properties of sintered Ti-10Nb-3Mo alloy. Mater. Lett. **140**, 55–58 (2015). https://doi.org/10.1016/j.matlet.2014.10.143
13. Mahboubi Soufiani, A., Karimzadeh, F., Enayati, M.H.: Formation mechanism and characterization of nanostructured Ti6Al4V alloy prepared by mechanical alloying. Mater. Des. **37**, 152–160 (2012). https://doi.org/10.1016/j.matdes.2011.12.044
14. Kearns, M.A., Murray, K., Davies, P.A., Ryabinin, V., Gonzalez, E.: Sintering and properties of MIM M2 high speed steel produced by prealloy and master alloy routes. Met. Powder Rep. **71**(3), 200–206 (2016). https://doi.org/10.1016/j.mprp.2016.04.085
15. Muley, A.V., Aravindan, S., Singh, I.P.: Nano and hybrid aluminum based metal matrix composites: an overview. Manuf. Rev. **2** (2015). https://doi.org/10.1051/mfreview/2015018
16. Samarov, V., Seliverstov, D., Froes, F.H.: Fabrication of near-net-shape cost-effective titanium components by use of prealloyed powders and hot isostatic pressing. In: Titanium Powder Metallurgy: Science, Technology and Applications. Elsevier Inc. (2015). https://doi.org/10.1016/B978-0-12-800054-0.00018-6
17. Abkowitz, S.M., Abkowitz, S., Fisher, H.: Breakthrough claimed for titanium PM. Met. Powder Rep. **66**(6), 16–21 (2011). https://doi.org/10.1016/S0026-0657(12)70015-2
18. Piotter, V., Zeep, B., Norajitra, P., Ruprecht, R., von der Weth, A., Hausselt, J.: Development of a powder metallurgy process for tungsten components. Fusion Eng. Des. **83**(10–12), 1517–1520 (2008). https://doi.org/10.1016/j.fusengdes.2008.06.022

19. Palacios, T., Pastor, J.Y.: Tungsten-vanadium-yttria alloys for fusion power reactors (II): mechanical characterization. Int. J. Refract Metal Hard Mater. **54**, 159–164 (2016). https://doi.org/10.1016/j.ijrmhm.2015.07.031
20. Fan, J., Li, J., Kou, H., Hua, K., Tang, B., Zhang, Y.: Microstructure and mechanical property correlation and property optimization of a near β titanium alloy Ti-7333. J. Alloys Compd. **682**, 517–524 (2016). https://doi.org/10.1016/j.jallcom.2016.04.303
21. Furukawa, M., Horita, Z., Nemoto, M., Langdon, T.G.: Processing of metals by equal-channel angular pressing. J. Mater. Sci. **36**(12), 2835–2843 (2001). https://doi.org/10.1023/A:1017932417043
22. Rusin, N.M.: The effect of temperature and equal-channel angular-pressing routes on the form of powders and structure in pressings. Russian J. Non-Ferrous Metals **50**(5), 529–533 (2009). https://doi.org/10.3103/S1067821209050186
23. Semenova, I.P., Valiev, R.Z., Yakushina, E.B., Salimgareeva, G.H., Lowe, T.C.: Strength and fatigue properties enhancement in ultrafine-grained Ti produced by severe plastic deformation. J. Mater. Sci. **43**(23–24), 7354–7359 (2008). https://doi.org/10.1007/s10853-008-2984-4
24. Zhu, Y.T., Langdon, T.G.: The fundamentals of nanostructured materials processed by severe plastic deformation. JOM **56**(10), 58–63 (2004). https://doi.org/10.1007/s11837-004-0294-0
25. Valiev, R.Z., Estrin, Y., Horita, Z., Langdon, T.G., Zehetbauer, M.J., Zhu, Y.: Producing bulk ultrafine-grained materials by severe plastic deformation: ten years later. JOM **68**(4), 1216–1226 (2016). https://doi.org/10.1007/s11837-016-1820-6
26. Gao, J., He, T., Huo, Y., Yao, T., Hong, H.: Effects of different deformation routes of ECAP on AA6063 mechanical properties and microstructure. Procedia Manuf. **50**(2019), 119–124 (2020). https://doi.org/10.1016/j.promfg.2020.08.022
27. Truong, V.K., Lapovok, R., Estrin, Y.S., Rundell, S., Wang, J.Y., Fluke, C.J., Crawford, R.J., Ivanova, E.P.: The influence of nano-scale surface roughness on bacterial adhesion to ultrafine-grained titanium. Biomaterials **31**(13), 3674–3683 (2010). https://doi.org/10.1016/j.biomaterials.2010.01.071
28. Estrin, Y., Kim, H., Pang, H., Lapovok, R., Ng, H.P., Jo, J.: Available from Deakin Research Online: Mechanical Strength and Biocompatibility of Ultrafine- Grained Commercial Purity Titanium, pp. 1–6 (2013). https://doi.org/10.1155/2013/914764
29. Langdon, T.G.: Processing of ultrafine-grained materials using severe plastic deformation: potential for achieving exceptional properties. Rev. Metal. (Madrid) **44**(6), 556–564 (2008). https://doi.org/10.3989/revmetalm.0838
30. Wu, H., Jiang, J., Liu, H., Sun, J., Gu, Y., Tang, R., Zhao, X., Ma, A.: Fabrication of an ultra-fine grained pure titanium with high strength and good ductility via ECAP plus cold rolling. Metals **7**(12) (2017). https://doi.org/10.3390/met7120563
31. ASM Handbook Volume 2: Properties and Selection: Nonferrous Alloys and Special-Purpose Materials. ASM International, October 01, 1990. ISBN: 978-0-87170-378-1
32. Romero, C., Yang, F., Bolzoni, L.: Fatigue and fracture properties of Ti alloys from powder-based processes—a review. Int. J. Fatigue **117**(August), 407–419 (2018). https://doi.org/10.1016/j.ijfatigue.2018.08.029
33. Molaei, R., Fatemi, A., Phan, N.: Significance of hot isostatic pressing (HIP) on multiaxial deformation and fatigue behaviors of additive manufactured Ti-6Al-4V including build orientation and surface roughness effects. Int. J. Fatigue **117**(July), 352–370 (2018). https://doi.org/10.1016/j.ijfatigue.2018.07.035

Chapter 8
Consolidation of Lightweight Alloy Powders: Overcoming the Problems During Pressing and Sintering of Low-Dense Alloy Powders like Aluminium, Magnesium, Titanium, and Beryllium Alloys

B. N. Sharath, R. Raghavendra Rao, K. P. Rakesh, and S. Pradeep

Abstract The process of consolidating lightweight alloy powders, which includes materials such as aluminium, magnesium, titanium, and beryllium, plays a crucial role in the production of high-performance components in many sectors. Nevertheless, the pressing and sintering processes of these alloys are accompanied by specific difficulties arising from their low density, reactivity, and particular material features. This chapter examines novel methodologies for overcoming these obstacles and improving the compaction of lightweight alloy powders. The study encompasses several aspects including powder modification techniques, innovative binder systems, additive manufacturing methods, examination of microstructure, evaluation of mechanical and thermal properties, as well as issues related to environmental and safety factors. Through a comprehensive analysis of existing scholarly works and the implementation of empirical investigations, the primary objective of this study is to offer significant perspectives on the advancement of effective consolidation methodologies. This endeavour has the potential to bring about transformative changes in sectors heavily reliant on lightweight alloys.

Keywords Lightweight alloys · Alloy powders · Titanium alloys

---

B. N. Sharath · R. Raghavendra Rao · S. Pradeep
Department of Mechanical Engineering, Malnad College of Engineering, Hassan, Affiliated to VTU, Belagavi, Karnataka 573202, India

K. P. Rakesh (✉)
Department of Radiation Chemistry, University of North Carolina at Chapel Hill, Chapel Hill, NC 27514, US
e-mail: rakeshasg@gmail.com

S. Rajendrachari (ed.), *Mechanically Alloyed Novel Materials*, Advanced Structured Materials 220, https://doi.org/10.1007/978-981-97-6504-1_8

8.1 Introduction

The use of lightweight alloy powders derived from various materials, including aluminium, magnesium, titanium, and beryllium, has emerged as a central area of interest within the fields of contemporary engineering and material science. These alloys possess a compelling amalgamation of strength, endurance, and weight efficiency, rendering them vital in a wide array of sectors, including aerospace, automotive, medical devices, and consumer electronics [1]. The appeal of these lightweight metals resides not alone in their exceptional characteristics but also in their capacity to transform product design and mitigate environmental consequences [2]. This study explores the diverse domain of consolidating lightweight alloy powders, with a specific emphasis on addressing the challenges faced during the processes of pressing and sintering. These processes play a crucial role in the conversion of loose powder into components that are completely dense, structurally sound, and mechanically durable [3]. The importance of this endeavour becomes evident when considering the crucial uses of these alloys, such as the production of aviation components that are both lightweight and durable, medical implants that are compatible with the human body, and vehicle parts that enhance fuel efficiency [4]. To explore the complexities associated with the consolidation of lightweight alloys and analyse the challenges and constraints that emerge throughout these procedures, while also highlighting novel approaches and remedies that are propelling advancements in this sector. This review seeks to offer a comprehensive comprehension of the consolidation of lightweight alloy powders, encompassing surface modifications, novel binder systems, cutting-edge additive manufacturing techniques, and in-depth microstructural analysis. The primary objective is to effectively address the challenges that arise in this context [5]. The pursuit of overcoming these obstacles associated with the consolidation of lightweight alloy powders extends beyond mere scientific inquisitiveness. It is a significant endeavour with wide-ranging ramifications for technological progress and societal improvement. The objective is to cultivate a more profound appreciation of the vast capabilities inherent in these alloys and elucidate the trajectory for effectively harnessing their extraordinary attributes in order to bring about transformative changes across all sectors. This study aims to provide valuable insights for academics, engineers, and manufacturers who are interested in harnessing the full capabilities of lightweight alloys and addressing the challenges that have impeded their extensive use [6].

8.1.1 Overview of Consolidation of Lightweight Alloy Powders and Its Significance in Material Science and Engineering

The consolidation of lightweight alloy powders is of paramount importance in the domain of materials science and engineering. The use of lightweight alloys, comprising elements such as aluminium, magnesium, titanium, and beryllium, has attracted considerable interest owing to their remarkable amalgamation of strength, low density, and other advantageous characteristics [7, 8]. These alloys possess significant promise for transforming many industries, such as aerospace, automotive, medical, and consumer electronics. In order to properly comprehend and use this potential, it is crucial to get a comprehensive understanding of the obstacles presented by the process of consolidation and the inventive strategies adopted to overcome them [9, 10]. The potential of lightweight alloys to decrease the mass of structural components while maintaining their strength and performance is a noteworthy achievement. In the context of aircraft and automobile industries, these factors correspond to improved fuel economy, decreased emissions, and increased cargo capacity [11–13]. Lightweight alloys frequently have exceptional mechanical characteristics, such as elevated strength-to-weight ratios, resistance to corrosion, and resilience against fatigue. These qualities are crucial in engineering materials (Fig. 8.1) that are capable of withstanding challenging environments [14–16]. The use of lightweight alloys plays a significant role in the reduction of fuel consumption and greenhouse gas emissions by effectively decreasing the weight of cars and buildings [17, 18]. These materials and production practises are in line with the worldwide movement towards sustainability. Certain lightweight alloys exhibit favourable characteristics for use in medical settings, particularly in cases where biocompatibility is of utmost importance. Titanium alloys are extensively employed in the field of orthopaedic implants owing to their exceptional biocompatibility and robust mechanical properties [19]. The advancement of innovative manufacturing processes, such as additive manufacturing or 3D printing, has been propelled by the use of lightweight alloys. This has facilitated the fabrication of intricate and lightweight structures with exceptional accuracy [20]. The aerospace sector derives advantages from the use of lightweight alloys, as they facilitate the development and production of aircraft that possess diminished mass while upholding their structural soundness [21–23]. As a consequence, there is an increase in the distance that can be covered during flights and a reduction in expenses [24]. The use of lightweight alloys plays a crucial role in enhancing the fuel efficiency of cars. Manufacturers boost the overall performance of vehicles by reducing weight through the substitution of heavy materials with lightweight alloys [25]. The consolidation of lightweight alloy powders presents issues because to their low density, material reactivity, and the inherent difficulties associated with obtaining complete density during the sintering process [26, 27]. Researchers and engineers have been actively engaged in developing inventive strategies, such as surface modification techniques, novel binder systems, and sophisticated production procedures, to surmount these challenges. The consolidation of lightweight alloy powders holds great importance in the field of materials science and engineering, as it exerts a substantial influence on several sectors. The capacity to use the remarkable characteristics of these alloys while surmounting their distinct obstacles is vital in propelling technology forward, mitigating environmental repercussions, and determining the trajectory of engineering and materials research. Researchers and engineers persist in investigating novel methodologies to enhance

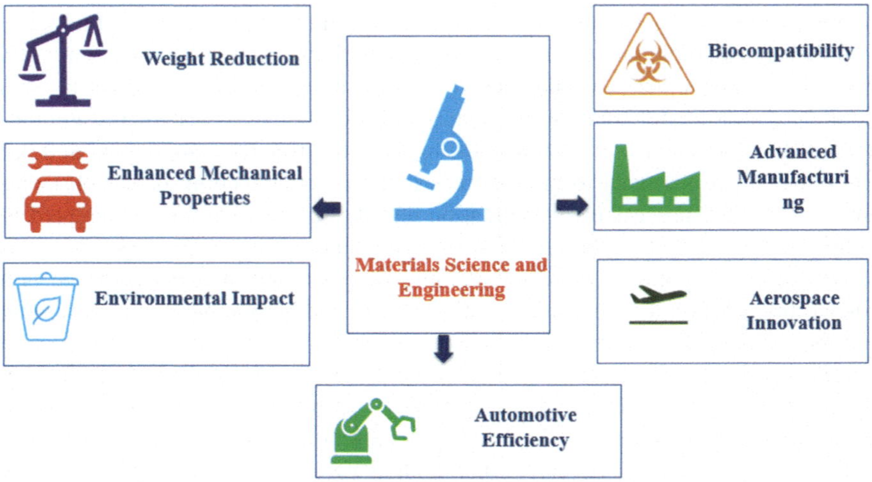

Fig. 8.1 Significance in materials science and engineering

the accessibility and dependability of these alloys, hence facilitating the progression towards a more promising and sustainable future.

8.2 Powder Preparation and Modification

The process of powder preparation and modification plays a crucial role in the consolidation of lightweight alloy powders, as it helps to address the difficulties encountered during pressing and sintering. The procedures encompass the manipulation and modification of the alloy powders in order to enhance their appropriateness for future consolidation. The process of powder preparation and modification involves a range of procedures and tactics that are designed to improve the flowability, manage reactivity, and enhance the adhesion between particles. The controlling of particle size distribution is a crucial factor in attaining consistent and compacted materials during the process of consolidation. Methods like as milling and mechanical alloying can be utilized to decrease particle size and attain a more uniform powder feedstock. Surface modifications can be implemented on lightweight alloy powders in order to modify their surface chemistry and improve their ability to flow [28]. This encompasses several procedures like as coating, chemical passivation, and oxidation reduction, which are employed to enhance the interparticle contact. The manipulation of alloy powder characteristics can be achieved by the integration of additives or reinforcements, such as nanoparticles or micro-scale reinforcements. The use of additives has the potential to augment the strength, wear resistance, or other desirable attributes of the ultimate component. The composition of the alloy can be modified in order to enhance its appropriateness for consolidation procedures. This process may entail the incorporation of additional parts in order to augment characteristics such as sinterability and control of reactivity. In some instances, it is possible for powder particles to have irregular shapes [29, 30]. The use of spheroidization techniques may be applied to generate particles that possess a higher degree of sphericity and uniformity. This, in turn, can enhance the flowability and packing density of powders. The consolidation process may be impeded by the agglomeration of powder particles [31]. Commonly employed techniques for the prevention or disintegration of agglomerates include ultrasonication and sifting. The selection of binders or temporary bonding agents plays a critical role in the consolidation process of lightweight alloy powders. The use of binders plays a crucial role in preserving the structural integrity of the densely packed powder prior to the sintering process, hence exerting a substantial influence on the ultimate quality of the product [32]. When there is a need to address the reactivity of alloys, the process of consolidation can be carried out in controlled atmospheres, such as inert gas settings. This is done in order to minimise any undesired chemical interactions that may occur between the particles. The use of nanoscale powders or processing processes has the potential to provide nanostructured materials that possess distinctive features, such as enhanced mechanical strength and wear resistance. Following the consolidation process, other post-processing treatments like as heat treatment or mechanical manipulation can be employed to further alter the microstructure and characteristics of the consolidated components [33]. The attainment of desirable properties in consolidated lightweight alloy components heavily relies on the utilisation of efficient powder preparation and modification methods. These approaches are not only crucial for tackling issues associated with low-density alloy powders but also have a significant impact on customising materials for specific applications in sectors such as aerospace, automotive, and other related fields. Scholars and practitioners persist in their exploration of novel methodologies within this domain in order to fully harness the inherent capabilities of lightweight alloys [34, 35] (Fig. 8.2).

8.3 Binder Systems

Binder systems play a crucial role in the consolidation of lightweight alloy powders, especially in applications such as powder metallurgy and additive manufacturing. These methods utilise temporary binding agents to facilitate the cohesion of powder particles during the first shaping process, such as pressing or 3D printing, prior to the ultimate consolidation achieved by sintering or other applicable procedures. The selection of a binder system significantly influences the consolidation process in terms of its quality, integrity, and efficiency. The selection of an appropriate binder is of utmost importance. Binders can exhibit organic, inorganic, or hybrid characteristics, contingent upon their intended use. Organic binders, such as waxes, polymers, and stearates, are frequently employed in many applications. On the other hand, inorganic binders, such as colloidal silica or alumina, are also commonly utilised. The composition of the binder can be customised to align with the specific alloy and

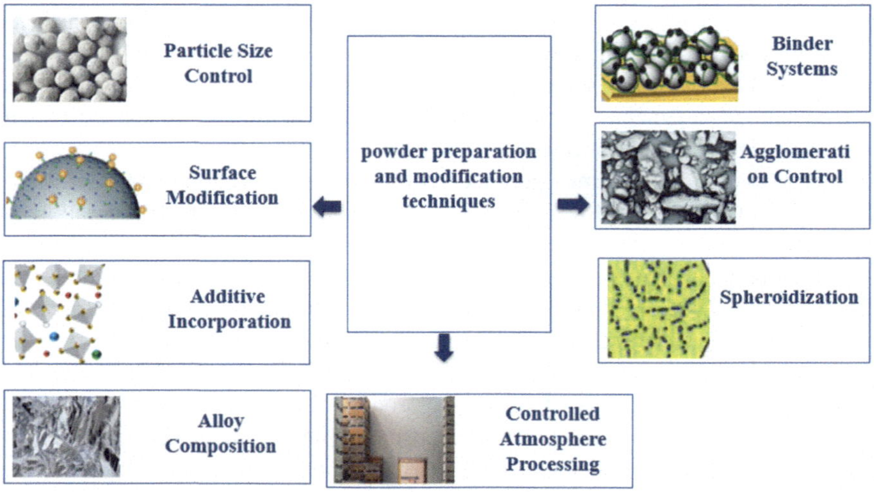

Fig. 8.2 Effective powder preparation and modification techniques

consolidation technique employed. For example, binders possessing a high carbon content may exhibit enhanced suitability for sintering operations, whereas water-based binders are commonly employed in green forming methodologies. In the process of shaping, binders are commonly used with alloy powder to generate a feedstock. The composition of the binder, encompassing its viscosity and incorporation of additives, plays a pivotal role in attaining desirable flow characteristics and adhesion properties. The flowability and formability of the binder-powder combination are determined by its rheological characteristics [36–38]. The precise management of binder rheology is crucial in order to achieve uniform and flawless compaction. In order to prevent contamination and facilitate the sintering process, it is necessary to remove binders after the shaping stage. The process of debinding typically entails a carefully regulated thermal procedure, occasionally incorporating successive phases of heating to progressively dissolve the binder substance. Certain binder systems may have sintering aids that serve to assist the process of sintering [39]. These additives have the potential to enhance the process of particle coalescence and densification when undergoing sintering. The cohesive force exerted by the binder in maintaining the integrity of the powder particles is a critical determinant. The material should possess sufficient strength to retain its form during the processing stage, while also exhibiting a degree of weakness that facilitates effortless debinding and sintering. The quality of the final consolidated component is influenced by the effectiveness of binder removal, which is commonly achieved using heat or chemical methods. The presence of residual binder traces has the potential to result in flaws and diminished mechanical characteristics [40]. The significance of environmental sustainability in the selection of binder systems is growing. Water-based or solvent-free binders are frequently favoured in order to reduce their environmental effect. The use of binder

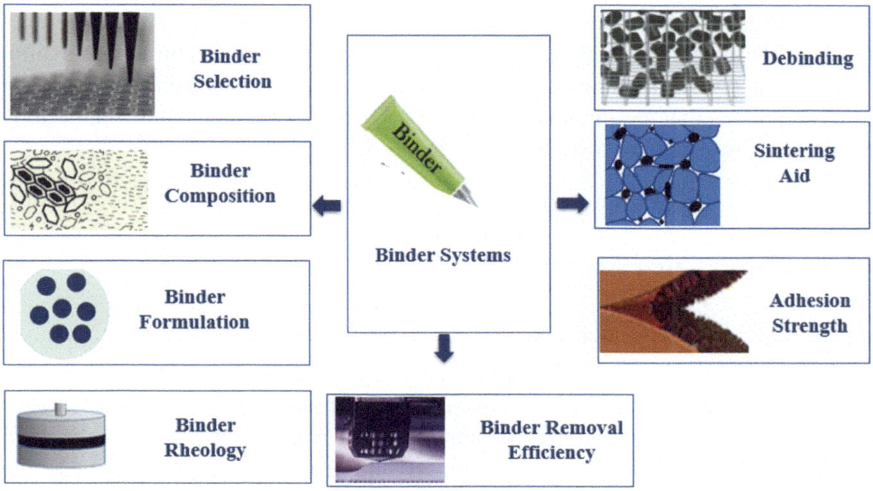

Fig. 8.3 Binder systems

systems can have a substantial influence on both the financial expenses and operational effectiveness of the consolidation procedure. The careful management of cost-effectiveness and performance is a crucial factor to be taken into account. The selection of an appropriate binder system is heavily contingent upon the particular alloy, technique of consolidation, and desired characteristics of the resulting component. In recent years, there has been a significant emphasis on the advancement of binder systems that exhibit enhanced sustainability, superior adhesion properties, and the ability to be customised to meet the specific demands of consolidating lightweight alloys. Binder systems play a crucial role in addressing the difficulties presented by low-density alloy powders and guaranteeing the manufacturing of superior components across diverse sectors [41] (Fig. 8.3).

8.4 Additive Manufacturing

The process of additive manufacturing, also known as 3D printing, has brought about a significant transformation in the consolidation of lightweight metal powders into intricate components. The utilisation of this technology has several benefits, such as the capacity to generate complex geometries, minimise material wastage, and fabricate lightweight structures that exhibit improved mechanical characteristics. There are two prominent additive manufacturing processes that are employed for the purpose of combining lightweight alloy powders, namely powder bed fusion and binder jetting [42, 43]. The process of additive manufacturing, also known as 3D printing, has brought about a significant transformation in the consolidation of lightweight metal powders into intricate components. The utilisation of this technology has several benefits, such as the capacity to generate complex geometries, minimise material wastage, and fabricate lightweight structures that exhibit improved mechanical characteristics. There are two prominent additive manufacturing processes that are employed for the purpose of combining lightweight alloy powders, namely powder bed fusion and binder jetting [44].

8.4.1 Powder Bed Fusion

Powder bed fusion is a widely utilised additive manufacturing technology employed for the consolidation of powders composed of lightweight alloys. The construction platform is coated with a homogeneous coating of metal powder. The thickness of the powder layer is often found to be below 100 μm. A highly energetic source, such as a laser or an electron beam, is accurately directed across the powder bed. The energy source exhibits the ability to discriminate between alloy powder particles, causing them to undergo sintering or melting processes, resulting in the formation of a cohesive and compacted layer [45]. Following the processing of each layer, the build platform undergoes a downward movement, thereby facilitating the application of a fresh coating of powder onto its surface. The aforementioned procedure is iteratively executed, with each layer being successively added until the entirety of the component is fabricated [46]. The utilisation of Powder Bed Fusion (PBF) enables the production of complex and lightweight structures that would present significant difficulties or even be unattainable using conventional manufacturing techniques. Powder bed fusion (PBF) is an additive manufacturing technique that selectively fuses or sinters essential materials, hence reducing material waste. This characteristic renders PBF a viable and sustainable alternative for the production of lightweight alloys. Powder bed fusion (PBF) technology allows for the fabrication of bespoke components that are specifically designed for certain purposes. This capability is especially advantageous in sectors such as aerospace and medicine [47].

8.4.2 Binder Jetting

Binder jetting is a commonly used additive manufacturing method that is utilised to consolidate powders of lightweight alloys. In an approach identical to Powder Bed Fusion (PBF), a slender coating of alloy powder is applied to the construction platform. Instead of using energy for the fusion of the powder, a liquid binding agent, commonly referred to as a binder, is administered selectively to the layer of powder. The binder serves to establish cohesive forces between the individual powder particles, resulting in their consolidation. The construction platform is systematically lowered, and the procedure is iteratively performed layer by layer, whereby each subsequent layer is selectively bonded by the liquid binder. Binder jetting is often acknowledged to possess superior speed and cost-effectiveness compared to other additive manufacturing methodologies, rendering it a highly appealing choice for the fabrication of lightweight alloys. The process of binder jetting demonstrates the ability to effectively handle a diverse array of alloy powders, hence offering considerable flexibility in the choice of materials. Complex geometries can be fabricated with little post-processing, hence minimising the requirement for supplementary machining or finishing operations. Although additive manufacturing has benefits in terms of geometric precision and efficient use of materials, it is sometimes necessary to perform additional procedures after the printing process, such as heat treatment or surface finishing, in order to get the appropriate material characteristics and surface quality [48]. The selection of alloy powder and binder is of utmost importance because of its significant impact on the sintering characteristics and ultimate qualities of the component. The use of additive manufacturing methods, such as powder bed fusion and binder jetting, has greatly broadened the potential for combining lightweight alloy powders. These approaches facilitate the development of intricate, low-weight components while minimising material wastage, hence creating opportunities for innovation in industries that heavily depend on such sophisticated materials. Researchers and engineers persist in investigating and enhancing these techniques to maximise their use in particular lightweight alloy systems and applications [49].

8.5 Microstructure Analysis

The examination of microstructure plays a pivotal role in comprehending the consolidation mechanism of lightweight alloy powders and its consequential effects on the ultimate material characteristics [50, 51]. The examination of microstructure evolution yields significant insights into the effect of several conditions, including temperature, pressure, and time, on the properties of consolidated components [52, 53]. The mechanical, thermal, and other qualities of a material are directly influenced by its microstructure. The optimisation of particular qualities for various applications may be achieved by comprehending the impact of consolidation factors on microstructure [54, 55]. The examination of microstructure has the potential to detect imperfections, including porosity, fractures, and inclusions, that must be addressed in order to guarantee the structural soundness of the consolidated components. Through the monitoring of microstructure evolution, producers are able to guarantee a consistent level of quality in their goods, hence adhering to industry norms and criteria. Microstructure analysis plays a crucial role in identifying the underlying reasons for component failure, hence facilitating the prevention of future failures and enhancing the efficacy of design and manufacturing processes [56–58].

8.5.1 Methods of Microstructure Analysis

Optical microscopy is a widely employed technique for the examination of the microstructure of solidified materials, particularly at low levels of magnification. This technique facilitates the examination of grain structure, porosity, and other macroscopic characteristics. The arrangement and size of grains in a metal or alloy have a crucial role in determining its mechanical characteristics, making it particularly vital in the field of metallurgy [59]. Through the analysis of grain boundaries, crystallographic orientation, and grain size, scholars are able to get enhanced insights into the mechanical response and performance of materials under many environmental circumstances. The identification and spatial arrangement of pores or spaces inside a material play a pivotal role in ascertaining its characteristics. In the field of ceramics, it has been shown that the mechanical strength and thermal conductivity of a material may be influenced by its porosity [60]. The utilisation of optical microscopy enables the quantification and analysis of material porosity. Optical microscopy may be utilised as a valuable tool for the characterization of several microstructural attributes, including inclusions, phases, and defects, in addition to grains and holes. This aspect has significant importance in the domains of quality control and research, particularly within businesses such as steel manufacture or concrete production. Although electron microscopes generally offer higher magnification than optical microscopes, the latter are well-suited for the examination of macroscopic features and microstructures on a larger scale. The use of Scanning Electron Microscopy (SEM) is a very effective imaging methodology that offers the capability to generate high-resolution visual representations of the microstructure of various materials. The utilisation of scanning electron microscopy (SEM) proves to be highly advantageous in the examination of intricate microstructural features such as grain boundaries, phase distribution, and microstructural flaws [61, 62]. The utilisation of scanning electron microscopy (SEM) is prevalent in the examination of grain boundaries inside various materials. Through the examination and evaluation of the topographical features and chemical composition of grain boundaries, scholars are able to get valuable knowledge on the characteristics and performance of materials. The comprehension of the mechanical characteristics of materials is of utmost importance [63, 64]. Scanning electron microscopy (SEM) is a powerful technique that can be utilised to analyse the distribution of various phases inside a material. This capability is particularly useful in the identification and quantification of specific components present in composite materials or multiphase alloys. Scanning electron microscopy (SEM) is a very effective technique for the identification and characterization of microstructural imperfections such as voids, fractures, dislocations, and inclusions. The acquisition of this information holds significant importance in the context of quality control and the comprehension of material failure processes [65]. The use of scanning electron microscopy (SEM) enables the investigation of several surface characteristics, including but not limited to roughness, texture, and contamination. This has significant importance across diverse industries, encompassing semiconductor fabrication and materials exhibiting specialised surface features. Transmission Electron Microscopy (TEM) is a very effective method employed in the investigation of intricate microstructural characteristics, including minute features at the nanoscale. The device has high-resolution capabilities and enables the examination of many microstructural attributes, including dislocations, precipitates, and other features at the nanoscale level. Transmission electron microscopy (TEM) is particularly advantageous in the investigation of nanoscale characteristics present in various materials. The utilisation of this technique enables the elucidation of the organisation and characteristics of nanoparticles, nanocrystals, and other diminutive structures, hence playing a crucial role in the field of nanomaterials investigation. Transmission electron microscopy (TEM) is a pivotal methodology employed in the investigation of dislocations inside various materials. The technique has the capability to furnish comprehensive data pertaining to the kind, concentration, and arrangement of dislocations, hence playing a crucial role in enhancing comprehension of the mechanical properties shown by materials. Electron microscopy (EM) is a valuable technique for investigating the characteristics of precipitates in materials, including their formation, size, distribution, and composition. The incorporation of this element is of utmost importance in the design and fabrication of high-strength alloys and materials that exhibit improved qualities. Transmission electron microscopy (TEM) facilitates the examination of crystallographic properties, such as the determination of a material's crystal structure, grain orientation, and phase identification. Electron microscopy (EM) is a widely utilised technique for the characterization of many types of defects, including point defects, stacking faults, and twinning, which has a significant impact on the characteristics and behaviour of materials. X-ray diffraction (XRD) and energy-dispersive X-ray spectroscopy, specifically Energy Dispersive X-ray Spectroscopy (EDS), represent a pair of vital analytical methodologies employed in the realm of materials science and several other disciplines. These procedures serve the purpose of furnishing valuable insights pertaining to the composition and structural characteristics of diverse materials [66]. X-ray diffraction (XRD) is utilised as a technique for ascertaining the crystallographic properties and phase composition of various materials. The utilisation of this technique proves to be highly advantageous in the detection of distinct phases inside a compacted substance, as well as in comprehending the processes of phase transitions. Energy-dispersive X-ray spectroscopy (EDS) is a technique employed for the purpose of examining and determining the elemental makeup of various materials. Frequently, it is used with scanning electron microscopy (SEM) or transmission electron microscopy (TEM) to get data on the elemental composition of distinct phases inside a microstructure [67] (Fig. 8.4).

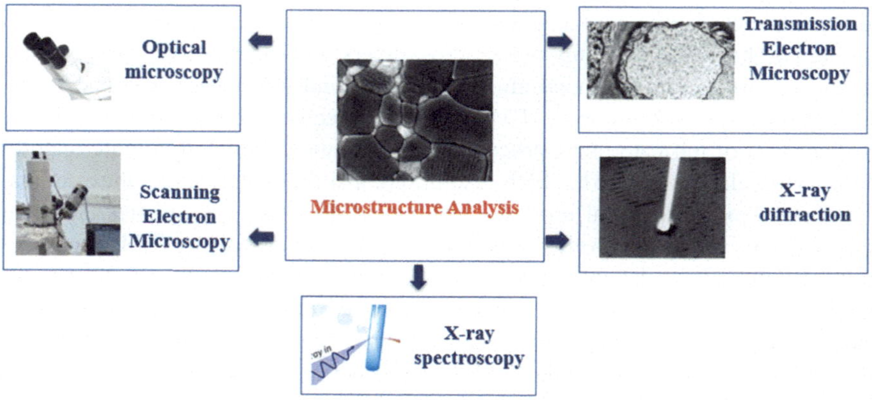

Fig. 8.4 Methods of microstructure analysis

8.6 Characterization of Mechanical and Thermal Properties

The assessment of the viability of consolidated lightweight alloy components for specific applications necessitates the characterization of their mechanical and thermal characteristics. These characteristics are influential factors in determining the performance, durability, and thermal behaviour of the material, and are crucial in ensuring that the components adhere to the necessary industry requirements. The evaluation of a material's capacity to endure different loads and situations relies on its mechanical qualities, which encompass strength, hardness, and ductility [68]. The comprehension of thermal characteristics, including thermal conductivity and expansion, is crucial in gaining insight into the behaviour of materials in response to variations in temperature. Ensuring compliance with mechanical and thermal requirements is of utmost importance in areas where safety and dependability are of fundamental concern, such as the aerospace and automotive sectors. The availability of precise property data allows engineers to enhance the design of components by carefully choosing materials that possess the specific mechanical and thermal attributes sought. Characterization plays a crucial role in ensuring the adherence of consolidated lightweight alloy components to prescribed quality requirements, hence mitigating the utilisation of inferior materials [69, 70].

8.6.1 Methods of Mechanical Property Characterization

Tensile tests are utilised to evaluate the mechanical characteristics of a material, encompassing its ultimate tensile strength, yield strength, and elongation. These tests offer valuable information on the behaviour of a material under axial loading

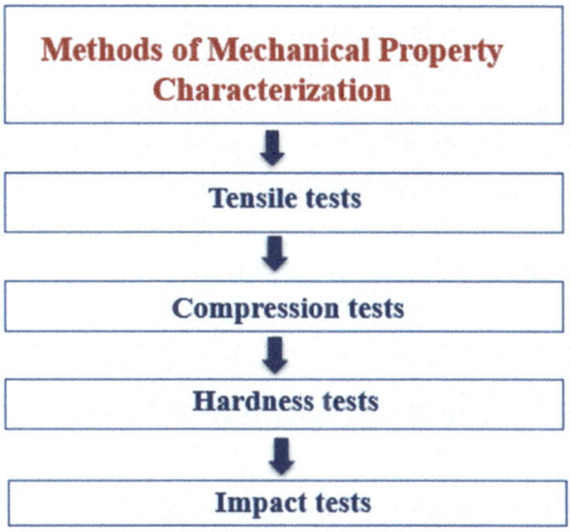

Fig. 8.5 Methods of mechanical property characterization

conditions. Compression tests are utilised to assess the ability of a material to withstand compressive pressures [71]. The aforementioned tests play a pivotal role in applications whereby components undergo crushing or compression. Hardness tests, such as the Rockwell or Brinell tests, are utilised to quantify the capacity of a material to withstand deformation or indentation. The evaluation of a material's capacity to withstand wear and maintain durability is crucial. Impact tests, such as the Charpy or Izod tests, are utilised to assess the ability of a material to withstand abrupt loading. This characteristic has significant relevance in situations when the capacity to endure shocks is of utmost importance [72–74] (Fig. 8.5).

8.6.2 Methods of Thermal Property Characterization

Thermal conductivity tests are employed to assess the heat conduction capacity of a certain material. This characteristic is crucial in scenarios where the dissipation of heat is of utmost importance. Thermal expansion tests are conducted to ascertain the alterations in the dimensions of a material in response to fluctuations in temperature [75, 76]. The attribute is of utmost importance for applications that undergo variations in temperature. Differential Scanning Calorimetry (DSC) is a technique used to quantify the heat exchange between a sample and its surroundings as a function of temperature. This method offers valuable insights into many thermodynamic properties, including phase transitions, heat capacity, and thermal stability. Thermogravimetric analysis (TGA) is a technique used to investigate the alterations in weight of a substance under controlled heating conditions. This source presents information pertaining to the behaviour of breakdown and oxidation. The utilisation of infrared and thermal imaging methodologies enables the visualisation and quantification of temperature distribution throughout the surface of a component, hence facilitating the detection of areas of elevated temperature and irregular thermal patterns. The acquisition of data pertaining to the mechanical and thermal properties of materials has significant importance in the processes of materials selection, quality control, and the assessment of suitability of lightweight alloy components for diverse applications [77–79]. These studies offer engineers and researchers useful data for making educated judgements on the appropriateness and effectiveness of materials, particularly in the context of lightweight alloys used in sectors such as aerospace, automotive, and manufacturing.

8.7 Environmental and Safety Considerations

The evaluation of the environmental and safety considerations associated with the consolidation methods employed in the production of lightweight alloy powders is of utmost importance in order to guarantee conscientious and sustainable manufacturing procedures. There are several crucial concerns that necessitate consideration, including the proper disposal of binders, the emission of volatile compounds, and the possible health risks related to the manipulation and treatment of lightweight alloy powders [80]. The utilisation of binders in consolidation procedures may give rise to environmental consequences. Certain binders have the potential to contain harmful substances or release volatile organic compounds (VOCs) throughout the manufacturing process. The appropriate disposal and recycling of these binders are necessary in order to mitigate the environmental consequences. It is vital to exert endeavours in order to mitigate the utilisation of binders and choose ecologically conscious alternatives [81, 82]. The use of waste reduction strategies and the optimisation of binder utilisation can make significant contributions towards achieving sustainability goals. In the context of consolidation procedures, particularly those using binders, it is imperative to exercise control over the emission of volatile organic compounds (VOCs) in order to mitigate the risk of air pollution. In order to effectively catch and treat these pollutants, it is imperative to implement appropriate ventilation and emission control systems [83]. The selection of binders can have a substantial impact on the release of volatile chemicals. The use of low-VOC binders or binders that can be readily removed during debinding processes serves to mitigate environmental and health hazards. The manipulation of lightweight alloy powders has the potential to create tiny particulate matter, hence presenting inhalation hazards to individuals in occupational settings. In order to effectively address health concerns, it is imperative to provide appropriate personal protective equipment, dust control measures, and comprehensive worker training. Certain lightweight alloys, such as those containing beryllium, exhibit pronounced toxicity and provide substantial health hazards. In order to ensure the safety of workers, it is imperative that handling, storage, and processing methods must follow rigorous safety rules. It is imperative for workers who are involved in the handling of lightweight alloy powders to utilise suitable respiratory protection to mitigate the risk of inhaling potentially hazardous particles [84]. To mitigate potential environmental and health hazards, it is imperative to adhere to local, regional, and national requirements when disposing of waste produced during consolidation procedures, including but not limited to old binders, unused powders, and hazardous substances. The use of sustainable practises in the selection of materials, design of processes, and management of waste is of utmost importance in order to mitigate the environmental impact of consolidation processes. Scientists and engineers are now investigating more ecologically sustainable options, including the use of water-based binders, the development of eco-friendly debinding techniques, and the implementation of responsible disposal procedures. Through a comprehensive evaluation and resolution of the environmental and safety factors associated with the amalgamation of lightweight alloy powders, producers and researchers may effectively safeguard the welfare of personnel, mitigate adverse ecological consequences, and uphold sustainability objectives. The adoption of green technologies and practises is an essential measure in promoting ethical and environmentally sensitive approaches to materials processing and production [85–87].

8.8 Comparative Analysis of Different Consolidation Methods

Conducting a comparative examination of various consolidation procedures pertaining to lightweight alloy materials is vital to choosing the most appropriate strategy for certain applications. The present study necessitates the consideration of many elements, encompassing cost, energy efficiency, and the overall sustainability of the process.

8.8.1 Powder Metallurgy

Powder metallurgy (PM) is a highly adaptable and extensively employed technique utilised in the fabrication of diverse materials, including lightweight alloys such as aluminium, magnesium, titanium, and beryllium alloys. The utilisation of this technology has several benefits in the manufacturing process of components and materials that possess distinct features. The initial stage involves the manufacturing of high-quality metal powders. These powders are commonly produced using techniques such as atomization, mechanical comminution (grinding), or chemical procedures. The blending of the powders may be conducted in order to get a consistent composition and uniform characteristics [88, 89]. This particular procedure holds significant importance in attaining the necessary material qualities. The mixture of powders is introduced into a die and exposed to elevated levels of pressure. The compaction process yields a "green compact" which maintains the finished part's shape while possessing a porous structure. The green compact undergoes sintering, a controlled environment process, wherein it is subjected to high temperatures. During the process of sintering, the metal particles undergo fusion, resulting in the elimination of porosity. The sintered component attains its ultimate density and mechanical characteristics. Additional processes, including machining, heat treatment, and surface treatments, may be carried out depending on the required requirements of the item. Powder metallurgy has the potential to offer cost-effective solutions, particularly in scenarios involving high-volume manufacturing. Tooling expenses are often incurred as a single, non-recurring expenditure, and the utilisation of materials is reasonably efficient. The energy efficiency of sintering processes is subject to variation based on the specific technology employed, such as conventional sintering or spark plasma sintering. The utilisation of lower temperatures during the sintering process has the potential to result in energy conservation. The sustainability of project management (PM) may be enhanced by utilising binders that are recyclable or have little environmental effect. The implementation of post-processing techniques for the recycling of surplus powders has the potential to mitigate waste generation [90, 91].

8.8.2 Additive Manufacturing

Additive manufacturing (AM) techniques such as powder bed fusion and binder jetting demonstrate suitability for the consolidation of lightweight alloys, particularly in scenarios involving intricate geometries and low-volume production. The initial capital investment required for acquiring additive manufacturing (AM) equipment might be substantial; nevertheless, this technology can prove to be economically advantageous when utilised for the production of bespoke or complex components. The reduction in material waste is seen. Additive manufacturing (AM) procedures have the potential for energy efficiency owing to their layer-by-layer additive nature, resulting in reduced energy usage when compared to subtractive manufacturing methods. Additive manufacturing (AM) techniques have the potential to generate reduced amounts of material waste and can contribute to environmental sustainability by utilising recyclable resources. Nevertheless, it is possible that post-processing will be necessary [92, 93].

8.8.3 Hot Isostatic Pressing

The Hot Isostatic Pressing (HIP) method is a materials processing technique that is employed to enhance the density and mechanical properties of various materials, with a special focus on metals, ceramics, and composites. This approach involves the simultaneous application of high temperature and high pressure. The use of Hot Isostatic Pressing (HIP) is implemented with the objective of enhancing the characteristics of materials, eliminating imperfections, and attaining increased density and uniformity. The hot isostatic pressing (HIP) technique is well-suited for the consolidation of lightweight alloy powders that possess intricate geometries, while also ensuring the preservation of near-net form integrity. The cost of HIP might be somewhat high as a result of the requirement for specialised equipment and consumables [94]. Nevertheless, the implementation of this approach leads to a reduction in waste and has the potential to be economically advantageous, particularly for crucial components. The process of hot isostatic pressing (HIP) often entails the use of elevated temperature and pressure, resulting in a considerable energy expenditure. The sustainability of HIP may be enhanced by the utilisation of recyclable materials and the use of waste reduction strategies. However, it is important to note that energy consumption is a significant obstacle to achieving sustainability goals [95].

8.8.4 Forging and Extrusion

Forging and extrusion are two distinct metalworking techniques that fulfil distinct objectives. The process of forging is predominantly employed for the purpose of fabricating solid components by subjecting them to compressive pressures. Conversely, extrusion is utilised to manufacture goods that possess a consistent cross-sectional form by compelling materials to pass through a die. Both processes play a crucial role in the manufacturing sector and have a wide range of applications across different sectors. The processes of forging and extrusion demonstrate favourable suitability for the consolidation of lightweight alloy powders, with a specific emphasis on enhancing mechanical qualities. The techniques have the potential to be economically efficient in the production of components with high strength. However, it is important to note that the expenses associated with tooling and equipment may be substantial. The energy efficiency of forging and extrusion operations can be quite high as a result of the localised heating and deformation mechanisms involved. Although the use of materials is effective, sustainability can be affected by energy-intensive heating procedures and tool wear [96, 97].

8.8.5 Investment Casting

The manufacturing method of investment casting, sometimes referred to as precision casting or lost-wax casting, is employed to fabricate precise and complicated components using a diverse array of materials, including lightweight alloys. This methodology provides a high level of precision in reproducing fine details, making it particularly suitable for components that require exact dimensional accuracy, surface quality, and intricate features. Investment casting demonstrates suitability for the production of intricate lightweight alloy components; nevertheless, its cost-effectiveness may not be as pronounced in comparison to alternative manufacturing techniques [98]. The cost of investment casting is elevated as a result of the requirement for tooling and the meticulousness involved in pattern fabrication. Low- to medium-volume manufacturing is considered optimal in this context. The energy efficiency of investment casting is contingent upon the dimensions and magnitude of the undertaking. Smaller, specialised facilities have the potential to exhibit greater efficiency. The sustainability of investment casting may be enhanced by the use of recyclable materials and minimising waste through the implementation of precision casting techniques. The selection of a consolidation process should be determined by the particular demands of the application, while considering limitations in cost, objectives for energy efficiency, and concerns for sustainability. Engineers and manufacturers may make well-informed judgements on the optimal consolidation technique for lightweight alloy materials, while simultaneously attaining the necessary material attributes and process efficiency, through the use of a thorough comparative study [99].

8.9 Consolidation of Lightweight Alloy Powders; Overcoming the Problems

The process of consolidating lightweight alloy powders, such as those composed of aluminium, magnesium, titanium, and beryllium alloys, can provide intricate difficulties owing to the distinctive obstacles arising from their low density and special material characteristics. The successful resolution of these issues encountered in the processes of pressing and sintering is of utmost importance in the production of components of superior quality, catering to diverse industrial sectors [100, 101].

8.9.1 Challenges During Consolidation

The low particle density of lightweight alloys presents a significant challenge in the attainment of high-density compacts through the processes of pressing and sintering. A decrease in density has the potential to result in increased porosity and diminished mechanical characteristics. The proposed solution involves the adjustment of pressing parameters, utilisation of suitable binders, and optimisation of sintering conditions in order to enhance the densification process. Certain lightweight alloys, such as magnesium and beryllium, have a high level of reactivity, which might result in unfavourable chemical interactions with the surrounding environment or binders during the consolidation process [102]. One potential solution is to conduct consolidation processes within controlled atmospheres, such as inert gas environments, in order to address and minimise any reactivity concerns. The flowability of lightweight alloy powders during the pressing process may be hindered, resulting in the formation of non-uniform compacts and an elevated occurrence of flaws. The proposed solution involves the use of appropriate powder modification techniques, such as surface treatments and particle size manipulation, with the aim of improving the flowability of the powder. Obtaining the intended microstructure and mechanical qualities in consolidated components might pose difficulties, particularly in alloys characterised by intricate phase transitions [103, 104]. The proposed solution involves acquiring a comprehensive understanding of the phase diagrams of the alloy in question and employing sophisticated heat treatment techniques throughout the sintering process in order to effectively regulate the microstructure. The selection of binders has significant importance, since the use of suboptimal binders may lead to inadequate compaction and unfavourable chemical reactions with the alloy. The proposed solution involves the careful selection of binders that are compatible with the specific alloy and sintering circumstances, while also prioritising environmental sustainability [105, 106].

8.9.2 Methods to Overcome Challenges

The enhancement of flowability, reactivity control, and densification in alloy powders may be achieved by employing several strategies, including but not limited to particle size manipulation, surface modification, and the incorporation of sintering aids. Optimise the sintering procedure by adjusting key parameters such as temperature, duration, and gas composition, to enhance the compactness of the material and regulate its microstructural characteristics. For instance, the utilisation of pressure-assisted sintering techniques like pressure-assisted sintering or spark plasma sintering has been seen to significantly augment the densification process [40, 107]. The process of consolidation under-regulated atmospheres, such as vacuum or inert gas environments, is employed to reduce the reactivity of extremely reactive alloys, such as magnesium or beryllium. Choose binders that are suitable for the specific alloy and intended use, with the aim of facilitating effective removal throughout the debinding and sintering processes. This study aims to investigate several additive manufacturing processes, including powder bed fusion and binder jetting, that have the potential to facilitate the accurate consolidation of lightweight alloys while mitigating some difficulties commonly encountered in conventional production approaches. Incorporate stringent quality control protocols to identify and rectify any concerns throughout the consolidation procedure, hence decreasing the likelihood of manufacturing flawed components [40]. Engaging in collaborative efforts with materials scientists and engineers that possess expertise in lightweight alloys can yield significant insights into the behaviour of alloys and offer viable resolutions to issues associated with consolidation. The resolution of challenges faced in the pressing and sintering of lightweight alloy powders necessitates the integration of materials knowledge, process refinement, and novel methodologies [108, 109]. By effectively overcoming these aforementioned obstacles, it becomes feasible to exploit the remarkable characteristics exhibited by lightweight alloys, rendering them appropriate for a wide range of applications in sectors such as aerospace, automotive, medical devices, and other related fields.

8.10 Applications Consolidation of Lightweight Alloy Powders

The process of consolidating lightweight alloy powders, such as those composed of aluminium, magnesium, titanium, and beryllium alloys, has significant implications across many sectors, notwithstanding the inherent difficulties encountered during pressing and sintering. The manufacturing of specialised applications can be enhanced through the resolution of challenges associated with low-density alloy powders. This can lead to the production of novel and high-performance components [4]. The use of consolidated lightweight alloy components has significant importance in the design of aircraft and spacecraft. Alloys of this nature find application in the construction of aeroplane frames, engine components, and spaceship constructions, primarily with the aim of minimising weight and augmenting fuel efficiency. The automotive sector employs lightweight alloys as a means to decrease the overall weight of cars, hence resulting in enhanced fuel efficiency and less emissions [110]. Consolidation procedures can yield advantages for many components such as engine blocks, wheels, and chassis pieces. Lightweight alloys, particularly titanium has favourable characteristics for the production of biocompatible implants, including orthopaedic implants, dental prosthesis, and surgical equipment. The consolidation of lightweight alloy powders may be employed to fabricate heat sinks, which are utilised in electronic equipment for the purpose of improving thermal management and mitigating the risk of overheating [111]. Lightweight alloys are extensively utilised in defence industries to fabricate aircraft, missiles, and other high-performance equipment, wherein the imperative need for enhanced mobility and efficiency necessitates the incorporation of decreased weight materials. Lightweight alloy consolidation can provide advantages for components like turbine blades and engine casings, as they necessitate a balance between high-temperature resistance and low weight. The sports sector uses lightweight alloys in the manufacturing of various equipment, including bicycle frames, golf club heads, and tennis racket frames [112]. The use of lightweight alloy materials in electrochemical devices has been shown to enhance both energy storage and conversion efficiency, rendering them highly appropriate for applications such as batteries, fuel cells, and capacitors. The utilisation of lightweight alloy components in the fabrication of machinery can mitigate wear and prolong the lifespan of tooling and dies, hence leading to cost reductions and enhanced operational efficacy. Beryllium alloys, renowned for their exceptional strength-to-weight ratio, find application in the oil and gas industry for the production of lightweight drilling and exploration equipment [113]. The use of lightweight alloys is of utmost importance in the fabrication of spacecraft and satellites, as the imposition of weight limitations plays a crucial role in both the expenses associated with launching and the overall effectiveness of operations. The utilisation of lightweight alloys in the fabrication of wind turbine components, including blades and nacelles, serves the purpose of diminishing total weight and enhancing operational efficiency. Lightweight alloys are utilised in consumer electronics with the purpose of attaining casings and components that are both sturdy and lightweight [114]. It is important to acknowledge and tackle the obstacles linked to the pressing and sintering processes of lightweight alloy powders in these particular applications, as this is crucial for guaranteeing that components adhere to the necessary benchmarks in terms of performance, strength, and longevity. With the continuous advancement of technology and materials science, there is a growing potential for enhanced efficiency and cost-effectiveness in the consolidation techniques employed for alloys. This progress holds promise for the facilitation of further innovation and the promotion of sustainability in many sectors.

8.11 The Identification of Prospective Research Opportunities and Areas for Advancement

The identification of prospective research avenues and areas for more improvement in the consolidation of lightweight alloy powders, while effectively addressing the obstacles encountered throughout the processes of pressing and sintering, holds the potential to make significant contributions to the fields of materials science, engineering, and diverse industries. Consolidation methods that may be improved upon in terms of efficiency, accuracy, and power consumption should be actively sought and developed [115]. Methods like powder bed fusion, binder jetting, and hybrid processes are just a few examples of how this may be done. The use of nanoparticles and nanostructures in lightweight alloys has yet to be explored to its full potential. Consolidation approaches may be used to take advantage of our growing understanding of how nanoscale characteristics affect material properties. Investigate the consolidation process of lightweight alloy powders when combined with other materials, such composites, ceramics, or polymers [116]. The development of multi-material components has the potential to provide improved characteristics and usefulness. Focusing on eco-friendly binders, debinding techniques, and energy savings, research consolidation procedures that are gentler on the planet. Sustainability can also benefit from studies on material reuse and recycling. To get the most out of your consolidation efforts, you should use digital twin and computational modelling tools [117]. Microstructural evolution and material characteristics may be predicted and manipulated with the use of these methods. The objective is to devise in-situ monitoring methodologies that offer instantaneous input on the consolidation process. The use of this approach can facilitate the identification and resolution of emerging difficulties, hence enhancing the overall quality of integrated components. Utilise materials informatics and data-driven methodologies to expedite the process of materials identification and enhance the optimisation of consolidation processes [118]. The utilisation of machine learning and artificial intelligence (AI) has the potential to contribute significantly to the design and analysis of sophisticated alloys, as well as the accurate prediction of their consolidation behaviour. Additive manufacturing strategies for the consolidation of lightweight alloys should be investigated and improved. Explore the potential of 3D printing to fabricate intricate geometries and components with specialised attributes. Create streamlining techniques that work well for rapid prototyping and low-volume manufacturing to serve sectors where quick design iterations and small-batch production are essential. Each of these lines of inquiry might lead to improvements in the functionality, durability, and versatility of lightweight alloys. These advancements would not be possible without the close cooperation of materials scientists, engineers, and researchers from a wide range of fields to overcome the particular obstacles presented by the compacting of lightweight alloy powders [119].

8.12 Conclusion

Lightweight alloy powder consolidation, including aluminium, magnesium, titanium, and beryllium alloys, is a key field of R&D with broad industrial applications. In order to make use of the remarkable features these low-density materials provide, such as lightweight, high strength, and corrosion resistance, it is necessary to find solutions to the problems connected with pressing and sintering them. The aerospace, automotive, and medical device sectors have benefited greatly from the evolution and standardisation of lightweight metals. To improve performance and get around pressing and sintering issues, scientists are always looking for novel alloy compositions and material alterations to try. Newer methods of consolidation, including additive manufacturing, have made it possible to create more complex component designs and reduce waste. The use of these developments has the potential to revolutionise the production of aluminium and magnesium alloys. The development of environmentally friendly binders, energy efficient sintering techniques, and recycling practises are all examples of the growing focus on sustainability in this area of study. These actions are in line with international targets for environmental protection. Lightweight alloys must be consolidated with the utmost accuracy and quality. To guarantee that these parts are up to snuff, the industry requires a high level of microstructure and material property control. The consolidation of lightweight alloy powders is more than a mere scientific pursuit; it serves as a transformational influence that profoundly influences the processes of construction, manufacturing, and technological progress in many sectors. Through unwavering commitment and innovative thinking, the forthcoming era presents an auspicious prospect of surmounting the formidable obstacles associated with pressing and sintering processes. This achievement will pave the way for novel advancements in the realm of lightweight alloy applications, therefore making significant contributions towards fostering a more sustainable and efficient global environment.

References

1. Sathyanarayana, K., Puttegowda, M., Rangappa, S.M., Siengchin, S., Shivanna, P., Nagaraju, S.B., Somashekara, M.K., Girijashankar, P.B., Girijappa, Y.G.: Metallic lightweight materials: properties and their applications. In: Lightweight and Sustainable Composite Materials, pp. 47–67. Woodhead Publishing (2023)
2. Shashanka, R., Chaira, D.: Optimization of milling parameters for the synthesis of nano-structured duplex and ferritic stainless steel powders by high energy planetary milling. Powder Technol. **1**(278), 35–45 (2015)
3. Wei, L.K., Abd Rahim, S.Z., Al Bakri Abdullah, M.M., Yin, A.T., Ghazali, M.F., Omar, M.F., Nemeş, O., Sandu, A.V., Vizureanu, P., Abdellah, A.E.: Producing metal powder from machining chips using ball milling process: a review. Materials **16**(13), 4635 (2023)
4. Prasad, S.S., Prasad, S.B., Verma, K., Mishra, R.K., Kumar, V., Singh, S.: The role and significance of magnesium in modern day research-a review. J. Magnes. Alloys **10**(1), 1–61 (2022)
5. Agarwal, J., Sahoo, S., Mohanty, S., Nayak, S.K.: Progress of novel techniques for lightweight automobile applications through innovative eco-friendly composite materials: a review. J. Thermoplast. Compos. Mater **33**(7), 978–1013 (2020)
6. Busarac, N., Adamovic, D., Grujovic, N., Zivic, F.: Lightweight materials for automobiles. IOP Conf. Ser. Mater. Sci. Eng. **1271**(1), 012010 (2022). IOP Publishing
7. Sharath, B.N., Venkatesh, C.V., Afzal, A., Aslfattahi, N., Aabid, A., Baig, M., Saleh, B.: Multi ceramic particles inclusion in the aluminium matrix and wear characterization through experimental and response surface-artificial neural networks. Materials **14**(11), 2895 (2021)
8. Sharath, B.N., Madhu, K.S., Venkatesh, C.V.: Experimental study on dry sliding wear behaviour of Al-B4C-Gr metal matrix composite at different temperatures. Appl. Mech. Mater. **2**(895), 96–101 (2019)
9. Sharath, B.N., Venkatesh, C.V.: Study on effect of boron carbide, aluminium oxide and graphite on dry sliding wear behaviour of aluminium based metal matrix composite at different temperature. Tribol. Finn. J. Tribol. **38**(1–2), 35–46 (2021)
10. Ghassemieh, E.: Materials in automotive application, state of the art and prospects. New trends and developments in automotive industry. **20**, 365–394 (2011)
11. Sharath, B.N., Jeevan, T.P., Baig, M.A., Ashrith, H.S., Afzal, A., Reddy, A.R.: Machinability studies on boron carbide and graphite reinforced aluminium hybrid composites. Mater. Today: Proc. **1**(46), 8734–8741 (2021)
12. Sharath, B.N., Karthik, S., Pradeep, D.G., Madhu, K.S., Venkatesh, C.V.: Machinability studies on boron carbide and graphite reinforced Al7029-based hybrid composites. In: Materials, Design and Manufacturing for Sustainable Environment: Select Proceedings of ICMDMSE 2022, pp. 511–522. Springer Nature, Singapore (2022)
13. Tisza, M., Czinege, I.: Comparative study of the application of steels and aluminium in lightweight production of automotive parts. Int. J. Light. Mater. Manuf. **1**(4), 229–238 (2018)
14. Sharath, B.N., Venkatesh, C.V., Afzal, A., Baig, M.A., Kumar, A.P.: Study on effect of ceramics on dry sliding wear behaviour of Al-Cu-Mg based metal matrix composite at different temperature. Mater. Today: Proc. **1**(46), 8723–8733 (2021)
15. Madhu, K.S., Venkatesh, C.V., Sharath, B.N., Karthik, S.: Effect of boron carbide on wear resistance of graphite containing Al7029 based hybrid composites and its dry sliding wear characterization through experimental, response surface method and ANOVA. Tribol. Finn. J. Tribol. **38**(3–4), 48–60 (2021)
16. Madhu, K.S., Sharath, B.N., Venkatesh, C.V., Pradeep, D.G.: Evaluation of mechanical properties of ceramic reinforced aluminium-7029 hybrid composite. IOP Conf. Ser. Mater. Sci. Eng. **1189**(1), 012019 (2021). IOP Publishing
17. Salonitis, K., Pandremenos, J., Paralikas, J., Chryssolouris, G.: Multifunctional materials used in automotive industry: a critical review. In: Engineering Against Fracture: Proceedings of the 1st Conference 2009, pp. 59–70. Springer Netherlands (2009)

18. Chaira, D.: Development of nano-structured duplex and ferritic stainless steels by pulverisette planetary milling followed by pressureless sintering. Mater. Charact. **1**(99), 220–229 (2015)
19. Nagaraju, S.B., Priya, H.C., Girijappa, Y.G., Puttegowda, M.: Lightweight and sustainable materials for aerospace applications. In: Lightweight and Sustainable Composite Materials, pp. 157–178. Woodhead Publishing (2023)
20. Ngo, T.D., Kashani, A., Imbalzano, G., Nguyen, K.T., Hui, D.: Additive manufacturing (3D printing): a review of materials, methods, applications and challenges. Compos. B Eng. **15**(143), 172–196 (2018)
21. Chaira, D.: Phase transformation and microstructure study of nano-structured austenitic and ferritic stainless steel powders prepared by planetary milling. Powder Technol. **1**(259), 125–136 (2014)
22. Sharath, B.N., Madhu, K.S., Pradeep, D.G., Venkatesh, C.V.: Tribological Suitability of aluminium hybrid composite above atmospheric temperature. IOP Conf. Ser. Mater. Sci. Eng. **1189**(1), 012018 (2021). IOP Publishing
23. Ballupete Nagaraju, S., Kodigarahalli Somashekara, M., Puttegowda, M., Manjulaiah, H., Kini, C.R., Channarayapattana Venkataramaiah, V.: Effect of B_4C/Gr on hardness and wear behavior of Al2618 based hybrid composites through Taguchi and artificial neural network analysis. Catalysts **12**(12), 1654 (2022)
24. Madhu, K.S., Venkatesh, C.V., Sharath, B.N., Karthik, S.: Characterization and evaluation of mechanical properties of Al–Zn based hybrid metal matrix composites. Appl. Sci. Eng. Progress. **16**(1), 5804 (2022)
25. Rajendrachari, S., Kamacı, Y., Taş, R., Ceylan, Y., Bülbül, A.S., Uzun, O., Karaoğlanlı, A.C.: Antimicrobial investigation of CuO and ZnO nanoparticles prepared by a rapid combustion method. Phys. Chem. Res. **7**(4), 799–812 (2019)
26. Rajendrachari, S.: Effect of sintering temperature on the pitting corrosion of ball milled duplex stainless steel by using linear sweep voltammetry. Anal. Bioanal. Electrochem. **10**(3), 349–361 (2018)
27. Shashanka, R., Chaira, D.: Effects of nano-Y_2O_3 and sintering parameters on the fabrication of PM duplex and ferritic stainless steels. Acta Metall. Sin. (English Lett.) **29**, 58–71 (2016)
28. Nayak, A.K., Shashanka, R., Chaira, D.: Effect of nanosize yittria and tungsten addition to duplex stainless steel during high energy planetary milling. IOP Conf. Ser. Mater. Sci. Eng. **115**(1), 012008 (2016). IOP Publishing
29. Gupta, S., Shashanka, R., Chaira, D.: Synthesis of nano-structured duplex and ferritic stainless steel powders by planetary milling: An experimental and simulation study. IOP Conf. Ser. Mater. Sci. Eng. **75**(1), 012033 (2015). IOP Publishing
30. Pavitra, V., Praveen, B.M., Nagaraju, G., Shashanka, R.: Energy storage, photocatalytic and electrochemical nitrite sensing of ultrasound-assisted stable Ta_2O_5 nanoparticles. Top. Catal. **8**, 1–4 (2022)
31. Shashanka, R.: Synthesis of nano-structured stainless steel powder by mechanical alloying-an overview. Int. J. Sci. Eng. Res. **8**, 588–594 (2017)
32. Shashanka, R., Chaira, D.: Effect of sintering temperature and atmosphere on nonlubricated sliding wear of nano-yttria-dispersed and yttria-free duplex and ferritic stainless steel fabricated by powder metallurgy. Tribol. Trans. **60**(2), 324–336 (2017)
33. Rajendrachari, S.: An overview of high-entropy alloys prepared by mechanical alloying followed by the characterization of their microstructure and various properties. Alloys **1**(2), 116–132 (2022)
34. Shashanka, R., Chaira, D., Chakravarty, D.: Fabrication of nano-yttria dispersed duplex and ferritic stainless steels by planetary milling followed by spark plasma sintering and non-lubricated sliding wear behaviour study. J. Mater. Sci. Eng. B **6**, 111–125 (2016)
35. Hoppentocht, M., Hagedoorn, P., Frijlink, H.W., De Boer, A.H.: Technological and practical challenges of dry powder inhalers and formulations. Adv. Drug Deliv. Rev. **30**(75), 18–31 (2014)
36. Khalyfa, A., Vogt, S., Weisser, J., Grimm, G., Rechtenbach, A., Meyer, W., Schnabelrauch, M.: Development of a new calcium phosphate powder-binder system for the 3D printing of patient specific implants. J. Mater. Sci. Mater. Med. **18**, 909–916 (2007)

37. Mehdipour, I., Khayat, K.H.: Effect of particle-size distribution and specific surface area of different binder systems on packing density and flow characteristics of cement paste. Cement Concr. Compos. **1**(78), 120–131 (2017)
38. Daniel, M.A.: Polyurethane binder systems for polymer bonded explosives. DSTO (2006)
39. Lenka, R.K., Patro, P.K., Mahata, T.: Shape forming and sintering of ceramics. In: Handbook on Synthesis Strategies for Advanced Materials: Volume-II: Processing and Functionalization of Materials, pp. 1–54 (2022)
40. Ziaee, M., Crane, N.B.: Binder jetting: a review of process, materials, and methods. Addit. Manuf. **1**(28), 781–801 (2019)
41. Lewis, R.M.: Powder binder interactions in 3D inkjet printing. Chalmers University of Technology (2014)
42. Wong, K.V., Hernandez, A.: A review of additive manufacturing. Int. Sch. Res. Notices (2012)
43. Frazier, W.E.: Metal additive manufacturing: a review. J. Mater. Eng. Perform. **23**, 1917–1928 (2014)
44. Gibson, I., Rosen, D.W., Stucker, B., Khorasani, M., Rosen, D., Stucker, B., Khorasani, M.: Additive Manufacturing Technologies. Springer, Cham, Switzerland (2021)
45. Sun, S., Brandt, M., Easton, M.J.: Powder bed fusion processes: an overview. Laser Addit. Manuf. **1**, 55–77 (2017)
46. Sutton, A.T., Kriewall, C.S., Leu, M.C., Newkirk, J.W.: Powder characterisation techniques and effects of powder characteristics on part properties in powder-bed fusion processes. Virtual Phys. Prototyp. **12**(1), 3–29 (2017)
47. Singh, D.D., Mahender, T., Reddy, A.R.: Powder bed fusion process: a brief review. Mater. Today Proc. **1**(46), 350–355 (2021)
48. Gibson, I., Rosen, D., Stucker, B., Khorasani, M., Gibson, I., Rosen, D., Stucker, B., Khorasani, M.: Binder jetting. In: Additive Manufacturing Technologies, pp. 237–52 (2021)
49. Li, M., Du, W., Elwany, A., Pei, Z., Ma, C.: Metal binder jetting additive manufacturing: a literature review. J. Manuf. Sci. Eng. **142**(9), 090801 (2020)
50. Shashanka, R., Uzun, O., Chaira, D.: Synthesis of nano-structured duplex and ferritic stainless steel powders by dry milling and its comparison with wet milling. Arch. Metall. Mater. **65**(1) (2020)
51. Pradeep, N.B., Hegde, M.R., Rajendrachari, S., Surendranathan, A.O.: Investigation of microstructure and mechanical properties of microwave consolidated TiMgSr alloy prepared by high energy ball milling. Powder Technol. **1**(408), 117715 (2022)
52. Sharath, B.N., Madhu, K.S., Pradeep, D.G., Madhu, P., Premkumar, B.G., Karthik, S.: Effects of tertiary ceramic additives on the micro hardness and wear characteristics of Al2618+ Si_3N_4-B_4C-Gr hybrid composites for automotive applications. J. Alloy. Metall. Syst. **31**, 100014 (2023)
53. Nagaraju, S.B., Sathyanarayana, K., Somashekara, M.K., Pradeep, D.G., Puttegowda, M., Verma, A.: Artificial neural networks for predicting mechanical properties of Al2219-B4C-Gr composites with multireinforcements. Proc. Inst. Mech. Eng. Part C: J. Mech. Eng. Sci. 09544062231196038 (2023)
54. Nagaraju, S.B., Somashekara, M.K., Govindaswamy, P.D., Puttegowda, M., Girijashankar, P.B., Sathyanarayana, K.: Mechanical characterization of B4C-Gr Al2618 based composites synthesized by stir casting method. Appl. Sci. Eng. Prog. **16**(3), 6579 (2023)
55. Pradeep, D.G., Nithin, H.S., Sharath, B.N., Madhu, K.S., Venkatesh, C.V.: Microstructure and wear behavior of microwave treated WC-10Co-4Cr composite coating on AISI 4140 alloy steel. IOP Conf. Ser. Mater. Sci. Eng. **1189**(1), 012012 (2021). IOP Publishing
56. Shashanka, R., Chaira, D.: Ball Milled Nano-Structured Stainless Steel Powders: Fabrication and Characterization. Educreation Publishing (2017)
57. Mahale, R.S., Shashanka, R.: Mechanical testing of spark plasma sintered materials: a review. AIP Conf. Proc. **2469**(1) (2022). AIP Publishing
58. Adimule, V., Keri, R., Rajendrachari, S., Kendrekar, P., Kulkarni, C.V.: Highly photoluminescence and wide band gap insulating metal hybrid nanoparticles array of samarium-doped SrO: CoO: synthesis, characterizations and sensor characteristics. J. Mater. Sci. Mater. Electron. **34**(5), 442 (2023)

59. Liang, X.K., Zhu, Q., Luo, M., Li, D.Q.: Application of polarized light microscopy on microstructural characterization of semisolid slurry. Rare Met. **4**, 1–8 (2018)
60. Nagaraju, S.B., Somashekara, M.K., Govindaswamy, P.D., Puttegowda, M., Shankar, P.B., Sathyanarayana, K.: Wear behaviour of hybrid (boron carbide-graphite) aluminium matrix composites under high temperature. J. Eng. Appl. Sci. **70**(1), 1–23 (2023)
61. Sharath, B.N., Karthik, S., Madhu, P., Madhu, K.S., Pradeep, D.G.: Predictive analysis of slurry erosion behaviour in aluminium-based hybrid metal matrix composites: experimental and machine learning approach. J. Bio Tribo Corros. **9**(4), 70 (2023)
62. Sharath, B.N., Madhu, P., Verma, A.: Enhancing tribological performance: a review of ceramic reinforced aluminium hybrid composites for high-temperature engineering applications. Hybrid Adv. **1**, 100094 (2023)
63. Mohammed, A., Abdullah, A.: Scanning electron microscopy (SEM): a review. In: Proceedings of the 2018 International Conference on Hydraulics and Pneumatics—HERVEX, vol. 2018, pp. 7–9. Băile Govora, Romania (2018)
64. Inkson, B.J.: Scanning electron microscopy (SEM) and transmission electron microscopy (TEM) for materials characterization. In: Materials Characterization Using Nondestructive Evaluation (NDE) Methods, pp. 17–43. Woodhead Publishing (2016)
65. Ma, H., Shieh, K.J., Qiao, T.X.: Study of transmission electron microscopy (TEM) and scanning electron microscopy (SEM). Nat. Sci. **4**(3), 14–22 (2006)
66. Ural, N.: The significance of scanning electron microscopy (SEM) analysis on the microstructure of improved clay: an overview. Open Geosci. **13**(1), 197–218 (2021)
67. Popović, S.: Quantitative phase analysis by X-ray diffraction—doping methods and applications. Crystals **10**(1), 27 (2020)
68. Praveena, B.A., Buradi, A., Santhosh, N., Vasu, V.K., Hatgundi, J., Huliya, D.: Study on characterization of mechanical, thermal properties, machinability and biodegradability of natural fiber reinforced polymer composites and its applications, recent developments and future potentials: a comprehensive review. Mater. Today: Proc. **1**(52), 1255–1259 (2022)
69. Rajendrachari, S., Adimule, V.M., Jayaprakash, G.K., Pandith, A.: Electrochemical oxidation of methylene blue dye in wastewater using mechanically alloyed high entropy alloy modified carbon paste electrode using cyclic voltammetry. Mater. Res. Exp. **10**(5), 054003 (2023)
70. Rajendrachari, S., Basavegowda, N., Vinaykumar, R., Narsimhachary, D., Somu, P., Lee, M.J.: Electrocatalytic determination of methyl orange dye using mechanically alloyed novel metallic glass modified carbon paste electrode by cyclic voltammetry. Inorg. Chem. Commun. **30**, 111010 (2023)
71. Hemker, K.J., Sharpe, W.N., Jr.: Microscale characterization of mechanical properties. Annu. Rev. Mater. Res. **4**(37), 93–126 (2007)
72. Sudha, G.T., Stalin, B., Ravichandran, M., Balasubramanian, M.: Mechanical properties, characterization and wear behavior of powder metallurgy composites-a review. Mater. Today: Proc. **1**(22), 2582–2596 (2020)
73. Shashanka, R., Chaira, D.: Optimization of Milling Parameters. Ball Milled Nano-Structured Stainless Steel Powders: Fabrication and Characterization. 32
74. Conventional and Spark Plasma Sintering. Ball Milled Nano-Structured Stainless Steel Powders: Fabrication and Characterization. 54
75. Kanaginahal, G.M., Mahale, R.S., Vasanth, S., Sharath, P.C., Shashanka, R.: Studies on dielectric, super-capacitive and photoluminescence properties of metal organic frameworks subjected to varying temperature and frequencies. Nano Hybrids Compos. **30**(39), 17–25 (2023)
76. Jiang, P., Qian, X., Yang, R.: Tutorial: time-domain thermoreflectance (TDTR) for thermal property characterization of bulk and thin film materials. J. Appl. Phys. **124**(16) (2018)
77. Bauld, R., Choi, D.Y., Bazylewski, P., Divigalpitiya, R., Fanchini, G.: Thermo-optical characterization and thermal properties of graphene–polymer composites: a review. J. Mater. Chem. C **6**(12), 2901–2914 (2018)
78. Prime, R.B., Bair, H.E., Vyazovkin, S., Gallagher, P.K., Riga, A.: Thermogravimetric analysis (TGA). In: Thermal Analysis of Polymers: Fundamentals and Applications, pp. 241–317

79. Chen, C., Ma, X., Liu, K.: Thermogravimetric analysis of microalgae combustion under different oxygen supply concentrations. Appl. Energy **88**(9), 3189–3196 (2011)
80. Iqbal, A., Zhao, G., Suhaimi, H., He, N., Hussain, G., Zhao, W.: Readiness of subtractive and additive manufacturing and their sustainable amalgamation from the perspective of Industry 4.0: a comprehensive review. Int. J. Adv. Manuf. Technol. **111**, 2475–2498 (2020)
81. Huang, S.H., Liu, P., Mokasdar, A., Hou, L.: Additive manufacturing and its societal impact: a literature review. Int. J. Adv. Manuf. Technol. **67**, 1191–1203 (2013)
82. Alami, A.H., Olabi, A.G., Alashkar, A., Alasad, S., Aljaghoub, H., Rezk, H., Abdelkareem, M.A.: Additive manufacturing in the aerospace and automotive industries: recent trends and role in achieving sustainable development goals. Ain Shams Eng. J. **14**(11), 102516 (2023)
83. Miller, S.A., Myers, R.J.: Environmental impacts of alternative cement binders. Environ. Sci. Technol. **54**(2), 677–686 (2019)
84. Jafer, H.M., Atherton, W., Sadique, M., Ruddock, F., Loffill, E.: Development of a new ternary blended cementitious binder produced from waste materials for use in soft soil stabilisation. J. Clean. Prod. **20**(172), 516–528 (2018)
85. Provis, J.L., Palomo, A., Shi, C.: Advances in understanding alkali-activated materials. Cem. Concr. Res. **1**(78), 110–125 (2015)
86. Revathi, T., Vanitha, N., Jeyalakshmi, R., Sundararaj, B., Jegan, M., Rajkumar, P.K.: Adoption of alkali-activated cement-based binders (geopolymers) from industrial by-products for sustainable construction of utility buildings-A field demonstration. J. Build. Eng. **15**(52), 104450 (2022)
87. Plati, C.: Sustainability factors in pavement materials, design, and preservation strategies: a literature review. Constr. Build. Mater. **30**(211), 539–555 (2019)
88. Danninger, H., Calderon, R.D., Gierl-Mayer, C.: Powder metallurgy and sintered materials. Addit. Manuf. **19**(4) (2017)
89. Gummadi, J., Alanka, S.: A review on titanium and titanium alloys with other metals for biomedical applications prepared by powder metallurgy techniques. Mater. Today: Proc. (2023)
90. Popov, V.V., Grilli, M.L., Koptyug, A., Jaworska, L., Katz-Demyanetz, A., Klobčar, D., Balos, S., Postolnyi, B.O., Goel, S.: Powder bed fusion additive manufacturing using critical raw materials: a review. Materials **14**(4), 909 (2021)
91. Anderson, K., Weritz, J., Kaufman, J.G.: Aluminum Powder Metallurgy Materials and Processes.
92. Altıparmak, S.C., Yardley, V.A., Shi, Z., Lin, J.: Challenges in additive manufacturing of high-strength aluminium alloys and current developments in hybrid additive manufacturing. Int. J. Lightweight Mater. Manuf. **4**(2), 246–261 (2021)
93. Najmon, J.C., Raeisi, S., Tovar, A.: Review of additive manufacturing technologies and applications in the aerospace industry. In: Additive Manufacturing for the Aerospace Industry, pp. 7–31 (2019)
94. Atkinson, H.V., Davies, S.: Fundamental aspects of hot isostatic pressing: an overview. Metall. Mater. Trans. A **31**, 2981–3000 (2000)
95. Bocanegra-Bernal, M.H.: Hot isostatic pressing (HIP) technology and its applications to metals and ceramics. J. Mater. Sci. **39**(21), 6399–6420 (2004)
96. Kumar, R., Gupta, M.D.: Manufacturing Processes. PHI Learning Pvt. Ltd. (2014)
97. Cao, J., Banu, M.: Opportunities and challenges in metal forming for lightweighting: review and future work. J. Manuf. Sci. Eng. **142**(11), 110813 (2020)
98. Prasath, M.K., Vignesh, S.: A review of advanced casting techniques. Res. J. Eng. Technol. **8**(4), 440–446 (2017)
99. Richard, C.T., Kwok, T.H.: Rapid investment casting: design and manufacturing technologies. In: International Design Engineering Technical Conferences and Computers and Information in Engineering Conference, vol. 59179, p. V001T02A022. American Society of Mechanical Engineers (2019)
100. Krasnowski, M., Gierlotka, S., Ciołek, S., Kulik, T.: Nanocrystalline NiAl intermetallic alloy with high hardness produced by mechanical alloying and hot-pressing consolidation. Adv. Powder Technol. **30**(7), 1312–1318 (2019)

101. Antolak-Dudka, A., Krasnowski, M., Kulik, T.: Nanocrystalline Ni3Al intermetallic produced by hot-pressing consolidation of mechanically alloyed powders. Intermetallics **1**(42), 41–44 (2013)
102. Viswanathan, V., Laha, T., Balani, K., Agarwal, A., Seal, S.: Challenges and advances in nanocomposite processing techniques. Mater. Sci. Eng. R. Rep. **54**(5–6), 121–285 (2006)
103. Samal, P.K., Newkirk, J.: Introduction to full density powder metallurgy. Powder Metall. 253–254 (2015)
104. Sharma, A.S., Yadav, S., Biswas, K., Basu, B.: High-entropy alloys and metallic nanocomposites: processing challenges, microstructure development and property enhancement. Mater. Sci. Eng. R. Rep. **1**(131), 1–42 (2018)
105. Suryanarayana, C.: Mechanical alloying: a novel technique to synthesize advanced materials. Research (2019)
106. Tuncer, N., Bose, A.: Solid-state metal additive manufacturing: a review. Jom **72**(9), 3090–3111 (2020)
107. Mostafaei, A., Elliott, A.M., Barnes, J.E., Li, F., Tan, W., Cramer, C.L., Nandwana, P., Chmielus, M.: Binder jet 3D printing—process parameters, materials, properties, modeling, and challenges. Prog. Mater. Sci. **1**(119), 100707 (2021)
108. Ahmadi, M., Tabary, S.B., Rahmatabadi, D., Ebrahimi, M.S., Abrinia, K., Hashemi, R.: Review of selective laser melting of magnesium alloys: advantages, microstructure and mechanical characterizations, defects, challenges, and applications. J. Mater. Res. Technol. **19**, 1537–1562 (2022)
109. Moghimian, P., Poirié, T., Habibnejad-Korayem, M., Zavala, J.A., Kroeger, J., Marion, F., Larouche, F.: Metal powders in additive manufacturing: a review on reusability and recyclability of common titanium, nickel and aluminum alloys. Addit. Manuf. **1**(43), 102017 (2021)
110. Kumar, A., Gupta, M.: An insight into evolution of light weight high entropy alloys: a review. Metals **6**(9), 199 (2016)
111. Moghadasi, K., Isa, M.S., Ariffin, M.A., Raja, S., Wu, B., Yamani, M., Muhamad, M.R., Yusof, F., Jamaludin, M.F., bin Ab Karim, M.S., bin Yusoff. N.: A review on biomedical implant materials and the effect of friction stir based techniques on their mechanical and tribological properties. J. Mater. Res. Technol. **17**, 1054–1121 (2022)
112. Dhakate, S.R., Singh, B.P., Gupta, B.K., Subhedar, K.M., Srivastava, S.K., Saravanan, M., Saini, P., Kumar, S., Prathap, P., Kumari, S., Singh, R.K.: Advanced materials for strategic and societal applications. In: Metrology for Inclusive Growth of India., pp. 811–879 (2020)
113. Behera, A., Behera, A.: Advanced semiconductor/conductor materials. In: Advanced Materials: An Introduction to Modern Materials Science, pp. 557–596 (2022)
114. Saini, A., Pabla, B.S., Prakash, C., Singh, G., Pramanik, A. (eds.): Manufacturing Engineering and Materials Science: Tools and Applications. CRC Press (2023)
115. Henriques, A.C., Winkler, I.: The advancement of virtual reality in automotive market research: challenges and opportunities. Appl. Sci. **11**(24), 11610 (2021)
116. Mukherjee, D., Lim, W.M., Kumar, S., Donthu, N.: Guidelines for advancing theory and practice through bibliometric research. J. Bus. Res. **1**(148), 101–115 (2022)
117. Gucciardi, D.F., Gordon, S., Dimmock, J.A.: Advancing mental toughness research and theory using personal construct psychology. Int. Rev. Sport Exerc. Psychol. **2**(1), 54–72 (2009)
118. Cummings, E.M., Davies, P.T.: Effects of marital conflict on children: Recent advances and emerging themes in process-oriented research. J. Child Psychol. Psychiatr. **43**(1), 31–63 (2002)
119. Drakvik, E., Altenburger, R., Aoki, Y., Backhaus, T., Bahadori, T., Barouki, R., Brack, W., Cronin, M.T., Demeneix, B., Bennekou, S.H., van Klaveren, J.: Statement on advancing the assessment of chemical mixtures and their risks for human health and the environment. Environ. Int. **1**(134), 105267 (2020)

Chapter 9
Properties and Novel Applications of Mechanical Alloyed Ferrous Materials

Uzwalkiran Rokkala, Borra N. Dhanunjayarao, and Tirupathi Kadam

Abstract Mechanical alloying is a versatile and innovative method for synthesizing materials with unique properties by subjecting elemental or pre-alloyed powders to high-energy ball milling. This process has been extensively employed to engineer ferrous materials with enhanced properties, such as improved hardness, wear resistance, and corrosion resistance. In this chapter, we explore the properties and novel applications of mechanical alloyed ferrous materials. We delve into the microstructural changes, phase transformations, and mechanical behavior exhibited by these materials as a result of the mechanical alloying process. Additionally, we highlight emerging applications of these materials in various industries, including automotive, aerospace, and energy storage. The ability to tailor the composition and microstructure of ferrous materials through mechanical alloying opens up exciting opportunities for designing advanced materials with improved performance characteristics, ultimately driving innovation in engineering and technology.

Keywords Ferrous materials · Phase transformation · Microstructure · Corrosion · Wear

U. Rokkala (✉)
Department of Mechanical Engineering, Aditya Engineering College (A), Surampaelm, Andhra Pradesh 533437, India
e-mail: uzwalkiranr@adityauniversity.in

B. N. Dhanunjayarao
Department of Mechanical Engineering, Vignan's Institute of Information Technology, Visakhapatnam, Andhra Pradesh 530049, India

T. Kadam
Department of Mechanical Engineering, RGUKT, Basar, Nirmal, Telangana, India

S. Rajendrachari (ed.), *Mechanically Alloyed Novel Materials*, Advanced Structured Materials 220, https://doi.org/10.1007/978-981-97-6504-1_9

9.1 Microstructure of Mechanically Alloyed Ferrous Materials

The microstructure of mechanically alloyed ferrous materials plays a crucial role in determining their mechanical properties and overall performance [1]. It is important to analyse and understand the microstructure of these materials to optimize their production processes and tailor their properties for specific applications [2]. By examining the microstructure, one can characterize factors such as grain size, phase composition, and distribution of reinforcing particles—all of which directly impact the material's strength, hardness, ductility, and wear resistance [3, 4]. Additionally, studying the microstructure can provide insights into the presence of defects or impurities that could affect the material's performance and durability. Understanding the microstructure of mechanically alloyed ferrous materials is essential for optimizing their properties and ensuring their successful application in various industries, such as automotive, aerospace, and construction [5]. The microstructure of mechanically alloyed ferrous materials refers to the arrangement and distribution of atoms, grains, and phases within the material [6]. It can be characterized using various techniques such as scanning electron microscopy, transmission electron microscopy, X-ray diffraction, and other advanced imaging and analytical methods [7–11]. Analyzing the microstructure of mechanically alloyed ferrous materials provides valuable information about their mechanical properties, including strength, hardness, ductility, and wear [12].

9.1.1 Mechanically Alloyed Ferrous Materials: An Examination of Microstructure Development

The development of microstructure in mechanically alloyed ferrous materials involves several stages. Initially, during the milling process, the powders undergo severe plastic deformation, causing their grain sizes to reduce significantly [13]. As the milling continues, the repeated impacts and collisions between the powders promote atomic mixing and diffusion, leading to the formation of solid solutions and new phases with different crystal structures [14–16]. This results in a fine and homogeneous distribution of microstructures, with nanocrystalline grains and intermetallic compounds being commonly observed [17]. Further milling and annealing processes can help in refining the microstructure and altering the phase composition [18]. Ultimately, the microstructure of mechanically alloyed ferrous materials plays a critical role in determining their mechanical and physical properties, making it a key area of research and development in materials science and engineering. Understanding the microstructure of mechanically alloyed ferrous materials is crucial for optimizing their properties and applications [19]. By studying the microstructure development of mechanically alloyed ferrous materials, researchers can gain insights into the underlying mechanisms and develop techniques to optimize these materials for specific applications, such as improving their strength, hardness, corrosion resistance, and other desired properties. Therefore, the microstructure of mechanically alloyed ferrous materials is a vital aspect that needs to be carefully examined and understood for effective design and utilization of these materials in various industries and applications.

9.1.2 Characteristics of Ferrous Materials Post-mechanical Alloying

Characteristics of ferrous materials post-mechanical alloying include refined grain size, homogenized microstructure, enhanced solid solution formation, and distribution of alloying elements. These characteristics are achieved through the process of mechanical alloying, which involves high-energy milling of elemental powders to create a fine and uniform mixture. During mechanical alloying, the repeated impact and deformation of particles result in a high density of defects such as dislocations, grain boundaries, and stacking faults. These defects act as nucleation sites for the formation of new phases, leading to solid solution strengthening and improved mechanical properties. Additionally, the mechanical alloying process promotes grain refinement by inducing severe plastic deformation and fragmentation of powder particles. This refining of grain size enhances the material's strength and hardness, making it more resistant to deformation and wear [20]. Furthermore, the mechanical alloying process helps to homogenize the microstructure by promoting the diffusion of alloying elements and eliminating compositional variations within the material. The resulting microstructure of mechanically alloyed ferrous materials is characterized by a fine, uniform distribution of grains and phases, with enhanced solid solution formation and excellent dispersion of alloying elements.

Additionally, the microstructure of mechanically alloyed ferrous materials may also exhibit features such as intermetallic compounds, amorphous phases, and nanocrystalline structures, depending on the specific alloy composition and processing parameters [21, 22]. These features can significantly influence the material's mechanical properties and performance. Therefore, analyzing the microstructure of mechanically alloyed ferrous materials using techniques such as electron microscopy, X-ray diffraction, and other advanced imaging and characterization methods is crucial for understanding the structure–property relationships and optimizing the material's performance in various applications [23, 24]. The microstructure of mechanically alloyed ferrous materials is characterized by features such as refined grain size, homogenized microstructure, enhanced solid solution formation and dispersion of alloying elements, as well as the potential presence of intermetallic compounds, amorphous phases, and nanocrystalline structures. Overall, the microstructure of mechanically alloyed ferrous materials is highly influenced by the repeated impact and deformation during the mechanical alloying process.

9.1.3 Understanding the Effects of Mechanical Alloying on Ferrous Microstructures

The microstructure of mechanically alloyed ferrous materials plays a crucial role in determining their properties and performance [25, 26]. It influences their strength, hardness, ductility, and resistance to wear and corrosion. Furthermore, the microstructure can also impact other aspects such as grain size, phase composition, and the presence of defects or impurities. He et al. [27] observed that the increase of Fe content in the alloy led to the change in the phase of alloy powders. The singe-phase FCC solid solution changes into the single-phase BCC solid solution which may be due to the increase of Fe content in the alloy. The mechanical alloying was also responsible for the high lattice strain and crystallinity By analyzing the microstructure, researchers can gain valuable insights into the mechanisms involved in mechanical alloying and develop new techniques to optimize these materials for various applications, including automotive components, aerospace materials, and structural steels [28]. The microstructure of mechanically alloyed ferrous materials is typically characterized by a fine and uniform distribution of constituent phases, such as nanocrystalline grains, intermetallic compounds, and amorphous phases [29–31]. These unique microstructural features are achieved through the process of mechanical alloying, which involves milling or grinding elemental powders together in a high-energy ball mill or attritor. During this process, the repeated impact and deformation of the powders result in a high degree of plastic strain, followed by strain-induced atomic mixing and diffusion, leading to the formation of solid solutions and new phase [32]. The microstructure of mechanically alloyed ferrous materials is a key factor in understanding their properties and performance. Understanding the microstructure of mechanically alloyed ferrous materials is crucial for discerning their properties and performance.

Mechanical alloying is a processing technique used to synthesize new materials by blending and grinding powders of different elements. During the mechanical alloying process, the powders undergo repeated deformation, cold welding, and fracturing, leading to the formation of nanostructured or fine-grained microstructures [33]. These microstructures often exhibit a non-equilibrium state, characterized by a high density of defects and distorted lattice structures.

9.2 Mechanical Properties of Alloyed Ferrous Materials

Mechanical alloying (MA) is a solid-state powder processing technique that produces materials with unique and often superior properties compared to their conventionally processed counterparts [34]. The mechanism of MA allows for the dispersion of insoluble phases and the addition of reactive alloying elements [35]. MA of ferrous materials, which contain iron either as the primary element or in combination with other elements [36]. As described in the abstract, which involves subjecting a mixture of elemental or pre-alloyed powders to severe plastic deformation, mixing, and alloying at the atomic level, comprises repeated cold welding, fracturing, and rewelding of powder particles in a high-energy ball milling [37]. It synthesizes material with unique properties, resulting in the production of metastable materials such as supersaturated solid solutions, intermediate phases, and nanocrystalline phases [38, 39]. This process leads to the formation of fine-grained, nanocrystalline microstructures with enhanced mechanical properties. The mechanical properties of alloyed ferrous materials play a crucial role in determining their performance and suitability for various applications [40].

MA can be used to produce high-strength steels with tensile strengths exceeding 1 GPa. These steels are characterized by fine-grained microstructures and the presence of intermetallic phases, such as Fe_3Al and FeMn. It can be used to produce tool steels with enhanced hardness, wear resistance, and hot hardness [41]. These steels are characterized by the presence of hard intermetallic phases, such as TiC and WC. It can also be used to produce stainless steel with improved corrosion resistance. This is achieved by adding alloying elements, such as chromium and molybdenum, which promote the formation of a passive oxide layer [42]. MA-processed ferrous materials are used in various automotive components, including engine parts, transmission components, and suspension components; aircraft components such as landing gear, structural components, and engine parts; energy components, such as turbine blades, geothermal energy components, and nuclear reactor components [43]. Their high strength, high strength-to-weight ratio, wear resistance, fatigue resistance, and corrosion resistance make them suitable for these applications. Alloying ferrous materials through MA offers several advantages over conventional melting techniques [44]. The ability to tailor the composition and microstructure of ferrous materials through mechanical alloying opens up exciting opportunities for designing advanced materials with improved performance characteristics [5].

9.2.1 Influencing Parameters

Mechanical alloying (MA) can influence the grain size, dislocation density, and other factors affecting the properties of alloyed ferrous materials [34, 43]. The following points highlight the impact of MA on the strength and ductility of ferrous materials:

- MA can lead to the formation of fine-grained microstructures, which can contribute to improved tensile properties.
- The process can increase the dislocation density, which can enhance the material's strength and ductility.
- MA can also introduce alloying elements that improve the corrosion resistance of ferrous materials, such as chromium, nickel, molybdenum, and copper.
- These alloying elements can contribute to the formation of protective oxides or other surface modifications that enhance corrosion resistance.

- Chromium additions, for example, can enhance the formation of a passive oxide layer that protects the material from corrosion [45].

Mechanical alloying can influence the tensile strength and ductility of alloyed ferrous materials by affecting factors such as grain size, dislocation density, and the presence of protective oxides or other surface modifications [46]. This process can lead to improved corrosion resistance and enhanced mechanical properties, making ferrous materials suitable for various applications in industries such as automotive, aerospace, and energy storage. Here are some key aspects of the mechanical properties of alloyed ferrous materials, those are: those are, high hardness, superior wear resistance, corrosion resistance, enhanced tensile strength, ductility, and impact toughness.

Svoboda et al. [47] worked on the development of high-performance materials for extreme environments which are crucial for advancements in various fields. Oxide dispersion-strengthened (ODS) nanocomposites have emerged as promising candidates due to their exceptional combination of strength, high-temperature resistance, and creep performance. Their study focused on optimizing the mechanical alloying (MA) process for the production of a Fe–10Al–4Cr–4Y_2O_3 ODS nanocomposite, known as FeAlOY. The objective of their work is to achieve the desired microstructure and exceptional high-temperature mechanical properties of FeAlOY while minimizing processing time and energy consumption. Figure 9.1a–f of their work presents representative metallographic cuts of the powders after mechanical alloying (MA) at different temperatures and times. The cuts are evaluated using an automatic image analyzer to provide the area distribution (A) at the particle cuts. The figure demonstrates a rather wide particle size distribution in the powders after MA. Figure 9.1g–h shows the transmission electron microscopy (TEM) analysis of the FeAlOY powder after mechanical alloying (MA). The microstructure of the powder is observed to have an ultrafine-grained (UFG) structure with a size of approximately 50 nm. The image also reveals a high density of dislocations within the grains and boundaries between grains that are not fully developed. The dislocation density is estimated to be around 10^{16} mm^3, with distances between dislocations of about 10 nm. These results suggest that the FeAlOY powder obtained through MA exhibits a fine-grained microstructure with a significant presence of dislocations. Figure 9.1i depicts the average ultimate strengths (Rm) of the FeAlOY nanocomposite under different mechanical alloying (MA) conditions. The results show that the strength of the FeAlOY becomes saturated after approximately 150 h of MA at both temperatures tested. Shorter times of MA lead to a significant decrease in strength, with MA at 33 °C being less effective and reliable compared to MA at 46 °C. The tensile tests conducted at 1100 °C and a strain rate of 10^{-6} s^{-1} resemble creep conditions. The figure provides valuable information about the optimal MA conditions for achieving the desired strength properties of the FeAlOY nanocomposite.

Kotan et al. [48] investigated the microstructural evolutions and mechanical properties of Fe–Ni–Zr alloys after equal channel angular extrusion (ECAE) processing at elevated temperatures. The authors specifically focused on understanding the strengthening mechanisms beyond the Hall–Petch grain size effect in these alloys.

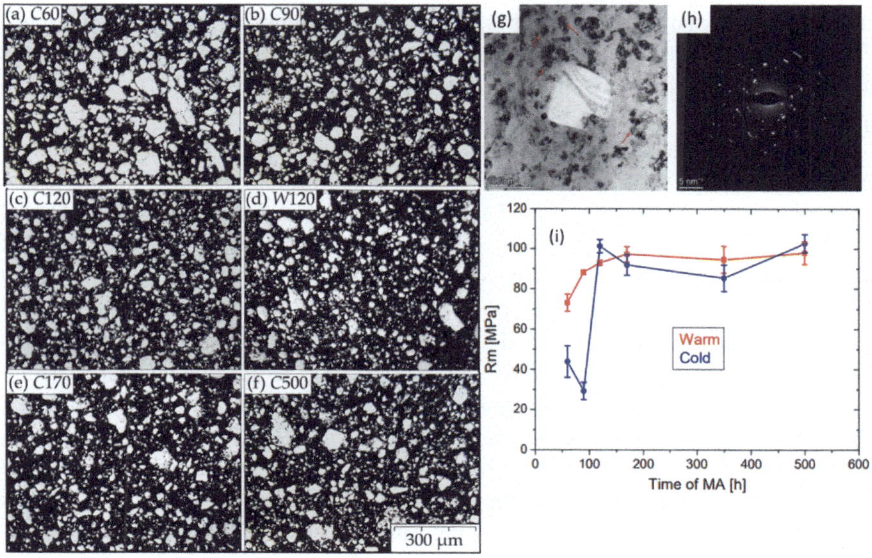

Fig. 9.1 **a–i** Illustrates metallographic cuts of the FeAlOY powders after mechanical alloying (MA) at various conditions, showing a wide particle size distribution and the area distributions at the particle cuts (**a**–**f**). Transmission electron microscopy (TEM) image of the FeAlOY powder after MA at 60 °C, revealing an ultrafine-grained (UFG) microstructure with a size of approximately 50 nm and a high density of dislocations (**g**–**h**). Average ultimate strengths (Rm) of the FeAlOY nanocomposite under different MA conditions, indicating that the strength becomes saturated after about 150 h of MA at both 33 and 46 °C. MA at 46 °C is found to be more effective and reliable than at 33 °C, with shorter MA times leading to a significant decrease in strength (**i**) [47]

The objective of the work is to compare the grain sizes of as-milled and ECAE-processed Fe–Ni–Zr alloys with previous research data, to determine the hardness values of ECAE-processed Fe–Ni–Zr alloys and assess the contribution of additional strengthening mechanisms and to emphasize the role of secondary particles in the strengthening mechanism of extruded Fe–Ni–Zr steels through Orowan strengthening. X-ray grain sizes in the as-milled state were estimated, grain sizes after extrusion processes were determined and compression tests were conducted at different temperatures. Microstructural evolutions were characterized using ion channeling contrast imaging (FIB-CCI) and scanning electron microscopy-electron backscatter diffraction (SEM-EBSD) mapping. The as-milled grain sizes matched closely with nanocrystalline regime data reported by previous researchers. There was a scatter in the data obtained after ECAE processing, indicating additional sources of strengthening beyond the Hall–Petch grain size effect. Hardness data for ECAE-processed Fe–Ni–Zr alloys corresponded to higher values than estimated from grain size alone, suggesting the presence of additional strengthening mechanisms. The role of secondary particles in the strengthening mechanism of extruded Fe–Ni–Zr steels through Orowan strengthening was emphasized. The study concludes that in addition to the Hall–Petch grain size effect, there are other sources of strengthening in ECAE-processed Fe–Ni–Zr alloys, such as the presence of secondary particles. These findings contribute to a better understanding of the microstructural evolutions and mechanical properties of these alloys, which can be beneficial for their practical applications.

Figure 9.2a–j sheds light on the microstructural evolution and grain size behavior of Fe–Ni-Zr alloys subjected to different consolidation temperatures during Equal Channel Angular Extrusion (ECAE). Scanning electron microscopy (SEM) and Focused Ion Beam (FIB) imaging confirmed full consolidation at temperatures of 700 °C and above. However, a stark contrast emerges as consolidation temperatures rise beyond 700 °C. In the $Fe_{91}Ni_8Zr_1$ alloy, grain growth becomes evident, evident in the grain sizes reaching 1 μm and 4 μm at 800 °C and 1000 °C, respectively. This trend is even more pronounced in the $Fe_{88}Ni_8Zr_4$ alloy, exhibiting an average grain size of 5 μm at 1000 °C. Interestingly, the $Fe_{92}Ni_8$ alloys containing Zr displayed a slower rate of grain coarsening during ECAE consolidation, highlighting the beneficial role of Zr addition in microstructural stabilization. While high-temperature exposure during ECAE promotes the formation of nano-sized superlattice clusters within the Fe–Ni–Zr alloys, a contrasting effect is observed on other microstructural features. The dominant factor in stabilizing the microstructure lies in the presence of precipitated second phases within the alloys. However, increasing the ECAE processing temperature beyond 700 °C leads to a double whammy—coarsening of the particle size and weakening of the pinning forces they exert on the grain boundaries. This combination explains the observed grain growth at ECAE processing temperatures of 800 °C and above.

Figure 9.3a–h unveil a complex interplay between composition, processing, and temperature in dictating the mechanical behavior of Fe–Ni–Zr alloys. While grain size plays a crucial role, the presence and distribution of secondary phases, particularly Zr-rich oxides, offer an additional layer of strengthening through Orowan mechanisms. Understanding these intricate relationships is key to tailoring the

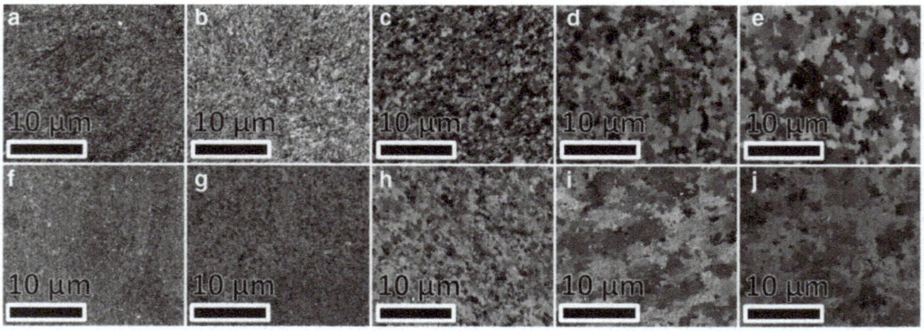

Fig. 9.2 **a–j** Illustration of focused ion beam channeling contrast images reveal the influence of alloy composition and equal channel angular extrusion (ECAE) consolidation temperature on the grain structure of Fe–Ni–Zr steel. Higher processing temperatures lead to a more pronounced coarsening of the grain structure, particularly evident in the steel with 4% Zr addition [48]

processing and composition of these alloys for optimal high-temperature performance. Figure 9.3i–j shows the compressive yield properties of Fe 91 Ni 8 Zr 1 and Fe 88 Ni 8 Zr 4 alloys at different processing and testing temperatures. The compressive yield strength decreases with increasing ECAE consolidation temperature due to coarsening of oxide-based precipitates and grain coarsening. Both alloys exhibit significant strength losses at high testing temperatures, with Fe 91 Ni 8 Zr 1 alloy showing a greater reduction in yield strength compared to Fe 88 Ni 8 Zr 4 alloy. Smaller grain sizes and particles are more sensitive to thermal softening, indicating that smaller precipitates and grain sizes contribute to increased strength at low temperatures. There is a limit to increasing and retaining high strength at elevated temperatures, suggesting that smaller precipitates and grain sizes are beneficial for strength at low temperatures. A study that examines the relationship between hardness and grain size in two types of steel alloys: Fe 88 Ni 8 Zr 4 and Fe 91 Ni 8 Zr 1. The relationship is summarized using a plot called the Hall–Petch plot, which is shown in Fig. 9.3k. The grain sizes used in the plot are determined using X-ray measurements in the as-milled state and from micrographs taken after extrusion processes at elevated temperatures. The plot also includes data from other researchers for comparison, including data for nano and coarse-grained iron alloys. The as-milled values in the plot closely match the data points reported by previous researchers for nanocrystalline regimes. However, there is a scatter in the data obtained after extrusion processes, indicating that the relationship between hardness and grain size is not as consistent. The hardness data for the Fe–Ni–Zr alloys processed at elevated temperatures show much higher hardness values than what would be expected based on grain size alone. This suggests that there are additional sources of strengthening beyond the grain size effect described by the Hall–Petch relationship. One possible source of strengthening is the presence of secondary particles, which play a role in the strengthening mechanism of the extruded Fe–Ni–Zr steels through a process called Orowan strengthening.

9.2.2 Higher Hardness

MA-processed ferrous materials exhibit significantly higher hardness compared to their conventionally processed counterparts. This is attributed to the formation of fine-grained microstructures and the presence of intermetallic phases [49]. The impact of the milling atmosphere on the structure and magnetic properties of mechanically alloyed $Fe_{40}Co_{30}Ni_{30}$ was explored [43]. Powders mechanically alloyed in argon and ambient atmospheres were micron-sized constituting nanocrystalline γ phase. The as-prepared powders exhibited phase transformation $\gamma \rightarrow \alpha$ (Fe) + γ (Fe deplete) between ~725 and ~800 K. Thermal treatment increased the magnetic saturation and decreased the coercivity and remanence [50]. A CoCrFeNiMn high-entropy alloy processed via cold rolling and post-deformation annealing showed strength-ductility synergy with high strengths of over ~1 GPa and uniform elongations of over 12%. Different deformation mechanisms such as dislocation slip, deformation

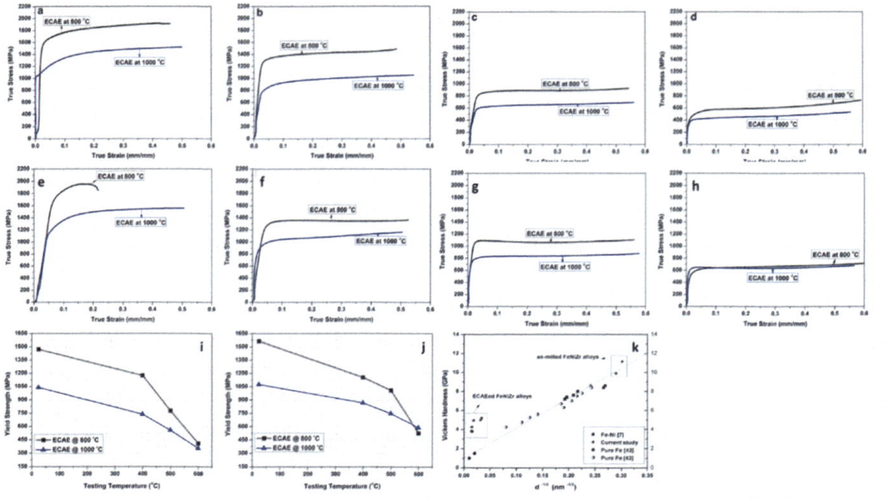

Fig. 9.3 **a**–**h** Illustrates the stress–strain curves for Fe 91 Ni 8 Zr 1 (**a**–**d**) and Fe 88 Ni 8 Zr 4 (**e**–**h**) alloys, respectively, at different testing temperatures. The compressive yield properties of Fe 91 Ni 8 Zr 1 (**i**) and Fe 88 Ni 8 Zr 4 (**j**) alloys at different processing and testing temperatures. The hardness data relationship is summarized using a plot called the Hall–Petch plot (**k**) [48]

twinning, and hetero-deformation-induced hardening were responsible for achieving this mechanical property [51].

9.2.3 Superior Wear Resistance

Alloyed ferrous materials produced through mechanical alloying frequently exhibit superior wear resistance compared to conventionally processed materials [6]. The fine-grained microstructure and intermetallic phases in MA-processed ferrous materials contribute to enhanced wear resistance. This makes them suitable for applications involving sliding contact and abrasion [52, 53]. The use of thermodynamic modeling in the production of wear-resistant coatings, such as titanium-based coatings, allows for the quantification and prediction of composition and properties, resulting in coatings with unique physical and chemical properties [54]. Coatings grown on DIN 1.2714 steel surfaces using a high impulse closed field unbalanced magnetron sputtering method, such as the TiCrNb–hBN layer, significantly increase hardness and improve wear performance [54]. Hardfacing techniques, such as thermal spraying, electric arc welding, and TIG welding, are cost-effective methods for extending the life of components exposed to wear, with Stellite6 being a suitable material due to its impact and abrasion resistance [55].

9.2.4 *Corrosion Resistance*

MA can be used to introduce alloying elements that improve the corrosion resistance of ferrous materials. The milling process can lead to the formation of protective oxides or other surface modifications that enhance corrosion resistance. For instance, chromium additions can enhance the formation of a passive oxide layer that protects the material from corrosion. The process of MA can accelerate diffusion in solids, leading to a decrease in grain size and an increase in micro strain in the grain [56–58] In the case of carbonitrided components formed of ferrous material, MA can be used to improve corrosion resistance by treating the components with a solution of a thermosetting organic synthetic resin and heat-treating them at a specific temperature. By reducing grain size and increasing micro strain, MA can enhance the corrosion resistance of ferrous materials, making them more resistant to degradation in corrosive environments [59].

In one study, the use of MA was effective in improving the corrosion protection efficacy of Ni–Mo coatings from a low-concentration bath [60]. Another study discussed the use of organic compounds and natural products as corrosion inhibitors for steel in CO_2 and CO_2/H_2S coexisting environments. The corrosion inhibition effectiveness of these compounds was evaluated using weight loss and electrochemical techniques [61]. Additionally, MA was used to produce oxide-dispersion-strengthened (ODS) steels, which are potential candidate structural materials for fusion reactors. The use of TiO_2 powder in the MA process successfully reduced the milling time and generated nanoparticles that maintained good coherent interfaces with the matrix, improving the corrosion resistance of the ODS alloys [62]. The effects of mechanical alloying (MA) on the corrosion resistance of ferrous materials have been studied in several papers. MA can accelerate diffusion in solids, leading to a decrease in grain size and an increase in micro strain. The microstructure and mechanical properties of a welded joint of 620-grade marine steel were found to be influenced by the MA process. The weld center area (WMmid) exhibited higher strength and hardness compared to the inner and outer welding surface regions (WMin and WMout) due to a higher volume fraction of martensite-austenite (MA) constituents [60]. Additionally, the corrosion resistance of the welded joint was found to vary across different regions, with the weld metal showing the highest corrosion resistance, followed by the heat-affected zone (HAZ) and base metal (BM) [63]. These findings suggest that MA can have a positive effect on the corrosion resistance of ferrous materials by altering their microstructure and surface properties [64].

9.2.5 *Tensile Strength and Ductility*

The tensile strength and ductility of alloyed ferrous materials can be investigated through various methods. Mechanical alloying is one such method that can influence the grain size, dislocation density, and other factors affecting the materials' tensile properties. Tensile strength refers to the amount of load a material can withstand before breaking, while ductility is the ability of a material to undergo significant plastic deformation before failure. Research has shown that the microstructure and properties of ferrous medium-entropy alloys prepared using powder metallurgy can be affected by the Fe content, leading to variations in strength and hardness. Ductility is often measured using a tensile test as a percentage of elongation, or the reduction in the cross-sectional area of the sample before failure. The effects of loading rate and test temperature on the impact energies and fracture toughness have been investigated for some ferrous alloys used in power transmission line hardware, highlighting the importance of these properties under different conditions. In summary, the tensile strength and ductility of alloyed ferrous materials are crucial mechanical properties that can be influenced by factors such as microstructure, Fe content, and testing conditions. Mechanical alloying is one of the methods that can affect these properties by influencing the material's microstructure [65].

9.2.6 Impact Toughness

The impact toughness of alloyed ferrous materials is a critical property for materials subjected to dynamic loading conditions. Dynamic fracture toughness is a measure of a material's ability to resist fracture under dynamic loading conditions [66]. Research has shown that the dynamic fracture toughness of materials, such as tungsten alloy and high-strength steel alloys, can increase with the increase of the loading rate. This property is particularly important for materials used in power transmission line hardware, as the effects of loading rate and test temperature on the impact energies and fracture toughness have been investigated for some ferrous alloys used in this application. The dynamic fracture toughness of materials is an important consideration for the design and selection of materials for applications where they will be subjected to dynamic loading conditions [67].

Mechanical alloying is a versatile and powerful technique for producing ferrous materials with enhanced mechanical properties and tailored microstructures. The ability to control the microstructure and composition of these materials at the atomic level opens up exciting opportunities for designing advanced materials with improved performance characteristics, ultimately driving innovation in engineering and technology.

9.3 Novel Applications of Mechanically Alloyed Ferrous Materials

One of the challenging uses of cutting-edge ferrous materials in conjunction with nanomaterials is in the field of chemosensors. Chemosensors [68], whether in the form of analytical devices or molecules, are engineered to identify and react to particular chemical species or alterations in the environment. Their significance spans diverse domains such as chemistry, biology, medicine, and environmental science. The fundamental operation of chemo sensors involves undergoing a distinct transformation when exposed to a specific target analyte. The resulting change can be quantified or observed, serving as an indicator of the concentration or presence of the analyte.

Another notable application of alloyed ferrous steels lies in the production of automobile parts through the powder metallurgy processing route. The creation of these alloys involves a combination of admixing, partial alloying, and pre-alloying techniques [69]. Automobile parts, such as cylinder heads, intake manifolds, wheels, and instrument panels, find significant applications in various aspects of vehicle manufacturing and functionality. Cylinder heads play a crucial role in the internal combustion engine, providing housing for the cylinders and facilitating the combustion process. Intake manifolds are responsible for distributing air to the engine cylinders for efficient fuel combustion. Wheels contribute to the vehicle's mobility and are integral to its overall performance and safety. Instrument panels serve as the control center for the driver, displaying critical information and providing a user interface for operating the vehicle. These components collectively contribute to the functionality, performance, and safety of automobiles.

Ferrous alloys also play a crucial role in water treatment processes [70]. Currently, ferrous sulfate is widely utilized in wastewater treatment, with studies suggesting a common Fe/P molar ratio ranging from 1.5 to 2.2. In the case of municipal wastewater containing 5 mg/L of phosphorus, the recommended ferrous dosage typically falls between 0.24 and 0.35 mM. This addition is instrumental in aiding biological phosphorus removal. However, it has been noted that higher ferrous dosages can negatively impact biological nitrogen removal. This observation underscores the need for careful consideration and optimization of ferrous dosage in water treatment processes to ensure the effectiveness of both phosphorus and nitrogen removal while minimizing adverse effects.

Ferrous alloys, particularly iron-manganese-silicon (Fe–Mn–Si) shape memory alloys, find notable applications in shape memory technology. The martensitic transformation in these alloys involves lattice deformation, resulting in a shift from the parent phase to the martensite phase. Unlike the conventional crystal slip-through dislocations, as depicted in the illustration in Fig. 9.4, the mechanism of deformation during martensitic transformation is distinct.

This process ensures that the nearest neighbouring atoms maintain their relative positions unchanged within the lattice. Martensite formation is achieved by impeding the diffusion of atoms within the crystal. This hindrance can be introduced through

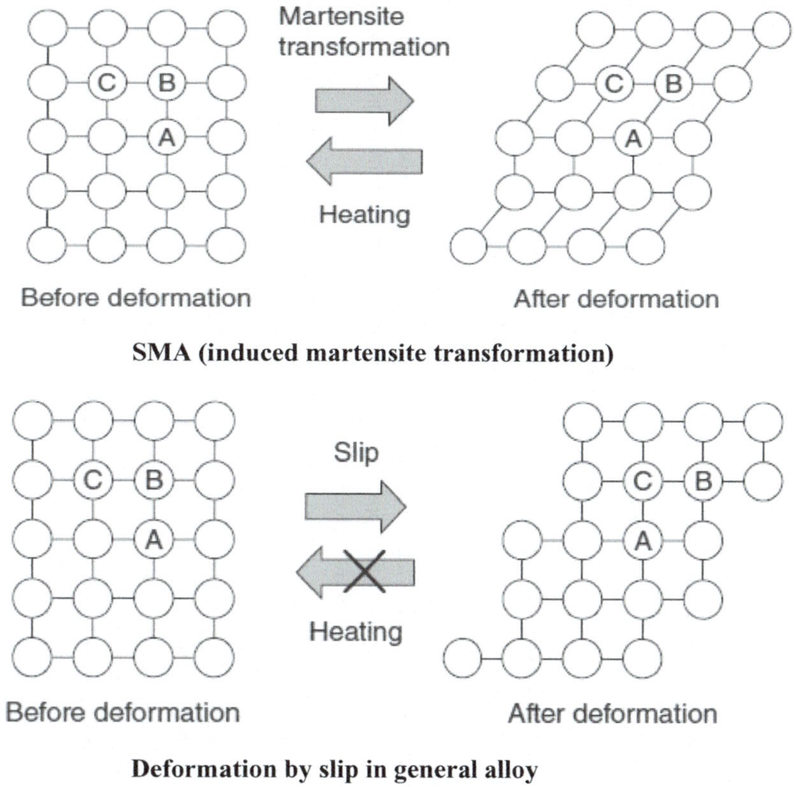

Fig. 9.4 Generation of shape memory effect by stress-induced martensite in a shape memory alloy and slip deformation in non-shape memory alloy

rapid quenching or sub-cooling of the material. In both instances, the transformation occurs while preserving the spatial arrangement of adjacent atoms, contributing to the unique properties of martensitic structures. These shape memory [71] alloys are particularly valued for their ability to revert to a predetermined shape upon the application of external stimuli, making them valuable in various applications, including actuators and medical devices.

Layered configurations incorporating ferrites [72] with piezoelectric oxides represent a promising frontier in magnetic field sensor technology. Their operational principle involves a two-step process that converts magnetic fields into electrical voltages. Initially, the magnetic field induces mechanical strain in the magnetic material, facilitated by its magnetostriction. This strain subsequently triggers the generation of a voltage within the piezoelectric layer. These sensors boast several advantages, including high sensitivity, compact size, and virtually negligible power consumption. They are versatile enough to be tailored for the detection of both alternating current (ac) and direct current (dc) magnetic fields, as well as ac and dc electric currents. Particularly noteworthy are sensors constructed with nickel ferrite, showcasing their potential for exceptional performance in diverse applications.

Ferrites play a dominant role in power applications [73], particularly in the power supplies for a wide range of devices. This includes computers, various peripherals, TV, and video systems, as well as various small and medium-sized instruments. The primary application lies in systems referred to as switched-mode power supplies.

References

1. Straffelini, G., Fontanari, V., Molinari, A.: Influence of microstructure on impact behaviour of sintered ferrous materials. Powder Metall. **38**, 45–51 (1995). https://doi.org/10.1179/pom.1995.38.1.45
2. Mara, D.D., Williams, D.J.A.: Influence of the microstructure of ferrous metals on the rate of microbial corrosion. Br. Corros. J. **7**, 139–142 (1972). https://doi.org/10.1179/000705972798323062
3. Charles, J.A.: Development and use of layered ferrous microstructure. Mater. Sci. Technol. **14**, 496–503 (1998). https://doi.org/10.1179/mst.1998.14.6.496
4. Patil, A., Nithin, S.K., Krishna, H., Vasanth, S., Mahale, R.S., Sharath, P.C., et al.: Investigation of wear behavior for three-body made of CA-40 steel and nickel hard cast iron by abrasion method. AIP Conf. Proc. **2901**, 100002 (2023). https://doi.org/10.1063/5.0178634
5. Pollock, T.M., Clarke, A.J., Babu, S.S.: Design and tailoring of alloys for additive manufacturing. Metall. Mater. Trans. A **51**, 6000–6019 (2020). https://doi.org/10.1007/s11661-020-06009-3
6. Yang, L., Zhao, C., Zhu, W., Cheng, Z., Wei, P., Ren, F.: Microstructure, mechanical properties, and sliding wear behavior of oxide-dispersion-strengthened FeMnNi alloy fabricated by spark plasma sintering. Metall. Mater. Trans. A **51**, 2796–2810 (2020). https://doi.org/10.1007/s11661-020-05748-7
7. Golinejad, S., Mirjalili, M.H.: Fast and cost-effective preparation of plant cells for scanning electron microscopy (SEM) analysis. Anal. Biochem. **609**, 113920 (2020). https://doi.org/10.1016/j.ab.2020.113920

8. Spurgeon, S.R., Ophus, C., Jones, L., Petford-Long, A., Kalinin, S.V., Olszta, M.J., et al.: Towards data-driven next-generation transmission electron microscopy. Nat. Mater. **20**, 274–279 (2021). https://doi.org/10.1038/s41563-020-00833-z
9. Begic Hadzipasic, A., Malina, J., Malina, M.: The influence of microstructure on hydrogen diffusion and embrittlement of fine-grained high strength dual-phase steels. Kov. Mater. **59**, 69–78 (2021). https://doi.org/10.4149/km_2021_1_69
10. Kiani, D.: X-ray diffraction (XRD) BT. In: Wachs, I.E., Bañares, M.A. (eds.) Springer Handbook of Advanced Catalyst Characterization, p. 519–539. Springer International Publishing, Cham (2023). https://doi.org/10.1007/978-3-031-07125-6_25
11. Raabe, D., Sun, B., Kwiatkowski Da Silva, A., Gault, B., Yen, H.-W., Sedighiani, K., et al.: Current challenges and opportunities in microstructure-related properties of advanced high-strength steels. Metall. Mater. Trans. A **51**, 5517–5586 (2020). https://doi.org/10.1007/s11661-020-05947-2
12. Simchi, A., Danninger, H., Weiss, B.: Microstructural modelling of electrical conductivity and mechanical properties of sintered ferrous materials. Powder Metall. **43**, 219–227 (2000). https://doi.org/10.1179/003258900665970
13. Fayazfar, H., Salarian, M., Rogalsky, A., Sarker, D., Russo, P., Paserin, V., et al.: A critical review of powder-based additive manufacturing of ferrous alloys: process parameters, microstructure and mechanical properties. Mater. Des. **144**, 98–128 (2018). https://doi.org/10.1016/j.matdes.2018.02.018
14. Scott, T.A.: The influence of microstructure on the mechanical properties of polycrystalline diamond: a literature review*. Adv. Appl. Ceram. **117**, 161–176 (2018). https://doi.org/10.1080/17436753.2017.1389462
15. Mahale, R.S., Vasanth, S., Sharath, P.C., Shashanka, R., Tambrallimath, V., Badari, A.: Effect of ball to powder ratios on the phase transformation of austenite and ferrite during ball milling of SAF-2507 super duplex stainless steel powders. J. Inst. Eng. Ser. D (2024). https://doi.org/10.1007/s40033-023-00624-1
16. Shashanka, R., Chaira, D.: Optimization of milling parameters for the synthesis of nano-structured duplex and ferritic stainless steel powders by high energy planetary milling. Powder Technol. **278**, 35–45 (2015). https://doi.org/10.1016/j.powtec.2015.03.007
17. Simchi, A.: Effects of lubrication procedure on the consolidation, sintering and microstructural features of powder compacts. Mater. Des. **24**, 585–594 (2003). https://doi.org/10.1016/S0261-3069(03)00155-9
18. Shashanka, R., Chaira, D.: Development of nano-structured duplex and ferritic stainless steels by pulverisette planetary milling followed by pressureless sintering. Mater. Charact. **99**, 220–229 (2015). https://doi.org/10.1016/j.matchar.2014.11.030
19. Shashanka, R., Chaira, D.: Phase transformation and microstructure study of nano-structured austenitic and ferritic stainless steel powders prepared by planetary milling. Powder Technol. **259**, 125–136 (2014). https://doi.org/10.1016/j.powtec.2014.03.061
20. Straffelini, G., Molinari, A.: Dry sliding wear of ferrous PM materials. Powder Metall. **44**, 248–252 (2001). https://doi.org/10.1179/003258901666419
21. Liu, S., Ma, L., Zhen, C., Wang, Y., Li, D., Guo, F.: Microstructural evolution of intermetallic compounds between crystalline/amorphous Cobalt-Phosphorous coatings and lead-free solders. J. Mater. Res. Technol. **19**, 2916–2929 (2022). https://doi.org/10.1016/j.jmrt.2022.06.035
22. Halfadji, A., Naous, M., Nadia Kharroubi, K., el zahraà Belmehdi, F., Rajendrachari, S.: An ultrasonic-assisted synthesis, characterization, and application of nano-Fe_3O_4/TiO_2 as nano-catalyst for the removal of organic dye by like-photo-fenton reactions. Inorg. Chem. Commun. **158**, 111686 (2023). https://doi.org/10.1016/j.inoche.2023.111686
23. Khan, H., Yerramilli, A.S., D'Oliveira, A., Alford, T.L., Boffito, D.C., Patience, G.S.: Experimental methods in chemical engineering: X-ray diffraction spectroscopy—XRD. Can. J. Chem. Eng. **98**, 1255–1266 (2020). https://doi.org/10.1002/cjce.23747

24. Kara, U., Susoy, G., Issa, S.A.M., Elshami, W., Yorgun, N.Y., Abuzaid, M.M., et al.: Scanning electron microscopy (SEM), energy-dispersive X-ray (EDX) spectroscopy and nuclear radiation shielding properties of [α-Fe3+O(OH)]-doped lithium borate glasses. Appl. Phys. A **126**, 506 (2020). https://doi.org/10.1007/s00339-020-03683-3
25. Xu, W., Ferry, M., Wang, Y.: Influence of alloying elements on as-cast microstructure and strength of gray iron. Mater. Sci. Eng. A **390**, 326–333 (2005). https://doi.org/10.1016/j.msea.2004.08.030
26. Boulifa, I., Hadji, A.: Study of the influence of alloying elements on the mechanical characteristics and wear behavior of a ductile cast iron. Frat. Ed. Integrità. Strutt. **15**, 74–83 (2021). https://doi.org/10.3221/IGF-ESIS.56.06
27. He, J., Ding, X., Li, Z., Chen, J., Wang, H., Zhang, M., et al.: Effects of Fe content on microstructure and properties of ferrous medium-entropy alloys prepared using powder metallurgy. Mater. Today Commun. **34**, 105344 (2023). https://doi.org/10.1016/j.mtcomm.2023.105344
28. Simchi, A., Danninger, H., Gierl, C.: Electrical conductivity and microstructure of sintered ferrous materials: iron-graphite compacts. Powder Metall. **44**, 148–156 (2001). https://doi.org/10.1179/003258901666293
29. López, D.A., Pérez, T., Simison, S.N.: The influence of microstructure and chemical composition of carbon and low alloy steels in CO_2 corrosion. A state-of-the-art appraisal. Mater. Des. **24**, 561–575 (2003). https://doi.org/10.1016/S0261-3069(03)00158-4
30. Moazzen, P., Toroghinejad, M.R., Cavaliere, P.: Effect of iron content on the microstructure evolution, mechanical properties and wear resistance of FeXCoCrNi high-entropy alloy system produced via MA-SPS. J. Alloys Compd. **870**, 159410 (2021). https://doi.org/10.1016/j.jallcom.2021.159410
31. Pan, S., Zeng, F., Su, N., Xian, Z.: The effect of niobium addition on the microstructure and properties of cast iron used in cylinder head. J. Mater. Res. Technol. **9**, 1509–1518 (2020). https://doi.org/10.1016/j.jmrt.2019.11.076
32. Alrubaiey, S.I., Anoon, E.A., Hanoon, M.M.: The influence of microstructure on the corrosion rate of carbon steels. Eng. Technol. J. **31**, 1825–1836 (2013). https://doi.org/10.30684/etj.31.10a2
33. Upadhyay, S., Saxena, K.K.: Effect of Cu and Mo addition on mechanical properties and microstructure of grey cast iron: an overview. Mater. Today Proc. **26**, 2462–2470 (2020). https://doi.org/10.1016/j.matpr.2020.02.524
34. Neikov, O.D.: Mechanical alloying. In: Neikov, O.D., Naboychenko, S.S., Yefimov, N.A. (eds.) Handbook of Non-Ferrous Metal Powders, pp. 91–124. Elsevier, Oxford (2019). https://doi.org/10.1016/B978-0-08-100543-9.00003-8
35. Suryanarayana, C.: Mechanical alloying: a novel technique to synthesize advanced materials. Research (2019). https://doi.org/10.34133/2019/4219812
36. Suñol, J.-J.: Mechanical alloying: processing and materials. Metals (Basel) **11**, 798 (2021). https://doi.org/10.3390/met11050798
37. Suryanarayana, C.: Mechanical alloying: a critical review. Mater. Res. Lett. **10**, 619–647 (2022). https://doi.org/10.1080/21663831.2022.2075243
38. Shuai, C., He, C., Peng, S., Qi, F., Wang, G., Min, A., et al.: Mechanical alloying of immiscible metallic systems: process, microstructure, and mechanism. Adv. Eng. Mater. **23**, 2001098 (2021). https://doi.org/10.1002/adem.202001098
39. Avar, B., Panigrahi, M., Soguksu, A.K., Rajendrachari, S., Gundes, A.: Photocatalytic activity of soft magnetic $Fe_{80-x}Co_xZr_{10}Si_{10}$ (x = 0, 40, and 80) nanocrystalline melt-spun ribbons. Top. Catal. (2022). https://doi.org/10.1007/s11244-022-01569-7
40. Suryanarayana, C., Al-Joubori, A.A., Wang, Z.: Nanostructured materials and nanocomposites by mechanical alloying: an overview. Met. Mater. Int. **28**, 41–53 (2022). https://doi.org/10.1007/s12540-021-00998-5
41. Novák, P., Vanka, T., Nová, K., Stoulil, J., Průša, F., Kopeček, J., et al.: Structure and properties of Fe–Al–Si alloy prepared by mechanical alloying. Materials (Basel) **12**, 2463 (2019). https://doi.org/10.3390/ma12152463

42. Haušild, P., Čech, J., Kadlecová, V., Karlík, M., Průša, F., Nová, K., et al.: Technological aspects of Fe–Al–Si intermetallic alloy preparation by mechanical alloying. Key Eng. Mater. **810**, 101–106 (2019). https://doi.org/10.4028/www.scientific.net/KEM.810.101
43. Rajendrachari, S.: An overview of high-entropy alloys prepared by mechanical alloying followed by the characterization of their microstructure and various properties. Alloys **1**, 116–134 (2022). https://doi.org/10.3390/alloys1020008
44. Průša, F., Proshchenko, O., Školáková, A., Kučera, V., Laufek, F.: Properties of FeAlSi-X-Y alloys (X, Y=Ni, Mo) prepared by mechanical alloying and spark plasma sintering. Materials (Basel) **13**, 292 (2020). https://doi.org/10.3390/ma13020292
45. Khalfallah, I.Y., Aning, A.O.: The effect of milling time on the structure and properties of mechanically alloyed high carbon iron-carbon alloys. Ref. Modul. Mater. Sci. Mater. Eng. (Elsevier) (2019). https://doi.org/10.1016/B978-0-12-803581-8.11383-9
46. Shashanka, R., Chaira, D.: Effects of nano-Y_2O_3 and sintering parameters on the fabrication of PM duplex and ferritic stainless steels. Acta Metall. Sin. (English Lett.) **29**, 58–71 (2016). https://doi.org/10.1007/s40195-015-0362-1
47. Svoboda, J., Gamanov, Š., Bártková, D., Luptáková, N., Bořil, P., Jarý, M., et al.: The optimization of mechanical alloying conditions of powder for the preparation of a Fe-10Al-4Cr-4Y_2O_3 ODS nanocomposite. Materials (Basel) 15 (2022). https://doi.org/10.3390/ma15249034
48. Kotan, H., Darling, K.A., Luckenbaugh, T.: High temperature mechanical properties and microstructures of thermally stabilized Fe-based alloys synthesized by mechanical alloying followed by hot extrusion. Met. Mater. Int. **27**, 1790–1797 (2021). https://doi.org/10.1007/s12540-019-00567-x
49. Paul, A.A., Rathi, A., Thotakura, G.V., Jayaraman, T.V.: Influence of milling atmosphere on the structure and magnetic properties of mechanically alloyed $Fe_{40}Co_{30}Ni_{30}$. Mater. Chem. Phys. **258**, 123897 (2021). https://doi.org/10.1016/j.matchemphys.2020.123897
50. Shahmir, H., Saeedpour, P., Mehranpour, M.S., Shams, S.A., Lee, C.S.: Hetero-deformation induced hardening in a CoCrFeNiMn high-entropy alloy. Crystals **13** (2023). https://doi.org/10.3390/cryst13050844
51. Alvaredo, P., Escribano, J., Ferrari, B., Sánchez-Herencia, A.J., Gordo, E.: Steel binder cermets processed by combination of colloidal processing and powder metallurgy. Int. J. Refract. Met. Hard Mater. **74**, 1–6 (2018). https://doi.org/10.1016/j.ijrmhm.2018.02.018
52. Narayanaswamy, B., Ghaderi, A., Hodgson, P., Cizek, P., Chao, Q., Safi, M., et al.: Abrasive wear resistance of ferrous microstructures with similar bulk hardness levels evaluated by a scratch-tester method. Metall. Mater. Trans. A **50**, 4839–4850 (2019). https://doi.org/10.1007/s11661-019-05354-2
53. Sert, Y., Küçükömeroğlu, T., Kahramanzade, H., Efeoğlu, İ.: Investigation of the wear behavior of forging tool by ball on disc and impact sliding tribometer BT. In: Maleque, M.A., Ahmad Azhar, A.Z., Sarifuddin, N., Syed Shaharuddin, S.I., Mohd Ali, A., Abdul Halim, N.F.H. (eds.) Proceeding of 5th International Conference on Advances in Manufacturing and Materials Engineering, pp. 451–459. Springer Nature Singapore, Singapore (2023)
54. Rojacz, H., Katsich, C., Kirchgaßner, M., Kirchmayer, R., Badisch, E.: Impact-abrasive wear of martensitic steels and complex iron-based hardfacing alloys. Wear **492–493**, 204183 (2022). https://doi.org/10.1016/j.wear.2021.204183
55. Singh, S., Kumar, R., Goel, P., Singh, H.: Analysis of wear and hardness during surface hardfacing of alloy steel by thermal spraying, electric arc and TIG welding. Mater. Today Proc. **50**, 1599–1605 (2022). https://doi.org/10.1016/j.matpr.2021.09.122
56. Zhang, C., Wang, J., Li, X., Wang, S., Zhu, S., Guan, S.: Rapid screening alloying elements for improved corrosion resistance on the Mg(0001) surface using first principles calculations. Phys. Chem. Chem. Phys. **23**, 26887–26901 (2021). https://doi.org/10.1039/D1CP03868K
57. Halfadji, A., Chougui, A., Djeradi, R., Ouabad, F.Z., Aoudia, H., Rajendrachari, S.: TiO_2-decorated by nano-γ-Fe_2O_3 as a catalyst for efficient photocatalytic degradation of orange G dye under eco-friendly white LED irradiation. ACS Omega **8**, 39907–39916 (2023). https://doi.org/10.1021/acsomega.3c06420

58. Rajendrachari, S., Adimule, V., Gulen, M., Khosravi, F., Somashekharappa, K.K.: Synthesis and characterization of high entropy alloy 23Fe-21Cr-18Ni-20Ti-18Mn for electrochemical sensor applications. Materials (Basel) **15** (2022). https://doi.org/10.3390/ma15217591
59. Zhang, M., Hector, L.G., Guo, Y., Liu, M., Qi, L.: First-principles search for alloying elements that increase corrosion resistance of Mg with second-phase particles of transition metal impurities. Comput. Mater. Sci. **165**, 154–166 (2019). https://doi.org/10.1016/j.commatsci.2019.04.018
60. Gonsalves, C.N., Hegde, A.C.: Development of corrosion-resistant Ni–Mo coatings from low-concentration bath: effect of magnetoconvection. Mater. Sci. Technol. **37**, 1187–1198 (2021). https://doi.org/10.1080/02670836.2021.1987696
61. Morita, Y., Kageyama, T.: Development of medium-frequency cavity loaded with multi-ring magnetic alloy cores cooled by chemically inert liquid. Nucl. Instrum. Methods Phys. Res. Sect. A Accel. Spectrometers, Detect. Assoc. Equip. **1010**, 165525 (2021). https://doi.org/10.1016/j.nima.2021.165525
62. Chikwanda, H., Mahlatji, L.: Mechanical alloying of Ti-based materials. Key Eng. Mater. **770**, 95–105 (2018). https://doi.org/10.4028/www.scientific.net/KEM.770.95
63. Xhanari, K., Wang, Y., Yang, Z., Finšgar, M.: A review of recent advances in the inhibition of sweet corrosion. Chem. Rec. **21**, 1845–1875 (2021). https://doi.org/10.1002/tcr.202100072
64. Rajendrachari, S., Adimule, V.M., Jayaprakash, G.K., Pandith, A.: Electrochemical oxidation of methylene blue dye in wastewater using mechanically alloyed high entropy alloy modified carbon paste electrode using cyclic voltammetry. Mater. Res. Express **10**, 54003 (2023). https://doi.org/10.1088/2053-1591/acd3d7
65. Gundlach, R.B.: Heat treatments to develop high-strength ferritic ductile iron. Int. J. Met. **14**, 1065–1077 (2020). https://doi.org/10.1007/s40962-020-00489-0
66. Kim, M.T., Park, T.M., Baik, K., Choi, W.S., Choi, P.-P., Han, J.: Crucial microstructural feature to determine the impact toughness of intercritically annealed medium-Mn steel with triplex-phase microstructure. Acta Mater. Mater. **164**, 122–134 (2019). https://doi.org/10.1016/j.actamat.2018.10.043
67. Duan, Q.Q., Qu, R.T., Zhang, P., Zhang, Z.J., Zhang, Z.F.: Intrinsic impact toughness of relatively high strength alloys. Acta Mater. Mater. **142**, 226–235 (2018). https://doi.org/10.1016/j.actamat.2017.09.064
68. Basaleh, A.S., Sheta, S.M.: Novel advanced nanomaterial based on ferrous metal – organic framework and its application as chemosensors for mercury in environmental and biological samples. Anal. Bioanal. Chem. (2020). https://doi.org/10.1007/s00216-020-02566-z
69. Cole, G.S., Sherman, A.M.: Light weight materials for automotive applications. Mater Charact **35**, 3–9 (1995). https://doi.org/10.1016/1044-5803(95)00063-1
70. Ji, B., Zhu, L., Wang, S., Qin, H., Ma, Y., Liu, Y.: A novel micro-ferrous dosing strategy for enhancing biological phosphorus removal from municipal wastewater. Sci. Total Environ. 135453 (2019). https://doi.org/10.1016/j.scitotenv.2019.135453
71. Maruyama, T., Kubo, H.: Ferrous (Fe-based) shape memory alloys (SMAs): properties, processing and applications (2011). https://doi.org/10.1533/9780857092625.2.141
72. Jacob, M., Wissel, K.: Novel applications of microwaves in the metallurgical processing of a nickel-ferrous laterite ore and an aluminum industry waste 1–7 (2016). https://doi.org/10.1088/1757-899X/123/1/012049
73. Corporation HP: Novel Applications of Ferrites Valenzuela (2012). https://doi.org/10.1155/2012/591839

Chapter 10
Consolidation of Tungsten-Heavy Alloys

Mayadhar Debata, Pradyut Sengupta, Shuvam Mandal, and Deepak Adhikari

Abstract Mechanical alloying is an important powder processing technique that is often utilized for developing novel alloys and composites. The use of mechanical alloying for the fabrication of tungsten-heavy alloys is explored by different research groups varying the milling conditions like milling time, speed, ball-to-powder ratio, milling media, etc. This chapter provides a detailed overview of the prior work on mechanical alloying and sintering conditions on densification, microstructural evolution and mechanical properties of ternary, quaternary as well as oxide dispersed tungsten-heavy alloys. In brief, this chapter summarizes the utility of mechanical alloying as a crucial processing step for the fabrication of tungsten-heavy alloys.

Keywords Tungsten-heavy alloys · Powder metallurgy · Mechanical alloying · Consolidation · Microstructure · Properties

10.1 Introduction

Tungsten-heavy alloys (WHAs) have received profound attention in the research community due to their outstanding properties like high density, superior corrosion resistance and mechanical properties [1]. Due to these excellent properties, WHAs find widespread application in strategic sectors, e.g., kinetic energy penetrators, vibration dampers, radiation shields, etc. Generally, WHAs are constituted by 88–98 wt.% W, and the remaining 2–12 wt.% is composed of metallic additives or binders of Ni–Fe, Ni–Cu, Ni–Co, Ni–Fe–Co, etc. [2, 3] Owing to the extremely high

M. Debata (✉) · P. Sengupta · S. Mandal · D. Adhikari
Advanced Materials Technology Department, CSIR – Institute of Minerals and Materials Technology, Bhubaneswar 751013, India
e-mail: mdebata@immt.res.in

P. Sengupta
Department of Metallurgical & Materials Engineering, Indian Institute of Technology Kharagpur, Kharagpur 721302, India

M. Debata · S. Mandal · D. Adhikari
Academy of Scientific and Innovative Research (AcSIR), Ghaziabad 201002, India

S. Rajendrachari (ed.), *Mechanically Alloyed Novel Materials*, Advanced Structured Materials 220, https://doi.org/10.1007/978-981-97-6504-1_10

melting temperature of W (~3410 °C), pure W or its alloys cannot be processed by melting/casting route. Hence, WHAs are prepared by powder metallurgy (P/M) route involving various processing stages like the blending of powders, compaction, debinding, sintering, and post-processing treatments. In the case of WHAs processed by the P/M route, almost fully dense alloys are obtained by virtue of the liquid phase sintering (LPS) route. Even though WHAs can be sintered to almost full density by solid-state sintering (SSS), LPS is preferred due to a better combination of structure and properties. As the melting temperature of W is extremely high, those metallic binders are utilized for facilitating the densification of WHAs at relatively low temperatures (1440–1530 °C). It is worth noting that the liquid phase sintered tungsten-heavy alloys offer two-phase microstructures where W spheroids are encapsulated in a matrix of Ni–Fe, Ni–Cu, Ni–Co, Ni–Fe–Co, etc. In the matrix phase, W is found to be dissolved to a certain extent, depending on the nature of the metallic binder, temperature of sintering, isothermal holding time, etc. However, in the case of conventionally sintered WHAs, growth of W grains takes place, which is, many a time, undesirable for enhanced mechanical properties. Of late, our research group has revealed the benevolence of NiB as a binder for developing newer heavy alloys for strategic applications [4–6]. In order to restrict the grain growth of W, many research groups have explored the possibility of adding dispersoids (ZrO_2, Y_2O_3, ZrB_2) [7–10] or other alloying elements (Re, Ta, Mo) [11, 12], etc. Several studies have been devoted to introducing mechanical alloying (MA) during the processing of WHAs [13–17]. In brief, mechanical alloying is a solid-state processing technique involving repetitive welding, re-welding, fracturing of powder particles in a confined chamber [18, 19]. Mechanical alloying has been extensively used for developing novel binary alloys, high entropy alloys, high entropy ceramics as well as composites [20–23]. The WHAs fabricated by MA and sintering route often exhibit a superior combination of density, microstructure, hardness, yield strength, ultimate tensile strength, impact strength, etc. [16, 17, 24] Additionally, the mechanically alloyed and sintered WHAs show refined structure and better bonding between the matrix and the W grains, which attribute for enhanced physical and mechanical properties.

The present chapter attempts to summarize the effect of mechanical alloying on the processing, densification, and structure-property evolution of WHAs consolidated by different techniques like solid state sintering, liquid phase sintering, two-step sintering, etc. The influence of starting powders, milling conditions, sintering parameters on microstructure and various properties is illustrated in the subsequent sections.

10.2 Mechanical Alloying of WHAs

Different research groups have carried out mechanical alloying of elemental powders to facilitate the fabrication of tungsten-heavy alloys of various compositions. The effect of several factors like milling time, milling speed, ball to powder ratio on mechanical alloying of WHAs is illustrated in literature [16, 25, 26]. Wu et al. [26]

performed mechanical alloying of 99.9% pure W powders for 60 h in a 99.999% pure N_2 atmosphere, varying the milling speed (300–700 rpm) and the content of process control agents (8–12%). The authors obtained median particle size (D_{50}) of 24 and 16 nm, respectively, after 30 and 60 h of milling at 700 rpm, using a 12% process control agent. Hence, optimization of milling parameters is crucial for refining the size of W powder. In previous studies, a wide variation in milling duration (e.g., 1–72 h) is noticed; however, the milling time depends on the type of ball mill, milling speed, starting powder size, ball to powder ratio, filling ratio, type of milling media, etc. During mechanical alloying of WHA powders, it is important to ensure that milled powders do not pick up atmospheric oxygen, hydrogen, nitrogen, etc. Therefore, wet milling in a controlled atmosphere is preferred over dry milling. As a consequence of high-energy ball milling, the crystallite size reduces, lattice strain increases, and the specific surface area is enhanced. Studies reveal that the mechanical alloying of different WHA systems like W–Ni–Fe, W–Ni–Co, W–Ni–Fe–Y_2O_3, W–Ni–Y_2O_3, W–Ni–Fe–Re, W–Ni–Co–Y_2O_3, etc. have been investigated under varying milling conditions [27–32]. The mechanically alloyed powders were consolidated by various techniques (as shown in Table 10.1). Overall, the effect of mechanical alloying condition and sintering condition on densification, microstructure and mechanical properties of mechanically alloyed WHAs are presented in Fig. 10.1. Xiang et al. [33] showed that after 5 h milling, the W, Fe and Ni are thoroughly mixed and indistinguishable (as depicted in Fig. 10.1a). After 10 h milling, the particle size distribution is broader (Fig. 10.1b). However, in 20 h and 30 h milled powders (Fig. 10.1c, d, respectively) the formation of agglomerated particles is noticed due to repeated cold welding and breaking. Finally, after 40 h milling, the particles again become finer and homogeneous, as shown in Fig. 10.1e.

10.2.1 Densification and Microstructure Evolution

Figure 10.2 shows the effect of milling time on the sintered density of different WHAs processed by the mechanical alloying route. The sintered density of mechanically alloyed WHAs depends on many factors like (a) particle size of milled powders, (b) ball milling condition (type of ball mill, milling time, ball to powder ratio, type of milling media, speed of ball mill, etc.), (c) contamination occurred during milling process, (d) type of sintering (solid state sintering, liquid phase sintering, spark plasma sintering, etc.), (e) sintering parameters (temperature, holding time, heating rate, cooling rate, sintering atmosphere, etc.), (f) post sintering treatments, etc. From Fig. 10.3, it is evident that the sintering condition and milling duration play a phenomenal role in final densification behaviour. However, prolonged milling duration is not always beneficial for enhanced densification of WHAs. For instance, Chen et al. [32] reported the formation of 96.12% dense 90W–7Ni–3Fe alloys prepared by high energy ball milling of >99.5% pure W, Ni and Fe powders (1–3 μm) for 5 h. In contrast, the authors found reduced densification (87.58% theoretical) when the milling duration was extended to 40 h, even though the samples were prepared by the

Table 10.1 The effect of starting material, processing route and milling parameters on sintered density, grain size and other relevant properties of tungsten-heavy alloys. For brevity, tensile strength, % elongation, ultimate tensile strength, yield strength, contiguity, hardness and elastic modulus are respectively abbreviated as TS, % el., UTS, YS, C_{WW}, H and E

System	Starting material	Processing route	Milling parameters			Consolidation		Sintered density	Grain size of W in sintered samples	Properties of sintered samples	Refs.
			Time	Speed	Ball-to-powder ratio	Consolida-tion process	Consolidation parameters				
90W–7Ni–3Fe	W (2 μm, irregular), Ni (4.3 μm, spiky), Fe (<10 μm, nearly spherical)	Mechanical alloying, sintering	12 h	1000 rpm	10:1	Solid-state sintering	T: 1250–1400 °C, t: 60 min, Vacuum	–	1.7–3.0 μm	*Max. TS*: 1204 MPa	[27]
90W–7Ni–3Fe	W (2.91 μm), Ni (2.66 μm), Fe (3.97 μm)	Mechanical alloying, liquid phase sintering	40 h	200 rpm	10:1	Liquid phase sintering	T: 1480 °C, t: 30 min, H_2	98.7% theoretical	40–60 μm	*TS*: 923 MPa, *% el.*: 8	[31]
89W–7Ni–3Fe–1Re	W (3–5 μm, cuboid), Ni (6 μm, spherical), Fe (5 μm, spherical), Re (<200 mesh, irregular)	Mixing, liquid phase sintering	–	–	–	Liquid phase sintering	T: 1480 °C, t: 2 h, followed by heat treatment	98% theoretical	25 ± 11	C_{WW}: 0.36 ± 0.07, *YS*: 602 MPa, *TS*: 890 MPa, *% el.*: 14, *Impact energy:* 22 ± 10 J	[36]

(continued)

Table 10.1 (continued)

System	Starting material	Processing route	Milling parameters			Consolidation		Sintered density	Grain size of W in sintered samples	Properties of sintered samples	Refs.
			Time	Speed	Ball-to-powder ratio	Consolida-tion process	Consolidation parameters				
		Mechanical alloying, liquid phase sintering	5 h (W-1Re)	670 rpm	5:1	Liquid phase sintering	T: 1480 °C, t: 2 h, followed by heat treatment	99.7% theoretical	18 ± 6	C_{WW}: 0.27 ± 0.06, *YS*: 629 MPa, *TS*: 952 MPa, % *el.*: 23, *Impact energy*: 60 ± 5 J	
			48 h (milled W-1Re + Ni–Fe)	–	1:1						
93W–5.6Ni–1.4Fe	W (2.5 μm), Ni (2.5 μm), Fe (3.5 μm)	Mechanical alloying, solid state sintering	72 h	75 rpm	20:1	Solid-state sintering	T: 1300 °C, t: 1 h, H_2	> 99% theoretical	3 μm	C_{WW}: 0.74, *YS*: 1100 MPa, % *el.*: 0.5	[37]
		Mixing, liquid phase sintering	–	–	–	Liquid phase sintering	T: 1485 °C, t: 1 h, H_2	–	40 μm	C_{WW}: 0.35, *YS*: 620 MPa, % *el.*: 25	[37]
93W–5.6Ni–1.4Fe	W (2.5 μm), Ni (2.5 μm), Fe (3.5 μm)	Mechanical alloying of W, Ni, Fe, followed by two-step sintering	72 h			Two-step sintering	For SSS, T: 1300 °C, H_2 For LPS: T: 1445–1485 °C t: 4–90 min, H_2			*TS*: 955 MPa, % *el.*: 16	[45]

(continued)

Table 10.1 (continued)

System	Starting material	Processing route	Milling parameters			Consolidation		Sintered density	Grain size of W in sintered samples	Properties of sintered samples	Refs.
			Time	Speed	Ball-to-powder ratio	Consolida-tion process	Consolidation parameters				
		Mechanical alloying of 30W–56Ni–14Fe, followed by blending with W to meet the composition, followed by two-step sintering	72 h			Two-step sintering			6 µm	*TS:* 997 MPa, *% el.*: 19	
93W–5.6Ni–1.4Fe	W (2.5 µm), Ni (2.5 µm), Fe (3.5 µm)	Mechanical alloying, liquid phase sintering	72 h	–	–	Liquid phase sintering	T: 1485 °C, H_2		5–18 µm	*Impact energy:* ~70 J	[14]
							T: 1470 °C, H_2			*Impact energy:* ~52 J	
							T: 1460 °C, H_2			*Impact energy:* ~20 J	
							T: 1445 °C, H_2			*Impact energy:* ~5 J	

(continued)

Table 10.1 (continued)

System	Starting material	Processing route	Milling parameters			Consolidation		Sintered density	Grain size of W in sintered samples	Properties of sintered samples	Refs.
			Time	Speed	Ball-to-powder ratio	Consolidation process	Consolidation parameters				
93W–5.6Ni–1.4Fe	W (2.5 μm), Ni (2.5 μm), Fe (3.5 μm)	Mechanical alloying + two-stage sintering	72 h	75 rpm	20:1	Solid-state sintering	T: 1300 °C, t: 60 min, Medium: H_2	> 99% theoretical	–	*TS*: ~1100 MPa, *% el.*: <1	[46]
						Liquid phase sintering	T: 1445–1485 °C, T: 4–90 min, Hydrogen, followed by annealing @ 1150 °C for 1 h in N_2				
93W–5.6Ni–1.4Fe		Ball milling in N_2, followed by cold isostatic pressing and liquid phase sintering				Liquid phase sintering	T: 1493 °C, t: 3 h		35 μm	C_{WW}: 0.35, *TS*: 940 MPa, *% el.*: 30, *H*: 29.6 HRC	[17]
		Mechanical alloying, solid state or liquid phase sintering	72 h			Solid-state sintering	T: 1445 °C		5 μm	C_{WW}: 0.70, *TS*: 893 MPa, *% el.*: 1, *H*: 28.7 HRC, Charpy impact energy: 5 J	

(continued)

Table 10.1 (continued)

System	Starting material	Processing route	Milling parameters			Consolidation		Sintered density	Grain size of W in sintered samples	Properties of sintered samples	Refs.
			Time	Speed	Ball-to-powder ratio	Consolida-tion process	Consolidation parameters				
						Liquid phase sintering	T: 1460 °C		13 μm	C_{WW}: 0.46, *TS*: 932 MPa, *% el.*: 12.2, *H*: 28.3 HRC, Charpy impact energy: 19 J	
						Liquid phase sintering	T: 1470 °C		16.8 μm	C_{WW}: 0.43, *TS*: 924 MPa, *% el.*: 13.8, *H*: 28.1 HRC, Charpy impact energy: 52 J	
						Liquid phase sintering	T: 1480 °C		18.2 μm	C_{WW}: 0.40, *TS*: 923 MPa, *% el.*: 18.5, *H*: 27.8 HRC, Charpy impact energy: 70 J	
		Mechanical alloying, solid-state sintering + liquid phase sintering				Solid-state sintering	T: 1300 °C, t: 1 h		10.7 μm		
						Liquid phase sintering	T: 1470 °C, t: 4 min				

(continued)

Table 10.1 (continued)

System	Starting material	Processing route	Milling parameters			Consolidation		Sintered density	Grain size of W in sintered samples	Properties of sintered samples	Refs.
			Time	Speed	Ball-to-powder ratio	Consolida-tion process	Consolidation parameters				
93W–4.9Ni–2.1Fe		Mechanical alloying, liquid phase sintering	12 h	200 rpm	8:1	Liquid phase sintering	T: 1520 °C, t: 90 min	99.1% theoretical	40 µm	C_{WW}: 0.32 ± 0.04, Compressive strength: 1400 MPa	[35]
93W–4.9Ni–2.1Fe	W (2.5 µm), Ni (3.5 µm)	Mechanical alloying, cold isostatic pressing, liquid phase sintering	72 h	–	–	Liquid phase sintering	T: 1493 °C, t: 3 h	99.9% theoretical	27 µm	C_{WW}: 0.26, *TS*: 940 MPa, *% el.*: 30, Max. shear stress: 910 MPa	[28]
92.5W–4.9Ni–2.1Fe–0.5ZrO_2 (Y)		Ion-level liquid–liquid doping, mechanical alloying, liquid phase sintering	12 h	200 rpm	8:1	Liquid phase sintering	T: 1520 °C, t: 90 min	99.2% theoretical	25 µm	C_{WW}: 0.26 ± 0.03, Compressive strength: 1515 MPa	[35]
W–4.9Ni–2.1Fe–0.1 wt.% Y_2O_3	W (2.5 µm), Ni (3.5 µm)	Mechanical alloying, cold isostatic pressing, liquid phase sintering	72 h	–	–	Liquid phase sintering	T: 1493 °C, t: 3 h	99.1% theoretical	27 µm	C_{WW}: 0.27, *TS*: 883 MPa, *% el.*: 18.4, Max. shear stress: 856 MPa	[28]

(continued)

Table 10.1 (continued)

System	Starting material	Processing route	Milling parameters			Consolidation		Sintered density	Grain size of W in sintered samples	Properties of sintered samples	Refs.
			Time	Speed	Ball-to-powder ratio	Consolida-tion process	Consolidation parameters				
94W–4.56Ni–1.14Fe–$0.3Y_2O_3$	W (3 μm), Ni (6 μm), Fe (3 μm), Y_2O_3 (10 μm)	Mechanical alloying, two-stage sintering	6 h	200 rpm	10:1	Solid-state sintering, followed by liquid phase sintering	For SSS: T: 1300–1450 °C, t: 1 h For LPS: T: 1465–1485 °C, t: 1 h	>97% theoretical (for SSS), >99% theoretical (for LPS)	2 μm (for SSS), 7 μm (for LPS)	*YS*: 639–952 MPa, *UTS*: 883–1000 MPa	[29]
W–7Ni–3Co	W, Ni and Co powders (20–50 μm), Y_2O_3 (~20 nm)	Mechanical alloying, liquid phase sintering	4, 8, 16 and 24 h	350 rpm	10:1	Liquid phase sintering	Temperature: 1450 °C, Holding time: 60 min			*H*: 5.43 ± 0.23 GPa, *E*: 304.3 ± 7.4 GPa	[30]
91W–7Ni–2Co										*H*: 5.60 ± 0.38 GPa, *E*: 318.3 ± 15.1 GPa	
90W–8Ni–2Co										*H*: 5.77 ± 0.33 GPa, *E*: 314.9 ± 13.9 GPa	
92W–7Ni–1Co										*H*: 6.67 ± 0.25 GPa, *E*: 375.7 ± 17.0 GPa	
90W–9Ni–1Co										*H*: 6.87 ± 0.69 GPa, *E*: 333.9 ± 26.9 GPa	

(continued)

Table 10.1 (continued)

System	Starting material	Processing route	Milling parameters			Consolidation		Sintered density	Grain size of W in sintered samples	Properties of sintered samples	Refs.
			Time	Speed	Ball-to-powder ratio	Consolida-tion process	Consolidation parameters				
90W–7Ni–3Fe–0.04Y_2O_3	W (2.91 μm), Ni (2.66 μm), Fe (3.97 μm), Y_2O_3 (5 μm)	Mechanical alloying, liquid phase sintering	40 h	200 rpm	10:1	Liquid phase sintering	T: 1480 °C, t: 30 min, H_2	99.3% theoretical	8 – 12 μm	*TS*: 1050 MPa, % *el*.: 30.8	[31]
95W–3.5Ni–1.5Fe	W (5.2 μm), Ni (<10 μm), Fe (3–5 μm),	Mixing, liquid phase sintering	–	–	–	Liquid phase sintering	T: 1500 °C, t: 1 h, H_2	98.6 ± 0.05	31.1 ± 9.4	*E*: 467 ± 9 GPa, *H*: 62.07 ± 1.60 HRA	[34]
		Mechanical alloying, liquid phase sintering	5 h	300 rpm	10:1			98.2 ± 0.06	28.9 ± 5.4	*E*: 456 ± 23 GPa, *H*: 63.65 ± 1.40 HRA	
			10 h	300 rpm	10:1			97.4 ± 0.10	24.9 ± 6.2	*E*: 434 ± 18 GPa, *H*: 63.80 ± 0.70 HRA,	
			15 h	300 rpm	10:1			97.1 ± 0.15	22.7 ± 7.6	*E*: 431 ± 21 GPa, *H*: 64.80 ± 1.41 HRA	
W–3.5Ni–0.3Y_2O_3	W, Ni (15–50 μm), Y_2O_3 (~20 nm)	Mechanical alloying, sintering	1 h	1080 rpm	10:1		T: 1350 °C, t: 30 min, Ar			*H*: 386 HV	[32]
			3 h	1080 rpm	10:1					*H*: 369 HV	
			6 h	1080 rpm	10:1					*H*: 409 HV	
			9 h	1080 rpm	10:1					*H*: 572 HV	

Fig. 10.1 Microstructure of 90W–7Ni–3Fe powders subjected to high-energy ball milling, **a** 5, **b** 10, **c** 20, **d** 30 and **e** 40 h [33]. Reproduced with permission from Elsevier

same sintering cycle in spark plasma sintering (SPS, temp.: 1050 °C, holding time: 8 min, applied pressure: 50 MPa). In another work, Ryu et al. [13] found that the density of 93W–5.6Ni–1.4Fe alloy showed a decreasing trend when the milling was performed up to 48 h and the samples were sintered at 1485 °C. In the same work, the authors reported enhanced densification up to 24 h of milling when the samples were sintered by SSS at 1300 and 1400 °C. Figure 10.3 shows the microstructure of LPSed 90W–7Ni–3Co–0.5Y_2O_3 prepared from milled powders. Debata et al. [34] studied the effect of milling time on the structure and properties of 95W–3.5Ni–1.5Fe alloys prepared by mechanical alloying and LPS route and found that the enhanced milling duration caused a slight reduction in sintered density from 98.6% (in case of unmilled) to 97.1% (in case of 15 h milled). Figure 10.4 shows the effect of LPS on mechanically alloyed 95W-3.5Ni-1.5Fe where it is evident that increased milling duration results in grain refinement. Xu et al. [35] reported the effect of mechanical alloying on 12 h-milled 93W–4.9Ni–2.1Fe processed by LPS at 1520 °C and found that the sintered density was 99.1%. However, the grain size was relatively high (40 μm) which may be attributed to higher sintering temperature (1520 °C) and holding time (1 h 30 in). However, upon substitution of 0.5 wt.% W with ZrO_2(Y), the resultant alloy system achieved a sintered density of 99.2% theoretical with reduced grain size (25 μm). Interestingly, the contiguity of 92.5W–4.9Ni–2.1Fe–0.5ZrO_2(Y) was slightly low (0.26 ± 0.03) and compressive strength was found to be high (1515 MPa). The microstructure of solid-state sintered WHA prepared by mechanical alloying can be found in the literature [38]. Figure 10.5 shows the microstructure and compositional analysis of 90W–7Ni–3Fe mechanically alloyed powder processed by SPS at 1200 °C for 8 min under applied pressure of 50 MPa.

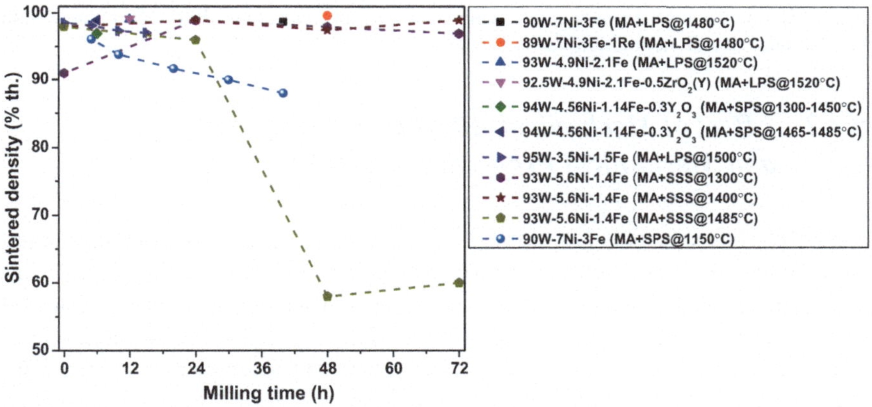

Fig. 10.2 Effect of milling duration on the sintered density of WHAs processed by mechanical alloying and sintering route. Data obtained from [13, 28, 29, 31, 33–37]

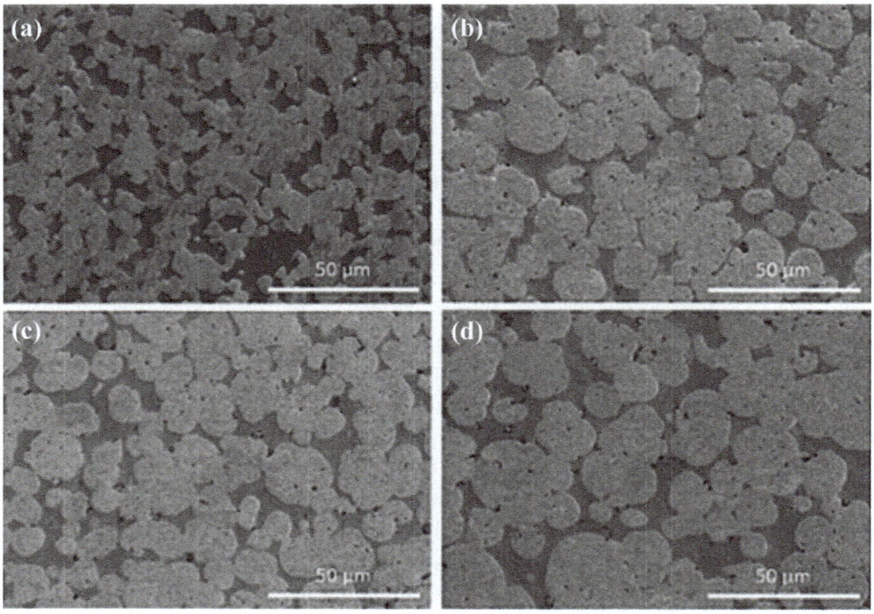

Fig. 10.3 The evolution of microstructure in 90W–7Ni–3Co–0.5Y_2O_3 alloy prepared from **a** 4, **b** 8, **c** 12 and **d** 16 h milled powders. The samples were consolidated by LPS 1450 °C for 60 min [30]. Reproduced with permission from Elsevier

10.2.2 Effect of Mechanical Alloying on Grain Size and Mechanical Properties of WHAs

Figure 10.6 shows the effect of milling time on the grain size of tungsten in sintered WHA. Of late, Antolak-Dudka et al. [40] fabricated W-ODS composites by mechanical alloying of W-(5, 10) wt.% Y_2O_3 powders for 50 h in argon atmosphere, followed by pulse plasma sintering at 1640 °C and found that Y_2O_3 incorporation improves the hardness of the resulted composites. Wang et al. [41] developed 95.1% dense W-Y_2O_3 alloy by mechanical alloying and spark plasma sintering. The authors reported dispersion of Y_2O_3 in the W matrix and found (a) BCC to FCC transformation and (b) significant grain size refinement of Y_2O_3 dispersoids owing to mechanical alloying. The resulted alloy exhibited a very good combination of hardness (625.7 ± 16.9 HV), compressive strength (1804.2 MPa) and bending strength (934 MPa).

In the past, Akhtar [27] performed solid-state sintering of 12 h mechanical alloyed 90W–7Ni–3Fe alloy at 1250–1400 °C for 60 min in vacuum. The grain size of W in sintered alloy was 1.7–3.0 μm. In another study, Fan et al. [31] performed solid state sintering of 72 h milled 93W–5.6Ni–1.4Fe alloy at 1300 °C for 1 h in an H_2 atmosphere and obtained a density of >99% theoretical and the grain size was 3 μm. The SSS led to a contiguous microstructure (contiguity: 0.74) and subsequently, the yield strength was very high (1100 MPa) and ductility was inferior (0.5%). In

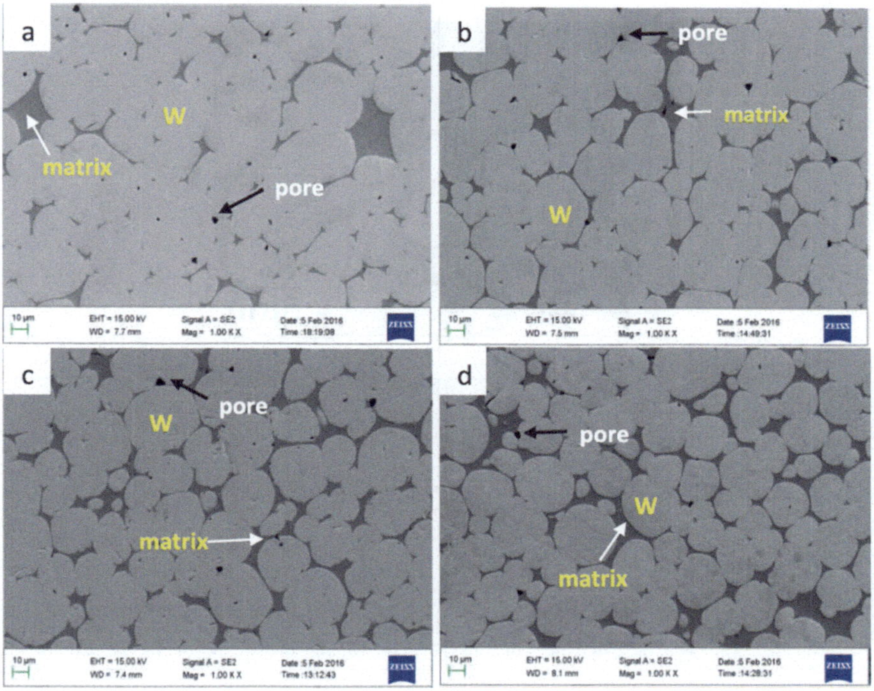

Fig. 10.4 The microstructure of liquid phase sintered 95W–3.5Ni–1.5Fe alloy prepared from **a** 0, **b** 5, **c** 10 and **d** 15 h milled powders [34] Reproduced with permission from Elsevier

the same study, the authors prepared 93W–5.6Ni–1.4Fe alloy by blending powders followed by LPS at 1485 °C for 1 h in an H_2 atmosphere and obtained a grain size of 40 μm. The enhancement in grain size is ascribed to liquid phase sintering which promotes the Ostwald ripening phenomenon. In the LPSed alloy, the contiguity was relatively low (0.35), which resulted in improved ductility (25%) in sintered alloy.

In another work, Kim et al. [17] prepared 93W–5.6Ni–1.4Fe by ball milling in an N_2 atmosphere, followed by Cold Isostatic Pressing (CIP) and LPS at 1493 °C for 3 h and obtained a grain size of 35 μm and the tensile strength and ductility was 940 MPa and 30%, respectively. When the same alloy was prepared by mechanical alloying for 72 h followed by SSS/LPS, the substantial reduction in grain size was observed. For instance, the minimum grain size of 5 μm and the highest contiguity of 0.70 was found in 1445 °C sintered alloy, which also exhibited extremely poor ductility (1%) and impact energy (5 J). Consequent upon consolidation by LPS at 1460 and 1470 °C, the % elongation was increased by 12.2 and 13.8%, respectively. Furthermore, the highest impact energy of 52 J was observed in 93W–5.6Ni–1.4Fe alloy prepared by mechanical alloying and LPS at 1470 °C. Compared to SSS, the grain size of LPS-ed WHA was relatively high. In the same study, the authors also reported the processing of WHA by mechanical alloying, followed by two-stage sintering. Park et al. [28] demonstrated the sintering behaviour and mechanical

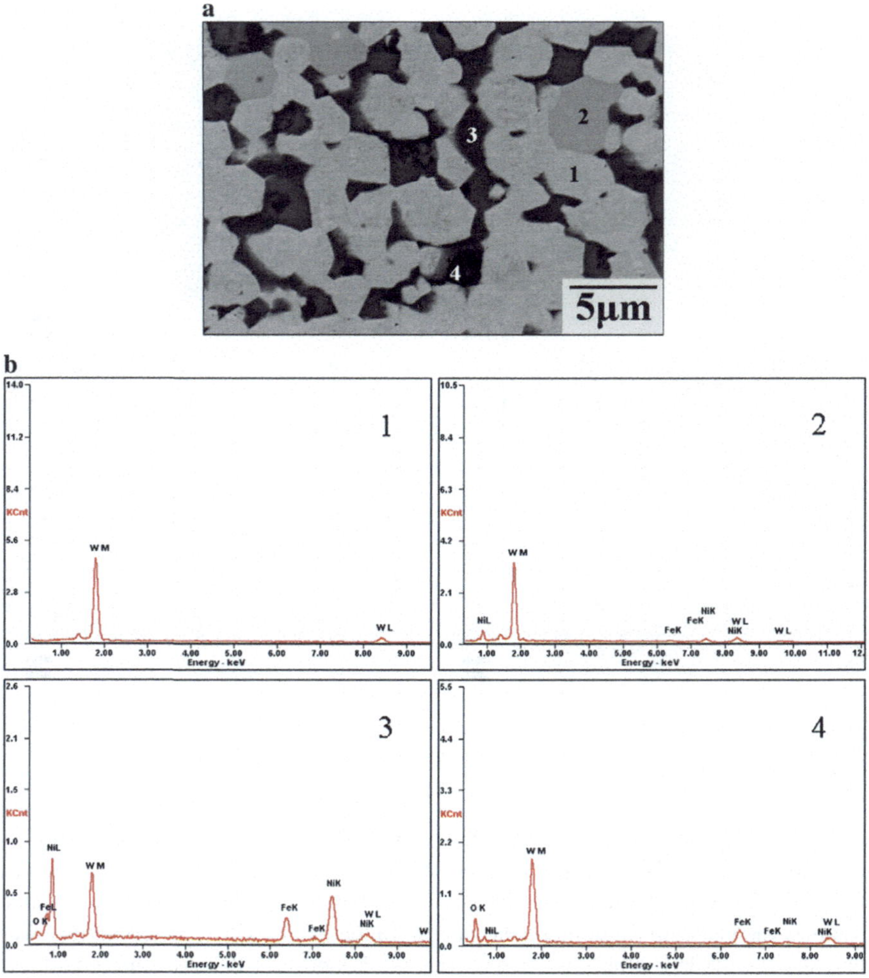

Fig. 10.5 **a** SEM image and **b** EDS mapping of 90W–7Ni–3Fe-heavy prepared by mechanical alloying for 40 h, followed by SPS at 1200 °C for 8 min under applied pressure of 50 MPa [39]. Reproduced with permission from Elsevier

properties of 93W–4.9Ni–2.1Fe and W–4.9Ni–2.1Fe–0.5Y_2O_3 systems prepared by mechanical alloying for 72 h. The compaction and consolidation were performed by CIP and LPS respectively. The Y_2O_3 incorporated WHA resulted in relatively low density (99.1%), tensile strength (883 MPa), ductility (18.4%) and shear stress (856 MPa). Lee et al. [29] prepared 94W–4.56Ni–1.14Fe–03Y_2O_3 by mechanical alloying and a two-stage sintering process. The respective density after SSS and LPS was >97% and >99%, while the grain size of W in SSS and LPSed samples was 2 μm and 7 μm, respectively. Chen et al. [30] investigated the effect of milling

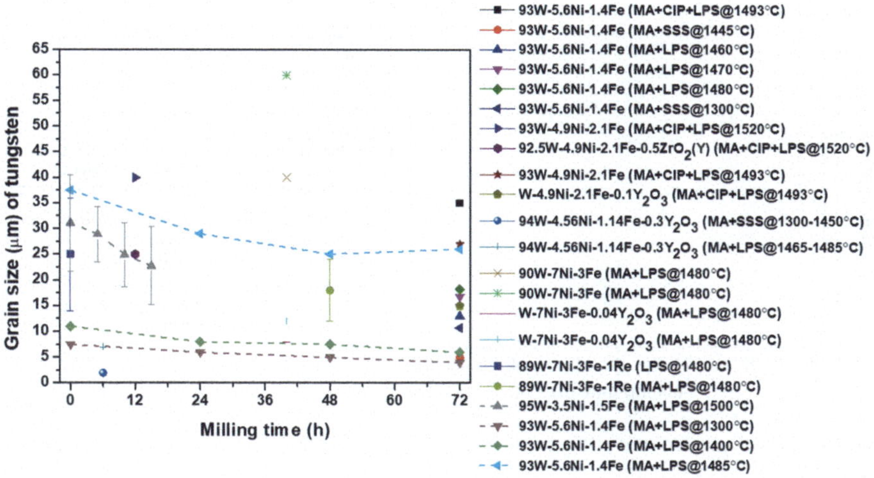

Fig. 10.6 The effect of milling time on grain size of tungsten in sintered WHAs. The processing condition is also mentioned against each composition. Data obtained from [13, 17, 28, 29, 31, 35, 42]

time (4–24 h) on the structure and properties of liquid phase sintered 90W–7Ni–3Co, 91W–7Ni–2Co, 92W–7Ni–1Co and 90W–9Ni–1Co alloys dispersed with 0.5 Wt.% Y_2O_3. The highest hardness of 6.87 ± 0.69 GPa was found in 90W–9Ni–1Co–0.5Y_2O_3 (Ni/Co = 9:1) alloy; while the maximum elastic modulus of 375.7 ± 17.0 GPa was recorded by 92W–7Ni–1Co–0.5Y_2O_3.

Fan et al. [31] prepared 90W–7Ni–3Fe and 90W–7Ni–3Fe–0.04Y_2O_3 alloys by mechanical alloying for 40 h and sintered the samples at 1480 °C for 30 min in H_2 atmosphere. Though the resulted density was almost similar, there was a substantial reduction in the grain size from 40 to 60 μm (in case of 90W–7Ni–3Fe) to 8–12 μm (in case of 90W–7Ni–3Fe–0.04Y_2O_3). The mechanically alloyed and LPSed 90W–7Ni–3Fe–0.04Y_2O_3 offered good combination of tensile strength (1050 MPa) and ductility (30.8%). The effect of Re addition and the processing condition on the structure and properties of W–Ni–Fe alloys is also investigated in the literature [36]. Initially, the milling of the W-1Re system was performed for 5 h, which was followed by mechanical alloying of W–1Re with Ni–Fe powder for 48 h. The grain size of mechanically alloyed 89W–7Ni–3Fe–1Re was 18 ± 6 μm, which was substantially low compared to the alloy processed by blending and LPS route. The YS, TS, % elongation and impact energy were found to be superior in mechanically alloyed systems. Figure 10.7 highlights the effect of mechanical alloying on the % elongation of sintered WHAs.

Mechanical properties of WHA are mostly dependent on microstructural factors. WHAs are made of predominantly BCC tungsten phase and the rest FCC matrix phase. The yield strength for liquid phase sintered WHA (80–90 wt.% W) is in the

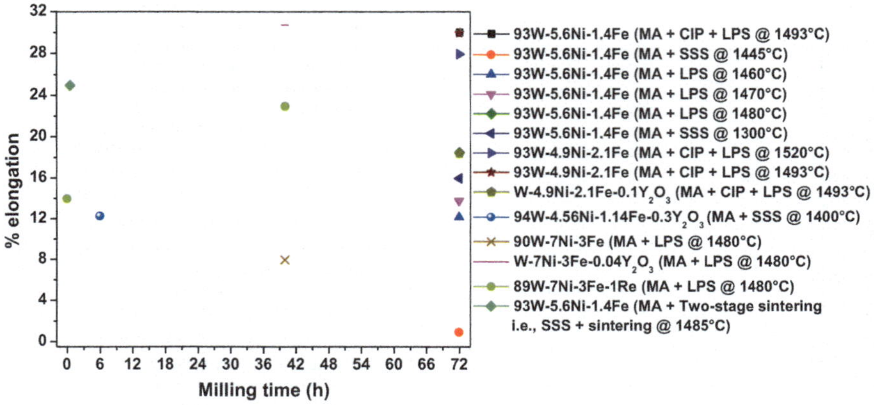

Fig. 10.7 The dependence of milling time and processing condition on % elongation behaviour of sintered WHAs. Data obtained from [17, 28, 29, 31, 36, 38]

range of 600–800 MPa, with a % elongation of 8–25%. Whereas, in the case of solid phase sintering the yield strength is in the range of 1100–1200 MPa with elongation of 0.5–2%. It is evident that solid-state sintered samples are much brittle and hard, this can be attributed to the microstructure of the solid-state sintered WHAs. The main mode of failure in case of solid-state sintered WHA is W-W intergranular cleavage, W-Matrix decohesion and matrix failure. Due to the inhomogeneous distribution of the matrix in solid-state sintered WHA, the matrix is more prone to failure [27]. The microstructure of liquid phase sintered WHA is comparatively homogeneous with rounded W grains evenly distributed in the matrix. In case of liquid phase sintering, the fracture mechanism is dominated by W–W cleavage and W-matrix decohesion and ductile failure of the matrix. The homogeneous distribution of the ductile matrix in the WHAs leads to increased elongation % and decreased yield strength.

The mechanical alloying of WHA changes the final properties of the sintered product drastically. This could be attributed to the different types of microstructure formed after the sintering of the mechanically alloyed, deformed powders. Wu et al. [26] studied extensively the mechanical alloying of WHA. They optimized various ball milling parameters and found that 700 RPM and 8% process control agent work most efficiently. However, it should be noted that the milling parameters may vary from instrument to instrument. Nokhrin et al. [25] found that mechanical alloying of WHA (90W–7Ni–3Fe) leads to the formation of the supersaturated solid solution of W in the FeNi matrix. They also studied the effect of SPS on the mechanically alloyed WHA, where it was established that mechanical alloyed WHA shows greater mechanical properties after SPS at higher temperatures (1450–1500 °C) [25]. They showed that increasing milling duration leads to smaller particle size which promotes diffusion along the nonequilibrium grain boundary, leading to increased creep resistance. Mechanical alloying generally promotes grain refinement and flattening of the starting powder particle [17, 25, 28]. After sintering (both solid state and liquid phase) the mechanically alloyed powder shows grain refinement and increase in contiguity [17]. Also, complete grain rounding of W may not be achieved during liquid sintering in the mechanically alloyed powder, leading to increased contiguity in the sample. The increased contiguity and uneven distribution of the W-matrix interface lead to the availability of higher crack nucleation sites and ultimately decrease the yield strength, tensile strength and fracture toughness [25].

Mechanical alloying is mostly adapted to disperse oxides in WHA. Mechanically alloyed oxide dispersed (Y_2O_3, Al_2O_3, La_2O_3, ZrO_2, etc.) WHAs are primarily studied for their superior dynamic deformation and penetration characteristics [28, 32, 43, 44]. Mechanical alloying is followed to obtain a uniform distribution of oxide nanoparticles in the WHA [32, 43, 44]. It is found that in case of ODS-WHA, sintering (both solid state and liquid phase) leads to a decrease in the yield strength, which can be attributed to the presence of oxides at the triple junction at W–W/matrix interface and acts as a crack initiation point [28]. The higher contiguity in case of SSS and LPS of the mechanically alloyed ODS-WHA is also responsible for the poor mechanical performance of the system [17]. However, mechanically alloyed ODS-WHA shows improved fracture toughness, impact resistance and creep resistance at high temperatures due to the prevention of W–W grain fracture by the dispersed oxides. Further to improve the mechanical strength and performance of the WHA prepared by mechanical alloying, two-stage sintering is followed by many researchers, where initially the mechanically alloyed WHAs are solid state sintered and subsequently liquid phase sintered. This type of two-step sintering is followed to achieve a comparatively lower contiguity and to allow grain growth.

10.3 Summary

Tungsten-heavy alloys have a high demand for various strategic applications due to their exceptional density and mechanical properties. These alloys are fabricated by conventional powder metallurgy routes mostly comprising of liquid phase sintering, due to distinct advantages over their solid-state sintered counterparts. The introduction of a mechanical alloying step in the processing of WHAs can assist in grain size refinement, which in turn improves the mechanical properties of consolidated heavy alloys. In addition to the refined microstructure, where the matrix phase is uniformly distributed encapsulating the tungsten spheroids, the mechanically alloyed and sintered WHAs exhibit better interfacial bonding between the W grain and the matrix phase; which improves the strength of the sintered alloys due to multiple factors like solid solution strengthening and grain boundary strengthening. However, the optimization of milling parameters and sintering conditions play a phenomenal role in defining the end properties of the mechanically alloyed and sintered WHAs for various intended applications.

References

1. German, R.M.: Sintered tungsten heavy alloys: review of microstructure, strength, densification, and distortion. Int. J. Refract. Met. Hard Mater. **108** (2022). https://doi.org/10.1016/j.ijrmhm.2022.105940
2. Lee, P.W., Trudel, Y., Iacocca, R., German, R.M., Ferguson, B.L., Eisen, W.B., Moyer, K., Madan, D., Sanderow, H.: ASM Handbook Volume 7: powder Metal Technologies and Applications. ASM International (1991)
3. Upadhyaya, A., G.S. Upadhyaya, Powder Metallurgy: science, Technology, and Materials. Universities Press (2011)
4. Sengupta, P., Debata, M.: Effect of partial and full substitution of Ni with NiB on densification, structure and properties of 90W–6Ni–2Fe–2Co heavy alloys. J. Alloys Compd. **774**, 145–152 (2019). https://doi.org/10.1016/j.jallcom.2018.09.368
5. Sengupta, P., Panigrahi, A., Indoria, B., Meher, P., Acharya, T.S., Basu, S., Debata, M.: Substitution of Ni with NiB prevents shape distortion of liquid phase sintered 90W–6Ni–2Fe–2Co heavy alloys. J. Alloys Compd. **840**, 155785 (2020). https://doi.org/10.1016/j.jallcom.2020.155785
6. Sengupta, P., Meher, P., Panigrahi, A., Mandal, S., Basu, S., Debata, M.: Microstructure, distortion characteristics and mechanical behaviour of NiB modified 90W–6Ni–4Co heavy alloys. J. Alloys Compd. **887**, 161404 (2021). https://doi.org/10.1016/j.jallcom.2021.161404
7. Lee, K.H., Cha, S.I., Ryu, H.J., Hong, S.H.: Effect of two-stage sintering process on microstructure and mechanical properties of ODS tungsten heavy alloy. Mater. Sci. Eng. A (2007). https://doi.org/10.1016/j.msea.2007.01.118
8. Ryu, H.J., Hong, S.H.: Fabrication and properties of mechanically alloyed oxide-dispersed tungsten heavy alloys. Mater. Sci. Eng. A (2003). https://doi.org/10.1016/S0921-5093(03)00641-5
9. Li, Z.B., Wang, Y., Zhang, H., Chen, B., Zhang, G.H., Chou, K.C.: Effect of ZrB2 addition on microstructure evolution and mechanical properties of 93 wt.% tungsten heavy alloys. Mater. Sci. Eng. A **825** (2021). https://doi.org/10.1016/j.msea.2021.141870
10. Mandal, S., Sengupta, P., Dey, S., Kumar, M., Basu, S., Debata, M.: Effect of Y2O3, La2O3 and ZrO2 dispersoid addition on ultra-high temperature stability of 95 W–3.5Ni–1.5Fe heavy alloy. Int. J. Refract. Met. Hard Mater. 106195 (2023). https://doi.org/10.1016/j.ijrmhm.2023.106195
11. Bose, A., Sadangi, R., German, R.M.: A review on alloying in tungsten heavy alloys. Suppl. Proc. Vol. 1 Mater. Process. Interfaces. 455–465 (2012). https://doi.org/10.1002/9781118356074.ch5
12. Bose, A., German, R.M.: Microstructural refinement of W-Ni-Fe heavy alloys by alloying additions. Metall. Trans. A **19**, 3100–3103 (1988)
13. Ryu, H.J., Hong, S.H., Back, W.H.: Mechanical alloying process of 93W–56Ni-14Fe tungsten heavy alloy. J. Mater. Process. Technol. **63**, 292–297 (1997)
14. Ryu, H.J., Hong, S.H., Kim, D.K., Lee, S.: Impact and dynamic deformation behavior of mechanically alloyed tungsten heavy alloy. Met. Mater. Int. **4**, 367–371 (1998). https://doi.org/10.1007/bf03187793
15. Lee, K.H., Cha, S.I., Ryu, H.J., Dilmore, M.F., Hong, S.H.: Effect of mechanical alloying process on microstructure and mechanical properties of ODS tungsten heavy alloys. J. Alloys Compd. **434–435**, 433–436 (2007). https://doi.org/10.1016/j.jallcom.2006.08.284
16. Chen, C.L., Huang, C.L.: Milling media and alloying effects on synthesis and characteristics of mechanically alloyed ODS heavy tungsten alloys. Int. J. Refract. Met. Hard Mater. (2014). https://doi.org/10.1016/j.ijrmhm.2014.01.007
17. Kim, D.K., Lee, S., Ryu, H.J., Hong, S.H., Noh, J.W.: Correlation of microstructure with dynamic deformation behavior and penetration performance of tungsten heavy alloys fabricated by mechanical alloying. Metall. Mater. Trans. A Phys. Metall. Mater. Sci. **31**, 2475–2489 (2000). https://doi.org/10.1007/s11661-000-0193-0

18. Suryanarayana, C.: Mechanical alloying and milling. Mech. Alloy. Milling. **46**, 1–472 (2004). https://doi.org/10.4150/kpmi.2006.13.5.371
19. Suryanarayana, C., Ivanov, E., Boldyrev, V.V.: The science and technology of mechanical alloying. Mater. Sci. Eng. A **304–306**, 151–158 (2001). https://doi.org/10.1016/S0921-5093(00)01465-9
20. Cantor, B., Chang, I.T.H., Knight, P., Vincent, A.J.B.: Microstructural development in equiatomic multicomponent alloys. Mater. Sci. Eng. A **375–377**, 213–218 (2004)
21. Mandal, S., Panigrahi, A., Rath, A., Bönisch, M., Sengupta, P., Debata, M., Basu, S.: Formation of L10 ordering in FeNi by mechanical alloying and field-assisted heat treatment: synchrotron XRD studies. ACS Omega **8**, 13690–13701 (2023). https://doi.org/10.1021/acsomega.2c07869
22. Salian, A., Sengupta, P., Vishalakshi Aswath, I., Gowda, A., Mandal, S.: A review on high entropy silicides and silicates: fundamental aspects, synthesis, properties. Int. J. Appl. Ceram. Technol. **20**, 2635–2660 (2023). https://doi.org/10.1111/ijac.14422
23. Sengupta, P., Debata, M., Jayasankar, K.: Properties of Immiscible Cu-20 wt.% Mo alloy prepared by high energy ball milling and cold isostatic pressing. Trans. Powder Metall. Assoc. India. **43**, 26–32 (2017)
24. Ishijima, Y., Kurishita, H., Yubuta, K., Arakawa, H., Hasegawa, M., Hiraoka, Y., Takida, T., Takebe, K.: Current status of ductile tungsten alloy development by mechanical alloying. J. Nucl. Mater. **329–333**, 775–779 (2004). https://doi.org/10.1016/j.jnucmat.2004.04.306
25. Nokhrin, A.V., Malekhonova, N.V., Chuvil'deev, V.N., Melekhin, N.V., Bragov, A.M., Filippov, A.R., Boldin, M.S., Lantsev, E.A., Sakharov, N.V.: Effect of high-energy ball milling time on the density and mechanical properties of W–7%Ni–3%Fe alloy. Metals (Basel) **13**, 1432 (2023). https://doi.org/10.3390/met13081432
26. Wu, Z., Liang, Y., Fu, E., Du, J., Wang, P., Fan, Y., Zhao, Y.: Effect of ball milling parameters on the refinement of tungsten powder. Metals (Basel). **8**, 281 (2018). https://doi.org/10.3390/met8040281
27. Akhtar, F.: An investigation on the solid state sintering of mechanically alloyed nano-structured 90W–Ni–Fe tungsten heavy alloy. Int. J. Refract. Met. Hard Mater. **26**, 145–151 (2008). https://doi.org/10.1016/j.ijrmhm.2007.05.011
28. Park, S., Kim, D.K., Lee, S., Kim, D.K., Ryu, H.J., Hong, S.H.: Dynamic deformation behavior of an oxide-dispersed tungsten heavy alloy fabricated by mechanical alloying. Metall. Mater. Trans. A Phys. Metall. Mater. Sci. **32**, 2011–2020 (2001). https://doi.org/10.1007/s11661-001-0013-1
29. Lee, K.H., Cha, S.I., Ryu, H.J., Hong, S.H.: Effect of two-stage sintering process on microstructure and mechanical properties of ODS tungsten heavy alloy. Mater. Sci. Eng. A **458**, 323–329 (2007). https://doi.org/10.1016/j.msea.2007.01.118
30. Chen, C.L., Ma, S.H.: Effects of Ni/Co ratio and mechanical alloying on characteristics and sintering behavior of W–Ni–Co tungsten heavy alloys. J. Alloys Compd. **711**, 488–494 (2017). https://doi.org/10.1016/j.jallcom.2017.04.037
31. Fan, J.L., Cheng, H.C., Huang, B.Y., Liu, T., Wang, D.L.: Fine-grained tungsten heavy alloy by mechanical alloying with Yttrium oxide addition. Mater. Sci. Forum **534–536**, 1261–1264 (2007). https://doi.org/10.4028/www.scientific.net/msf.534-536.1261
32. Chen, C.L., Huang, C.L.: Mechanical alloying of ODS tungsten heavy alloys and microstructure development of intermetallic compounds. Met. Mater. Int. **19**, 1047–1051 (2013). https://doi.org/10.1007/s12540-013-5018-y
33. Xiang, D.P., Ding, L., Li, Y.Y., Chen, G.B., Zhao, Y.W.: Preparation of fine-grained tungsten heavy alloys by spark plasma sintered W–7Ni–3Fe composite powders with different ball milling time. J. Alloys Compd. **562**, 19–24 (2013). https://doi.org/10.1016/j.jallcom.2013.02.014
34. Debata, M., Acharya, T.S., Sengupta, P., Acharya, P.P., Bajpai, S., Jayasankar, K.: Effect of high energy ball milling on structure and properties of 95W–3.5Ni–1.5Fe heavy alloys. Int. J. Refract. Met. Hard Mater. **69**, 170–179 (2017). https://doi.org/10.1016/j.ijrmhm.2017.08.007
35. Xu, L., Xiao, F., Wei, S., Zhou, Y., Pan, K., Li, X., Li, J., Liu, W.: Development of tungsten heavy alloy reinforced by cubic zirconia through liquid-liquid doping and mechanical alloying

methods. Int. J. Refract. Met. Hard Mater. **78**, 1–8 (2019). https://doi.org/10.1016/j.ijrmhm.2018.08.009
36. Ravi Kiran, U., Panchal, A., Prem Kumar, M., Sankaranarayana, M., Nageswara Rao, G.V.S., Nandy, T.K.: Refractory metal alloying: a new method for improving mechanical properties of tungsten heavy alloys. J. Alloys Compd. **709**, 609–619 (2017). https://doi.org/10.1016/j.jallcom.2017.03.174
37. Ryu, H.J., Hong, S.H., Baek, W.H.: Microstructure and mechanical properties of mechanically alloyed and solid-state sintered tungsten heavy alloys. Mater. Sci. Eng. A **A291**, 91–96 (2000)
38. Ryu, H.J., Hong, S.H.: Effects of sintering conditions on mechanical properties of mechanically alloyed tungsten heavy alloys. Met. Mater. Int. **7**, 221–226 (2001). https://doi.org/10.4028/www.scientific.net/kem.183-187.1291
39. Ding, L., Xiang, D.P., Li, Y.Y., Li, C., Li, J.B.: Effects of sintering temperature on fine-grained tungsten heavy alloy produced by high-energy ball milling assisted spark plasma sintering. Int. J. Refract. Met. Hard Mater. **33**, 65–69 (2012). https://doi.org/10.1016/j.ijrmhm.2012.02.017
40. Antolak-Dudka, A., Oleszak, D., Zielinski, R., Kulik, T.: W-Y2O3 composites obtained by mechanical alloying and sintering. Adv. Powder Technol. **32**, 390–397 (2021). https://doi.org/10.1016/j.apt.2020.12.017
41. Wang, M., Sun, H., Pang, B., Xi, X., Nie, Z.: Structure evolution of Y2O3 and consequent effects on mechanical properties of W-Y2O3 alloy prepared by ball milling and SPS. Mater. Sci. Eng. A **832**, 142448 (2022). https://doi.org/10.1016/j.msea.2021.142448
42. Kiran, U.R., Panchal, A., Kumar, M.P., Sankaranarayana, M., Rao, G.V.S.N., Nandy, T.K.: Refractory metal alloying: a new method for improving mechanical properties of tungsten heavy alloys. J. Alloys Compd. (2017). https://doi.org/10.1016/j.jallcom.2017.03.174
43. Lee, K.H., Cha, S.I., Ryu, H.J., Hong, S.H.: Effect of oxide dispersoids addition on mechanical properties of tungsten heavy alloy fabricated by mechanical alloying process. Mater. Sci. Eng. A **452–453**, 55–60 (2007). https://doi.org/10.1016/j.msea.2006.10.155
44. Prabhu, G., Arockia Kumar, R., Nandy, T.K.: Effect of yttrium oxide dispersion on the microstructure and properties of tungsten heavy alloys. Def. Sci. J. **68**, 406–411 (2018). https://doi.org/10.14429/dsj.68.12255
45. Ryu, H.J., Hong, S.H., Lee, S., Baek, W.H.: Microstructural control of and mechanical properties of mechanically alloyed tungsten heavy alloys. Met. Mater. Int. **5**, 185–191 (1999). https://doi.org/10.1007/BF03026051
46. Hong, S.H., Ryu, H.J.: Combination of mechanical alloying and two-stage sintering of a 93W–5.6Ni–1.4Fe tungsten heavy alloy. Mater. Sci. Eng. A **344**, 253–260 (2003). https://doi.org/10.1016/S0921-5093(02)00410-0

Chapter 11
Fabrication of Cermets by Mechanical Alloying

Emre Altas and Bugra Ates

Abstract Today a wide variety of studies are carried out to discover new materials and develop production methods, and solutions to various technological needs are sought. Thanks to scientific and technological advances, we have reached a point where materials can be designed according to the needs of the product to be manufactured. In this regard, composite materials, in which a new material with better properties is obtained by combining the properties of its components, offer many technological innovations and developments and are the subject of scientific examination and research. It is possible to achieve the desired properties by manufacturing composites with various production techniques using the powder metallurgy technique. Therefore, it is important to reduce foreign dependency in this field and ensure academic accumulation by developing such materials. Considering the abovementioned situations, this study aims to develop cermet materials with high specific strength and density by mechanical alloying, one of the production techniques of powder metallurgy.

Keywords Mechanical alloying · Cermets · Sintering

11.1 Introduction

One of the impossibilities of today's technology is the difficulty in the production of some metal alloys. For example, it is very difficult to alloy a material with a high melting temperature with another material with a low melting temperature using old methods. In this case, the two metals can form a solution in the liquid state, but when it reaches a lower melting point, the metal separates as it cools and solidifies. At the

E. Altas
Department of Mechanical Engineering, Faculty of Engineering, Architecture and Design, Bartın University, Bartın, Turkey
e-mail: emrealtas@bartin.edu.tr

B. Ates (✉)
Department of Mechanical Engineering, Graduate School, Bartın University, Bartın, Turkey
e-mail: bugraatess92@gmail.com

S. Rajendrachari (ed.), *Mechanically Alloyed Novel Materials*, Advanced Structured Materials 220, https://doi.org/10.1007/978-981-97-6504-1_11

same time, the increasing demand for more durable, lighter and harder materials than those that can be produced with generally used manufacturing methods has made it necessary to design and develop new materials [1, 2].

Scientific research on materials science has continuously been directed towards improving the properties and performance of materials. Significant changes in mechanical, chemical and physical properties were achieved through chemical modifications and convectional thermal, mechanical and thermo-mechanical processing methods. However, increasing demands for materials that are stronger, harder and lighter than traditional materials have led to the design and development of advanced materials [3]. High-tech industries have contributed to this path. In recent years, material technology has been in a very rapid development and change process. Considering the conditions required by contemporary life in many main industrial branches such as automotive, energy, defense, communication and space-aircraft industries, reliability, economic difficulties and environmental concerns increase the need for new materials with superior properties beyond the existing ones. It is known that advanced technology systems cannot be achieved with the inherent properties of traditional materials such as metals, ceramics and polymers. Composite materials have been developed as alternatives to conventional classical materials to eliminate the limiting properties of metals, such as limited high-temperature properties, brittle characteristics of ceramics and low melting points of polymers [4–6].

Cermets are among the composite materials covering a very wide area; They are composite materials consisting of ceramic and metal materials, generally created by combining the hardness and high-temperature resistance of ceramic materials and the plastic deformability properties of metals. These materials are structures formed by metals such as nickel, molybdenum or cobalt, which have a high melting point and high wettability of ceramics, and ceramics in the form of oxide, boride, nitride or carbide dispersed in the matrix, and they stand out with their high wear and corrosion resistance [7]. Tungsten carbide is the most widely used cermet produced by sintering with a cobalt binder phase. In addition to the effect of the binder phase on the plastic deformation of the WC–Co system, which was developed as a cutting tool tip in the 1920s, it has a constantly rigid WC skeleton structure, maintains its stability up to 600 °C and attracts attention with its high wear and corrosion resistance. The production of WC–Co group materials was done with traditional powder metallurgy techniques until recently, but today it is done with powder metallurgy techniques such as hot pressing, spark plasma sintering, electric-current-assisted sintering, which allow sintering under pressure in order to obtain high mechanical properties [8]. The electric-current-assisted sintering technique is a powder metallurgy method based on the principle of sintering pre-shaped or unshaped powders with the Joule effect by passing an electric current through the mold and exposing the powders to be produced to electric current. This technique, which has a very low initial investment cost compared to other powder metallurgy methods, allows material production in a short processing time and high-density products at low costs [9, 10].

Cermets are composite materials formed by combining ceramic and metallic materials, combining the high-temperature properties and high hardness values of ceramics with the good plastic deformation properties of metals. These materials, which stand out with their high wear and corrosion resistance, are structures formed by ceramics in the form of carbides, borides, nitrides or oxides distributed within the matrix, and the matrix formed by metals such as Ni, Co or Mo, which have high wettability and melting points [11]. Tungsten Carbide Cobalt (WC–Co), the most well-known and widely used among cermet composites, was developed as a cutting tool in the 1920s, used in surface hardening applications in the 1970s, is generally produced by traditional sintering methods, and can maintain its stability up to 600 °C. It is a cermet known for its high wear and corrosion resistance. Since these materials combine high hardness and toughness, they are widely used in industrial applications such as cutting tools, drill bits, wire drawing dies, sealing rings for machine parts, and as candidate materials in applications where excellent mechanical and corrosion resistance are simultaneously required, such as seawater mining is considered. Cermets can be obtained by bonding ceramics such as Al_2O_3 and TiC with metals such as Mo, Cr, Fe, and Ni. The most important feature of cermets compared to ceramics is that they have good impact and bending strength [5, 12].

Powder metallurgy has a very important place among metal-forming methods and its history dates back to ancient times. The technology of modern powder metallurgy started in the 1920s with the production of tungsten carbides. While it was initially used due to its low production cost, today it is used because high quality, complex shaped and homogeneous parts can be manufactured easily and cheaply. Powder metallurgy provides many advantages when compared to known conventional production methods. Serial production of parts with complex geometry can be achieved. Products close to the final shape can be produced with almost no need for machining [13]. While parts with a wide composition range can be easily manufactured according to the desired mechanical properties, material losses during production are also very low. For such reasons, powder metallurgy is a highly preferred production technique. Metal powders used in powder metallurgy are produced by chemical, physical or mechanical methods, and the produced powders are mixed and blended together in various ways. The mixed powders are compressed at specified pressures and shaped. Finally, after pressing, the compressed powders are sintered to gain strength. When necessary, various secondary processes are applied to obtain the final product [14].

For the production of cermets, powder metallurgy is preferred because it offers many advantages. The biggest advantage of the powder metallurgy method for cermets is the good distribution of the reinforcement material within the matrix. Nowadays, the production of hard metals by powder metallurgy method has brought the mechanical properties to a higher level than the traditional method. The ease of flexible production, the fact that the production can be controlled at any time, and the strength of the parts, sometimes being twice as good as the traditional method, put powder metallurgy one step ahead in hard metal production. Powder metallurgy consists of three main stages: mixing of powders, pressing and sintering [15, 16].

The recent development of new methods used in joining metals has arisen from the limitations of alloying with old methods. One of the methods is mechanical alloying. However, mechanical alloying, a powder metallurgy method, is preferred in the production of cermets with better mechanical properties compared to other particle-reinforced composite production methods. With this technique, it is possible to produce alloys that are impossible and difficult to manufacture with traditional metallurgical techniques by properly alloying metal oxides and metal. Therefore, with the development of these types of materials, foreign exchange in this field is important to reduce dependency and ensure academic accumulation. In this study, it was aimed to develop cermet materials with high specific strength and density by mechanical alloying, which is one of the production techniques of powder metallurgy [17].

11.2 Cermet Metals

Cermets are structural materials composed of ceramic and metal materials, in which approximately equal fine-grained ceramic hard phase particles are embedded in a metal or alloy binder matrix. It is generally designed as an optimal combination of the high-temperature resistance and high hardness properties of a ceramic and the formability properties of a metal. In the formation of composite materials with cermet, a metal is used as a binder for ceramics in the form of oxide, boride, nitride or carbide. The metals generally used are nickel, molybdenum or cobalt, which have a high melting point and have high wettability of such ceramics. In addition to these metals, other metals within the solubility limits can also be used for alloying [12].

The most common use of cermets is cutting tool tips. Cermets are used in saws and other grinding devices instead of tungsten carbide (WC) thanks to their high wear and corrosion resistance. Titanium nitride (TiN), titanium carbonitride (TiCN), titanium carbide (TiC) and similar structures can be used in grinding and grinding like tungsten carbide if they are prepared properly. More complex materials are called Cermet II and are used in cutting tool tips if they have a longer working life in both grinding and grinding, such as tungsten carbide [18]. Cermets are used as ceramic-metal joint gaskets, as sealing elements in vacuum tubes, in systems that convert chemical, nuclear or thermo-ionic energy into electricity and in solid fuel cells. They are used as filling and prosthesis materials in dentistry. Since they are materials resistant to wear and friction, they are preferred in brake and lining mechanisms. Some types of cermets are also used as space shuttle shields because they are more resistant to high-speed impacts from orbital debris and micrometers than traditional space shuttle materials such as aluminum and other metals. In addition, cermets are among the preferred material groups in the production of resistors (especially potentiometers), capacitors and electronic components exposed to high temperatures [19–22]. The most well-known and widely used cermet composite material is tungsten carbide cobalt. WC–Co materials have mechanical properties that vary depending on the amount of cobalt they contain but also exhibit very high hardness and wear resistance. When the historical process of WC–Co is examined, it emerges with its introduction as a cutting tool tip in 1923, its use in surface hardening applications in the 1970s, and the production of nanocrystalline cemented carbides in 1994. The most important disadvantage of WC–Co is the oxidation of Co at high temperatures (around 600 °C) resulting from high cutting speeds and the degradation of WC and the deterioration of the integrity of WC–Co, followed by sudden breakage and loss of function of the abrasive tip. The disadvantages of these materials are eliminated by alloy additions that allow sintering at low temperatures [23, 24].

Because these materials combine toughness and high hardness, they are widely used in various industrial areas such as wire drawing dies, cutting and drilling bits, and sealing rings in machine parts. They are also candidate materials for logging, seawater mining and similar applications. Today, in applications where high wear resistance in corrosive environments comes to the fore, the materials to be used are required to have excellent mechanical and corrosion resistance as well as low cost. WC–Co/Ni and its additive combinations are thought to be capable of responding to these study demands [25, 26].

11.2.1 Physical and Mechanical Properties of Cermet Metals

Cermets are widely used in cutting tool tip and metal extrusion mold applications due to their high hardness, wear resistance and high-temperature resistance properties. Cermet tools are preferred in cutting tool tip applications due to their longer life and superior mechanical properties compared to high-speed steels. In this direction, the industrial and scientific studies carried out consist of new designs depending on the tip geometry and modifications made in the material composition and surface properties [27, 28].

The starting point for the industrial use of the WC–Co system was the search for an alternative to diamond-based wire drawing dies. WC–Co group sintered materials have very high hardness and wear resistance. Diamond, TiC, TiN and Al_2O_3 coatings on WC–Co are some of the studies carried out in recent years. The rapid development in coating technologies has also been reflected in the search for alternatives in this material, but the improvement of bulk materials has always maintained its importance. Binder and carbide phase modifications in the WC–Co system constitute the most studied topics [29–31].

WC–Co group materials are frequently sintered under pressure in order to obtain high mechanical properties. Today, the most common pressure sintering methods applied to this type of material are hot pressing and spark plasma sintering. Spark plasma sintering is a much newer technology compared to hot pressing, and in this method, superior mechanical properties of sintered materials can be obtained in shorter sintering times. There is a process starting from WC–Co cutting tool tips, developing with coating technologies, and then extending to the use of nano-sized powder today. It is noteworthy that in this development process, there was a search for alternative materials, especially tungsten, and binder and carbide phase modifications were made for this purpose. Carbide phase additions to the WC–Co system progressed over time to the use of tungsten-free carbides. TiC bonded with nickel-molybdenum is the most studied alternative, but the WC–Co system has never lost its universality in the application [23, 32].

The basic components of cermets are very hard and brittle WC grains and the tough cobalt-based binder phase that holds these grains together. In this structure, the binding phase is continuous. However, this is not the case for the carbide phase. According to one hypothesis, the carbide grains are dispersed so that they are embedded in the binder phase and there are thin binder films between them. According to this hypothesis, only the binder phase has an effect on the plastic deformation of WC–Co. There are studies supporting this, and the region that controls strength is the part that lacks the binder. Cermets are generally quite resistant to abrasive wear. This feature makes cemented carbides preferred in a wide range of usage areas. Abrasive wear is the most understandable type of wear for cermets and has been studied extensively. It occurs when a hard particle breaks away from a surface. In general, the particle exposed to abrasive effect is harder, but there are exceptional cases. Abrasive wear is a phenomenon that occurs especially in metal cutting processes and is closely related to the hardness of materials [33–35].

11.2.2 Production of Cermet Metals

Cermet metal production is basically carried out in three steps. The first step is the preparation of powders, where the necessary grinding and mixing processes are carried out. In the second step, the powders are shaped by pressing or extrusion. In the last step, hard metal production is carried out by sintering. Cemented carbides are produced by classical powder metallurgy methods. Powder mixtures are prepared in the desired composition and grain size and are shaped by compression with the help of metal molds. Compacts that gain raw strength are sintered with sintering methods selected depending on the properties expected from the material [12].

11.2.3 Preparation of Cermet Metal Powders

In the first stage, the basic materials to be used (WC, other carbides, grain growth inhibitors, carbon, binder metal) and the necessary additives (organic solvent, lubricant) are placed in the containers to be used for mixing and grinding processes. At this stage, generally, ball, vibrating or spindle grinders or attritors are used. The purpose of grinding powders may be to separate agglomerated particles by mixing or to reduce particle size. The type of grinder to be used, rotation speed, ball diameter, powder-ball ratio and hard metal content are optimized according to the purpose. At the end of the grinding process, the particle size can be reduced to 0.5–1 mm in industrial grinders and to 100–150 nm in more special grinders. Although a wide particle size distribution is achieved by grinding, this process ensures that the binder metal used is completely dispersed in the mixture and the binder, which causes porosity formation during sintering, is prevented from mixing and forming grains [36, 37].

Production methods of metallic powders are directly related to determining the properties of the starting powders and, therefore the final product. WC powders used in cermets are generally produced as a result of the carbothermal reduction of tungsten oxides with carbon heat. The resulting powders are subjected to grinding processes to bring them to the desired grain size and grain size distribution. Grain size and size distribution of powder-based materials are of great importance for the final properties of the material. Therefore, the production stage of powders is an important parameter for the quality of the final product. The resulting tungsten carbide powders are mixed with a binder and made ready for sintering. Due to its good wetting properties, cobalt is almost unbeatable as a binder choice. If carbide-based reinforcement additives are to be added to the system, these materials are added to the powder mixture during the powder preparation process [38, 39].

11.2.4 Cermet Powder Production by Mechanical Alloying Method

The mechanical alloying process is a method used to produce powders with smaller average grain size and smaller crystal structure, with homogeneous internal structure, thanks to the high energy generated in the system. Mechanical alloying process can obtain mixtures with homogeneous internal structures that do not form alloys with each other. In this method, matrix and reinforcement powders, WC, zirconia, sapphire, agate, hardened steel materials, etc. Balls made of different materials are placed in a container made of the same type of material. In the first stage, the powders crumble thanks to the high energy generated, and powder with a lower average powder size than the initial one is formed. In the second stage, the powders are combined with cold welding, and in the final stage, the combined powders are broken again. These stages continue continuously and thus, powders with a homogeneous and fine internal structure and increased dislocation density are obtained. Powders reinforced with this method are preferred in the production of metal matrix composite materials because they are distributed more homogeneously and the agglomeration rate is less. In addition, since the strength increases with the increasing amount of deformation, it ensures that defects in the final product are eliminated [12, 40]. Figure 11.1 shows a ball mill that grinds particles in a short time by releasing high energy with extreme friction and impact.

Mechanical alloying is defined as a high-energy milling process in which various metal powders are repeatedly cold welded, crushed and reworked. Thanks to its advantages such as low cost, ease of application, and simple technological equipment requirement, it has been accepted as an effective and practical way to produce various advanced nanocomposite powders. As mechanical energy is transferred to the powder particles, particle and grain sizes become smaller, diffusion distances decrease and a slight increase in temperature during grinding leads to the alloying of the blended

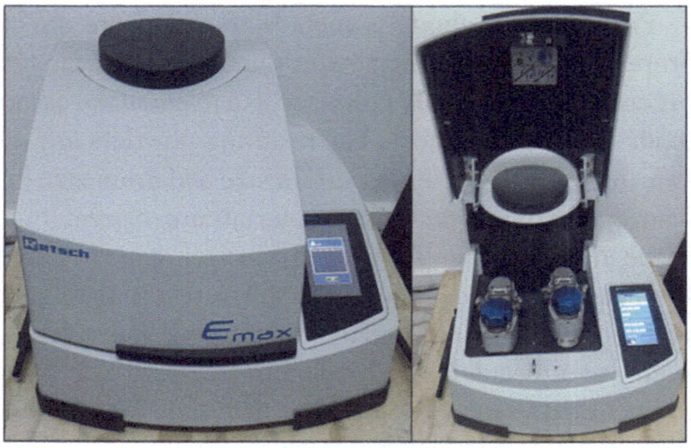

Fig. 11.1 Image of high-energy ball grinder

powders. A very important feature of mechanical alloying is that it triggers solid-state reactions and the synthesis of intermetallic phases such as boride, carbide, nitride and silicide. The extraordinary unique properties of transition metal carbides enable them to be used in very interesting and important industrial applications. For example, tungsten carbide (WC) has a high melting point, very high hardness, low coefficient of friction, low reactivity, high oxidation resistance and good thermal and electrical conductivity. This superiority enables WC to be used in cutting tools and wear-resistant parts, usually in the form of WC–Co hard materials, scratch-resistant materials, erosion-resistant coatings for aerospace parts, resistive parts in wire drawing, extrusion and pressing dies, platinum for polymer electrode membrane fuel cells. This makes it suitable for a variety of applications, such as the production of catalytic materials and thin-film diffusion barriers in microelectronics [40–44]. The reduction of grain sizes is carried out in a high-energy planetary-type ball milling device (Fig. 11.2).

We can summarize the advantages of materials obtained by mechanical alloying over materials manufactured according to other manufacturing techniques as follows:

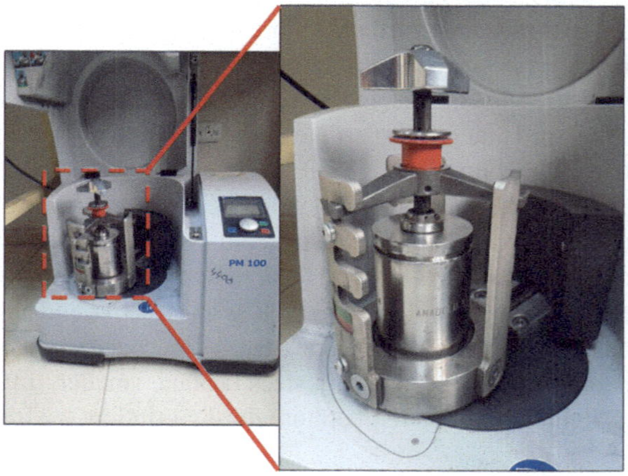

Fig. 11.2 Image of planetary ball mill

It is possible to manufacture alloys with very good corrosion resistance, to obtain intermetallic compounds from materials with very different melting points, to obtain parts that are difficult to manufacture with normal production methods such as casting, to ensure homogeneity throughout the structure of powders produced by mechanical alloying, to have complex shapes and sensitive tolerances [45, 46]. The advantages of this technique can be summarized as the ability to obtain parts and products that can operate at temperatures close to the melting point. The disadvantages of mechanical alloying are the long mixing time, high cost and oxygen-containing impurities. To eliminate the foreign matter problem, we can use an inert atmosphere and carefully select the tooling materials to be used, or we can coat the mixer with the same material to be ground. Long mixing time is a typical feature, as in other mixing processes. In ball mixers, the time for each charge (amount of mixed powder) is between 10 and 50 h. Mixing time can even exceed several hundred hours in some cases. If we look at the usage properties of materials manufactured with mechanical alloys, it is observed that they are not actually much higher than other alloys [47].

11.2.5 Forming Cermet Powders

Since the properties of the final product depend on density, uniform density must be maintained throughout the material. In the traditional compression method, pressure is applied from one direction, resulting in an anisotropic density distribution. The shaping process of cermets is done by applying 100–300 MPa pressure with semi-automatic or automatic mechanical or hydraulic presses. Since the pressure is applied from one direction, a slightly anisotropic density distribution is observed in the pressed part. Since carbide powders do not deform during shaping, the maximum density that can be achieved is 65% of the theoretical density [48, 49].

11.2.6 Sintering of Cermet Materials

The sintering process is done under hydrogen or vacuum. Alloys containing high amounts of TiC and TaC are sintered in vacuum to prevent residual gas porosity. In the first stage, the binder removal process is carried out by increasing the temperature to 400–500 °C. After this, the temperature is gradually increased to 1350–1600 °C, taking into account the hard metal content and grain size. During heating, carbon monoxide gas, which causes the formation of residual oxygen, is removed by keeping it at certain temperatures. Isothermal liquid phase sintering is a 1–1.5 h process [5, 50].

During sintering, condensation occurs below the liquidus temperature. Particles are rearranged during solid-state sintering, which is used for submicron or very fine-grained structures. For WC–Co hard metal, at these temperatures, Co diffuses onto WC surfaces by capillary forces. Partial dissolution of WC in Co activates this

Fig. 11.3 Image of high-temperature furnace for sintering powders

process. Partial melting occurs around 1300–1340 °C and WC continues to dissolve in Co up to its eutectic concentration (54% Co and 46% WC). WC–Co binary eutectic temperature is 1310 °C, and W–Co–C triple eutectic temperature is 1275–1280 °C. Increasing the temperature further causes the binder phase to melt completely and the WC to dissolve slightly more. At this stage, condensation occurs faster and a non-porous structure is obtained [39]. Sintering of the resulting powders is carried out in a high-temperature tube furnace environment (Fig. 11.3).

11.3 Conclusion

In this research, an overview of the mechanical alloying method, one of the most important powder metallurgy production methods for cermet materials, and an indication of its important properties are presented. It is possible to achieve the desired properties by manufacturing composites with various production techniques using the powder metallurgy technique. Therefore, it is important to reduce foreign dependency in this field and ensure academic accumulation by developing such materials. Considering the above-mentioned situations, this study aims to develop cermet materials with high specific strength and density by mechanical alloying, one of the production techniques of powder metallurgy.

References

1. Zhu, G., Liu, Y., Ye, J.: Fabrication and properties of Ti (C, N)-based cermets with multi-component AlCoCrFeNi high-entropy alloys binder. Mater. Lett. **113**, 80–82 (2013)
2. Dasdemir, U., Altas, E.: Metal based additive manufacturing. In: Rajendrachari, S. (ed.) Practical Implementations of Additive Manufacturing Technologies, pp. 63–87. Springer Nature Singapore (2024)
3. Dios, M., Kraleva, I., González, Z., Alvaredo, P., Ferrari, B., Gordo, E., Bermejo, R.: Mechanical characterization of Ti (C, N)-based cermets fabricated through different colloidal processing routes. J. Alloys Compd. **732**, 806–817 (2018)
4. Mhadhbi, M., Dağ, İE., Avar, B., Khitouni, M., Bousnina, M.A., Schoenstein, F., Jouini, N.: Manufacturing of novel nanostructured TiCrC Carbides using mechanical alloying and Spark Plasma Sintering. Metals **13**(6), 1040 (2023)
5. Aramian, A., Razavi, N., Sadeghian, Z., Berto, F.: A review of additive manufacturing of cermets. Addit. Manuf. **33**, 101130 (2020)
6. Nayak, A., Shashanka, R., Chaira, D.: Effect of nanosize yittria and tungsten addition to duplex stainless steel during high energy planetary milling. IOP Conf. Ser.: Mater. Sci. Eng. 012008 (IOP Publishing) (2016)
7. Shashanka, R., Chaira, D.: Phase transformation and microstructure study of nano-structured austenitic and ferritic stainless steel powders prepared by planetary milling. Powder Technol. **259**, 125–136 (2014)
8. Xu, X., Zheng, Y., Zhang, J., Ke, Z., Wu, H., Yang, Z.: Evolution of microstructure and interfacial characteristics of complete solid-solution Ti (C, N)-based cermets fabricated by mechanical activation and subsequent in situ carbothermal reduction. Ceram. Int. **47**(12), 16786–16793 (2021)
9. Real, C., Alcala, M., Trigo, I., Fombella, I., Cordoba, J.: Fabrication and characterization of FeCoNiCrMn, (Al) high entropy alloy based (Ti, Ta, Nb)(C, N) cermet. Int. J. Refract. Metals Hard Mater. **101**, 105694 (2021)
10. Shashanka, R., Chaira, D.: Optimization of milling parameters for the synthesis of nano-structured duplex and ferritic stainless steel powders by high energy planetary milling. Powder Technol. **278**, 35–45 (2015)
11. Yadav, S., Zhang, Q., Behera, A., Haridas, R.S., Agrawal, P., Gong, J., Mishra, R.S.: Role of binder phase on the microstructure and mechanical properties of a mechanically alloyed and spark plasma sintered WC-FCC HEA composites. J. Alloys Compd. **877**, 160265 (2021)
12. Kübarsepp, J., Juhani, K.: Cermets with Fe-alloy binder: a review. Int. J. Refract. Metals Hard Mater. **92**, 105290 (2020)
13. Shashanka, R., Chaira, D., Chakravarty, D.: Fabrication of nano-yttria dispersed duplex and ferritic stainless steels by planetary milling followed by spark plasma sintering and non-lubricated sliding wear behaviour study. J. Mater. Sci. Eng. B **6**, 111–125 (2016)
14. Xu, X., Zheng, Y., Zhao, Y., Lu, X., Ke, Z., Wu, H., Yang, M., Liang, H.: Influence of TaC content on microstructure and mechanical performance of Ti (C, N)-based cermets fabricated by mechanical activation and subsequent in situ carbothermal reduction. Ceram. Int. **48**(3), 3826–3832 (2022)
15. Zheng, Z., Lin, N., Zhao, L., Li, X., He, Y.: Fabrication and wear mechanism of Ti (C, N)-based cermets tools with designed microstructures used for machining aluminum alloy. Vacuum **156**, 30–38 (2018)
16. Shashanka, R.: Synthesis of nano-structured stainless steel powder by mechanical alloying-an overview. Int. J. Sci. Eng. Res. **8**, 588–594 (2017)
17. Grigoriev, S., Tarasova, T., Gusarov, A., Khmyrov, R., Egorov, S.: Possibilities of manufacturing products from cermet compositions using nanoscale powders by additive manufacturing methods. Materials **12**(20), 3425 (2019)
18. Plucknett, K.: Cermets and hardmetals, ed., MDPI, p. 963 (2018)
19. Perepezko, J., Bero, J., Sakidja, R., Talmy, I., Zaykoski, J.: Oxidation resistant coatings for refractory metal cermets. Surf. Coat. Technol. **206**(19–20), 3816–3822 (2012)

20. Shashanka, R., Uzun, O., Chaira, D.: Synthesis of nano-structured duplex and ferritic stainless steel powders by dry milling and its comparison with wet milling. Arch. Metall. Mater. **65**(1) (2020)
21. Altaş, E., Erdogan, A., Koçyiğit, F.: A comparative study on the high temperature dry sliding wear behavior of TiN and AlTiN/TiSiN coatings fabricated by PVD technique. Ind. Lubr. Tribol. **71**(7), 861–868 (2019)
22. Shashanka, R.: Non-lubricated dry sliding wear behavior of spark plasma sintered nano-structured stainless steel. J. Mater. Environ. Sci. **10**(8) (2019)
23. Ghasali, E., Orooji, Y., Tahamtan, H., Asadian, K., Alizadeh, M., Ebadzadeh, T.: The effects of metallic additives on the microstructure and mechanical properties of WC–Co cermets prepared by microwave sintering. Ceram. Int. **46**(18), 29199–29206 (2020)
24. Erdoğan, A., Altaş, E.: Experimental study on micro-abrasion behavior of hard coatings: the role of load, sliding distance and abrasive particle size. Mater. Res. Express **6**(11), 116430 (2019)
25. Wang, H., Song, X., Liu, X., Gao, Y., Wei, C., Wang, Y., Guo, G.: Effect of carbon content of WC–Co composite powder on properties of cermet coating. Powder Technol. **246**, 492–498 (2013)
26. Kübarsepp, J., Juhani, K., Tarraste, M.: Abrasion and erosion resistance of cermets: a review. Materials **15**(1), 69 (2021)
27. Rajabi, A., Ghazali, M., Daud, A.: Chemical composition, microstructure and sintering temperature modifications on mechanical properties of TiC-based cermet—A review. Mater. Des. **67**, 95–106 (2015)
28. Rajendrachari, S., Adimule, V.M., Jayaprakash, G.K., Pandith, A.: Electrochemical oxidation of methylene blue dye in wastewater using mechanically alloyed high entropy alloy modified carbon paste electrode using cyclic voltammetry. Mater. Res. Express **10**(5), 054003 (2023)
29. Rajendrachari, S.: An overview of high-entropy alloys prepared by mechanical alloying followed by the characterization of their microstructure and various properties. Alloys **1**(2), 116–132 (2022)
30. Caliskan, H., Altas, E., Panjan, P.: Study of nanolayer AlTiN/TiN coating deposition on cemented carbide and its performance as a cutting tool. J. Nano Res. **47**, 1–10 (2017)
31. Caliskan, H., Altas, E.: The effect of cutting conditions on cutting forces in milling of Ti6Al4V alloy. Int. J. Eng. Technol. Nat. Sci. **13**, 13–22 (2015)
32. Miranda-Hernández, J.G., Rocha-Rangel, E., Díaz de la Torre, S.: Synthesis, microstructural analysis and mechanical properties of alumina-matrix cermets. Epa.-J. Silic. Based Compos. Mater. **2010**(1) (2010)
33. Hussainova, I., Antonov, M., Jasiuk, I., Du, X.: Characterisation of microstructure and mechanical properties of cermets at micro-and nanoscales. Int. J. Mater. Prod. Technol. **40**(1–2), 58–74 (2011)
34. Nahvi, S., Jafari, M.: Microstructural and mechanical properties of advanced HVOF-sprayed WC-based cermet coatings. Surf. Coat. Technol. **286**, 95–102 (2016)
35. Altas, E., Gokkaya, H., Karatas, M.A., Ozkan, D.: Analysis of surface roughness and flank wear using the Taguchi method in milling of NiTi shape memory alloy with uncoated tools. Coatings **10**(12), 1259 (2020)
36. Leal, E., Gomes, U., Alves, S., Costa, F.A.D.: The influence of powder preparation condition on densification and microstructural properties of WC-Co-Al2O3 cermets. Int. J. Refract. Metals Hard Mater. **92**, 105275 (2020)
37. Mahale, R.S., Vasanth, S., Chikkegouda, S.P., Rajendrachari, S., Narsimhachary, D., Basavegowda, N.: Electrochemical determination of ascorbic acid by mechanically alloyed super duplex stainless steel powders. Metals **13**(8), 1430 (2023)
38. Miazga, A., Konopka, K., Gizowska, M., Szafran, M.: Preparation of Al2O3–Ni cermet composites by aqueous gelcasting. Powder Metall. Met. Ceram. **52**, 567–571 (2014)
39. Basavarajappa, P.N., Hegde, M.M.R., Rajendrachari, S., Surendranathan, A.O.: Investigation of structural and mechanical properties of nanostructured TiMgSr alloy for biomedical applications (2023)

40. Stanciu, V.I., Vitry, V., Delaunois, F.: Tungsten carbide powder obtained by direct carburization of tungsten trioxide using mechanical alloying method. J. Alloys Compd. **659**, 302–308 (2016)
41. Chen, C.-S., Yang, C.-C., Chai, H.-Y., Yeh, J.-W., Chau, J.L.H.: Novel cermet material of WC/ multi-element alloy. Int. J. Refract. Metals Hard Mater. **43**, 200–204 (2014)
42. Aslibeiki, B., Kameli, P.: Structural and magnetic properties of Co/Al2O3 cermet synthesized by mechanical ball milling. Ceram. Int. **46**(12), 20116–20121 (2020)
43. Altas, E., Altin Karatas, M., Gokkaya, H., Akinay, Y.: Surface integrity of NiTi shape memory alloy in milling with cryogenic heat treated cutting tools under different cutting conditions. J. Mater. Eng. Perform.
44. Altas, E., Erkan, O., Ozkan, D., Gokkaya, H.: Optimization of cutting conditions, parameters, and cryogenic heat treatment for surface roughness in milling of NiTi shape memory alloy. J. Mater. Eng. Perform. 1–13
45. Pradeep, N., Hegde, M.R., Rajendrachari, S., Surendranathan, A.: Investigation of microstructure and mechanical properties of microwave consolidated TiMgSr alloy prepared by high energy ball milling. Powder Technol. **408**, 117715 (2022)
46. Rajendrachari, S., Adimule, V., Gulen, M., Khosravi, F., Somashekharappa, K.K.: Synthesis and characterization of high entropy alloy 23Fe–21Cr–18Ni–20Ti–18Mn for electrochemical sensor applications. Materials **15**(21), 7591 (2022)
47. Gogebakan, M., Kursun, C., Eckert, J.: Formation of new Cu-based nanocrystalline powders by mechanical alloying technique. Powder Technol. **247**, 172–177 (2013)
48. Cramer, C.L., Preston, A.D., Ma, K., Nandwana, P.: In-situ metal binder-phase formation to make WC–FeNi Cermets with spark plasma sintering from WC, Fe, Ni, and carbon powders. Int. J. Refract. Metals Hard Mater. **88**, 105204 (2020)
49. Mahale, R.S., Rajendrachari, S., Vasanth, S., Krishna, H., Kapanigowda, N.S., Chikkegowda, S.P., Patil, A.: Technology and challenges in additive manufacturing of duplex stainless steels (2022)
50. Shashanka, R., Chaira, D.: Effects of nano-Y2O3 and sintering parameters on the fabrication of PM duplex and ferritic stainless steels. Acta Metall. Sinica (English Letters) **29**, 58–71 (2016)

Chapter 12
Ball Milling of Copper and Zinc Alloys Followed by Their Consolidation and Application

Krutika L. Routray and Sunirmal Saha

Abstract Ball milling is a mechanical process that involves the grinding of solid materials into small, fine particles using balls as the grinding medium. In the context of copper and zinc alloys, this process can be used to prepare fine powders of these metals or their alloys, which can then be consolidated into various forms for different applications. Copper and zinc alloys, commonly known as brass, are versatile materials with a wide range of applications due to their unique combination of properties. Here's a general overview of the process and its potential applications. This article explores the synthesis of copper and zinc alloy powders through ball milling, followed by consolidation processes and the diverse range of applications for these materials. The controlled production of alloy powders using ball milling, combined with various consolidation techniques, opens opportunities for tailored material properties suitable for industries such as electronics, architecture, and health care.

Keywords Powder particle · Mechanical process · Copper and zinc alloys · Consolidation processes

12.1 Introduction

The ball milling process is a versatile and widely used technique in materials science and engineering, offering a valuable means of modifying and synthesizing materials with enhanced properties [1–3]. It involves the use of metallic or ceramic balls in a rotating drum or container to effect mechanical alloying or milling of various materials [4–7]. This process has found applications in a multitude of fields, including materials synthesis, nanoparticle production, pharmaceuticals, and more [2–6]. The mechanical alloying/milling process is a solid-state powder process where the powder particles are subjected to high energetic impact by the balls in a vial [4, 8, 9]. As the powder particles in the vial are continuously impacted by the balls, cold welding

K. L. Routray (✉) · S. Saha
C.V. Raman Global University, Bhubaneswar, India
e-mail: krutika.routray@cvrp.edu.in; rkruttika@gmail.com

S. Rajendrachari (ed.), *Mechanically Alloyed Novel Materials*, Advanced Structured Materials 220, https://doi.org/10.1007/978-981-97-6504-1_12

between particles and fracturing of the particles takes place repeatedly during the ball milling process. The entire process includes blending the powder mixture prior to the ball milling, vacuuming and/or filling with protective gases to prevent oxidation and contamination, and the ball milling process itself [10–14]. Post-milling processes after mechanical alloying include canning, degassing, and finally plastic deformation by extrusion or rolling [15].

The ball milling process is a mechanical method for crushing and grinding materials into fine powders or blending different substances that have evolved as a fundamental technique in materials science, chemistry, and engineering. This versatile and efficient process employs the use of a rotating drum filled with grinding media, such as metallic or ceramic balls, to carry out a myriad of functions, including mechanical alloying, particle size reduction, and materials modification [16–18]. The impact and friction generated by the milling balls in the rotating drum result in a wide range of effects on the materials, leading to the creation of novel materials, the synthesis of nanoparticles, and the improvement of material properties [19]. The ball milling process has been indispensable in the development of advanced materials, serving as a bridge between the micro and nanoscale worlds. It offers the ability to control and tailor the structure and properties of materials with precision, making it a powerful tool for researchers and engineers. This process has become an essential technique for manufacturing and engineering advanced materials with enhanced characteristics for applications in electronics, energy storage, pharmaceuticals, and various other fields [20–24].

It is a fact that ball milling is a prominent mechanical processing technique that has been instrumental in the development and modification of materials with diverse applications across several industries [2–4]. One intriguing area of application is the ball milling of copper and zinc alloys, a process that plays a pivotal role in tailoring the properties of these alloy systems. Copper-zinc alloys, also known as brass, exhibit a unique combination of characteristics, such as good electrical conductivity, corrosion resistance, and excellent machinability, making them vital components in various engineering and manufacturing domains. Ball milling of copper-zinc alloys, often performed under controlled conditions, presents a powerful avenue for refining these alloys, enhancing their microstructural properties, and optimizing their performance in specific applications. This process can be utilized to manipulate the grain size, distribution, and alloy composition, thereby achieving a wide range of material properties [20, 25–28].

Ball milling and consolidation are two fundamental processes in materials science and engineering that, when combined, offer a powerful approach to tailoring the properties of copper and zinc alloys [16]. These alloys, commonly referred to as brass, are renowned for their exceptional combination of electrical conductivity, corrosion resistance, and machinability. However, to unlock their full potential and extend their applications, it is essential to fine-tune their microstructure and mechanical properties [29]. Ball milling is a mechanical processing technique that involves the use of milling balls and a high-energy ball mill to induce a series of mechanical and chemical reactions within a powdered material. It can be precisely controlled to modify the particle size, shape, and distribution, while also altering the alloy's composition at the nanoscale [3–7]. This process is instrumental in enhancing the properties of materials and is particularly advantageous for copper-zinc alloys [30, 31]. Consolidation refers to the techniques used to take the modified powders from ball milling and transform them into a solid, functional material [21, 32]. Common consolidation methods include sintering, spark plasma sintering, and hot pressing. These methods facilitate the compaction of the milled powders, resulting in dense and well-integrated structures [33, 34].

In this article, we embark on a journey into the realm of ball milling with a focus on copper and zinc alloys. We explore the fundamental principles of ball milling, delving into the mechanisms that underpin this technique, and discuss its application to brass alloys. Moreover, we examine the various improvements and advancements that ball milling has brought to the field of copper-zinc alloys, such as increased hardness, strength, and durability. As we delve into this topic, we aim to highlight the potential and versatility of ball milling as a valuable tool for materials engineers, researchers, and industries looking to harness the unique properties of copper and zinc alloys for innovative and improved products. We examine the principles and applications of the ball milling process, discussing its significant role in materials science, nanoparticle synthesis, and the development of advanced materials with tailored properties. We explore the mechanisms that underlie this technique and highlight its broad spectrum of applications in current research and industrial processes. We also explore the dynamic synergy between ball milling and consolidation, focusing on their application to copper and zinc alloys. We discover the underlying principles of these processes, the mechanisms at play during ball milling, and the critical aspects of consolidation. Furthermore, we examine the implications of combining these processes on the microstructure and properties of brass alloys, such as improved mechanical strength, hardness, and thermal stability. This work highlights the potential of leveraging ball milling and consolidation as a route to engineer advanced copper and zinc alloy materials, offering unprecedented opportunities for innovation in various industries, from electronics to automotive manufacturing.

12.2 Concept and Significance of Ball Milling of Copper and Zinc Alloys

Ball milling, at its core, is the process of reducing large particles or crystalline solids into smaller, fine particles using the energy generated by milling balls. This technique relies on the mechanical energy imparted by the milling balls as they collide with and crush the material. The primary mechanisms at play in ball milling include abrasion, impact, and cold welding, all of which contribute to altering the material's structure, size, and properties [35–38].

The significance of ball milling extends across various scientific and industrial domains.

Ball milling is instrumental in the production of nanomaterials, which are materials with unique properties due to their nanoscale dimensions. This is particularly relevant in the development of advanced materials for applications in electronics, optics, and catalysis. It can alter the structural, mechanical, and chemical properties of materials [39, 40]. It is used to refine grains, change crystalline structures, and introduce defects, thus enhancing or tailoring the materials' characteristics to specific applications [41–46]. In the pharmaceutical industry, ball milling is used to reduce the particle size of drug substances, improving their solubility and bioavailability. This technique is vital for drug formulation [47]. Ball milling plays a role in the development of materials for energy storage applications, such as batteries and supercapacitors. It can enhance the electrochemical performance of electrode materials [20, 48]. It also enables chemical reactions to occur under mechanical force, a field known as mechanochemistry. This has implications for green chemistry, as it reduces the need for hazardous solvents and reagents. Ball milling is a key step in the powder metallurgy process, helping to mix and blend metal powders for subsequent sintering or pressing into components [49]. It allows the creation of advanced composites, alloys, and materials with unique combinations of properties. Recent advances have seen ball milling applied to additive manufacturing processes, such as 3D printing, offering the possibility to engineer novel materials with precise microstructures [50, 51]. Hence, ball milling is a transformative mechanical processing technique with a profound impact on materials science and industry. Its applications are diverse, ranging from the synthesis of nanomaterials to materials engineering, drug formulation, and green chemistry. This article will delve into the principles, applications, techniques, and recent advancements of ball milling, shedding light on the multifaceted significance of this mechanical processing method [2–4].

Ball milling, a mechanical processing technique, is employed in the field of metallurgy and materials science to refine, modify, and engineer the properties of copper and zinc alloys. This process entails the use of milling balls to apply mechanical energy, enabling the reduction of large particles or crystalline structures into smaller, fine particles [52–54]. In the context of copper and zinc alloys, ball milling can significantly influence their microstructure, mechanical properties, and performance. It is considered to be a powerful tool for reducing the grain size and refining the microstructure of copper and zinc alloys. Smaller grain sizes often lead to enhanced mechanical properties and increased strength [55, 56]. In copper-zinc alloy systems, ball milling promotes the formation of solid solutions with improved homogeneity and uniform distribution of alloying elements. This results in alloys with superior properties [57–59]. Through ball milling, dislocations, defects, and grain boundaries are introduced, which contribute to the strengthening of the alloys. This process enhances hardness and tensile strength. Ball milling can alter the thermal stability of copper and zinc alloys, making them suitable for high-temperature applications [60–62]. Fine-tuning thermal properties is essential in industries like aerospace and automotive manufacturing. The modification of alloy microstructures through ball milling can lead to improved corrosion resistance, which is crucial in applications exposed to harsh environments or corrosive substances. Copper and zinc alloys subjected to ball milling often exhibit increased wear resistance, making them suitable for use in bearings, gears, and other components requiring resistance to friction and abrasion. Ball milling can help engineer alloys with tailored tribological properties, including reduced friction and improved lubrication behavior [4–7]. Ball milling facilitates the development of advanced copper and zinc alloy materials with unique combinations of properties, making them highly adaptable to various industries [63–65]. The process is versatile, efficient, and amenable to scaling, allowing for the production of copper and zinc alloy powders with improved properties for subsequent applications in additive manufacturing, powder metallurgy, and more [66].

Hence, ball milling is a pivotal mechanical processing technique in the domain of copper and zinc alloys, offering opportunities to tailor their microstructure and properties for a wide range of applications. Its significance extends to industries like aerospace, automotive, and manufacturing, where finely tuned alloys with enhanced properties are of paramount importance [67]. This article will delve into the principles, methods, applications, and recent advancements in the context of ball milling in copper and zinc alloys, emphasizing its critical role in materials science and engineering.

12.3 Principles of Ball Milling of Copper and Zinc Alloys

Ball milling is a mechanical processing technique widely utilized in materials science, chemistry, and various industries for the refinement, modification, and engineering of materials at the nanoscale. It involves the use of milling balls, typically made of stainless steel, ceramic, or other materials, in a ball mill to apply mechanical energy and reduce larger particles or crystalline structures into smaller, fine particles [68–70]. The principles of ball milling encompass several key aspects that make this technique a versatile and essential tool for materials research and development. At its core, ball milling is a size-reduction process. It enables the reduction of large particles, aggregates, or bulk materials into finer particles with sizes ranging from micrometers to nanometers. This size reduction is critical for various applications, including the production of fine powders, dispersion of particles in a matrix, and the synthesis of nanomaterials. Ball milling relies on the input of mechanical energy into the system [35, 71, 72]. As the milling balls collide with the material to be processed, they create stress fields, generate fractures, and induce deformation in the crystalline structure of the material. This mechanical energy results in the reduction of particle size and changes in the material's properties.

Ball milling operates based on two primary mechanisms: impact and attrition. During the process, the milling balls impact the material, causing it to fracture and break into smaller particles. Additionally, the balls exert attrition forces, leading to abrasion, grinding, and refinement of the material. The principles of ball milling offer control and precision over the processing conditions. Researchers can adjust parameters such as milling time, rotation speed, ball-to-powder ratio, and milling environment to tailor the size, morphology, and properties of the resulting materials. This control allows for the customization of materials for specific applications [73–75]. Ball milling is instrumental in the technique known as mechanical alloying [1]. In this process, elements or compounds are mixed and milled together to form alloys or composite materials. Mechanical alloying can create materials with unique properties, such as enhanced strength, improved wear resistance, and altered phase compositions. The principles of ball milling are pivotal in the synthesis of nanomaterials. By reducing materials to the nanoscale, researchers can achieve unique properties and behaviors that do not present in bulk materials. Nanomaterials have a wide range of applications in fields like electronics, catalysis, energy storage, and biomedicine. Ball milling has found applications in various industries, including pharmaceuticals, ceramics, materials science, and chemical engineering. It is equally valuable in research laboratories for materials development and analysis. The efficiency of ball milling in terms of energy consumption and material utilization is a crucial aspect. Researchers continually seek ways to improve efficiency, reduce processing times, and minimize energy consumption during the ball milling process.

Ball milling of copper and zinc alloys is a mechanical processing technique that follows the same fundamental principles as general ball milling but is tailored for specific applications involving these metals. The principles guiding ball milling of copper and zinc alloys are as follows [74–77]:

Alloy Synthesis: Ball milling is commonly used to synthesize copper-zinc alloys or composite materials. It involves the mechanical mixing of copper and zinc powders. The principles applied here encompass size reduction and alloying. Through repeated collisions of milling balls with the powder mixture, the elemental components are homogenously mixed on a nanoscale.

Elemental Composites: Copper and zinc are two distinct elements, and alloying them through ball milling allows for the creation of composite materials with unique properties. The principles involved include mechanical alloying and the formation of solid solutions. During ball milling, the intimate mixing of copper and zinc powders occurs, resulting in the formation of a uniform and well-dispersed composite.

Particle Size Reduction: One of the primary objectives of ball milling in copper-zinc alloy synthesis is to reduce the particle size of the constituent powders. The principles of size reduction apply to break down the bulk copper and zinc powders into smaller, more reactive particles. This is crucial for achieving a uniform distribution of the two elements.

Homogenization and Microstructure Control: Ball milling enables precise control over the microstructure of copper-zinc alloys. The principles governing homogenization are crucial here, as the process ensures that the copper and zinc components are evenly distributed. This control over microstructure influences the resulting material's properties, such as mechanical strength, electrical conductivity, and corrosion resistance.

Phase Formation: The ball milling of copper and zinc powders can lead to the formation of various phases, depending on processing conditions and alloy compositions. The principles related to phase transformation and transition are important. Researchers aim to achieve specific phases in the resulting alloy, such as brass, which is a well-known copper-zinc alloy.

Energy Input and Milling Parameters: The principles of ball milling also encompass controlling various parameters. These parameters include milling time, rotational speed, ball-to-powder ratio, and milling media. Researchers manipulate these parameters to achieve specific outcomes in terms of particle size, phase composition, and microstructure.

Enhanced Properties: The primary objective in ball milling of copper and zinc alloys is to enhance material properties. By creating well-mixed, fine-grained alloys, researchers can achieve improved mechanical, electrical, and thermal properties. Tailoring the alloy's microstructure can result in materials with specific characteristics suitable for various applications.

The choice of milling equipment depends on the nature of the copper and zinc alloys, the desired particle size, and the processing goals [69]. For instance, if you are working with copper-zinc alloys to develop materials with improved mechanical properties or other tailored features, you may opt for a high-energy mill like a planetary ball mill. On the other hand, if you are focused on creating homogeneous mixtures of copper and zinc alloy powders, vibratory ball mills can be an excellent choice. Some common types of ball mills and milling equipment suitable for copper and zinc alloy processing [69, 78, 79]:

Planetary Ball Mill: Planetary ball mills are versatile and can be used for mechanical alloying, blending, and nanoparticle synthesis. They are suitable for high-energy milling and offer excellent control over milling conditions. This precision makes them well-suited for research involving copper and zinc alloys, particularly for developing advanced materials.

Vibratory Ball Mill: Vibratory ball mills are ideal for the rapid processing of powders and are suitable for mixing, blending, and milling. The gentle, oscillating motion can help ensure a uniform mixture of copper and zinc alloy powders.

Conventional Ball Mill: Conventional ball mills are widely used for grinding and mixing applications. They are suitable for both wet and dry milling processes. These mills can be used for homogenizing copper and zinc alloy powders before further processing.

Rod Mill: While not a ball mill, rod mills can be used for coarse grinding of copper and zinc alloys. The long rods tumble and grind the alloy to the desired size.

Semi-Autogenous Grinding (SAG) Mill: In mining and metallurgy, SAG mills are used for the primary grinding of copper and zinc ores. They can handle large volumes of material and are often equipped with steel balls to assist with milling.

Customized Mills: For specific applications or research needs, custom-designed mills can be manufactured to accommodate the unique requirements of copper and zinc alloy processing. These can be tailored in terms of size, materials, and other parameters.

In many cases, researchers and engineers may use a combination of milling techniques to achieve their specific goals, as different milling equipment offers varying levels of energy input, milling speed, and control over milling parameters.

12.4 Consolidation Processes of Ball Milling of Copper and Zinc Alloys

The consolidation of ball-milled copper and zinc alloys involves the transformation of the milled powders into a solid material with the desired shape, density, and properties. Several consolidation processes can be employed, depending on the specific application and requirements. Some common consolidation techniques for ball-milled copper and zinc alloys [76, 80–82]:

Powder Compaction: In this process, ball-milled alloy powders are loaded into a die cavity and subjected to high mechanical pressure, typically using a hydraulic press. The pressure compacts the powders into a green compact with the desired shape, which can be cylindrical or any other form depending on the die used. The green compact is typically porous and requires further processing to enhance density and mechanical properties.

Sintering: Sintering is a widely used consolidation method for copper and zinc alloy powders. The green compact obtained through powder compaction is heated in a controlled atmosphere (usually in a furnace) at temperatures below the alloy's melting point. During sintering, the particles diffuse and bond at contact points, reducing porosity and increasing density. Sintering also promotes grain growth and recrystallization, leading to changes in the microstructure and mechanical properties. The choice of sintering temperature and atmosphere can be tailored to achieve specific properties.

Hot Isostatic Pressing (HIP): HIP is used to consolidate ball-milled powders with the goal of achieving high density and enhanced mechanical properties. In HIP, the powders are enclosed in a canister or capsule and subjected to high temperature and isostatic pressure. The combination of high temperature and pressure leads to densification and can eliminate residual porosity. HIP is often used for complex-shaped parts or when a high degree of densification is required.

Spark Plasma Sintering (SPS): SPS is a rapid consolidation technique that involves the application of pulsed direct current and uniaxial pressure. SPS can rapidly densify copper and zinc alloy powders at lower temperatures and shorter processing times compared to traditional sintering methods. The pulsed current and pressure facilitate particle rearrangement and bonding. SPS can lead to the preservation of fine microstructures obtained through ball milling.

Extrusion: Extrusion is employed when complex shapes, such as rods or tubes, are needed. Ball-milled powders can be mixed with a binder to form a paste, which is then extruded through a die at elevated temperatures. The extruded material is subsequently sintered to remove the binder and achieve the desired properties.

Additive Manufacturing: Additive manufacturing methods, such as selective laser melting (SLM) or binder jetting, can be used to consolidate ball-milled powders. These methods involve layer-by-layer deposition and sintering of powders, enabling the creation of complex geometries and customized components. The choice of consolidation method depends on factors such as the alloy composition, desired properties, part geometry, and production volume. The goal is to achieve high-density, well-bonded materials with tailored microstructures and properties suitable for specific applications in fields like aerospace, electronics, and manufacturing.

12.5 Mechanical and Chemical Effects

Ball milling of copper and zinc alloys can induce both mechanical and chemical effects on the materials. These effects are essential for understanding the changes in structural, morphological, and chemical properties. Here are the mechanical effects of ball milling on copper and zinc alloys. Ball milling leads to a significant reduction in the particle size of copper and zinc alloy powders [77, 83]. As the milling balls collide with the alloy particles, they are subjected to mechanical forces that break down larger particles into smaller ones. Ball milling helps in achieving better mixing and homogenization of alloy components. The repeated impacts and shear forces promote the distribution of zinc in the copper matrix, resulting in a more uniform structure. The high-energy impacts during ball milling can lead to the deformation of particles and the occurrence of cold welding, where alloy particles adhere to each other. This effect can promote grain refinement. The mechanical energy imparted during milling introduces microstrain and dislocations in the crystal lattice [84–89]. These defects can lead to changes in the structural properties of the alloy. In some cases, extended ball milling can induce partial or complete amorphization of the alloy, transforming the crystalline structure into an amorphous phase [89, 90].

The chemical effects of ball milling on copper and zinc alloys include oxidation, solid-state alloying, surface functionalization, chemical mixing, surface defects, hydrogen uptake. Ball milling is often carried out under an inert atmosphere or with the addition of process control agents to prevent oxidation. However, in the presence of oxygen or moisture, oxidation of the alloy surface may occur, leading to the formation of oxide layers. The repeated mechanical deformation and strain can promote solid-state alloying between copper and zinc atoms [89]. This can result in the formation of copper-zinc solid solutions or intermetallic compounds, depending on the milling conditions. Ball milling can introduce changes in the chemical functional groups on the alloy surface. These changes can influence the alloy's reactivity and interaction with other materials. Intimate mixing of alloy constituents occurs during ball milling, facilitating chemical reactions between copper and zinc. This mixing can lead to the formation of complex chemical compounds within the alloy [87]. Ball milling can create surface defects and imperfections, making the alloy more reactive and accessible for subsequent chemical modifications or surface treatments. In the presence of hydrogen gas, ball milling can facilitate hydrogen uptake in the alloy, leading to the formation of metal hydrides.

The balance between these mechanical and chemical effects depends on various factors, including milling time, ball-to-powder ratio, milling speed, milling media, and the presence of process control agents. Understanding and controlling these effects are crucial for tailoring the properties of ball-milled copper and zinc alloys for specific applications.

12.6 Characterization Techniques of Ball Milling of Copper and Zinc Alloys

Characterization techniques play a crucial role in evaluating the effects of ball milling on copper and zinc alloys. These techniques provide insights into the structural, morphological, and chemical changes induced by the milling process. Here are some common characterization techniques for ball-milled copper and zinc alloys [89, 91–93]:

X-ray Diffraction (XRD): XRD is used to analyze changes in the crystal structure, phase composition, and lattice parameters resulting from ball milling. It helps identify crystallographic changes, the formation of new phases, and crystallite size reduction.

Scanning Electron Microscopy (SEM): SEM allows for the observation of changes in the surface morphology, grain size, and particle distribution of ball-milled copper and zinc alloys. It provides visual insights into the microstructure.

Transmission Electron Microscopy (TEM): TEM offers high-resolution imaging, enabling the examination of the nanoscale features of milled materials. It helps visualize structural defects, dislocations, and particle size distributions.

Energy-Dispersive X-ray Spectroscopy (EDS or EDX): EDS, combined with SEM or TEM, helps determine changes in elemental composition and distribution resulting from ball milling.

X-ray Photoelectron Spectroscopy (XPS): XPS is used to analyze changes in the chemical composition and oxidation states of the elements present in ball-milled alloys.

Fourier-Transform Infrared Spectroscopy (FTIR): FTIR can detect changes in the chemical bonds and functional groups in ball-milled copper and zinc alloys. It provides information about surface chemistry.

Differential Scanning Calorimetry (DSC): DSC measures changes in heat flow associated with thermal events, such as phase transformations, which can occur during ball milling.

Thermogravimetric Analysis (TGA): TGA measures changes in weight as a function of temperature and time, helping to understand the thermal stability and any decomposition or oxidation of milled alloys.

Particle Size Analysis: Particle size distribution is assessed using techniques like laser diffraction or dynamic light scattering to quantify the size reduction resulting from ball milling.

Magnetic Testing: Magnetic hysteresis loop measurements and magnetic susceptibility tests can provide information about the magnetic properties affected by ball milling.

Mechanical Testing: Changes in mechanical properties, such as hardness, tensile strength, and impact resistance, can be evaluated to assess the impact of ball milling on alloy properties.

Electron Backscatter Diffraction (EBSD): EBSD helps study changes in crystallography, grain orientation, and texture induced by ball milling.

Raman Spectroscopy: Raman spectroscopy can identify changes in vibrational modes and lattice vibrations in ball-milled materials.

Ultrasonic Testing: Ultrasonic techniques can help assess changes in the acoustic properties and detect defects or changes in the material structure.

Electrochemical Testing: Electrochemical techniques, such as cyclic voltammetry and potentiostatic/galvanostatic tests, can be used to investigate the electrochemical behavior of ball-milled alloys, particularly in corrosion and electrocatalysis studies.

The choice of characterization techniques depends on the specific aspects and properties being investigated. Researchers often use a combination of these techniques to gain a comprehensive understanding of the effects of ball milling on copper and zinc alloys.

12.7 Recent Advances and Innovations

The field of ball milling and consolidation processes for copper and zinc alloys has seen continuous advancements and innovations in recent years. Researchers and engineers are exploring new techniques and refining existing methods to improve the efficiency, quality, and properties of materials [93]. Researchers are exploring mechanochemical synthesis, which combines mechanical milling with chemical reactions. Introducing reactive additives during ball milling, this approach can lead to the formation of complex alloys and compounds that are challenging to obtain through traditional methods [94]. High-energy ball milling is being optimized to reduce processing times and enhance the uniformity of alloying and mixing. Innovations in milling equipment, such as high-speed planetary ball mills, enable more controlled and efficient alloy production. In situ alloying during ball milling is gaining attention, allowing the simultaneous production and consolidation of alloys. This approach minimizes the need for subsequent consolidation steps and can lead to the formation of alloys with unique properties [20, 95]. Hybrid processes that combine ball milling with other techniques like spark plasma sintering (SPS) or additive manufacturing are being explored. These approaches offer precise control over microstructures and properties, leading to tailored materials for specific applications. Researchers are focusing on the development of copper and zinc nanocomposites through ball milling. By adding nanoparticles, like carbon nanotubes or graphene, the mechanical and electrical properties of the materials can be significantly enhanced [37, 96].

Innovations in the consolidation process aim to reduce energy consumption and environmental impact. Sustainable sintering methods, such as microwave sintering or flash sintering, are being investigated [97–100]. Recent advances in characterization techniques, such as in situ transmission electron microscopy (TEM), allow real-time observation of microstructural changes during ball milling and consolidation. These techniques provide valuable insights into the processing mechanisms [101]. Tailoring copper and zinc alloy compositions for specific applications, such as electrical contacts, heat exchangers, or antimicrobial materials, is an ongoing area of innovation. Alloy design and optimization are being conducted through computational methods. Advances in data-driven materials science, including machine learning and artificial intelligence, are being applied to optimize processing parameters and predict material properties [102]. Copper and zinc alloys are being adapted for additive manufacturing techniques, enabling the production of complex geometries and customized components for various industries [103, 104].

These innovations are driven by the demand for materials with enhanced properties, reduced production costs, and improved sustainability [105, 106]. As the field continues to evolve, the development of copper and zinc alloys through ball milling and consolidation processes is expected to play a crucial role in various industrial applications [80].

12.8 Challenges and Future Prospects

The field of ball milling and consolidation processes for copper and zinc alloys presents various challenges and holds promising future prospects. Understanding these challenges and opportunities is essential for advancing research and applications in this domain. Achieving a homogeneous alloy composition while avoiding contamination remains a challenge. Addressing issues related to materials compatibility and tool wear during ball milling is crucial. Ball milling can be an energy-intensive process, and finding more energy-efficient methods or optimizing milling parameters to reduce energy consumption is a priority. In the case of complex alloy systems, obtaining the desired microstructure and properties through mechanical alloying can be challenging. Researchers need to explore innovative alloying techniques. Achieving uniform consolidation, especially in complex or large components, is a challenge. Reducing defects and improving the consolidation process's reliability is essential. While ball milling can provide fine microstructures, fine-tuning the control over grain size, distribution, and texture remains a challenge for specific applications [77]. Transitioning from lab-scale ball milling to industrial-scale production while maintaining material quality and properties is a significant challenge.

Reducing the environmental impact of the entire process, from milling to consolidation, is becoming increasingly important. Sustainable practices and materials are needed.

The future prospects of ball milling and consolidation processes for copper and zinc alloys include the utilizing of computational methods, such as molecular dynamics simulations and finite element analysis, which will enable researchers to gain a deeper understanding of the underlying mechanisms and optimize processing conditions [107]. The development of new alloy compositions tailored for specific applications will continue to expand. These alloys may have enhanced properties, including improved electrical conductivity, corrosion resistance, or antimicrobial features. Integrating nanoparticles or nanotubes into copper and zinc alloys through ball milling will result in advanced nanocomposite materials with unique mechanical, electrical, and thermal properties [19]. Combining ball milling with advanced manufacturing techniques, like 3D printing and laser processing, offers opportunities to produce complex, high-performance components. The incorporation of sustainable and green processing methods, such as energy-efficient sintering and eco-friendly milling techniques, will be a focus of future research. The use of machine learning and artificial intelligence to predict material properties, optimize processing parameters, and reduce energy consumption will continue to grow [108]. Tailoring materials to meet the specific needs of industries like aerospace, electronics, and health care will drive the development of customized copper and zinc alloys. Collaborations between academic researchers and industry professionals will accelerate the translation of research findings into practical applications. Copper and zinc alloys with antimicrobial properties are of interest for applications in healthcare settings [109]. Researchers will explore their potential in preventing infections and enhancing patient care. Addressing global challenges such as climate change, resource scarcity, and health care demands will drive innovation in copper and zinc alloy development for sustainable solutions. As researchers and engineers tackle these challenges and explore these prospects, the field of ball milling and consolidation processes for copper and zinc alloys is poised to make significant contributions to various industries and address pressing societal needs [108–110].

12.9 Applications

Ball milling of copper and zinc alloys followed by their consolidation offers a diverse range of applications due to the unique properties and microstructures that can be achieved [30]. Some of the notable applications are presented in this chapter. Copper and its alloys are renowned for their excellent electrical conductivity. Ball milling and consolidation techniques can be used to refine copper-zinc alloy powders, leading to materials with enhanced conductivity [69]. These materials find applications in electrical wiring, connectors, and electronic components. Copper and copper alloys exhibit intrinsic antimicrobial properties. The finely powdered copper-zinc alloys obtained through ball milling can be consolidated into coatings or surfaces for applications in healthcare settings, public transportation, and food processing to reduce the spread of infectious agents. Copper and its alloys possess high thermal conductivity. Ball-milled and consolidated copper-zinc composites can be employed in heat sinks and thermal management solutions for electronic devices to enhance heat dissipation [111]. Copper and zinc alloys can serve as catalysts in various chemical reactions. The high surface area and unique microstructures achieved through ball milling can improve their catalytic activity, making them valuable in chemical processes and environmental applications. Consolidated copper-zinc alloys with tailored mechanical properties are used in the manufacturing of gears, bearings, bushings, and other mechanical components. The improved wear resistance and strength of these materials extend their service life [112]. The combination of biocompatible copper-zinc alloys and the antimicrobial properties obtained through ball milling can be advantageous for biomedical implants. These materials can help reduce the risk of post-surgical infections while maintaining the mechanical integrity of the implants. Ball milling can be used to incorporate nanoparticles (e.g., carbon nanotubes, graphene) into copper-zinc matrices. These nanocomposites have applications in lightweight and high-strength materials for aerospace and automotive industries [113, 114]. Copper and zinc alloys, when used as catalysts or adsorbents, can facilitate environmental remediation processes. They are applied in the degradation of pollutants in water treatment, the removal of volatile organic compounds, and catalytic converters in automobiles [115].

Consolidated copper-zinc alloys are used for structural applications in various industries, including construction, where their corrosion resistance and mechanical properties make them an ideal choice for architectural and structural elements [116]. Copper-zinc alloys obtained through ball milling can be combined with other materials such as polymers, ceramics, or composites to create hybrid materials tailored for specific applications. These hybrids have applications in aerospace, automotive, and construction. The electrical conductivity and piezoelectric properties of copper-zinc alloys make them suitable for sensor applications. They are used in pressure sensors, strain gauges, and accelerometers. Consolidated copper-zinc materials can be utilized as electrode materials in energy storage devices, including batteries and supercapacitors, to improve energy efficiency and cycle life [117].

The applications of ball milling and consolidation of copper and zinc alloys are vast and continue to expand as researchers explore new possibilities and refine the properties of these materials to meet the evolving demands of various industries. The versatility of these processes ensures that copper and zinc alloys remain at the forefront of materials science and engineering [118].

12.10 Conclusion

In conclusion, the ball milling and consolidation processes for copper and zinc alloys are versatile and promising techniques that offer numerous advantages and have found applications across various industries. Through this research journey, we have explored the principles, methodologies, challenges, and future prospects associated with these processes. Here is a summary of the key takeaways: Ball milling is a mechanical processing technique that involves the use of ball mills and milling equipment to refine, alloy, and homogenize copper and zinc powders. It enables the generation of finely powdered materials with controlled microstructures. Different types of ball mills and milling equipment, including planetary ball mills and attritor mills, offer flexibility in terms of processing conditions and are adapted to various applications. Ball milling imparts mechanical and chemical effects on materials, leading to alloying, refinement of grains, and the formation of metastable phases. The consolidation of ball-milled powders is crucial to obtain functional components. Techniques such as spark plasma sintering (SPS) and hot pressing are employed to create dense and homogeneous structures. Ball milling and consolidation offer a platform for enhancing the properties of copper and zinc alloys, including improved mechanical strength, electrical conductivity, and corrosion resistance. Researchers continually innovate by exploring new alloys, nanocomposites, and hybrid manufacturing techniques, tailoring copper and zinc alloys to meet the needs of specific applications. Challenges include materials compatibility, energy consumption, achieving uniform consolidation, and addressing sustainability concerns in the entire process.

In the ever-evolving landscape of materials science and engineering, the processes of ball milling and consolidation for copper and zinc alloys are pivotal. They enable the production of advanced materials with tailored properties, supporting innovations across a wide range of applications. As we look ahead, interdisciplinary research, knowledge sharing, and sustainable practices will shape the direction of this field, fostering creativity and discovery. With a focus on solving real-world challenges and improving our quality of life, the ball milling and consolidation processes for copper and zinc alloys have a bright and transformative future.

References

1. Kumar, M., Xiong, X., Wan, Z., Sun, Y., Tsang, D.C., Gupta, J., Gao, B., Cao, X., Tang, J., Ok, Y.S.: Ball milling as a mechanochemical technology for fabrication of novel biochar nanomaterials. Biores. Technol. **312**, 123613 (2020)
2. Amrute, A.P., De Bellis, J., Felderhoff, M., Schüth, F.: Mechanochemical synthesis of catalytic materials. Chem. A Eur. J. **27**(23), 6819–6847
3. Jamkhande, P.G., Ghule, N.W., Bamer, A.H., Kalaskar, M.G.: Metal nanoparticles synthesis: an overview on methods of preparation, advantages and disadvantages, and applications. J. Drug Deliv. Sci. Technol. **53**, 101174 (2019)
4. Suryanarayana, C.: Mechanical alloying and milling. Prog. Mater. Sci. **46**(1–2), 1–184 (2001)

5. El-Eskandarany, M.S.: Mechanical Alloying: for Fabrication of Advanced Engineering Materials. William Andrew (2001)
6. R. Shashanka, D.C.: Phase transformation and microstructure study of nanostructured austenitic and ferritic stainless steel powders prepared by planetary milling. Powder Technol. **259**, 125–136 (2014)
7. Shashanka, R., Chaira, D.: Development of nano-structured duplex and ferritic stainless steel by pulverisette planetary milling followed by pressureless sintering. Mater. Charact. **99**, 220–229 (2015)
8. Lü, L., Lai, M.O.: Mechanical Alloying. Springer Science & Business Media (1997)
9. Suryanarayana, C.: Mechanical alloying. In: Pergamon Materials Series, vol. 2, pp. 49–85. Pergamon (1999)
10. Wang, W.: Modeling and simulation of the dynamic process in high energy ball milling of metal powders (Doctoral dissertation, The University of Waikato) (2000)
11. Othman, A.R., Sardarinejad, A., Masrom, A.K.: Effect of milling parameters on mechanical alloying of aluminum powders. Int. J. Adv. Manuf. Technol. **76**, 1319–1332 (2015)
12. Angelo, P.C., Subramanian, R., Ravisankar, B.: Powder Metallurgy: science, Technology and Applications. PHI Learning Pvt. Ltd. (2022)
13. Zarader, C.: Preparation of Polyetheretherketone-Nickel composite particles using ball milling for cold sprayed coatings on fiber reinforced plastics (2023)
14. Shashanka, R., Chaira, D.: Optimization of milling parameters for the synthesis of nanostructured duplex and ferritic stainless steel powders by high energy planetary milling. Powder Technol. **278**, 35–45 (2015)
15. Benjamin, J.S.: Mechanical alloying—A perspective. Metal Powder Rep. **45**(2), 122–127 (1990)
16. Gaffet, E., Le Caër, G.: Mechanical processing for nanomaterials. In: Encyclopedia of Nanoscience and Nanotechnology, vol. 5, no. 129, pp. 91–129. American scientific publishers, Stevenson Ranch, calif, USA
17. Gorrasi, G., Sorrentino, A.: Mechanical milling as a technology to produce structural and functional bio-nanocomposites. Green Chem. **17**(5), 2610–2625 (2015)
18. Wei, L.K., Abd Rahim, S.Z., Al Bakri Abdullah, M.M., Yin, A.T.M., Ghazali, M.F., Omar, M.F., Nemeş, O., Sandu, A.V., Vizureanu, P., Abdellah, A.E.H.: Producing metal powder from machining chips using ball milling process: a review. Materials **16**(13), 4635
19. Yadav, T.P., Yadav, R.M., Singh, D.P.: Mechanical milling: a top down approach for the synthesis of nanomaterials and nanocomposites. Nanosci. Nanotechnol. **2**(3), 22–48 (2012)
20. Joy, J., Krishnamoorthy, A., Tanna, A., Kamathe, V., Nagar, R., Srinivasan, S.: Recent developments on the synthesis of nanocomposite materials via ball milling approach for energy storage applications. Appl. Sci. **12**(18), 9312 (2022)
21. El-Eskandarany, M.S., Al-Hazza, A., Al-Hajji, L.A., Ali, N., Al-Duweesh, A.A., Banyan, M., Al-Ajmi, F.: Mechanical milling: a superior nanotechnological tool for fabrication of nanocrystalline and nanocomposite materials. Nanomaterials **11**(10), 2484 (2021)
22. Shashanka, R., Chaira, D., Kumara Swamy, B.E.: Electrocatalytic response of duplex and Yittria dispersed duplex stainless steel modified carbon paste electrode in detecting folic acid using cyclic voltammetry. Int. J. Electrochem. Sci. **10**, 5586–5598 (2015)
23. Gupta, S., Shashanka, R., Chaira, D.: Synthesis of nano-structured duplex and ferritic stainless steel powders by planetary milling: an experimental and simulation study. In: 4th National Conference on Processing and Characterization of Materials, IOP Conference on Series: materials Science and Engineering, vol. 75, p. 012033 (2015)
24. Shashanka, R., Chaira, D.: Effects of nano-Y2O3 and Sintering parameters on the fabrication of PM duplex and ferritic stainless steels. Acta Metall. Sin. (Engl. Lett.) **29**, 58–71 (2016)
25. Dasgupta, R.: A look into Cu-based shape memory alloys: present scenario and future prospects. J. Mater. Res. **29**(16), 1681–1698 (2014)
26. Hesse, F.A.B., Verissimo, N.C., Soyama, J., Bertazzoli, R.: High-energy ball milling of intermetallic Ti–Cu alloys for the preparation of oxide nanoparticles. Adv. Powder Technol. **32**(12), 4609–4620 (2021)

27. Saravanan, R., Rani, M.P.: Metal and Alloy Bonding-an Experimental Analysis: charge Density in Metals and Alloys. Springer Science & Business Media
28. Chaudhari, A.K., Singh, V.B.: A review of fundamental aspects, characterization and applications of electrodeposited nanocrystalline iron group metals, Ni–Fe alloy and oxide ceramics reinforced nanocomposite coatings. J. Alloy. Compd. **751**, 194–214 (2018)
29. Azimi, A., Shokuhfar, A., Zolriasatein, A.: Nanostructured Al–Zn–Mg–Cu–Zr alloy prepared by mechanical alloying followed by hot pressing. Mater. Sci. Eng., A **595**, 124–130 (2014)
30. Costa da Silva, F., Kazmierczak, K., Edil da Costa, C., Milan, J.C.G., Torralba, J.M.: Zamak 2 alloy produced by mechanical alloying and consolidated by sintering and hot pressing. J. Manuf. Sci. Eng. **139**(9), 091011 (2017)
31. Houshmand, A., 2019. Phase selection and transformation kinetics of copper-zinc powder mixture subjected to ultrasonic powder consolidation. Northeastern University. College of Engineering. Department of Mechanical and Industrial Engineering
32. Raihanuzzaman, R.M., Xie, Z., Hong, S.J., Ghomashchi, R.: Powder refinement, consolidation and mechanical properties of cemented carbides—An overview. Powder Technol. **261**, 1–13 (2014)
33. Yurlova, M.S., Demenyuk, V.D., Lebedeva, L.Y., Dudina, D.V., Grigoryev, E.G., Olevsky, E.A.: Electric pulse consolidation: an alternative to spark plasma sintering. J. Mater. Sci. **49**(3), 952–985 (2014)
34. Oguntuyi, S.D., Johnson, O.T., Shongwe, M.B.: Spark plasma sintering of ceramic matrix composite of TiC: microstructure, densification, and mechanical properties: a review. Int. J. Adv. Manuf. Technol. **116**(1–2), 69–82 (2021)
35. Šepelák, V., Bégin-Colin, S., Le Caer, G.: Transformations in oxides induced by high-energy ball-milling. Dalton Trans. **41**(39), 11927–11948 (2012)
36. Samal, P., Newkirk, J.: Milling of brittle and ductile materials. In: Powder Metall, pp. 77–87. ASM International
37. Delogu, F., Gorrasi, G., Sorrentino, A.: Fabrication of polymer nanocomposites via ball milling: present status and future perspectives. Prog. Mater. Sci. **86**, 75–126 (2017)
38. Neikov, O.D., Naboychenko, S.S., Yefimov, N.A., Neikov, O.D.: Mechanical alloying. In: Neikov, O.D., Naboychenko, S.S., Murashova, IV., Gopienko, V.G., Frishberg, IV., Lotsko, D.V.S. (eds.) Handbook of Non-ferrous Metal Powders, pp. 63–79 (2019)
39. Lines, M.G.: Nanomaterials for practical functional uses. J. Alloys Compd. **449**(1–2), 242–245 (2008)
40. Gauthier, M., Mazouzi, D., Reyter, D., Lestriez, B., Moreau, P., Guyomard, D., Roué, L.: A low-cost and high performance ball-milled Si-based negative electrode for high-energy Li-ion batteries. Energy Environ. Sci. **6**(7), 2145–2155 (2013)
41. Xing, T., Sunarso, J., Yang, W., Yin, Y., Glushenkov, A.M., Li, L.H., Howlett, P.C., Chen, Y.: Ball milling: a green mechanochemical approach for synthesis of nitrogen doped carbon nanoparticles. Nanoscale **5**(17), 7970–7976 (2013)
42. Ahmadi, M., Zabihi, O., Li, Q., Fakhrhoseini, S.M., Naebe, M.: A hydrothermal-assisted ball milling approach for scalable production of high-quality functionalized MoS2 nanosheets for polymer nanocomposites. Nanomaterials **9**(10), 1400 (2019)
43. Gusev, A.I., Kurlov, A.S.: Production of nanocrystalline powders by high-energy ball milling: model and experiment. Nanotechnology **19**(26), 265302 (2008)
44. Nayak, A.K., Shashanka, R., Chaira, D.: Effect of nanosize yittria and tungsten addition to duplex stainless steel during high energy planetary milling. In: 5th National Conference on Processing and Characterization of Materials, IOP Conference on Series: materials Science and Engineering, vol. 115, 012008 (2016)
45. Shashanka, R., Chaira, D., Chakravarty, D.: Fabrication of nano-yttria dispersed duplex and ferritic stainless steels by planetary milling followed by spark plasma sintering and non-lubricated sliding wear behaviour study. J. Mater. Sci. Eng. B **6**(5–6), 111–125 (2016)
46. Shashanka, R., Chaira, D.: Effect of sintering temperature and atmosphere on nonlubricated sliding wear of nano-yttria dispersed and yttria free duplex and ferritic stainless steel fabricated by powder metallurgy. Tribol. Trans. **60**, 324–336 (2017)

47. Khadka, P., Ro, J., Kim, H., Kim, I., Kim, J.T., Kim, H., Cho, J.M., Yun, G., Lee, J.: Pharmaceutical particle technologies: an approach to improve drug solubility, dissolution and bioavailability. Asian J. Pharm. Sci. **9**(6), 304–316 (2014)
48. Muñoz-Batista, M.J., Rodriguez-Padron, D., Puente-Santiago, A.R., Luque, R.: Mechanochemistry: toward sustainable design of advanced nanomaterials for electrochemical energy storage and catalytic applications (2018)
49. Mateti, S., Mathesh, M., Liu, Z., Tao, T., Ramireddy, T., Glushenkov, A.M., Yang, W., Chen, Y.I.: Mechanochemistry: a force in disguise and conditional effects towards chemical reactions. Chem. Commun. **57**(9), 1080–1092 (2021)
50. Lee, J.Y., An, J., Chua, C.K.: Fundamentals and applications of 3D printing for novel materials. Appl. Mater. Today **7**, 120–133 (2017)
51. Ngo, T.D., Kashani, A., Imbalzano, G., Nguyen, K.T., Hui, D.: Additive manufacturing (3D printing): a review of materials, methods, applications and challenges. Compos. B Eng. **143**, 172–196 (2018)
52. Hu, Y., Li, B., Yu, C., Fang, H., Li, Z.: Mechanochemical preparation of single atom catalysts for versatile catalytic applications: a perspective review. Mater. Today (2023)
53. Kabashin, A.V., Singh, A., Swihart, M.T., Zavestovskaya, I.N., Prasad, P.N.: Laser-processed nanosilicon: a multifunctional nanomaterial for energy and healthcare. ACS Nano **13**(9), 9841–9867 (2019)
54. Neikov, O.D., Yefimov, N.A., Naboychenko, S.: Handbook of Non-ferrous Metal Powders: technologies and Applications. Elsevier (2009)
55. Şap, S., Uzun, M., Usca, Ü.A., Pimenov, D.Y., Giasin, K., Wojciechowski, S.: Investigation on microstructure, mechanical, and tribological performance of Cu base hybrid composite materials. J. Market. Res. **15**, 6990–7003 (2021)
56. Casati, R., Vedani, M.: Metal matrix composites reinforced by nano-particles—A review. Metals **4**(1), 65–83 (2014)
57. The Luong, N., Okumura, H., Yamasue, E., Ishihara, K.N.: Structure and catalytic behaviour of CuO–CeO2 prepared by high-energy ball milling. Royal Soc. Open Sci. **6**(2), 181861 (2019)
58. Lu, L., Lai, M.O.: Formation of new materials in the solid state by mechanical alloying. Mater. Des. **16**(1), 33–39 (1995)
59. Shashanka, R.: Synthesis of nano-structured stainless steel powder by mechanical alloyingan overview. Int. J. Sci. Eng. Res. **8**, 588–594 (2017)
60. Shashanka, R., Chaira, D., Kumara Swamy, B.E.: Effect of Y2O3 nanoparticles on corrosion study of spark plasma sintered duplex and ferritic stainless steel samples by linear sweep voltammetric method. Arch. Metall. Mater. **63**, 749–763 (2018)
61. Scattergood, R.O., Koch, C.C., Murty, K.L., Brenner, D.: Strengthening mechanisms in nanocrystalline alloys. Mater. Sci. Eng. A **493**(1–2), 3–11 (2008)
62. Mokdad, F., Chen, D.L., Liu, Z.Y., Xiao, B.L., Ni, D.R., Ma, Z.Y.: Deformation and strengthening mechanisms of a carbon nanotube reinforced aluminum composite. Carbon **104**, 64–77 (2016)
63. Kumar, K., Dixit, S., Prakash, A., Vatin, N.I., Ul Haq, M.Z., Tummala, S.K., Bobba, P.B., Sobti, R., Kalpana, K., 2023. Understanding composites and intermetallic: Microstructure, properties, and applications. In: E3S Web of Conferences, vol. 430, p. 01196. EDP Sciences
64. Sasikumar, Y., Indira, K., Rajendran, N.: Surface modification methods for titanium and its alloys and their corrosion behavior in biological environment: a review. J. Bio- Tribo-Corros. **5**, 1–25 (2019)
65. Manivasagam, G., Mudali, U.K., Asokamani, R., Raj, B.: Corrosion and microstructural aspects of titanium and its alloys as orthopaedic devices. Corros. Rev. **21**(2–3), 125–160 (2003)
66. Ziaee, M., Crane, N.B.: Binder jetting: A review of process, materials, and methods. Addit. Manuf. **28**, 781–801 (2019)
67. Basheer, A.A.: Advances in the smart materials applications in the aerospace industries. Aircr. Eng. Aerosp. Technol. **92**(7), 1027–1035 (2020)

68. Sathish, T., Saravanan, R., Kumar, A., Prakash, C., Shahazad, M., Gupta, M., Senthilkumar, N., Pandit, B., Ubaidullah, M., Smirnov, V.A.: Influence of synthesizing parameters on surface qualities of aluminium alloy AA5083/CNT/MoS2 nanocomposite in powder metallurgy technique. J. Mater. Res. Technol. (2023)
69. Everhart, J.L.: Copper and Copper Alloy Powder Metallurgy: properties and Applications, vol. 129. Metal Powder Industries Federation, Princeton, NJ (1975)
70. Mechiri, S.K., Vasu, V., Venu Gopal, A.: Investigation of thermal conductivity and rheological properties of vegetable oil based hybrid nanofluids containing Cu–Zn hybrid nanoparticles. Exp. Heat Transfer **30**(3), 205–217 (2017)
71. Rubinstein, M.H., Gould, P.: Praticle size reduction in the ball mill. Drug Dev. Ind. Pharm. **13**(1), 81–92 (1987)
72. Nabiyouni, G., Fesharaki, M.J., Mozafari, M., Amighian, J.: Characterization and magnetic properties of nickel ferrite nanoparticles prepared by ball milling technique. Chin. Phys. Lett. **27**(12), 126401 (2010)
73. DeCastro, C.L., Mitchell, B.S.: Nanoparticles from mechanical attrition. In: Synthesis, Functionalization, and Surface Treatment of Nanoparticles, vol. 5 (2002)
74. Santhanam, P.R., Dreizin, E.L.: Predicting conditions for scaled-up manufacturing of materials prepared by ball milling. Powder Technol. **221**, 403–411 (2012)
75. Koch, C.C.: The synthesis and structure of nanocrystalline materials produced by mechanical attrition: a review. Nanostruct. Mater. **2**(2), 109–129 (1993)
76. Bahmanpour, H., Youssef, K.M., Scattergood, R.O., Koch, C.C.: Mechanical behavior of bulk nanocrystalline copper alloys produced by high energy ball milling. J. Mater. Sci. **46**, 6316–6322 (2011)
77. Rajeshkumar, L., Suriyanarayanan, R., Hari, K.S., Babu, S.V., Bhuvaneswari, V. and Karunan, M.J.: Influence of boron carbide addition on particle size of copper zinc alloys synthesized by powder metallurgy. In: IOP Conference Series: materials Science and Engineering, vol. 954, no. 1, p. 012008. IOP Publishing (2020)
78. Bor, A., Jargalsaikhan, B., Lee, J., Choi, H.: Effect of different milling media for surface coating on the copper powder using two kinds of ball mills with discrete element method simulation. Coatings **10**(9), 898 (2020)
79. Tyler, D.E., Black, W.T.: Introduction to copper and copper alloys (1990)
80. Bahmanpour, H., Yousseff, K.M., Khoshkhoo, M.S., Scudino, S., Eckert, J., Scattergood, R.O., Koch, C.C., Khoshkhoo, M.S., Freudenberger, J., Zehetbauer, M.J.: 5.0. in situ consolidation of Cu and Cu–Zn alloys via room temperature ball milling
81. Camurri, C., Ortiz, M., Carrasco, C.: Hot consolidation of Cu–Li powder alloys: a first approach to characterization. Mater. Charact. **51**(2–3), 171–176 (2003)
82. Queudet, H., Lemonnier, S., Barraud, E., Guyon, J., Ghanbaja, J., Allain, N., Gaffet, E.: One-step consolidation and precipitation hardening of an ultrafine-grained Al–Zn–Mg alloy powder by Spark Plasma Sintering. Mater. Sci. Eng. A **685**, 227–234 (2017)
83. Shashanka, R.: Non-lubricated dry sliding wear behavior of spark plasma sintered nanostructured stainless steel. J. Mater. Environ. Sci. **10**(8), 767–777 (2019)
84. Shashanka, R., Orhan Uzun, D. Chaira, Synthesis of nano-structured duplex and ferritic stainless steel powders by dry milling and its comparison with wet milling. Arch. Metall. Mater. **65**(1), 5–14 (2020)
85. Rayappa, S.M., Shamanth, V., Sharath, P.C., Shashanka, R., Hemanth, K.: A review on spark plasma sintering of duplex stainless steels. Mater. Today: Proc. **45**, 138–144 (2021)
86. Mahale, R.S., Rajendrachari, S., Vasanth, S., Krishna, H., Kapanigowda, N.S., Chikkegowda, S.P., Patil, A.: Technology and challenges in additive manufacturing of duplex stainless steels. Biointerface Res. Appl. Chem. **12**(1), 1110–1119 (2022)
87. Clinktan, R., Senthil, V., Ramkumar, K.R., Sivasankaran, S., Al-Mufadi, F.A.: Effect of boron carbide nano particles in CuSi4Zn14 silicone bronze nanocomposites on matrix powder surface morphology and structural evolution via mechanical alloying. Ceram. Int. **45**(3), 3492–3501 (2019)

88. Ramkumar, K.R., Sivasankaran, S., Alaboodi, A.S.: Effect of alumina content on microstructures, mechanical, wear and machining behavior of Cu-10Zn nanocomposite prepared by mechanical alloying and hot-pressing. J. Alloy. Compd. **709**, 129–141 (2017)
89. Stalin, B., Ravichandran, M., Jasper, S., Kannan, C.R.: Synthesis and characterization of brass—AlN composites synthesized by ball milling. Mater. Today: Proc. **22**, 2573–2581 (2020)
90. Pradeep, N.B., Hegde, M.R., Patel, G.M., Giasin, K., Pimenov, D.Y., Wojciechowski, S.: Synthesis and characterization of mechanically alloyed nanostructured ternary titanium based alloy for bio-medical applications. J. Market. Res. **16**, 88–101 (2022)
91. Andrade-Gamboa, J., Gennari, F.C., Larochette, P.A., Neyertz, C., Ahlers, M., Pelegrina, J.L.: Stability of Cu–Zn phases under low energy ball milling. Mater. Sci. Eng. A **447**(1–2), 324–331 (2007)
92. Abdolhoseinzadeh, A., Sheibani, S.: Enhanced photocatalytic performance of Cu2O nano-photocatalyst powder modified by ball milling and ZnO. Adv. Powder Technol. **31**(1), 40–50 (2020)
93. Guo, D., Kazasidis, M., Hawkins, A., Fan, N., Leclerc, Z., MacDonald, D., Nastic, A., Nikbakht, R., Ortiz-Fernandez, R., Rahmati, S., Razavipour, M.: Cold spray: over 30 years of development toward a hot future. J. Therm. Spray Technol. **31**(4), 866–907 (2022)
94. Capuzzi, S., Timelli, G.: Preparation and melting of scrap in aluminum recycling: a review. Metals **8**(4), 249 (2018)
95. Rajendrachari, S., Adimule, V.M., Jayaprakash, G.K., Pandith, A.: Electrochemical oxidation of methylene blue dye in wastewater using mechanically alloyed high entropy alloy modified carbon paste electrode using cyclic voltammetry. Mater. Res. Express **10**, 054003 (2023). https://doi.org/10.1088/2053-1591/acd3d7
96. Tjong, S.C.: Recent progress in the development and properties of novel metal matrix nanocomposites reinforced with carbon nanotubes and graphene nanosheets. Mater. Sci. Eng. R. Rep. **74**(10), 281–350 (2013)
97. Orru, R., Licheri, R., Locci, A.M., Cincotti, A., Cao, G.: Consolidation/synthesis of materials by electric current activated/assisted sintering. Mater. Sci. Eng. R. Rep. **63**(4–6), 127–287 (2009)
98. Dancer, C.E.: Flash sintering of ceramic materials. Mater. Res. Express **3**(10), 102001 (2016)
99. Bykov, Y.V., Egorov, S.V., Eremeev, A.G., Kholoptsev, V.V., Plotnikov, I.V., Rybakov, K.I., Sorokin, A.A.: On the mechanism of microwave flash sintering of ceramics. Materials **9**(8), 684 (2016)
100. Heidary, D.S.B., Lanagan, M., Randall, C.A.: Contrasting energy efficiency in various ceramic sintering processes. J. Eur. Ceram. Soc. **38**(4), 1018–1029 (2018)
101. Ross, F.M., Minor, A.M.: In situ transmission electron microscopy. In: Springer Handbook of Microscopy, pp. 101–187 (2019)
102. Himanen, L., Geurts, A., Foster, A.S., Rinke, P.: Data-driven materials science: status, challenges, and perspectives. Adv. Sci. **6**(21), 1900808 (2019)
103. Yang, L., Hsu, K., Baughman, B., Godfrey, D., Medina, F., Menon, M., Wiener, S.: Additive Manufacturing of Metals: the Technology, Materials, Design and Production (2017)
104. Mobarak, M.H., Islam, M.A., Hossain, N., Al Mahmud, M.Z., Rayhan, M.T., Nishi, N.J., Chowdhury, M.A.: Recent advances of additive manufacturing in implant fabrication—A review. Appl. Surf. Sci. Adv. **18**, 100462 (2023)
105. Mohanty, A.K., Vivekanandhan, S., Pin, J.M., Misra, M.: Composites from renewable and sustainable resources: challenges and innovations. Science **362**(6414), 536–542 (2018)
106. Nidumolu, R., Prahalad, C.K., Rangaswami, M.R.: Why sustainability is now the key driver of innovation. Harv. Bus. Rev. **87**(9), 56–64 (2009)
107. Whang, S.H. (ed.).: Nanostructured Metals and Alloys: processing, Microstructure, Mechanical Properties and Applications. Elsevier (2011)
108. Pimenov, D.Y., Mia, M., Gupta, M.K., Machado, Á.R., Pintaude, G., Unune, D.R., Khanna, N., Khan, A.M., Tomaz, Í., Wojciechowski, S., Kuntoğlu, M.: Resource saving by optimization and machining environments for sustainable manufacturing: a review and future prospects. Renew. Sustain. Energy Rev. **166**, 112660 (2022)

109. Konieczny, J., Rdzawski, Z.: Antibacterial properties of copper and its alloys. Arch. Mater. Sci. Eng. **56**(2), 53–60 (2012)
110. Meyers, M.A., Mishra, A., Benson, D.J.: Mechanical properties of nanocrystalline materials. Prog. Mater Sci. **51**(4), 427–556 (2006)
111. Carro, G., Muñoz, A., Monge, M.A., Savoini, B., Galatanu, A., Galatanu, M., Pareja, R.: Thermal conductivity and diffusivity of Cu-Y alloys produced by different powder metallurgy routes. Fusion Eng. Des. **124**, 1156–1160 (2017)
112. Myagkov, L.L., Mahkamov, K., Chainov, N.D., Makhkamova, I.: Advanced and conventional internal combustion engine materials. In: Alternative Fuels and Advanced Vehicle Technologies for Improved Environmental Performance, pp. 370–408e. Woodhead Publishing (2014)
113. Mondal, S.: Aluminum or its alloy matrix hybrid nanocomposites. Met. Mater. Int. **27**, 2188–2204 (2021)
114. Sayyadi, R., Khodabakhshi, F., Javid, N.S., Khatibi, G.: Influence of graphene content and nickel decoration on the microstructural and mechanical characteristics of the Cu/Sn–Ag–Cu/Cu soldered joint. J. Market. Res. **9**(4), 8953–8970 (2020)
115. Khin, M.M., Nair, A.S., Babu, V.J., Murugan, R., Ramakrishna, S.: A review on nanomaterials for environmental remediation. Energy Environ. Sci. **5**(8), 8075–8109 (2012)
116. Jester, T.C. (ed.). Twentieth-Century Building Materials: history and Conservation. Getty Publications (2014)
117. Thomas, P., Lai, C.W., Johan, M.R.B.: Recent developments in biomass-derived carbon as a potential sustainable material for super-capacitor-based energy storage and environmental applications. J. Anal. Appl. Pyrol. **140**, 54–85 (2019)
118. Raabe, D.: The materials science behind sustainable metals and alloys. Chem. Rev. **123**(5), 2436–2608 (2023)

Chapter 13
Processing of Magnesium Alloys by Mechanical Alloying

Kumar Debajyoti Jena and Peng Cao

Abstract This chapter overviews the mechanical alloying (MA) process applied to produce magnesium (Mg) alloys, often called MA-Mg alloys. It discusses the influence of MA on the microstructures and mechanical properties of Mg alloys. The chapter also highlights recent advancements over the past 5 years, emphasising potential applications of MA-Mg alloys.

Keywords Material properties · Magnesium alloys · Mechanical alloying · Nanostructured alloys · Microstructure

13.1 Background

Mg, an alkaline earth metal, exhibits a shiny, silvery-white appearance, and remarkable reactivity [1]. While it is not naturally found in its pure form, Mg exists in many chemical compounds. Its recognition as an element can be credited to Joseph Black, with contributions from Friedrich Hoffman, who identified it in 1754 [2]. The production of pure Mg was later accomplished by Antoine-Alexander Bussy in 1828, following Humphry Davy's discovery of Mg metal in 1808 [3, 4].

Throughout history, Mg has played a significant role in military applications and warfare. It was employed as a weapon during World War II, finding use in constructing incendiary bombs, flares, and ammunition, which led to widespread devastation [1, 5]. However, in the post-war era, its availability and unique properties piqued interest in various industries. Today, Mg is extensively utilised in engineering applications, particularly in the automotive, aerospace, and consumer electronics sectors. Its exceptional mechanical properties, including a high strength-to-weight

K. D. Jena · P. Cao (✉)
Department of Chemical and Materials Engineering, The University of Auckland, Auckland 1142, New Zealand
e-mail: p.cao@auckland.ac.nz

P. Cao
MacDiarmid Institute of Advanced Materials and Nanotechnology, Victoria University of Wellington, PO Box 600, Wellington 6140, New Zealand

S. Rajendrachari (ed.), *Mechanically Alloyed Novel Materials*, Advanced Structured Materials 220, https://doi.org/10.1007/978-981-97-6504-1_13

ratio and machinability, make it an ideal material for manufacturing engine blocks, transmission cases, steering wheels, and structural components in the automotive industry. In aerospace sectors, Mg alloys contribute to weight reduction and increase fuel efficiency. Mg also finds applications in organic chemistry, pharmaceuticals, and the production of general-purpose items like sporting goods, household products, and office equipment. Furthermore, its biodegradable and biocompatible nature makes it a promising biomaterial, attracting significant interest from research and industry communities. This has spurred the growth of the global Mg market, with projections of 1.6 million metric tons by 2027 [2, 6, 7].

13.2 Physical Properties of Mg

Mg is a lightweight metal with a 1.74 g/cm^3 density and a relatively low melting point of around 650 °C. It typically has a room-temperature hexagonal close-packed (HCP) crystal structure. Mg exhibits good thermal conductivity (156 W/m K), a specific heat capacity of 1024 J/kg K, Young's modulus of 45 GPa, and a shear modulus of 17 GPa. Its Poisson ratio is approximately 0.29. These properties make Mg suitable for applications, where lightweight, heat conduction and specific stiffness are important factors. Pure polycrystalline Mg is brittle and prone to fracturing along shear bands. However, its malleability improves when alloyed with small amounts of other metals, such as 1% aluminium. While Mg tarnishes slightly when exposed to air, unlike heavier alkaline earth metals, it does not require an oxygen-free environment for storage. This is because Mg is naturally protected by a thin, impermeable oxide layer that is challenging to remove. Alloying is a well-established technique for improving the applicability of Mg. In Mg alloys, various intermetallic phases can be observed, and the mechanical performance of these materials can be enhanced through solid solution hardening and/or precipitation hardening. Given the ongoing developments and increasing interest in Mg, providing a comprehensive understanding of its properties and applications is crucial. This chapter presents recent progress and advancements in the field, specifically focusing on the engineering applications of Mg and its alloys and composites. Table 13.1 shows the common Mg alloys and their applications.

13.3 Extraction of Mg from Its Ores to the Final Product

Mg ores, primarily dolomite and magnesite, are found globally in regions such as China, Russia, North America (USA and Canada), North Korea, Australia, Brazil, South Africa, several European countries, Asia, and the Middle East [13]. The extraction of Mg from its primary ores involves several essential steps. First, Mg-bearing minerals like dolomite and magnesite are mined and processed to concentrate the

Table 13.1 Common Mg alloys and their applications [8–12]

Alloy designation	Alloying additives	Uses	Basic properties and applications
AZ91	9.0% Al, 0.7%Zn 0.13%Mn	General cast alloy	Good castability, good mechanical properties at T < 150 °C
AM60	6.0% Al, 0.15%Mn	High-pressure die-cast alloy	Greater toughness and ductility than AZ91, with slightly lower strength. Often preferred for automotive structural applications
AM 50	Mg–Al system	General cast alloy	Good strength, ductility, energy absorption properties, and castability
AE44	Mg–4%Al-4%rare earth	General cast alloy	Better creep behaviour and castability than AE42
AE42	Mg–4%Al-2% rare earth	General cast alloy	Low castability, good creep behaviour
AS41	4.2%Al, 1.0%Si	General cast alloy	Better creep resistance than AZ91 at elevated temperatures but lower strength
ZE41	4.2%Zn, 1.2%RE, 0.7% Zr	Specialist cast alloy	Rare earth addition improves creep strength at elevated temperatures. Pressure tight
AZ31	3.0% Al, 1.0Zn, 0.2% Mn	Wrought Mg products	Good extrusion alloy
AM20	Mg–Al system	Cast alloy	High ductility, toughness, poor die-castability
MRI 153 M	Mg–Al–Ca–Sr system	Cast alloy	For high-temperature applications up to 150 °C
MRI 230D	Mg–Al–Ca–Sr system	Cast alloy	For high-temperature applications up to 190 °C
AS 21	Mg–Al–Si system	Cast alloy	For use at temperatures over 120 °C
AJ62	Mg–Al–Sr system	High-pressure die-casting (HPDC)	Good thermal and mechanical strength, superior castability, corrosion resistance, and creep behaviour

Mg content. Subsequently, the concentrated ore undergoes high-temperature calcination processes, which drive off carbon dioxide and leave Mg oxide as the primary product. The reduction of Mg oxide to Mg metal can be achieved through two main methods: electrolysis and thermal reduction. Historically, electrolysis was predominant, involving the dissociation of Mg chloride into Mg metal and chlorine gas in electrolytic cells. Figure 13.1 shows the electrolysis process of $MgCl_2$. On the other hand, thermal reduction methods, such as the Pidgeon and Magnetherm processes, reduce Mg oxide with a reducing agent like silicon. After reduction, refining and purification steps are employed to ensure high-purity Mg [14–16]. Depending on the applications, alloying elements may be added, and the purified Mg is shaped

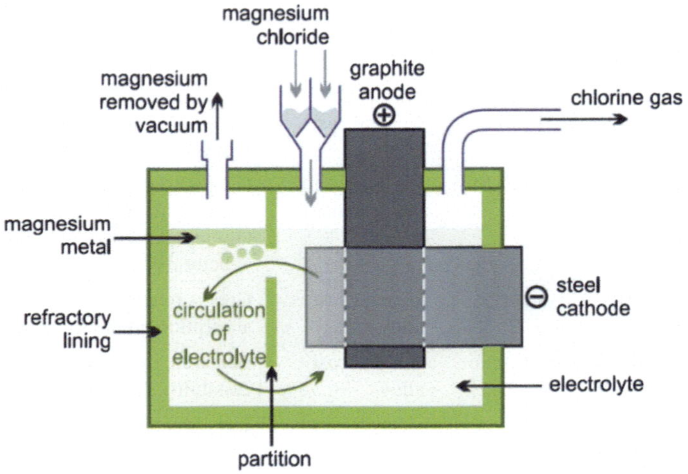

Fig. 13.1 Schematic illustration of electrolysis of Mg chloride. Copyrights: Department of Chemistry, University of York, 2016. *Source* www.essentialchemicalindustry.org/metals/magnesium

into various forms through processes like casting, extrusion, or forging. Due to their lightweight and corrosion-resistant properties, the resulting Mg and Mg alloys are utilised in aerospace, automotive and manufacturing industries. Ongoing research aims to improve the energy efficiency and environmental sustainability of Mg extraction methods.

13.4 Mg Alloys and Processing Techniques

Mg alloys offer unique mechanical behaviour, including plasticity and stiffness, which is advantageous compared to bioresorbable polymers [17]. These alloys have found applications in medical, transportation, and aerospace industries. In the physiological environment, Mg alloys play a crucial role in corrosion and are promising candidates for biodegradable tissue implants [18]. Mg alloys interact and interface with temporary biomaterials in the biological environment [19]. Mg alloys degrade in vivo three times faster than in vitro [20]. However, the applicability of Mg alloys is limited by their inherent formability challenges. Microstructural optimisation has proven beneficial, enhancing formability by reducing the grain size [21]. Techniques like electron backscattering diffraction analysis have been employed to investigate the microstructure and grains of Mg alloys, shedding light on the influence of grain size and microstructure on formability [22]. Despite exhibiting less ductility and poor formability at room temperature, Mg alloys can be efficiently processed using rolling methods at elevated temperatures, achieving significant reduction levels ranging from 18 to 50% [23]. This, coupled with their high strength-to-weight ratio, good damping behaviour, and high creep resistance, positions Mg alloys as essential materials in the automotive and aerospace industries.

Various techniques are employed to prepare Mg alloys, each tailored to specific applications. Casting methods, including sand casting and die casting, are utilised for producing large and intricate components. Mg alloy products for automotive applications are typically manufactured through casting processes. As-cast Mg–Cu alloys have shown promise in promoting osteogenesis, bone fixation, cell viability, and bacterial infection prevention [24]. AZ31 alloy, commonly used for its commercial availability and good tensile characteristics, faces limitations in its as-cast condition due to low yield strength and casting defects. Utilising grain refinement through secondary processes like rolling, extrusion, and severe plastic deformation (SPD) offers promise for improving yield strength. Achieving ultra-fine grain (UFG) or nano-sized grains in AZ31 alloy can activate grain boundary sliding at room temperature, resulting in quasi-superplastic elongation. Extrusion is commonly employed to create continuous profiles like rods and tubes while rolling reduces thickness to form sheets and foils. Powder metallurgy involves compacting Mg alloy powders, followed by processes like sintering. Wrought processing includes techniques such as forging and swagging at elevated temperatures. Ingot metallurgy produces ingots that can be further processed through rolling and extrusion. Thixomolding involves injecting semi-solid Mg alloy into moulds suitable for complex components. Additive manufacturing, including 3D printing, offers design flexibility. The selection of a technique depends on factors such as the desired product, mechanical properties, and cost considerations, with each method catering to specific needs in Mg alloy processing.

Powder metallurgy (PM) techniques have been employed to develop Mg alloys with increased strength, improved impact toughness, and enhanced corrosion resistance. The PM-Mg alloys have been explored particularly in biomedical applications where Mg-based alloys are combined with Ti, Fe, and Mn. These Mg-based biomedical alloys exhibit corrosion behaviour, mechanical properties, and cytotoxicity that ensure mechanical integrity in both in vivo and in vitro environments [25, 26]. This process also leads to smaller grain sizes compared to ingot metallurgy [27]. Mg–$x$$Al_2O_3$ ($x = 5.5$, 10.7 wt%) micro-composites were successfully fabricated through PM and microwave sintering, followed by extrusion [28]. The addition of 5.5 and 10.7 wt% Al_2O_3 enhanced microhardness, tensile yield strength, and ultimate tensile strength in Mg, with the trade-off of reduced ductility, reaching a minimum of approximately 2.8% with 10.7 wt% Al_2O_3 reinforcement. Mg–10Ti with graphite (Gr) micro-composite was also prepared via PM route, exhibiting enhanced strength and ductility compared to pure Mg [29]. Mg alloys processed through PM exhibited a high strain rate of superplasticity, especially at low temperatures.

The ductility and stability of Mg alloys often need to be improved due to a coarse grain structure and intermetallic compounds. Friction stir processing has been proposed to enhance the microstructure and mechanical properties of composite materials such as Mg alloys, titanium alloys, and steel [30]. Mg alloys demonstrate significant accuracy in punching and spring-back stamping processes. In bending, Mg alloys tend to exhibit neutral layer shifting towards the direction of the external tension zone [31]. Laser surface melting in an alternating magnetic field has been employed for enhancing the mechanical properties of Mg alloys. Studies have explored the wear resistance of Mg alloys, particularly those incorporating Mn, demonstrating favourable performance under slow sliding speed conditions [32]. However, adding Mn can negatively affect microstructure refinement in Mg–Mn alloys, leading to poor performance at room temperature due to intensified basal appearance and hardened microstructure.

Barani et al. demonstrated the effectiveness of chitosan coating agents on AZ91 Mg alloy to address corrosion challenges and improve bioactivity in Mg alloys [33, 34]. Mg alloys exhibit desirable mechanical and corrosion properties, particularly in high-solute alloys [35]. The corrosion behaviour of Mg alloys is influenced by various parameters such as alloy composition, microstructure, and film properties, which can be affected by exposure to the surrounding environment [36]. Various coating methods, including metal coatings, chemical conversion coatings, synthesised polymer coatings, and ceramics, have been explored to decrease degradation rates and increase corrosion resistance in Mg alloys. Spin coating techniques have proven crucial in achieving uniform coatings with fast processing and high uniformity. Xu et al. implemented cathodic protection on Mg alloys to enhance their corrosion resistance [37]. Stannate is commonly used in chemical conversion treatments to form a compact conversion film on Mg alloys. Plasma electrolytic oxidation is another method employed to create composite coatings on Mg alloys by treating them with an aqueous suspension, enhancing their wettability.

Wrought Mg alloys require elevated temperatures to dissolve the precipitates and activate additional slide systems [38]. The laser surface melting method modifies the Mg matrix layer. It creates a uniform distribution of the β phase, resulting in improved corrosion resistance and the formation of a barrier effect [32]. Mg is susceptible to the influence of high metal impurities, which can negatively impact corrosion resistance and sediment formation during alloy melting. Different temperature and time treatments are applied to Mg alloys to enhance their mechanical properties [39]. Mg alloys are essential for human metabolism and are non-toxic [40]. Carboneras et al. demonstrated methods to improve the poor corrosion behaviour of Mg alloys through surface modifications and undercoating techniques [41]. Mg alloy hydride plays a critical role in commercial applications, and efforts are made to develop new alloys and efficient production methods that take advantage of its effects. In the biomedical field, Mg alloys offer a promising solution for improved performance in simulated body fluids [42]. The production of Mg alloys involves slab casting, hot rolling, cold rolling, and multiple intermediate heat treatment processes [43].

The microstructures and properties of Mg alloys prepared via mechanical alloying (MA) depend on alloying elements. They play a crucial role in enhancing the properties of Mg alloys since as-cast Mg typically exhibits poor strength and rapid degradation [37]. When chemically active Mg interacts with alloying compounds, it forms intermetallic compounds, becoming integral to the microstructure of Mg alloys and subsequently impacting their material properties [44]. Alloying elements directly strengthen the mechanical properties of these alloys through mechanisms such as precipitation hardening, grain refinement, and solid solution strengthening [36]. To reinforce the Mg matrix effectively, it is essential to select alloying elements with high temperature-dependent solubility, considering factors like the atomic size relative to Mg and valence [35]. The HCP crystal structure of Mg, characterised by a c/a ratio of 1.624 and an atomic diameter of 0.320 nm, facilitates the formation of solid solutions with various alloying elements [35]. The impact of each alloying element can be linked to the number of elements present in the composition and the percentage or concentration of each alloying element. For a quick overview, Table 13.2 provides a summary of some common alloying elements and their respective effects on Mg alloys [45].

In summary, traditional methods like casting and extrusion offer simplicity but often need more control over microstructure and face challenges in producing intricate shapes. PM provides an avenue for shaping through compaction and sintering, yet achieving full density can be challenging. In contrast, mechanical alloying or ball milling, has emerged as a preferred technique due to several key advantages.

Table 13.2 A summary of some common alloying elements and their respective effects on Mg alloys

Alloying element	Influence on material properties of Mg alloys
Aluminum (Al)	Increases hardness, strength, and castability
Beryllium (Be)	Reduces surface melt oxidation
Calcium (Ca)	Improves corrosion resistance, creep resistance, and grain refinement
Cerium (Ce)	Increases elongation and work hardening rate. Reduces yield strength
Copper (Cu)	Increases ambient temperature and high-temperature strength. Reduces ductility
Iron (Fe)	Harmful to corrosion resistance
Lithium (Li)	Increases ductility and decreases strength
Manganese (Mn)	Improves saltwater corrosion resistance
Molybdenum (Mo)	Increases hardness, Young's modulus, and ductility
Nickel (Ni)	Improves YS and UTS. Reduces corrosion resistance
Neodymium (Nd)	Increases strength
Silicon (Si)	Improves fluidity
Silver (Ag)	Improves high-temperature strength and creep resistance
Strontium (Sr)	Enhances creep and corrosion resistance, grain refinement. Increases bone mass
Thorium (Th)	Enhances creep strength up to 370 °C. Increases weldability with zinc in alloys
Tin (Sn)	Ductility increases with Al in Mg. Enhanced corrosion resistance, compressive strength
Titanium (Ti)	Increased ductility and yield strength
Yttrium (Y)	Improved creep resistance and high-temperature resistance
Zinc (Zn)	Enhances yield stress and corrosion resistance
Zirconium (Zr)	Grain refinement

This process generates fine and homogeneous alloy powders, facilitates solid-state synthesis without needing high-temperature liquid phases, and allows for precise microstructure and phase composition control. MA also offers the preparation of non-equilibrium phases, supersaturated solid solutions, and intermetallics. The mechanical properties in Mg alloys are improved through two main approaches: solid solution strengthening and second phase strengthening. Chen et al. highlighted alloying as the most effective method for enhancing the mechanical properties of metals [46]. While MA demands energy-intensive milling, its capacity to produce complex alloys with controlled properties, fine grain sizes, and metastable phases makes it an excellent choice for advanced Mg alloy preparation.

13.5 Mechanical Alloying of Mg

MA is a solid-state powder processing technique that transforms materials into metastable by adjusting their free energy and then rapidly quenching them, resulting in unique chemical compositions and microstructures. Particularly valuable for producing challenging or difficult-to-obtain alloy powders, MA involves placing powders of a specific composition in a high-energy ball mill with large milling balls, leading to intense deformation, welding, and fracturing of the powders (Fig. 13.2) [47]. This process creates homogeneous materials with non-equilibrium phases, including supersaturated solid solutions, intermetallic phases, quasicrystalline phases, amorphous alloys, and high-entropy alloys. The dislocations within the grains multiply and slip during ball milling, causing plastic deformation. FCC crystals are more prone to plastic deformation than BCC and HCP due to the availability of slip systems. As ball milling time increases, dislocations accumulate, undergo reactions, and initiate cracks, leading to powder refinement through dislocation glide, multiplication, slip, reactions, stress concentration, and crack propagation. Temperature regulation during high-energy ball milling is critical as it affects powder morphology, particle size, and sintering behaviour [48–50]. To prevent oxidation, milling is typically performed in an inert gas atmosphere. The MA process significantly increases crystalline defects and dislocations, promoting the diffusion of alloying elements and resulting in Mg alloys with enhanced mechanical properties, such as increased strength, ductility, creep resistance, and toughness. These mechanically alloyed magnesium (MA-Mg) alloys find applications in aerospace, automotive, biomedical, and sports equipment, where their improved mechanical properties make them highly desirable materials.

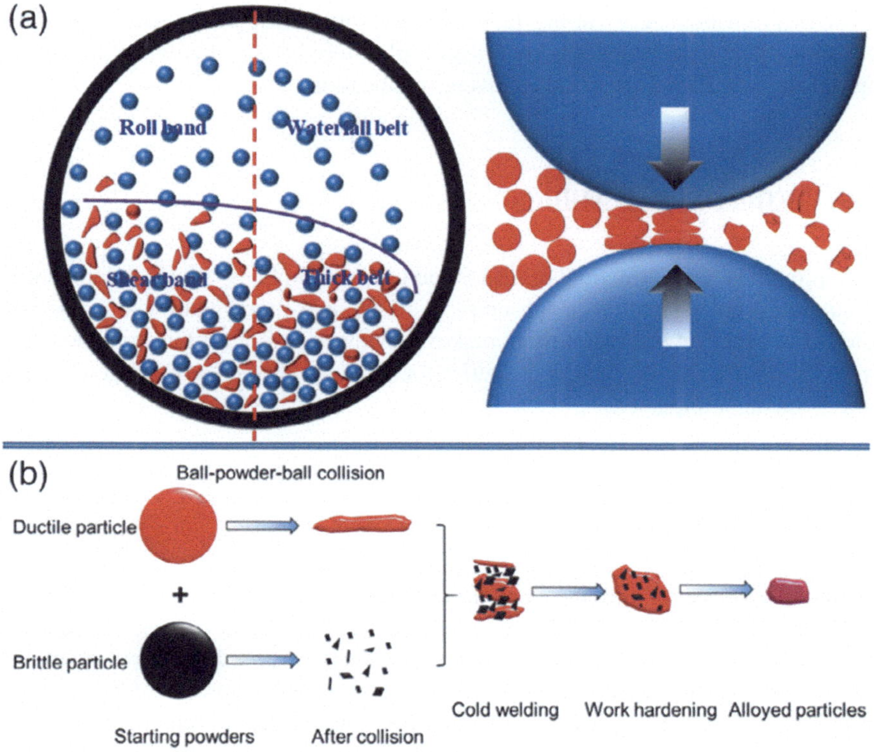

Fig. 13.2 **a** Schematic diagrams depict ball–powder–ball collision during MA. **b** Deformation processes of initial particles after collision. Taken from Ref. [47] with permission

13.6 Consolidation

Consolidating milled powders is crucial in determining the final properties of Mg alloys, ensuring enhanced performance and structural integrity. Spark plasma sintering (SPS), press-and-sinter, and hot rolling are the primary consolidation techniques for converting milled powders into semi-finished products. SPS has effectively consolidated Mg-10 wt% Al alloy prepared by high-energy ball milling (HEBM). Even after processing at 350 °C, the alloy maintains its grain size below 100 nm, and increasing SPS temperature correlates with higher alloy density [51]. The combination of HEBM treatment and SPS results in the fabrication of AZ31 Mg anodes with ultrafine grains, showcasing the versatility of this approach in achieving nanostructured materials [52]. HEBM has successfully synthesised Mg–Si alloys with varying grain sizes in sintered alloys [53]. PM-produced Mg–SiC composites exhibit improved mechanical properties, including enhanced density, hardness, compressive strength, and impact strength [54]. Hot pressing is another effective consolidation method to achieve ultrafine-grained Mg–3Al–Zn alloy through mechanical milling [55]. With an average grain size of 485 nm, the dominant deformation mechanism involves grain boundary sliding facilitated by slip controlled by grain boundary diffusion. This underscores the significance of tailored consolidation techniques in optimising Mg alloy properties for diverse applications.

13.7 Effect of Ball Milling

MA promotes the formation of nanostructures, amorphous alloys, intermetallic compounds, solid solutions, high entropy alloys, and precipitates within the Mg alloy. The versatility of ball milling makes it a powerful tool for tailoring the properties of Mg alloys, offering opportunities for applications such as hydrogen storage and biomedical implants.

13.7.1 Metastable Alloys

MA provides superior alloying capabilities to traditional ingot processing and rapid solidification methods. It tackles the thermodynamic challenges posed by mutual insolubility in alloy systems. It circumvents limitations posed by phase diagrams, enabling the creation of homogeneous alloys featuring supersaturated solid solutions of titanium. This is achieved by subjecting powder particles to repetitive deformation, welding, and fracturing in a high-energy ball mill. Notably, the limited mutual solubility between Ti and Mg, typically less than 2 at%, can be overcome through MA, which effectively extends terminal solid solubility and triggers chemical reactions at substantially lower temperatures than conventional methods. Ochoa et al. successfully synthesised a metastable FCC structure in a $Ti_{50}Mg_{50}$ alloy through 25 h of ball milling, using stearic acid as the process control agent [56]. The transmission electron microscopic (TEM) and selected area electron diffraction (SAED) image of the metastable $Ti_{50}Mg_{50}$ alloy is depicted in Fig. 13.3. The phase transformation from hexagonal HCP to FCC in the Ti–Mg alloy was attributed to stacking faults generated during the milling process. Various metastable alloys, including Mg–10Zn, Mg–25Sn, Mg–50In, and $Mg_{55}Co_{45}$, have been successfully prepared via ball milling, showcasing enhanced properties [57–61]. The recent report highlights the synthesis of the $Mg_{64}Zn_{30}Ca_4Er_1$ alloy through MA, showcasing the versatility of this method in producing novel alloy compositions [62].

13.7.2 Supersaturated Solid Solution

The $Mg_{95}Sn_3Zn_2$ alloy, synthesised via MA, promotes the prolonged solubility of Sn in the Mg lattice, resulting in the development of a supersaturated Mg (Sn, Zn) ternary solid solution during the MA process [63]. This solid solution demonstrates enhanced

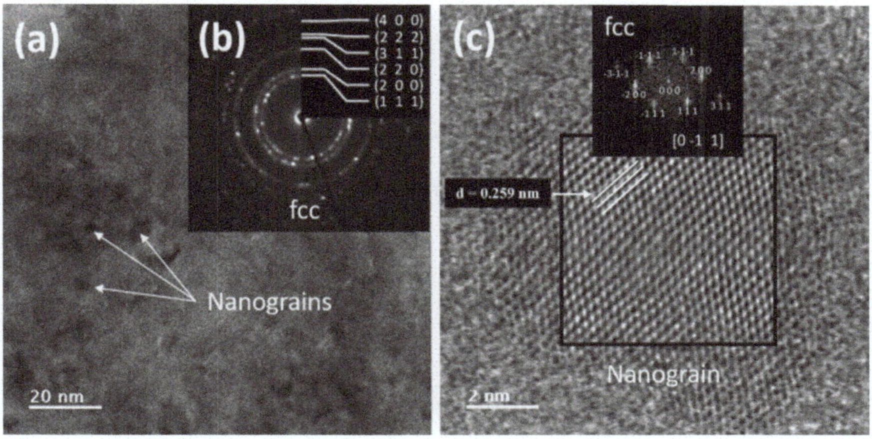

Fig. 13.3 TEM image of the $Ti_{50}Mg_{50}$ alloy processed by ball milling for 25 h: **a** bright field image, **b** indexed SAED pattern, and **c** high resolution image. Taken from Ref. [56] with permission

kinetics, with lower hydrogen absorption and desorption activation energies and a reduced dehydriding enthalpy. The MA process facilitates the incorporation of Zn into Mg, forming the Mg65–Zn30–Ca5 and Mg70–Zn25–Ca5 alloys. These alloys demonstrate enhanced corrosion resistance, credited to the high Zn content, which plays a role in forming a protective film on the alloy surface [64]. HEBM of Mg–50 at.% Ti powder mixtures increased solid solubility, creating an FCC and BCC matrix of Ti solid solution in Mg with small amounts of an HCP Ti-rich phase after milling for 48 and 72 h, respectively, at 800 rpm [65]. Twinning in Ti-rich crystallites indicated strain-induced martensitic transformation and solid solution formation in the Mg–Ti system, with crystallite boundaries as preferential sites for nucleation and lattice strain energy release facilitating these processes. Ball milling effectively produced a $Mg_2Si_{0.4}Sn_{0.6}$ solid solution. The formation mechanism involves initial Sn-rich Mg_2(Si, Sn) formation, followed by slow Si incorporation, as shown in Fig. 13.4. The milling process leads to the welding of Mg and Sn but the fracturing of Si. The fractured Si diffuses into the ductile matrix with longer milling times, resulting in a phase mixture close to the nominal starting composition after 35 h of milling. However, achieving a single-phase pure material required a subsequent sintering step, making the synthesis of Mg_2(Si, Sn) a two-step process. Interestingly, the study observed no significant influence of milling time on thermoelectric (TE) properties, and just 2 h of milling followed by compaction were sufficient to obtain a pellet with optimal properties. These works highlight the robustness and reproducibility of the MA technique for synthesising high-performance Mg-based solid solutions.

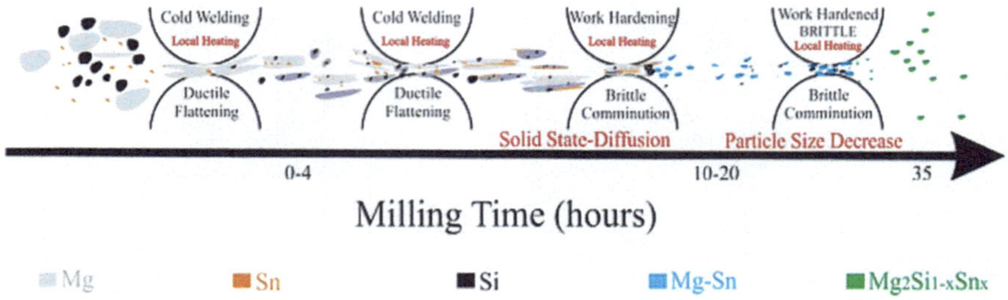

Fig. 13.4 Formation mechanism of $Mg_2Si_{1-x}Sn_x$ phase through HEBM process. Taken from Ref. [146] with permission

13.7.3 Amorphous and Nanostructured Alloys

MA exerts severe plastic deformation on Mg alloys, leading to profound microstructural changes. This process induces grain refinement, breaking down large grains into smaller fragments and resulting in nanostructured Mg alloys with reduced grain sizes into the nanometre scale. The intense mechanical forces during ball milling can also generate amorphous phases in the material, characterised by the absence of a regular crystalline structure. This process is driven by the lower free energy of the amorphous phase compared to a mixture of pure crystalline compounds, elucidating the formation of amorphous phases through the milling of a blend of pure components. For example, HEBM of Mg-10 wt% Al alloy achieved an average grain size of 44 ± 10.3 nm [51]. The ultrafine-grained Mg–3Al–Zn alloy, achieved through mechanical milling, hot-press consolidation, and extrusion, features an average grain size of 485 nm, as shown in Fig. 13.5 [55].

A combination of amorphous and nanocrystalline structures was achieved in Ce–Mg–Ni-based $CeMg_{12}$-type alloys synthesised through 10–60 h of ball milling (Fig. 13.6), which exhibited improved hydrogen storage kinetics [66]. The alloys demonstrated enhanced electrochemical kinetics performance with increased milling duration, attributed to a high active surface area.

Both amorphous and nanocrystalline structures were also obtained in ball-milled $Mg_{1-x}Ce_xNi_{0.9}Al_{0.1}$ ($x = 0$–0.08) + 50 wt% Ni alloys, $Mg_{50-x}Ti_xNi_{45}Al_3Co_2$ ($x = 0$, 1) + 50 wt% Ni, and $Mg_{50-x}V_xNi_{45}Fe_3Zn_2$ ($x = 0, 1, 2, 3, 4$) + 50 wt% Ni composite alloys [67–69]. Increasing the ball milling time from 20 to 80 h in $Mg_{80}Ni_5La_9Ce_1$ + 200 wt% Ni alloy with 0–4 wt.% graphite (Gr) also results in nanocrystalline and amorphous structures, depicted by the broad, amorphous peak in the XRD pattern in Fig. 13.7c [70]. The inclusion of graphite effectively reduces powder agglomeration, improves ball-milling efficiency, and facilitates structure formation, significantly enhancing electrochemical properties at room temperature.

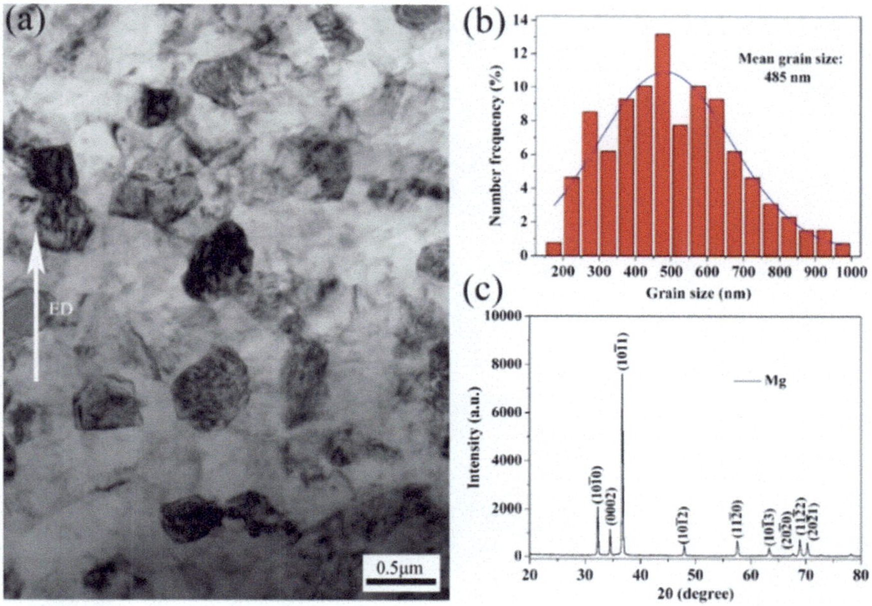

Fig. 13.5 The microstructure of Mg–3Al–Zn alloy extruded at 473 K: **a** TEM bright-field micrograph, **b** the grain size statistical distribution and **c** XRD pattern. Take from Ref. [55] with permission

13.7.4 Intermetallics

MA facilitates the creation of intermediate crystalline phases, often resembling equilibrium phases seen in phase diagrams. The intense plastic deformation during milling distorts both the shape and size of the unit cell, giving rise to entirely novel intermediate phases, such as Mg_2Sn, Mg_3Bi_2, MgZn, Mg_2Ni, and Mg_3Al [53, 63, 71–75]. Nevertheless, additional heat treatment is required to achieve these phases directly from the as-mechanically alloyed state. Mg-based alloys, incorporating low-melting-point additives like Bi, Sn, In, and Zn through ball milling, undergo intermetallic formation as these metals react with Mg during milling. The XRD patterns shown in Fig. 13.8 illustrate the formation of various intermetallics of Mg and alloying elements during ball milling. The resulting intermetallic compounds enhance Mg's hydrolysis reaction, facilitating rapid reactions with seawater and methanol even at room temperature [75]. MA proves effective in developing challenging dispersion-strengthened Mg alloys, such as Mg–*x*Al. These alloys exhibit increased hardness, dependent on the Al content. Specifically, the hardness of ball-milled Mg starts at 80 HV and rises to 183 HV for Mg-20 wt% Al. This hardness improvement is linked to grain refinement below 100 nm and the uniform dispersion of Al in the form of a solid solution, along with fine $Mg_{17}Al_{12}$ intermetallic dispersoids distributed uniformly, acting as effective dislocation barriers in the matrix [76].

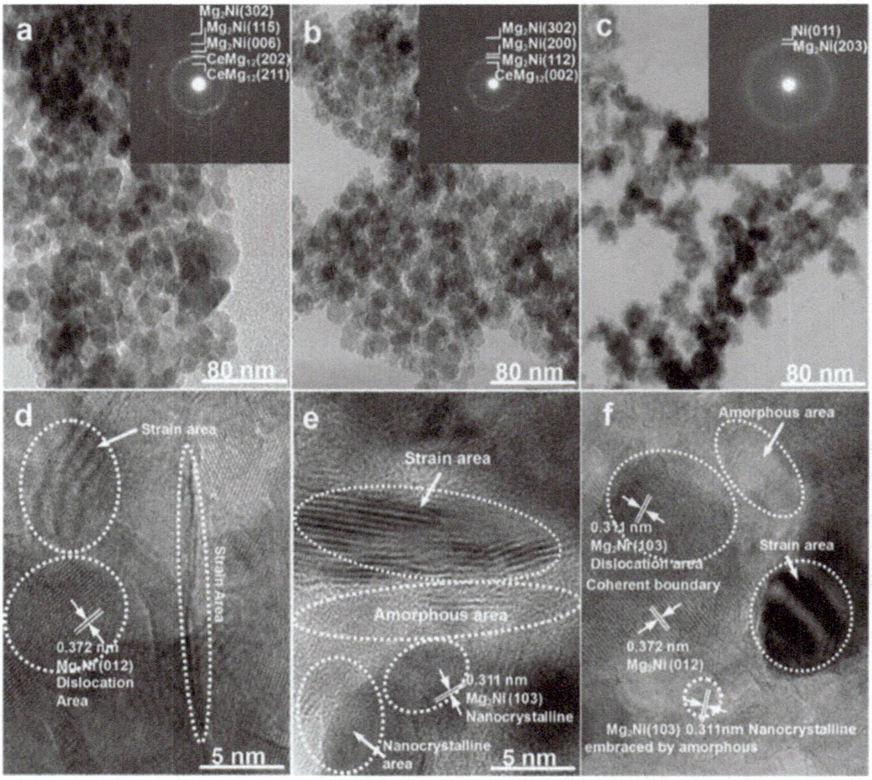

Fig. 13.6 TEM and SAED patterns of the as-milled CeMg11Ni + x wt.% Ni ($x = 100, 200$) alloys: **a** and **d** $x = 100$ alloy milled for 10 h; **b** and **e** $x = 200$ alloy milled for 10 h; **c** and **f** $x = 200$ alloy milled for 60 h. Taken from Ref. [66] with permission

13.7.5 High-Entropy Alloys

Forming single-phase lightweight high-entropy alloys (HEAs) poses challenges due to the tendency of light metal elements to form intermetallic compounds in binary systems. However, MA plays a crucial role in overcoming this difficulty, enabling the formation of solid solutions and facilitating the handling of light metal elements. For example, $LiMg_{0.5}AlScTi_{1.5}$ synthesised through MA exhibited an FCC single-phase structure [77]. MA has shown promise in preparing Mg-containing HEAs for solid-state hydrogen storage applications. The $Mg_{35}Al_{15}Ti_{25}V_{10}Zn_{15}$ alloy synthesised through MA comprised a BCC phase containing Mg, exhibiting a hydrogen storage capacity of 2.33 wt% for the initial absorption/desorption cycle. In contrast, the reactive milling (RM)-synthesised alloy formed an FCC hydride mixed with MgH_2, showing a hydrogen storage capacity of 2.75 wt% after the first desorption [78]. Quaternary LiMgAlTi HEAs displayed an HCP

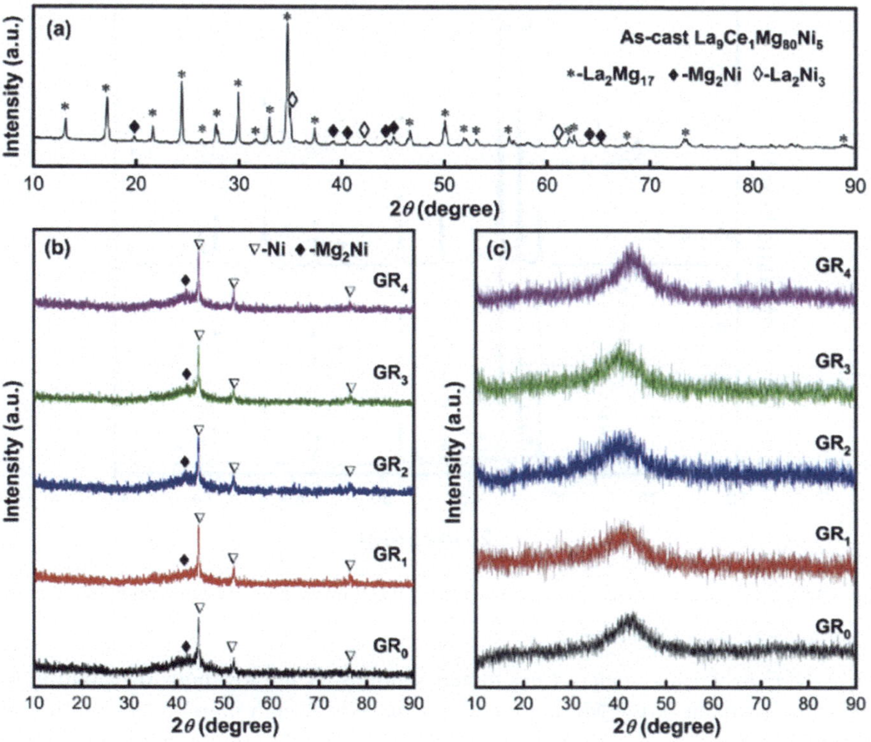

Fig. 13.7 XRD profiles of samples: **a** As-cast $La_9Ce_1Mg_{80}Ni_5$ alloy, **b** 20 h milled experimental alloys, **c** 80 h milled experimental alloys. Taken from Ref. [70] with permission

single-phase structure, while the Li–Mg–Al–Ti–V system achieved a BCC single-phase structure with appropriate composition adjustments [79]. Additionally, various lightweight alloys, including ternary, quaternary, and quinary systems, MgVAl, MgVCr, MgVNi, $Mg_{0.68}TiNbNi_{0.55}$, MgVAlCr, MgVAlNi, MgVCrNi, $AlMg_2TiZn$, and MgVAlCrNi alloys were successfully produced using ball milling, showcasing potential applications in hydrogen storage [80–82].

13.7.6 Mg Composites

Ball milling has emerged as a versatile technique for preparing metal matrix composites, offering improvements in various properties. For example, Mg–TiO_2 composites, crafted through ball milling and PM, demonstrated augmented characteristics, including increased density, controlled porosity, and uniform distribution of TiO_2 in the matrix [83]. This resulted in notable enhancements in hardness (74 BHN) and compressive strength (160 MPa), especially evident with the addition of 12

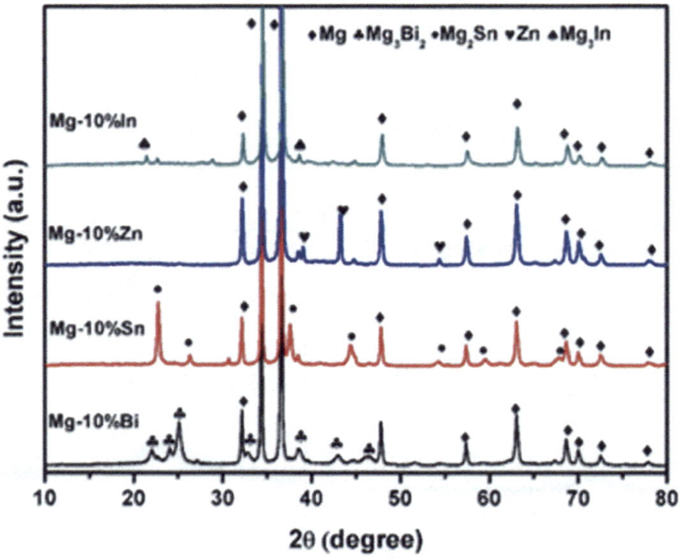

Fig. 13.8 XRD patterns of the activated Mg alloys. Taken from Ref. [75] with permission

wt.% TiO_2 in the Mg matrix for sintered samples. A homogenous distribution of SiC reinforcement in the Mg matrix was achieved through high-energy milling of Mg–SiC blends for 25 h [84]. This extended milling duration resulted in notable decreases in the crystallite size, reaching 12.73 nm. Mg–SiC–Graphite (Gr) hybrid composites, fabricated using MA and consolidated via PM route, showed increased hardness compared to the base material due to hard SiC. However, a slight reduction in hardness was observed compared to Mg–SiC composites, attributed to the inclusion of soft Gr particles. Nevertheless, the wear resistance of these composites significantly improved over the Mg matrix, benefiting from the combined effects of both reinforcements [85]. Additionally, HEBM was employed to create a Mg-based biodegradable nanocomposite incorporating Zr and graphene nanoplates, exhibiting a remarkable 68% reduction in corrosion alongside increased strength [86]. To address the inherent strength-ductility challenge in Mg-based composites, ball milling facilitated the synthesis of Mg–SiC nanocomposites, achieving a unique combination of high yield strength and ductility [87]. Mg matrix composites were successfully fabricated through MA, reinforced with chemically oxidised carbon nanotubes and in situ Mg2Si particles. The resulting composites demonstrated a yield strength of 213 MPa, ultimate tensile strength of 271 MPa, and elongation of 6.3%, attributed to load transfer strengthening mechanisms [88]. Moreover, MA was employed to develop nanosized NiTi-reinforced Mg–3Zn–0.5Ag–xNiTi ($x = 0$–45 wt%) nanocomposites, leading to notable improvements in compressive strength, elongation, and microhardness [89]. Figure 13.9 demonstrates the EDS elemental mapping of the composites, showing uniform distributions of constituent elements. While a slight increase in corrosion rate was observed with 15 wt% NiTi, higher NiTi content (30 and 45 wt%)

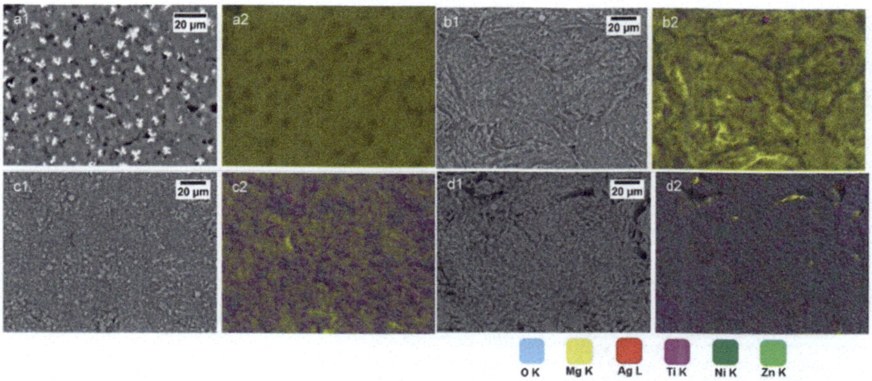

Fig. 13.9 SEM-EDS elemental mapping of Mg, Zn, Ag, Ni, Ti, and O for **a1**, **a2** Mg–3Zn–0.5Ag Matrix, **b1**, **b2** Matrix-15NiTi, **c1**, **c2** Matrix-30NiTi, and **d1**, **d2** Matrix-45NiTi. Taken from Ref. [89] with permission

resulted in a significant corrosion rate increase. Encouragingly, in vitro biocompatibility tests revealed no improper toxicity, underscoring the potential suitability of the Mg–3Zn–0.5Ag nanocomposite strengthened by 15 wt% NiTi for implant applications.

13.8 Microstructure Evolution of MA-Mg Alloys

MA, mainly through high-energy ball milling, is a transformative process that significantly enhances the properties of Mg alloys by inducing microstructural modifications. The resulting changes in Mg alloy microstructures lead to superior mechanical, thermal, and corrosion resistance properties, making them highly advantageous for diverse applications. An illustrative example is observed in the SEM micrograph of ball-milled Mg–6Sn–1Na powders, where different milling durations reveal the evolution of secondary phases in the Mg matrix [71]. The SEM micrographs of the alloy at different milling times are shown in Fig. 13.10. Notably, the 12 h milling duration demonstrates improved mixing, presenting homogeneously dispersed secondary phases with smaller sizes in the Mg matrix. This indicates the efficacy of MA in achieving a desirable microstructure. Additionally, the presence of Mg_2Sn intermetallic phases and the precipitation of Na elements along grain boundaries suggest the potential for precipitation hardening, contributing to enhanced mechanical properties in Mg alloys.

MA has proven crucial in producing a nanocrystalline and amorphous structure in $Mg_{1-x}Ce_xNi_{0.9}Al_{0.1}$ ($x = 0$–0.08) + 50 wt.%Ni alloys [90]. Figure 13.11 illustrates both the nanocrystalline Mg_2Ni and Ce_2Ni_7 and the amorphous matrix, highlighting the effectiveness of this method in achieving the desired structural characteristics in the alloy system. This approach resulted in a significant reduction in grain size

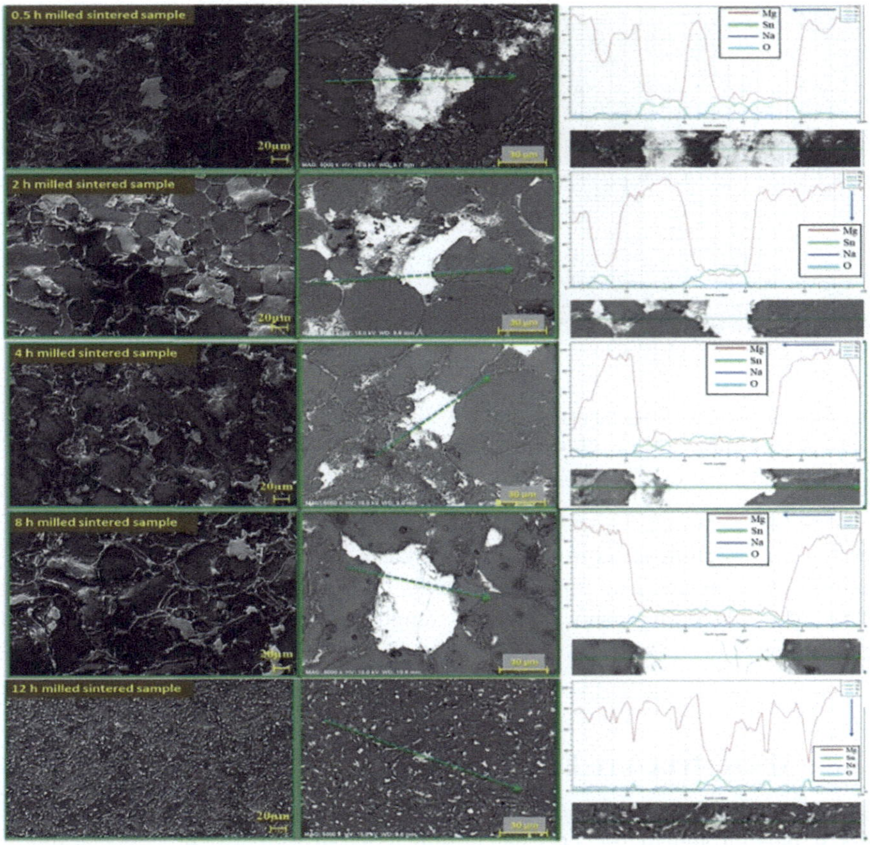

Fig. 13.10 SEM micrograph Mg–Na–Sn ternary alloy as a function of milling time. Taken from Ref. [71] [Open access]

due to Ce substitution for Mg. The electrochemical behaviour of these alloys was noteworthy, showcasing efficient hydrogen absorption and desorption. The enhanced discharge capacity observed in connection with Ce substitution indicates improved performance. Moreover, the cyclic stability of the alloys demonstrated a substantial improvement with an increase in Ce content, emphasising the positive impact of ball milling on the electrochemical properties of Mg alloys.

Utilising V as a novel reinforcement in AZ31 accelerated the refinement rate of Mg crystallite size during milling, achieving an average size of 25 nm, as can be seen in Fig. 13.12 [91].

The ball-milled Mg-10 wt.% Al alloy exhibited compositional homogeneity after cold compaction but revealed a heterogeneous microstructure post-SPS consolidation [51]. The microstructure featured contrasting bright and dark regions, with visible porosity and interparticle boundaries in samples after cold compaction and SPS at 200 °C (Fig. 13.13). These features disappeared in samples produced by SPS at or

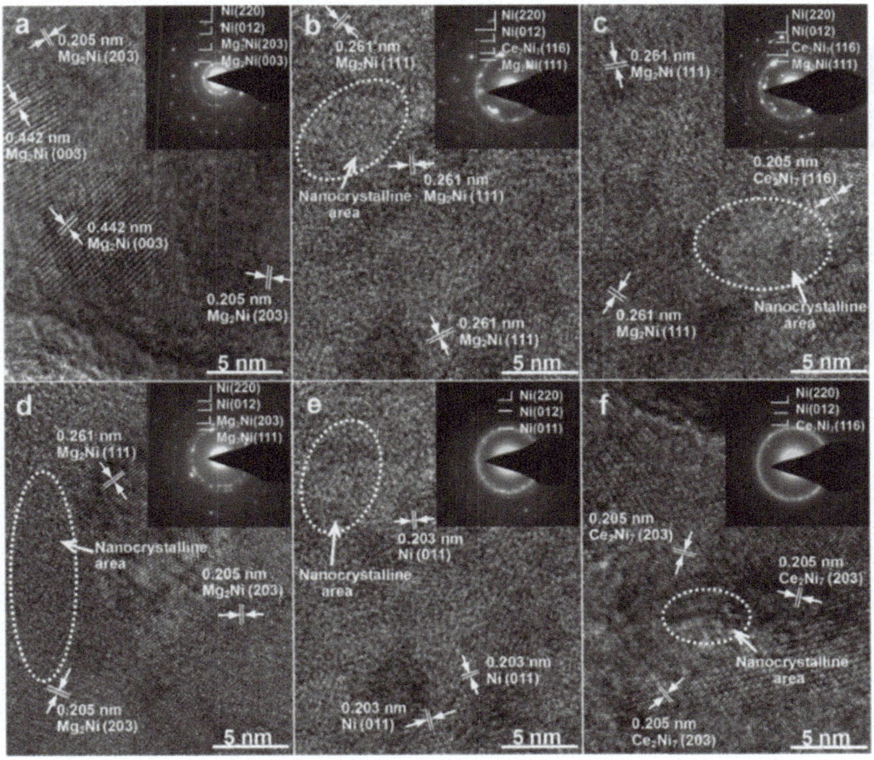

Fig. 13.11 HRTEM micrographs and SAED patterns of the as-milled $Mg_{1-x}Ce_xNi_{0.9}Al_{0.1}$ ($x =$ 0–0.08) + 50 wt.% Ni alloys: **a–c** Ce_0, $Ce_{0.02}$ and $Ce_{0.08}$ milled for 5 h; **d–f** Ce_0, $Ce_{0.02}$ and $Ce_{0.08}$ milled for 20 h. Taken from Ref. [90] with permission

above 250 °C. SPS at 200 °C induced elemental diffusion, leading to dispersed bright particles among dark regions. The size and coarseness of these particles increased at higher SPS temperatures, indicating coalescence and coarsening. Energy-dispersive X-ray Spectroscopy (EDXS) analysis confirmed Al-rich bright phases and Al-lean dark phases, with the size of the Al-rich phase increasing with SPS temperature.

Mg–6Si alloys were synthesised through PM, employing high-energy ball milling, leading to a phase fraction of 92.3% Mg and 7.7% Mg_2Si [53]. The formation of Mg_2Si through solid-state diffusion was evident by the presence of grey particles, and microstructure analysis unveiled a primary Mg_2Si phase surrounded by α-Mg dark halos, which can be observed in Fig. 13.14. Excess silicon reacts with Mg, forming Mg_2Si intermetallic compounds in grey particles, which increase with an increase in Si addition. A noteworthy observation was the decrease in the average grain size of Mg_2Si with an increase in the weight per cent of silicon, suggesting an inhibitory effect on grain growth.

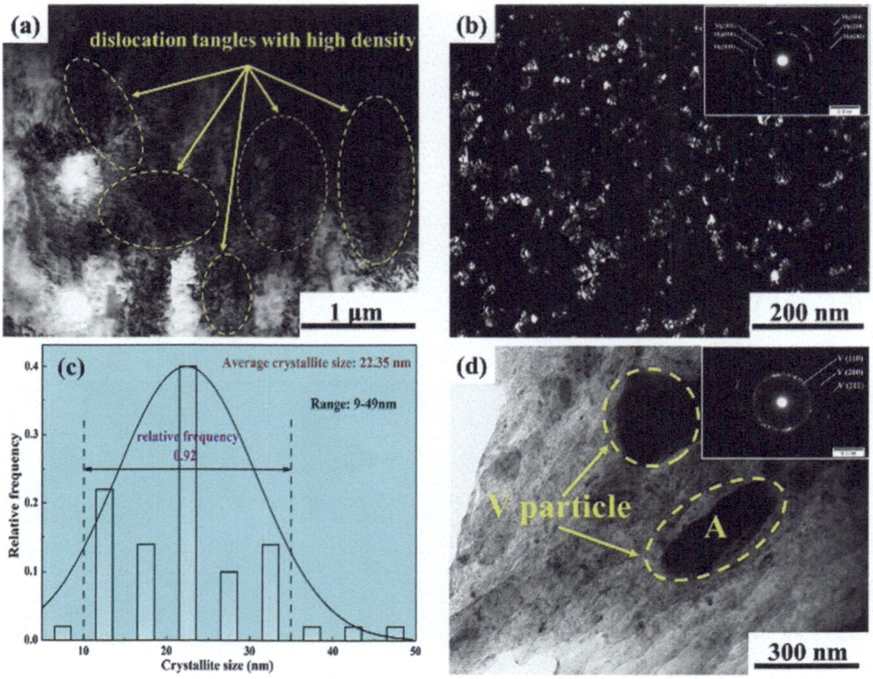

Fig. 13.12 TEM images of AZ31-2.5 wt% V composite powders after milling for different times: **a** bright-field image of 10 h, **b** dark-field image of 90 h, **c** crystallite size distribution in (**b**), and **d** bright-field image of 90 h. Taken from Ref. [91] [Open Access]

Ball milling was crucial in achieving uniform dispersion of elements within the Mg matrix in synthesising Mg2.3Y0.1Ni $+ x$ wt.% graphene ($x = 1, 3, 5, 7$) hydrogen storage alloys [92]. Figure 13.15 illustrates the EDS elemental mapping of the alloy.

Ball milling of MgH_2 with doping catalytic species is a simple and efficient method that facilitates the homogeneous distribution of catalytic additives on the surface of nanosized MgH_2 powders. The ball milling process involving the hydrogenolysis of nickelocene ($NiCp_2$) resulted in the in-situ generation of Ni nanoparticles, as depicted by HRTEM and SAED images in Fig. 13.16 [93]. The MgH_2–16.1 wt% $NiCp_2$ composite, obtained after 15 h of ball milling under a 4 MPa hydrogen atmosphere, displayed a uniform morphology with in situ formed Ni nanoparticles (~8 nm) widely dispersed within the MgH_2 matrix. These nanoparticles served as catalysts for MgH_2, demonstrating superior hydrogen desorption kinetics and exceptional cycling stability.

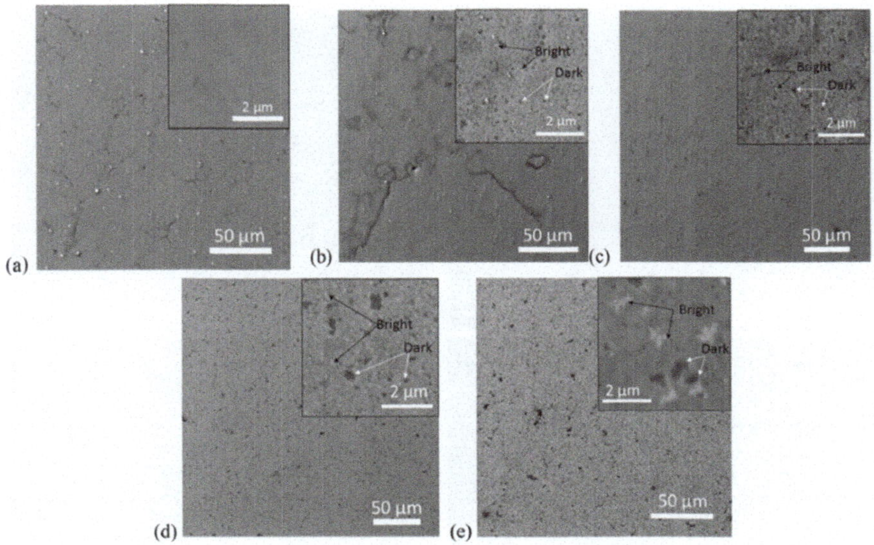

Fig. 13.13 Microstructure of MAed Mg-10 wt% Al alloy after cold compacted condition (**a**), and after SPS at 200, 250, 300, and 350 °C in (**b**), (**c**), (**d**) and (**e**), respectively. Typical bright regions and dark regions in b-e have been marked with black and white arrows, respectively. Taken from Ref. [51] [Open access]

13.9 Mechanical Properties

Mg alloys play a vital role in various applications due to their excellent mechanical properties and high corrosion resistance. As-cast Mg, however, suffers from rapid deterioration and weak strength, prompting the need for alloying with specific elements or compounds. Alloying elements react with Mg, forming intermetallic compounds that influence microstructure and material properties. These elements enhance alloy strength through precipitation hardening, grain refinement, and solid solution strengthening. An illustration of enhancement in mechanical properties due to the incorporation of a second phase is shown in Fig. 13.17 [89]

Recent studies showcase the effectiveness of the MA method in improving Mg alloy performance. For instance, mechanically alloyed Mg–Fe alloy demonstrated exceptional strength (1380 MPa), a Vickers hardness of 424.3 VHN, and an improved biodegradation rate [94] Seth et al. reported the successful formation of Mg–6%Si alloy using a combined approach of ball milling and PM [53] The alloy exhibited high hardness, primarily due to the abundant presence of a fine and irregular-shaped Mg_2Si phase. This phase forms because of the reaction between Mg and silicon atoms during the sintering process, resulting in the precipitation of the intermetallic compound Mg_2Si. The Al-50 vol% Mg composite, subjected to 12 h of ball milling at 200 rpm, exhibited the presence of intermetallic compounds, notably Al_3Mg_2 and $Al_{12}Mg_{17}$, contributing to a Vickers hardness of 189 HV [95] The intermetallic

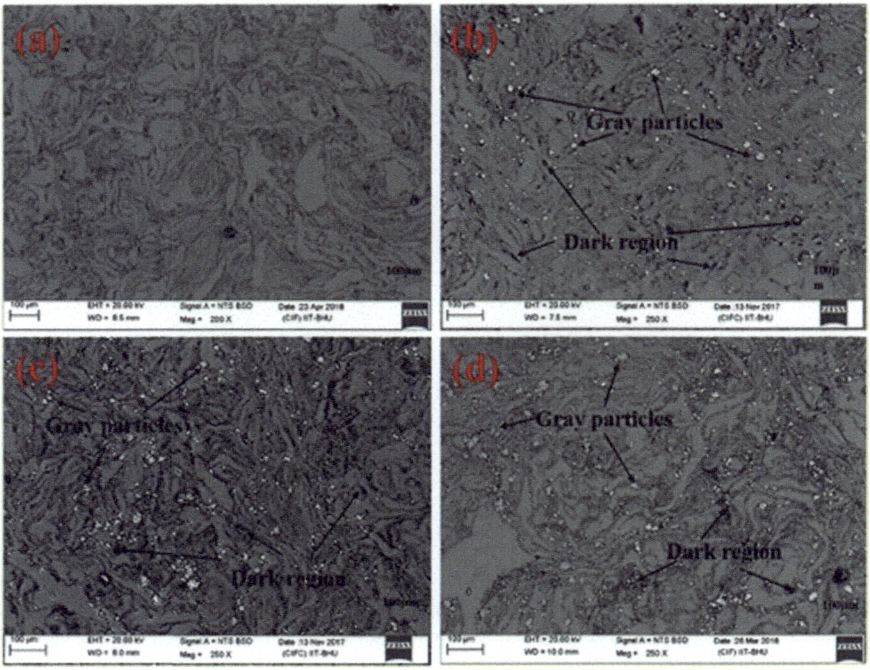

Fig. 13.14 SEM Micrographs of sintered. **a** Mg. **b** Mg–1%Si, **c** Mg–4%Si, and **d** Mg–6%Si alloy specimens. Taken from Ref. [53] with permission

compounds act as effective barriers hindering dislocation mobility within the matrix. Mg-10 wt% Al alloy, prepared through HEBM and consolidated by SPS at 300 °C, demonstrated a hardness of 235 HV [51]. Despite increasing grain size with higher SPS consolidation temperatures, the hardness exhibited an unexpected rise contrary to the Hall-Petch relationship, attributed to the enhanced solid solubility of Al in Mg due to elevated processing temperatures. The fine and uniformly dispersed submicron V in AZ31 alloy after 90 h milling induced a significant Orowan strengthening effect, thereby enhancing the mechanical properties of the matrix [91]. The microhardness of $Mg_{65}Zn_{30}Ca_4Gd_1$ powders, prepared via HEBM, showed a dependency on particle size, demonstrating an increase in hardness as the particle size decreased (Fig. 13.18) [96].

Zhou et al. successfully synthesised nanocrystalline AZ31 Mg alloy powders reinforced with submicron TiB_2 particles through mechanical milling [97]. This process led to accelerated milling due to TiB_2-induced fracture and welding. The TiB_2 particles were refined and distributed in the Mg matrix, reaching a submicron scale after 110 h of milling. The matrix grain sizes were significantly reduced to 53, 37, and 23 nm for AZ31–2.5 wt% TiB_2, AZ31–5 wt% TiB_2, and AZ31–10 wt % TiB_2 composites. The combined effect of the nanocrystalline matrix and well-dispersed submicron TiB_2 particles led to increased microhardness in the AZ31/TiB_2 composites, with the AZ31–10 wt% TiB_2 composite reaching 153 $HV_{0.5}$. The microhardness

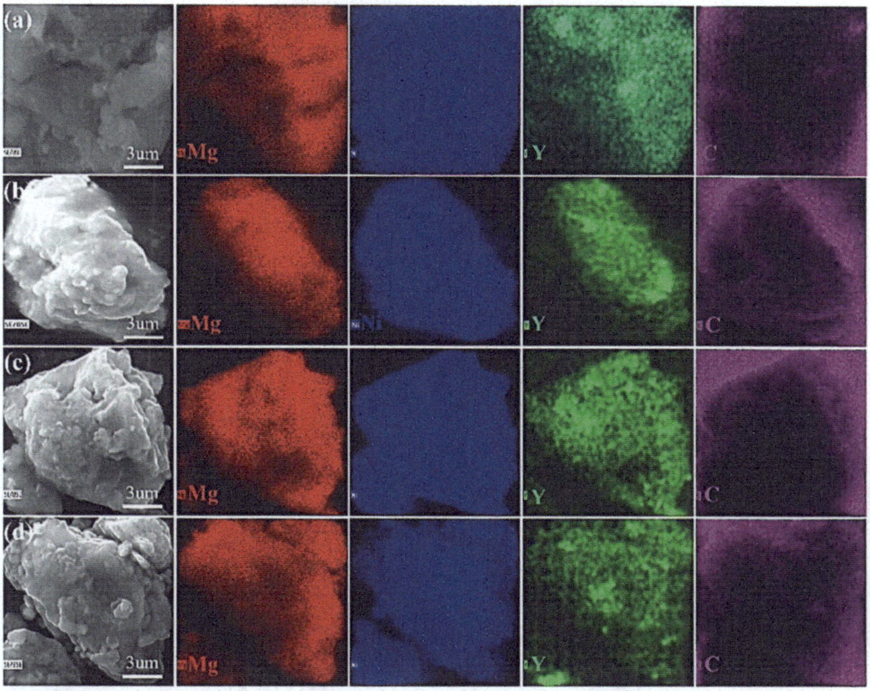

Fig. 13.15 Distribution of EDS elements of ball-milled Mg2.3Y0.1Ni + *x* wt.% graphene alloys compounded with different contents of graphene: **a** 1 wt.%; **b** 3 wt.%; **c** 5 wt.%; **d** 7 wt.%. Taken from Ref. [92] with permission

increased with higher TiB_2 additions. The precipitation of Mg_2Sn is advantageous for enhancing the strength and creep resistance of Mg alloys. However, a coarse Mg_2Sn phase can make the resulting alloy susceptible to denting. Fang et al. successfully produced an Mg_2Sn-nanophase reinforced Mg–Mg_2Sn composite, which exhibited significant improvements in hardness and compressive strength while maintaining high ductility [60, 61]. The obtained phase, microstructure, and mechanical properties of Mg–Sn alloy are shown in Fig. 13.19.

HEBM facilitates a homogeneous dispersion of 0.5 wt% zirconium (Zr) and 0.1 wt% graphene nanoplatelets (GNPs) in Mg powders, resulting in a Mg matrix nanocomposite (MMNC) [86]. This MMNC exhibited a substantial 91% improvement in compressive yield strength compared to pure Mg. The OM micrographs and SEM images of sintered samples are shown in Fig. 13.20. The addition of Zr and GNPs significantly improves mechanical properties through synergetic strengthening mechanisms, demonstrating higher compressive yield strength, ultimate compressive strength, and hardness compared to pure Mg and Mg–Zr samples. Zirconium (Zr) contributes a significant grain-refining effect to the MMNC. At the same time, GNPs enhance strength through synergistic mechanisms such as grain-refinement (GR), difference in thermal expansion (DTE), difference in elastic

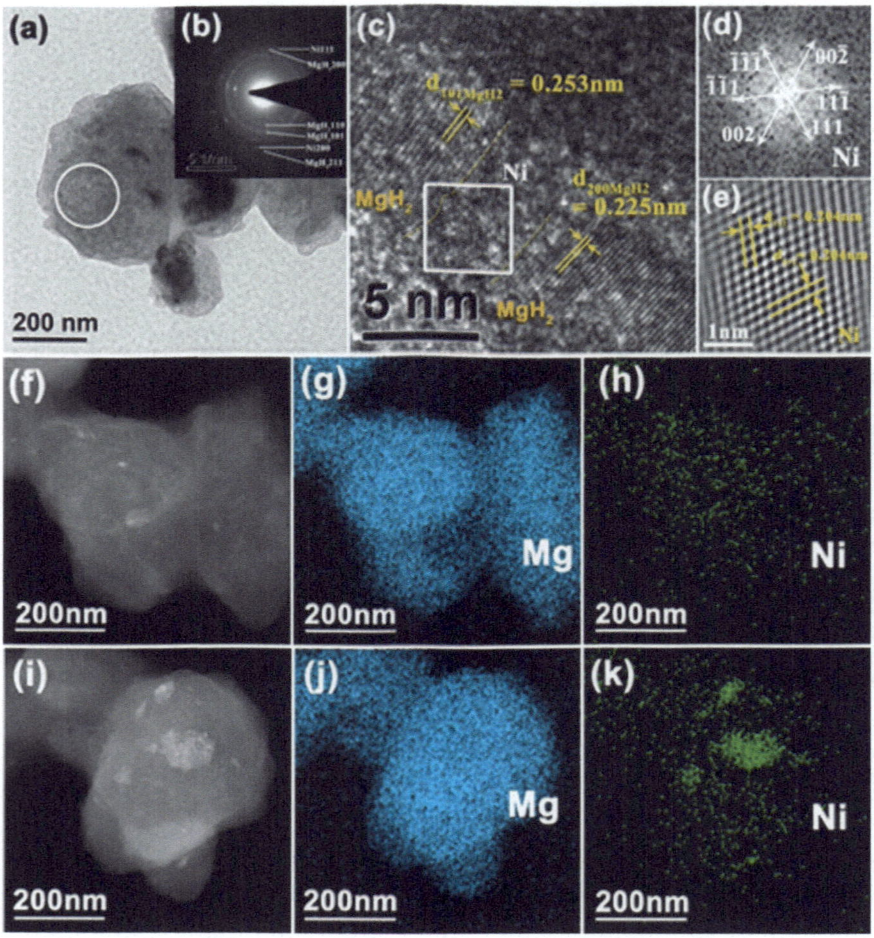

Fig. 13.16 **a** TEM image, **b** corresponding SAED pattern, **c** HRTEM image, **d** FFT pattern, and **e** IFFT image of the MgH_2-16.1 wt% $NiCp_2$ sample after ball milling for 15 h. HAADF-STEM images and corresponding EDX elemental mappings of the ball milled MgH_2-16.1 wt% $NiCp_2$ (**f**, **g** and **h**) and MgH_2-5.62 wt% Ni (**i**, **j**, and **k**) samples. Taken from Ref. [93] with permission

modulus (DEM), geometric mismatch (GM), load-transfer (LT) strengthening, and precipitation strengthening (PS).

Recent studies indicate that adding nano-sized reinforcements can enhance Mg-composites' strength and ductility. This effect is attributed to a more effective nano-sized particle hardening mechanism than micron-sized particles, with the nano-sized reinforcements activating non-basal slip systems in the Mg matrix at room temperature, thereby increasing ductility. However, challenges in achieving a uniform distribution of nanoparticles, particularly at higher contents, have hindered the expected property enhancement. A successful synthesis of Mg–SiC_n nanocomposite powders through mechanical milling addressed these challenges, reducing particle

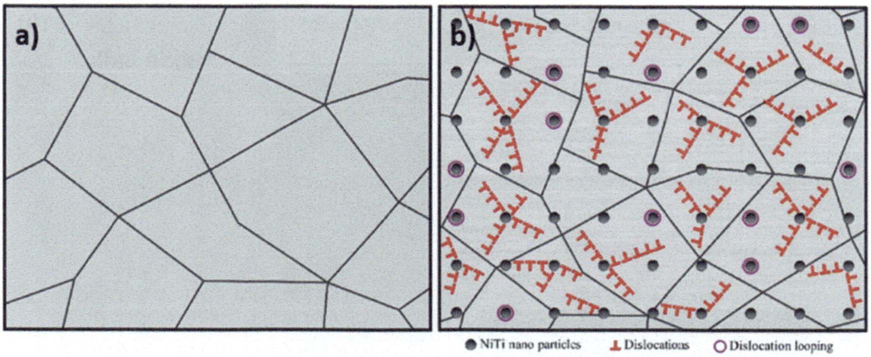

Fig. 13.17 Schematic illustration of the enhancement of mechanical properties due to incorporation of the second phase, **a** Mg alloy matrix and **b** particle-incorporated nanocomposite. Taken from Ref. [89] with permission

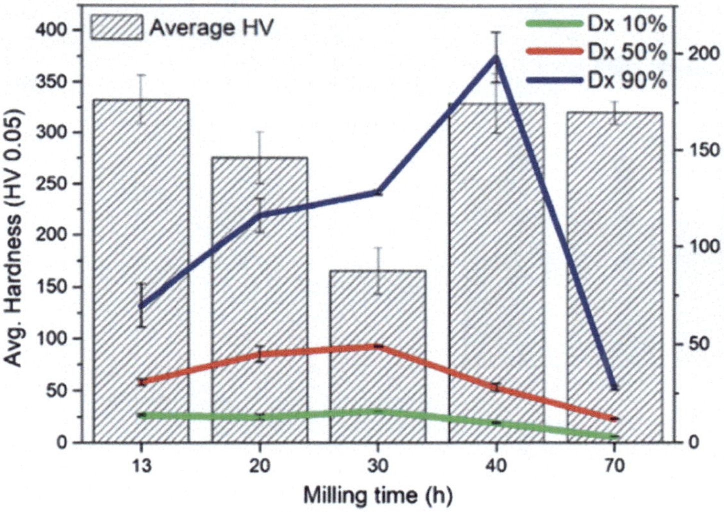

Fig. 13.18 Average hardness values as a function of particle size for $Mg_{65}Zn_{30}Ca_4Gd_1$ powder alloy samples milled for 13, 20, 30, 40, and 70 h, respectively. Taken from Ref. [96] [Open Access]

size, achieving a more uniform distribution, and enhancing mechanical properties [87]. The study demonstrated a uniform distribution of SiC nanoparticles in the Mg matrix, even at high volume fractions (up to 10%), achieved through severe plastic deformation during mechanical milling, leading to nanocrystalline Mg grains surrounding SiC particles (Fig. 13.21). Mg-(0, 4, 8, and 12 wt%) SiC nanocomposite was prepared via high energy ball milling for 6 h and consolidated by powder metallurgical route [54]. Figure 13.21 shows the microstructures of M10Sn nanocomposite powder. The density and hardness increased with higher weight percentages (4, 8, and 12 wt%) of SiC compared to pure Mg. The compressive strength and impact

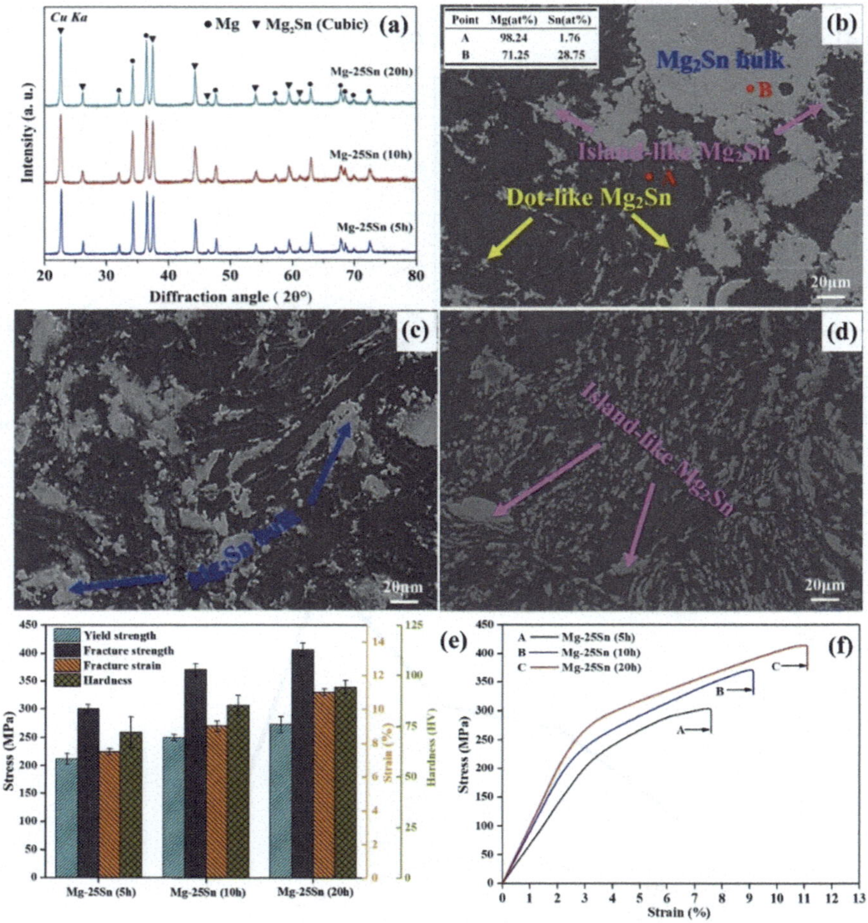

Fig. 13.19 **a** XRD patterns, hot-press sintered microstructure of **b** 5 h, **c** 10 h, **d** 20 h milled mixture, **e** mechanical properties diagram of as-sintered Mg–25Sn alloys, and **f** compression engineering stress–strain curves. Taken from Ref. [61] with permission

strength of the 4, 8, and 12 wt% SiC-reinforced Mg composites were found to be 474 MPa, 524 MPa, and 548 MPa, and 2 J, 2.8 J, 3.8 J, and 5 J, respectively.

The nanostructured Mg-7.4% Al alloy, featuring a 300 nm grain size, consists of solid solutions, intermetallic phases (γ-$Al_{12}Mg_{17}$), and MgO dispersion. Following uniaxial hot pressing consolidation, it achieves an exceptional compressive stress of 690 MPa and exhibits an impressive plastic strain exceeding 9% [98]. A ternary Mg–3Al–Zn alloy exhibited tensile yield stress of 459 MPa, an ultimate tensile strength of 519 MPa, an elongation to failure of 3.2%, and a low tension-compression yield asymmetry of 1.01 [99]. The exceptional strength of the material is attributed to the synergistic effects of a partially nanocrystallised structure, the presence of

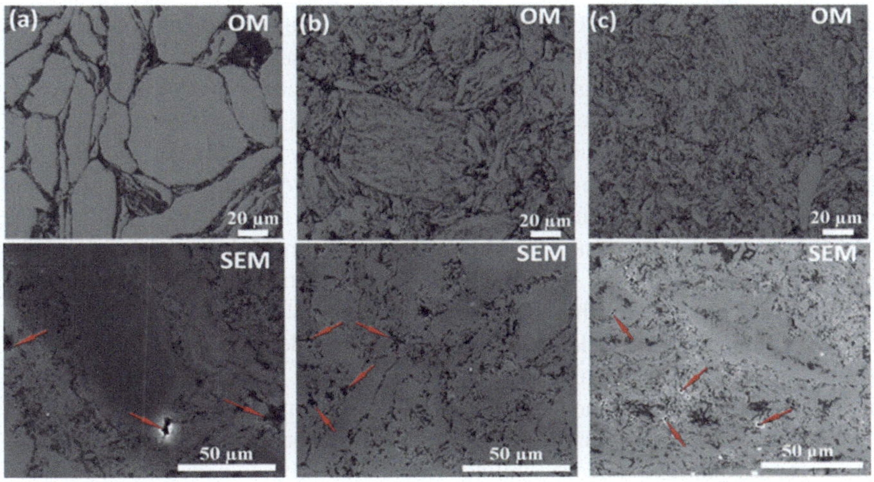

Fig. 13.20 OM micrographs and SEM images of sintered samples: **a** pure Mg; **b** Mg–0.5Zr; **c** Mg–0.5Zr-0.1 graphene nanoplates. Taken from Ref. [86] with permission

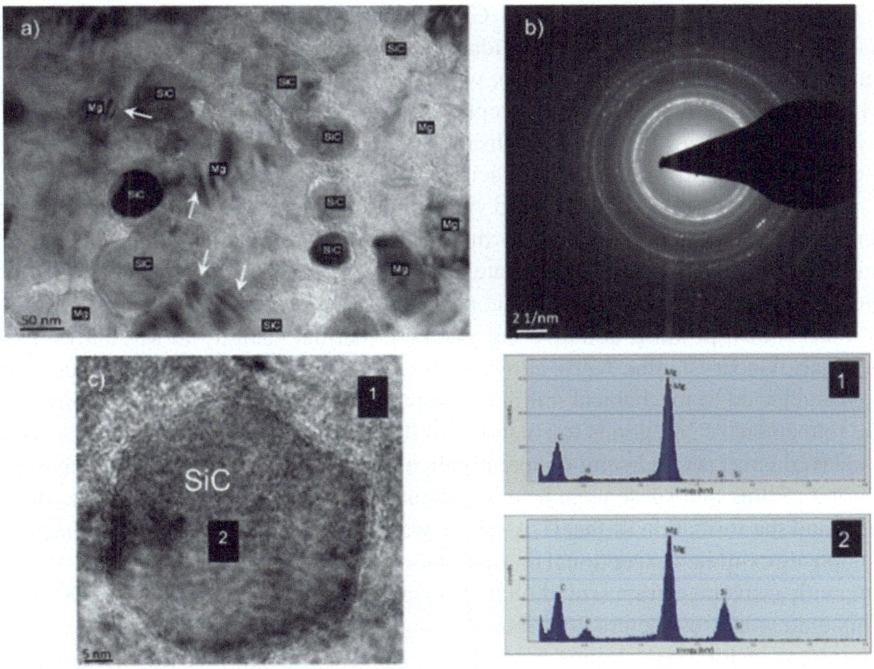

Fig. 13.21 Microstructures of M10Sn nanocomposite powder. **a** TEM micrograph of distribution of SiC nanoparticle in Mg matrix, **b** the corresponding SAED patterns, and **c** HRTEM image of the interface between SiC nanoparticle and Mg matrix supported by EDS analysis. Taken from Ref. [87] with permission

nanoscale and coherent MgO precipitates, and the formation of ultra-fine recrystallised Mg grains. These factors collectively lead to ultra-high-strength and reduced tension-compression yield asymmetry, further influenced by a weak basal texture. Mg–6Sn–1N alloy was made by ball-milling for varying durations (0.5, 2, 4, 8, and 12 h) to achieve a solid solution and MA in the ternary system (Mg–Sn–Na). The milling process initially formed a platelet structure, evolving into small spherical particles after 12 h. XRD data indicated Sn dissolution and the intermetallic phase (Mg_2Sn) formation after 8 h. The 12-h milling condition achieved a homogenous secondary phase distribution, resulting in the highest relative density (95%), modulus of elasticity (34.5 GPa), and hardness (89 HV) values in this ternary material system.

Cryomilling for 8 h yields Mg–Al–Zn alloy powders with an average grain size of 40 nm [100]. Deformation twinning induced in cryo-milled AZ80 powder enhances the ductility of nanocrystalline Mg AZ80 materials. The resulting bimodal microstructure, featuring increased grain size at the particle interface during SPS while retaining internal nanocrystalline structure showcases nano-sized $Mg_{17}Al_{12}$ precipitates within coarse grains. This unique structure leads to exceptional properties, including a maximum microhardness of 146 HV, compressive yield strength of 442.3 MPa, and ultimate strength of 546 MPa, surpassing conventional wrought Mg AZ80 alloys significantly. $Mg_{65}Cu_{25}Gd_{10}$ metallic glasses were produced through ball milling and subsequently consolidated under high pressure (500 MPa) at various temperatures (443, 433, and 423 K), which exhibited a microhardness of 436 HV. Multiple factors, including the disruption of surface oxides, structural relaxation within the milled powder, and the utilisation of high pressure during the SPS process, facilitated the improvement in interfacial bonding among amorphous powders.

V-reinforced AZ31 alloy prepared via ball milling displayed a fine-grained structure and superior mechanical performance, showcasing notable enhancements in microhardness, yield strength, ultimate tensile strength, and elongation compared to the as-cast AZ31 Mg alloy [91]. MA contributes to heightened mechanical properties and has shown promise in delivering excellent thermal stability in Mg alloys. For instance, nanocrystalline AZ61 alloy powders containing submicron Ti particulates were subjected to mechanical milling, leading to the cracking of Ti dispersions and the refinement of Mg grains to a nanocrystalline state [101]. The resulting material displayed coherent and semi-coherent lattice interfaces between Mg and titanium. During milling and subsequent heat treatment, the dissolution of Al into Ti particulates and the formation of the Ti_3Al phase were observed. This nanocrystalline Mg composite exhibited exceptional thermal stability, retaining its nanocrystalline structure with an average grain size of 68.6 nm even after annealing at 673 K for 600 min. This unique microstructure contributed to outstanding mechanical properties, with a hardness of 146.6 HV. Impressively, the annealed composite also maintained high hardness (134.7 HV), underscoring its excellent thermal stability.

13.10 Effect of Milling Time

The duration of high-energy ball milling profoundly affects Mg (Mg) alloys. As milling progresses, the most immediate consequence is the reduction in particle size, driven by processes like cold welding and fracturing. Simultaneously, microstructural changes occur, introducing defects such as vacancies, dislocations, and grain boundaries. This transformative process contributes to increased hardness because of repeated cold welding and fracturing cycles. Furthermore, prolonged milling promotes the homogenisation of alloy composition, ensuring a uniform distribution of elements through the alloy. The activation of alloying elements during extended milling enhances reactivity, a crucial aspect in applications like Mg-based hydrogen storage materials. In specific alloy systems, longer milling times may induce the formation of intermetallic compounds. Grain refinement is another effect associated with extended milling time, influencing the mechanical and thermal properties of Mg alloys. However, an optimal milling time exists, beyond which further milling may yield diminishing returns or even detrimental effects. The specific impacts depend on factors like alloy composition, milling conditions, and the intended application, necessitating experimentation for precise characterisation and property optimisation.

For instance, after 3 h of ball milling, the morphology of Mg powder transforms from an irregular semi-elliptical three-dimensional structure (Fig. 13.22a) to irregular thin sheet-like lamellae (Fig. 13.22c) [53].

The average grain size of the ball-milled powder was observed to be lower after sintering compared to the powder without ball milling (Fig. 13.23). Specifically, the average grain size of pure Mg with ball milling and without ball milling was 38 μm and 110 μm, respectively [53].

The particle size of V (Grey regions in Fig. 24a–e) in AZ31 alloy continuously reduces with increasing milling time, accompanied by a more uniform distribution in the Mg matrix [91]. Cold welding and cracking of composite powders allow V to gradually enter the Mg matrix, leading to a homogeneous distribution after 90 h of milling (Fig. 13.24e). The average particle size of V has been refined to 880 nm, primarily influenced by the fracture mechanism.

In the case of $Mg_{65}Zn_{30}Ca_4Gd_1$ powders prepared by HEBM, the particle size varies with milling time [96]. After 13 h, the average size is 31 μm, increasing to 45–49 μm after 30 h. Subsequently, a sharp decrease is observed after 40 and 70 h,

Fig. 13.22 SEM micrographs of **a** Mg powder, **b** Si powders, and **c** ball-milled Mg–1Si powder. Taken from Ref. [53] with permission

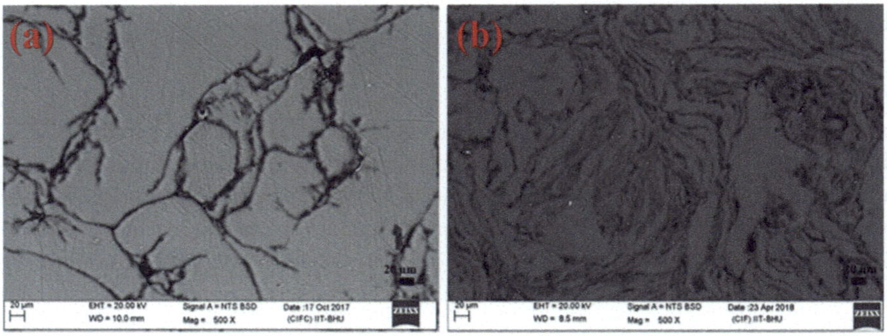

Fig. 13.23 SEM micrographs of sintered pure Mg samples using powders, **a** without ball milling and **b** after 3 h ball milling. Taken from Ref. [53] with permission

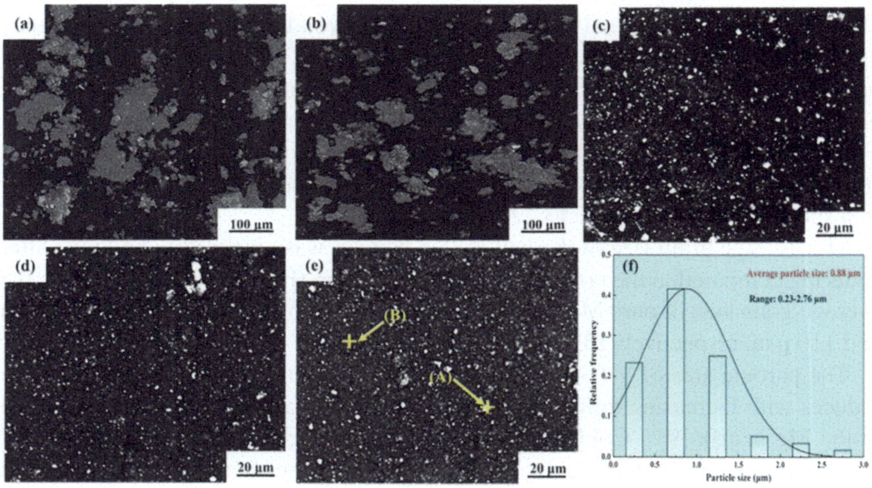

Fig. 13.24 The evolution of V in AZ31 alloy after milling for different times: **a** 10 h, **b** 30 h, **c** 50 h, **d** 70 h, **e** 90 h, **f** average particle size distribution map after milling for 90 h. Taken from Ref. [91] [Open Access]

reaching 28 μm and 12 μm, respectively, significantly influencing hardness. For the ternary material system of (Mg–Sn–Na), the initial powder mixture retains its spherical particle morphology after a 0.5-h milling process [71]. A 2 h ball-milling treatment efficiently transforms particles into flake structures through extreme plastic deformation. The 4-h milling process triggers significant cold welding, forming a platelet structure with an increased particle size of up to 120 μm. Subsequent milling up to 8 h leads to particle breakage into smaller forms, creating a bimodal particle size distribution. After a certain milling duration, the fracture mechanism becomes dominant, possibly due to extreme work hardening or loss of function of the processing

Fig. 13.25 SEM image of ball-milled Mg–6Sn–1Na powders at a particular milling duration. Taken from Ref. [71] [Open Access]

agent. With an 8-h milling process, particles have experienced severe plastic deformation and accumulated excessive stress concentration, affecting the sinterability of the material system. Further milling up to 12 h transforms the flake morphology into smaller fragmented particle formations, achieving a monomodal particle size distribution below 10 μm, well below the initial particle size. The final microstructure formation suggests that a 12-h milling duration is sufficient for proper mixing and MA in this ternary material system. Figure 13.25 demonstrates the SEM images of ball-milled Mg–6Sn–1Na powders at different milling duration.

Other reported impacts of extended milling time include forming nanocrystalline and amorphous structures, increased density and microhardness, and improved electrochemical kinetic properties and cycling stability in various Mg-based alloys, addressing issues in Ni–MH batteries. For instance, extended milling time consistently reduces the grain size of Mg–Fe alloy to 13–30 nm, enhancing hydrogen storage capacity by generating fresh Mg/MgH_2 surfaces [94, 102–104]. The electrochemical kinetic properties of $Mg_{1-x}Ce_xNi_{0.9}Al_{0.1}$ ($x = 0, 0.02, 0.04, 0.06, 0.08$) + 50 wt% Ni alloys demonstrate significant improvement with longer milling durations [67]. The cycling stability of the alloy increases from 47 to 63% with an extension of milling time from 5 to 30 h, reaching 82% for the 30-h milled alloy. Prolonging the milling time of $Mg_{50-x}V_xNi_{45}Fe_3Zn_2$ ($x = 0, 1, 2, 3, 4$) + 50 wt% Ni alloy also addressed cycle stability issues in Ni–MH batteries [68]. Zhang et al. observed that increasing ball milling time from 20 to 80 h in $Mg_{80}Ni_5La_9Ce_1$ + 200 wt% Ni alloy with 0–4 wt.% graphite resulted in nanocrystalline and amorphous structures [70]. The density and microhardness of Mg–Zn alloy were reported to increase with prolonged milling time [58].

13.11 Applications

MA-Mg alloys find diverse applications across various industries due to their advantageous properties. These alloys are notably used in hydrogen storage, structural engineering, biomedical applications, and aerospace. Mg alloys are explored for their potential to absorb and release hydrogen, contributing to solid-state hydrogen storage technology advancements. The lightweight nature of Mg alloys makes them attractive for structural applications, where weight reduction is crucial for enhancing fuel efficiency and overall performance. For instance, the Mg–3Al–Zn alloy demonstrates exceptional strength with a tensile yield stress of 459 MPa, an ultimate tensile strength of 519 MPa, and reduced tension-compression yield asymmetry, owing to its partially nanocrystallized structure, nanoscale coherent MgO precipitates, and ultra-fine recrystallised Mg grains, making it well-suited for structural applications [99]. In biomedical fields, Mg alloys are employed in developing biocompatible implants and medical devices, benefitting from their biodegradability and compatibility with biological tissues. Additionally, the aerospace industry utilises Mg alloys in manufacturing lightweight components, contributing to the overall efficiency and fuel economy of aerospace vehicles. The versatility of Mg alloys in these diverse applications underscores their significance in advancing technology and addressing various engineering challenges.

13.12 Hydrogen Storage

Mg-based alloys are promising candidates for solid-state hydrogen storage, boasting high capacity, reversibility, and cost-effectiveness. However, challenges such as slow absorption/desorption rates and high operating temperatures impede their widespread use. Recent research has focused on leveraging ball milling techniques to enhance hydrogen storage properties, introducing element substitutions and additives to optimise alloy performance. Ball milling has proven instrumental in significantly improving the hydrogenation and dehydrogenation kinetics of Mg alloys [74, 105, 106]. For example, the SmMg11Ni alloy undergoes a notable enhancement in kinetic and electrochemical properties, resulting in increased hydrogen storage capacity [107]. Additionally, ball-milled $Mg_{1-x}Ce_xNi_{0.9}Al_{0.1}$ ($x = 0$–0.08) + 50 wt% Ni alloys showcase exceptional electrochemical hydrogen storage performance with high cycling stability, displaying maximum discharge capacities of 378.8 and 545.3 mAh g^{-1} for 5 and 20 h of milling, respectively [67, 90]. The amorphisation induced by ball milling in Ce–Mg–Ni-based $CeMg_{12}$-type alloys contributes to a notable hydrogen storage capacity increase of 6.16% [66]. Ball milling of $La_9Ce_1Mg_{80}Ni_5$ + 200 wt% Ni alloy with 3 wt.% graphite results in enhanced discharge capacity from 288.1 to 315.4 mAh g^{-1} for the 20 h milled specimen and from 314.5 to 344.9 mAh g^{-1} for the 80 h milled specimens, as well as cyclic stability and electrochemical kinetics [70].

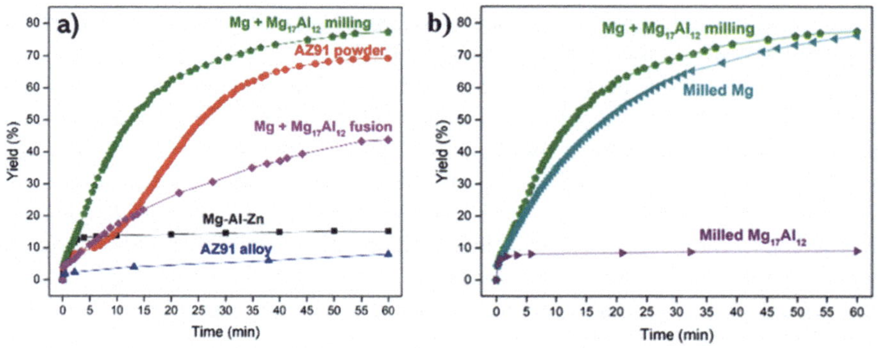

Fig. 13.26 Hydrogen generation of **a** Mg–Al–Zn, AZ91 powder, AZ91 alloy, Mg + $Mg_{17}Al_{12}$ fusion and Mg + $Mg_{17}Al_{12}$ milling, **b** Hydrogen generation of milled Mg, milled $Mg_{17}Al_{12}$ and Mg + $Mg_{17}Al_{12}$ milling. Taken from Ref. [110] with permission

Mg-based materials exhibit dual functionality, acting as hydrogen carriers and enabling hydrolysis for hydrogen production under mild conditions, thereby enhancing overall productivity [108]. Ball milling Mg alloy waste with graphite and $AlCl_3$ for 4 h results in favourable hydrolysis performance [102, 103]. Ball milling MgH_2 for 5 h and subjecting nanostructured Mg to dehydrogenation at 355 °C after 2.5 h of ball grinding improves hydrolysis performance, with nanometric Mg in a 0.6 M $MgCl_2$ solution yielding 910 ml/g of H_2 [109]. The milled Mg-10%In alloy exhibits excellent hydrolysis properties, with a maximum hydrogen generation rate of 7.4 mL g^{-1} s^{-1} and a hydrogen conversion yield of 93% at 30 °C. Additionally, Mg-10%In demonstrates a fast hydrogen generation rate in methanol, offering a straightforward method for preparing Mg methoxide [75]. Milling Mg with $Mg_{17}Al_{12}$ reveals a weak bond between Mg and $Mg_{17}Al_{12}$, with enhanced microstructural defects due to the brittleness of the intermetallic. The galvanic coupling between Mg and $Mg_{17}Al_{12}$ improves hydrolysis performance, achieving 80% of the theoretical capacity of hydrogen production in 60 min (Fig. 13.26) [110].

The ball milled $Mg_{50-x}Ti_xNi_{45}Al_3Co_2$ ($x = 0, 1$) + 50 wt% Ni composites show an increased discharge capacity from 343.6 to 468.8 mAh/g and the capacity retention from 41 to 57% with the milling time changing from 5 to 20 h [69] Mg2.3Y0.1Ni + x wt.% graphene (x = 1, 3, 5, 7) hydrogen storage alloys, synthesised through ball milling, exhibit enhanced hydrogen absorption and release due to graphene distribution at the alloy edges, promoting increased phase boundaries and facilitating H_2 diffusion. The alloy with 3 wt.% graphene shows optimal performance, achieving absorption and release of 3.251 wt.% and 3.389 wt.% of H_2, respectively, at 573 K, with the lowest activation energy and enthalpy changes of generation and release (−54.8 kJ/mol and 67.5 kJ/mol, respectively) [92]. The stability of the γ-MgH_2 phase, synthesised through high-energy ball milling, has also been explored. From an application perspective, the potential utilisation of the γ-phase for hydrogen storage technology becomes feasible if the hydrogen absorption/desorption process can be conducted at a temperature of 250 °C or lower [111].

Fujiwara and colleagues synthesised a BCC Mg–V–Cr alloy using HEBM, which displayed a relatively low hydrogen storage capacity of 0.9 mass % at room temperature [112]. The diminished storage capability is linked to lattice defects resulting from the ball milling process. In contrast, ball milling-induced lattice distortion and grain size refinement in $Mg_{91.4}Ni_7Y_{1.6}$ alloy delivers improved hydrogen absorption and desorption properties [113]. The $Mg_{91.4}Ni_7Y_{1.6}$ alloys, characterised by a higher number of phase boundaries and fine particles, demonstrate heightened hydrogen absorption capacity, reduced activation energy, and improved absorption rates. Notably, these alloys exhibit superior kinetic and thermodynamic properties, achieving hydrogen absorption contents of 4.64 wt.% and 5.78 wt.% at 300 and 350 °C under 3 MPa in just 5 min, with an activation energy of 25.4 kJ/mol H_2. Ball milling effectively reduces the nucleation barrier for hydrogen absorption and desorption in $Mg_{90}Y_{1.5}Ce_{1.5}Ni_7$ samples, lowering the optimal hydrogenation temperature to 200 °C [72]. Upon hydrogenation, these samples undergo a transformation from nanocrystalline and amorphous phases to a fully nanocrystalline structure, demonstrating a reversible hydrogen storage capacity of approximately 5.6 wt%. This enhancement in kinetics is attributed to particle refinement and the in-situ formation of $CeH_{2.73}$ and Mg_2NiH_4 nanocrystals.

The addition of catalysts during ball milling further enhances surface characteristics, introducing defects that improve channels for hydrogen diffusion and elevate activation and kinetic processes [114]. For instance, catalysing MgH_2 with highly dispersed Ni nanoparticles through in-situ hydrogenolysis of nickelocene ($NiCp_2$) during ball milling creates an MgH_2–16.1 wt% $NiCp_2$ composite, displaying superior hydrogen desorption kinetics and excellent cycling stability, maintaining a high-capacity retention rate of ~96% after 50 cycles [93]. Similarly, $Mg_{85}Ni_{10}Zn_5 + x$ wt% CeO_2 ($x = 0, 2, 4, 8$) composites synthesised through HEBM exhibit reversible hydrogen absorption/desorption, with the addition of CeO_2 nanoparticles enhancing hydrogen storage dynamics by promoting phase interfaces and grain boundaries. They exhibit reversible hydrogen absorption/desorption above 320 °C with hydrogen storage capacities of 5.28, 5.22, 5.05, and 4.92 wt% H_2 for $x = 0, 2, 4$, and 8 [73]. The activation energy for dehydrogenation is reduced to 82.93 kJ/mol H_2 with 4 wt% CeO_2. The incorporation of Sn and Zn into Mg through ball milling results in a nanocomposite comprising a supersaturated Mg (Sn, Zn) solid solution and Mg_2Sn. This solid solution decomposes into MgH_2, Sn, and $MgZn_2$ during hydrogenation with insitu-formed nanostructures Mg_2Sn and $MgZn_2$, positively influencing the hydrogen absorption and desorption kinetics of Mg. The alloy exhibits lower hydrogen absorption and desorption activation energies of 38.1 kJ/mol and 86.6 kJ/mol, respectively, and a reduced dehydriding enthalpy of 67.0 ± 1.9 kJ/(mol·H_2) [63]. Additionally, the addition of 5 wt.% $CeNi_5$ to MgH_2 via ball milling could achieve a desorption hydrogen capacity of 6.16 wt% at 300 °C [115].

In conclusion, ball milling has emerged as a versatile and effective technique for tailoring the properties of MAed Mg alloys, addressing challenges in solid-state hydrogen storage. From enhanced kinetics to improved electrochemical properties and reversible hydrogen storage capacities, the diverse applications of ball milling showcase its pivotal role in advancing Mg-based hydrogen storage materials.

Continued research in this direction holds the potential to unlock further innovations, making Mg alloys even more viable for hydrogen storage applications.

13.13 Aerospace

Mg has a historical legacy in military applications, including incendiary bombs, pyrotechnics, and aircraft construction during both the World Wars [6, 116]. Its entry into the civil aerospace industry was driven by notable advantages, such as exceptional castability, machinability, and ductility, offering superior performance compared to conventional metallic materials. Replacing aluminium alloys with Mg alloys in aircraft, as demonstrated with a Boeing 747 model, could achieve a significant weight reduction of 60.4 tonnes, enhancing fuel efficiency and reducing emissions [117]. The potential benefits of using Mg in the aerospace industry have led to numerous historical experiments where aircraft were constructed entirely with Mg or used as the primary material. Notable examples include military aircraft such as the Northrop XP-56, Lockheed F-80, Convair B-36, and Convair XC 99 [118]. Despite inherent limitations like flammability and corrosion resistance, Mg is currently used in specific aerospace components, but challenges have limited its widespread use in structures. Ongoing research addressing issues like ignition and mechanical strength shows promise for Mg's expanded role in the aerospace industry, with advancements in alloys like Elektron 21 and Elektron 675 offering improved strength and weight characteristics [118–121]. MA is key in tailoring the structure and intended properties of Mg alloys for various applications.

13.14 Automotive

The integration of Mg and its alloys into automotive applications traces back to 1918 when Mg-based racing engine pistons were introduced for the Indy 500 [120]. Widespread adoption occurred in the 1930s when Volkswagen incorporated a Mg-based engine block in its Beetle series [4]. Over the years, Mg has found extensive use in powertrain, chassis, and body structures, making it the third most used metallic material in the automotive sector [122]. Its growth trajectory in automotive applications mirrors that of the aerospace sector, emphasising usage benefits over overall market share. Despite being a relatively small contributor to general metallic materials, Mg has garnered attention for its potential to reduce vehicle weight, save energy, and minimise environmental impact [123]. Figure 13.27 illustrates the utilisation of Mg-based materials within the automotive sector [124].

Cast Mg alloys, especially the Mg–Al-based series like AZ and AM alloys, Mg-rare earth-based alloys such as WE43 and E21, and ZK alloys, are the most prevalent types for automotive applications. They are employed in critical components like instrument panel structures, cross-car beams, roof frames, and seat frames, primarily

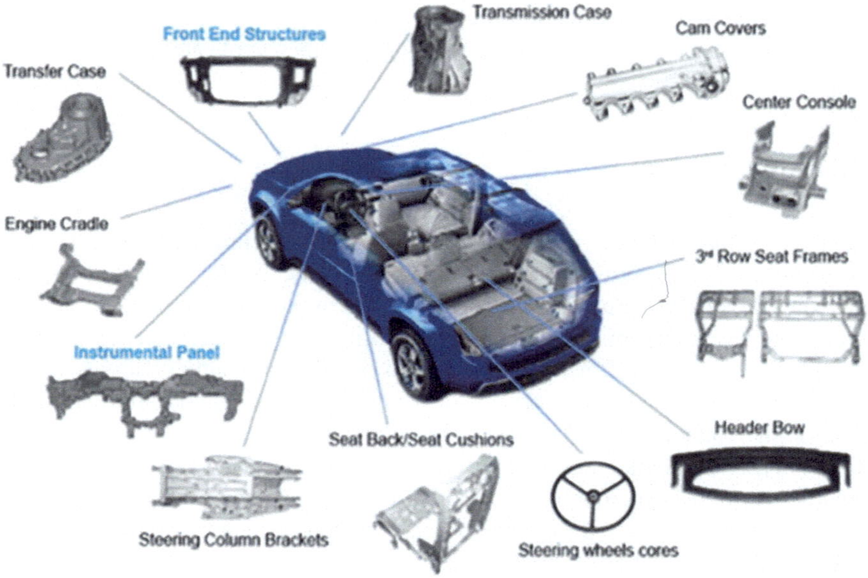

Fig. 13.27 Mg-based materials within the automotive sector. Taken from Ref. [124] with permission

in the front and top portions of automobiles where weight reduction is crucial [122]. While high-strength Mg alloys and alloy nanocomposites are considered for elevated temperatures due to their superior thermal and dimensional stability, other forms of Mg, such as sheet Mg and extruded Mg, have limited or no reported applications in the modern automotive industry [125]. Incorporating B_4C into the Mg matrix via MA enhances the wear resistance and strength [126]. The specific wear rate increases under varying loads, transitioning from abrasion to adhesion and delamination wear. Hard ceramic nanoparticles delay the transition to severe wear, indicating the potential for improved wear resistance in nano B_4C-reinforced AZ91 composites, suitable for wear resistance parts in automotive sectors. Addressing materials, processes, and performance challenges is crucial for promoting greater adoption of Mg in the automotive industry.

13.15 Biomedical

There has been a notable transition in the field of biomaterials towards making materials that function as temporary structures (3rd generation biomaterials). These materials are engineered to undergo gradual degradation, allowing for the seamless integration of native tissue with the implant and, eventually, the substitution of the implant for the body's tissue [127]. Mg alloys and composites have gained attention as potential alternatives to conventional implant materials due to their attractive biodegradation and mechanical properties [128, 129]. Table 13.3 compares the mechanical properties of Mg materials to bone and other metals.

However, one major challenge with Mg-based implants is their extremely high degradation rate in physiological conditions. This rapid degradation may lead to a loss of mechanical integrity before the complete healing of fractured bone tissue and result in the release of a large amount of hydrogen gases, which can delay the healing process. The most used Mg alloys in biomedical applications are Mg–Sn, Mg–Ca, Mg–Zn, Mg–Si, Mg–Sr, and Mg–Zr, which are non-toxic in a physiological environment. Additionally, Mg-based MMCs demonstrate appreciable mechanical strength and corrosion resistance [130–136].

The Mg65–Zn30–Ca5 and Mg70–Zn25–Ca5 alloys display non-cytotoxic behaviour towards human fibroblast cells, indicating their suitability for biomedical applications [64]. A biodegradable Mg-6.5 wt% Zn (Mg–Zn) alloy and Mg-6.5 wt% Zn reinforced with 10 wt% hydroxyapatite (HAP) composite are prepared by MA [137]. In vitro studies in Hank's balanced salt solution (HBSS) show decreased Mg ion (Mg^{2+}) dissolution, from 75.74 μg/ml for pure Mg to 46.88 μg/ml for Mg–Zn/HAP composite, indicating a progressive deceleration of degradation with HAP and Zn addition. The Mg–Zn/10 wt% HAP composite exhibits the highest corrosion resistance, with a more positive corrosion potential than pure Mg. After 7 days in HBSS, the Mg–Zn/HAP composite showed a compressive strength of 278.46 MPa, surpassing cortical bone strength (100–230 MPa), highlighting its potential for biomedical applications. An Mg-70 at.% Fe alloy, produced through ball milling and consolidated using SPS (773 K, 50 K/min, 10 min holding), exhibits exceptional strength (1380 MPa), elevated hardness, density, and enhanced biodegradability [94]. Figure 13.28 shows various Mg-based implants for orthopaedic, maxillo-facial, and vascular surgery applications [138].

One challenge with Mg material is its relatively high corrosion rate and the release of hydrogen gas (H_2), previously studied in the mid-20th century. Compositing Mg with hydroxyapatite (HAP) has been found to reduce the evolution of hydrogen gas (H_2) and minimise pH variations during the degradation of Mg in biomedical implants

Table 13.3 The comparison of mechanical properties of Mg composites to bone and other metals

Material/Tissue	Density (g/cm^3)	UTS (MPa)	Yield stress (MPa)	Elastic modulus (GPa)
Cortical bone	1.8–2.0	35–283	104.9–114.3	5–23
Cancellous bone	1.0–1.4	1.5–38	–	0.01–1.57
Ti–6Al–4V	4.43	830–1025	760–880	114
316 L stainless steel	8.0	450–650	200–300	190
Pure Mg, annealed	1.74	160	90	45
AZ91D Mg alloy	1.81	240–250	160	45

Data taken from Ref. [130]

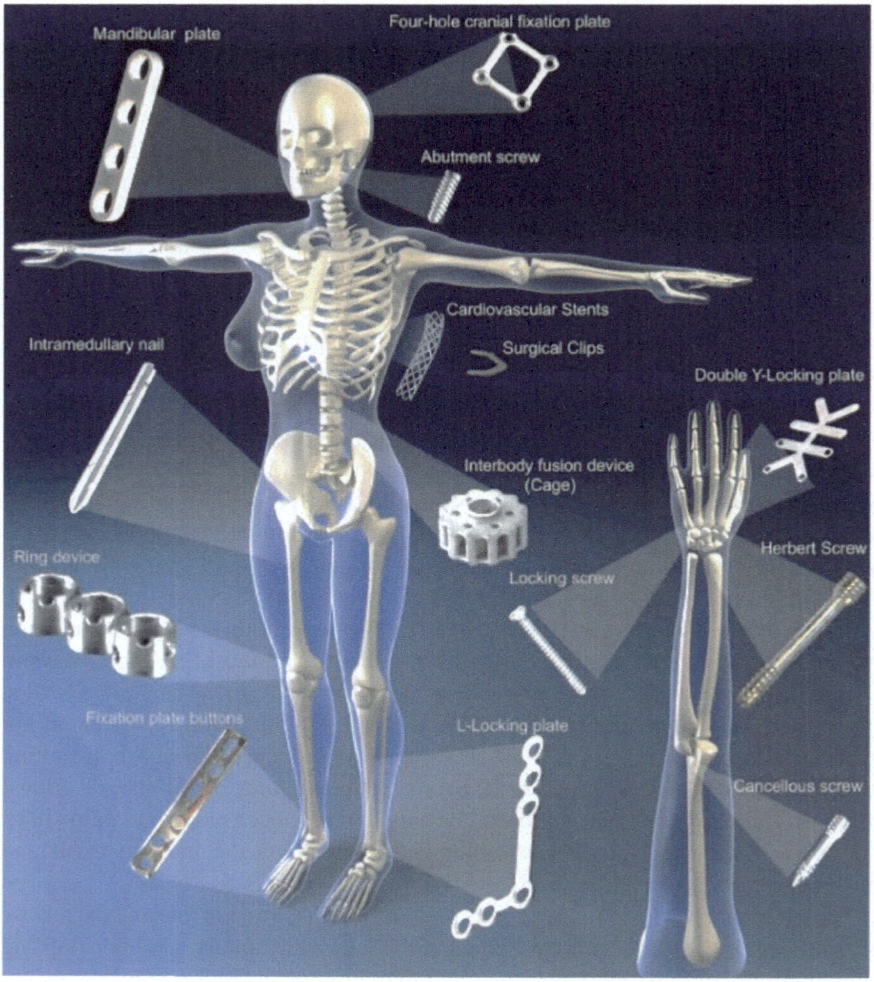

Fig. 13.28 Different Mg-based implants reported for orthopaedic, maxillo-facial, and vascular surgery applications. Taken from Ref. [138] with permission

[139, 140]. This Mg-HAP composite exhibits outstanding cytocompatibility, bioactivity, and enhanced mechanical properties, making it a highly suitable material for biomedical implants [141]. SiO_2 nanoparticle-reinforced Mg-based composites, produced through ball milling, exhibit a uniform reinforcement distribution and form a Mg_2Si intermetallic phase [142]. This composite demonstrates improved mechanical properties, including density and hardness, while maintaining an elastic modulus similar to the underlying Mg alloy, indicating compatibility with cortical bone. Notably, the composite with 3 wt% SiO_2 shows enhanced corrosion resistance in potentiodynamic tests, attributed to the Mg_2Si phase hindering hydroxide formation and delaying degradation. This suggests that Mg + 3 wt% SiO_2 can be a promising alternative biomaterial for implants and fixation devices, offering improved mechanical strength and corrosion resistance. Overall, using metallic materials, especially Mg composites, has shown great potential in advancing biomedical applications, particularly in developing orthopaedic implants and temporary devices, paving the way for safer and more effective medical solutions. Researchers continue improving these materials and addressing any associated challenges to ensure their successful integration into medical practice.

13.16 Electronic and Power-Tool Applications

In addition to the aerospace and automotive sectors, Mg-based alloys have found promising opportunities in the power tools and electronics industries. One of its standout advantages is its exceptional heat dissipation capability, which is approximately 100 times better than plastics. Moreover, Mg's recyclable nature, excellent vibration-damping characteristics, electromagnetic shielding properties, and ease of machinability make it highly valuable in the realm of electronics [143]. The applications of Mg in the consumer industry encompass a wide range of products, including audio–video players, mobile devices, cameras, laptops, radars, and rechargeable batteries [144]. The AZ31 Mg anode with ultrafine grains (667.28 $\pm$ 291.35 nm) was fabricated through SPS of alloy powder treated with HEBM [52]. The resulting ultrafine-grained AZ31 anode delivered a high voltage of 1.211 $\pm$ 0.004 V at 10 mA cm^{-2}, with a capacity of 1530 $\pm$ 20 mAh g^{-1} and an efficiency of 69.1 $\pm$ 0.9%, surpassing other Mg alloys. The enhanced performance is attributed to the increased grain boundaries in the ultrafine-grained AZ31, leading to stronger activity and suppressed side hydrogen evolution during battery discharge. Mg's combination of strength and low density makes it compelling for manufacturers to incorporate it into consumer electronic components. This usage helps in making these devices not only thin and lightweight but also highly portable and efficient [145]. Die-cast Mg products offer outstanding durability and ease of handling due to their lightweight designs. As a result, they are well-suited for power tool applications, particularly in demanding environments like construction sites, where they need to withstand extended working shifts [145].

13.17 Future Outlook

The versatility and distinctive properties of Mg alloys position them as highly attractive materials for a broad spectrum of engineering and biomedical applications. The alloying of Mg offers a pathway to tailor and enhance its properties, making it particularly valuable in industries such as aerospace and automotive, where the pursuit of lightweight materials carries significant environmental and economic advantages. The exploration of MA-Mg alloys, documented in this study, presents a promising avenue for the future of materials science and engineering. The synergistic approach of alloy composition and innovative processing techniques, including MA and SPS, can advance the structural integrity and performance of Mg alloys. Microstructure evolution and ongoing research into the effects of ball milling and milling time are vital aspects, offering insights into the refinement and customisation of material properties.

In the biomedical field, Mg possesses remarkable attributes such as biocompatibility, biodegradability, and bioabsorbability, in addition to its inherent engineering strengths. Its elastic modulus and natural antimicrobial properties further enhance its appeal. Consequently, Mg is predominantly employed in musculoskeletal and orthopaedic applications, with extensive research exploring its potential in cardiovascular medicine. However, its rapid biodegradability has presented challenges, hindering its widespread adoption in engineering and biomedical applications. Nonetheless, ongoing research and technological advancements are focused on developing strategies to mitigate Mg's excessively rapid degradation. Mg remains a promising material, as market research indicates a projected increase in demand. However, the current production of primary Mg is energy-intensive and inefficient, prompting exploration into additive manufacturing of Mg alloys and using solar energy for carbothermic reduction processes. Recycling Mg is another avenue to ensure a circular supply chain and ongoing investigations aim to determine the most commercially viable and environmentally sustainable recycling methods.

References

1. Shand, M.A.: The Chemistry and Technology of Magnesia. Wiley (2006). https://doi.org/10.1002/0471980579
2. Dobrzański, L.A., Totten, G.E., Bamberger, M.: The importance of magnesium and its alloys in modern technology and methods of shaping their structure and properties. In: Dobrzanski, L.A., Bamberger, M., Totten, G.E. (eds.) Magnesium and its Alloys, pp. 1–28. CRC Press (2019). https://doi.org/10.1201/9781351045476-1
3. Luthringer, B.J.C., Feyerabend, F., Willumeit-Römer, R.: Magnesium-based implants: a mini-review. Magnes. Res. **27**, 142–154 (2014)
4. Czerwinski, F.: Magnesium and its alloys. In: Magnesium Injection Molding, pp. 1–79. Springer, Boston, MA (2008). https://doi.org/10.1007/978-0-387-72528-4_1
5. Magnesium | Elements | RSC Education. https://edu.rsc.org/elements/magnesium/2020016.article
6. Polmear, I.J.: Magnesium alloys and applications. Mater. Sci. Technol. **10**, 1–16 (1994)
7. Holywell, G.C.: Magnesium: the first quarter millennium. JOM **57**, 26–33 (2005)
8. Aghion, E., Bronfín, B., Von Buch, F., Schumann, S., Friedrich, H.: Newly developed magnesium alloys for powertrain applications. JOM **55**, 30–33 (2003)
9. Baril, E., Labelle, P., Pekguleryuz, M.: Elevated temperature Mg–Al–Sr: creep resistance, mechanical properties, and microstructure. JOM **55**, 34–39 (2003)
10. Pekguleryuz, M.O., Kaya, A.A.: Creep resistant magnesium alloys for powertrain applications. Adv. Eng. Mater. **5**, 866–878 (2003)
11. Friedrich, H., Schumann, S.: Research for a "new age of magnesium" in the automotive industry. J. Mater. Process. Technol. **117**, 276–281 (2001)

12. Kuo, J.-L., Sugiyama, S., Hsiang, S.-H., Yanagimoto, J.: Investigating the characteristics of AZ61 magnesium alloy on the hot and semi-solid compression test. Int. J. Adv. Manuf. Technol. **29**, 670–677 (2006)
13. Kramer, D.A.: Magnesium, its Alloys and Compounds. U.S. Geological Survey Publication. (2001). https://pubs.usgs.gov/publication/ofr01341
14. Pidgeon, L.M.: New methods for the production of magnesium. Trans. Can. Inst. Min. Met **47**, 16–18 (1944)
15. Wu, L., Han, F., Liu, G.: Overview of magnesium metallurgy. In Comprehensive Utilization of Magnesium Slag by Pidgeon Process. SpringerBriefs in Materials, pp. 1–44. Springer, Singapore (2021). https://doi.org/10.1007/978-981-16-2171-0_1
16. Wu, L., Han, F., Liu, G.: Comprehensive Utilization of Magnesium Slag by Pidgeon Process. Springer Singapore (2021). https://doi.org/10.1007/978-981-16-2171-0
17. Hou, L., et al.: In vitro and in vivo studies on biodegradable magnesium alloy. Prog. Nat. Sci. Mater. Int. **24**, 466–471 (2014)
18. Xin, Y., Hu, T., Chu, P.K.: In vitro studies of biomedical magnesium alloys in a simulated physiological environment: a review. Acta Biomater. **7**, 1452–1459 (2011)
19. Witte, F.: The history of biodegradable magnesium implants: a review☆. Acta Biomater. **6**, 1680–1692 (2010)
20. Gu, X.N., et al.: The effect of tensile and fluid shear stress on the in vitro degradation of magnesium alloy for stent applications. Bioact. Mater. **3**, 448–454 (2018)
21. Panigrahi, S.K., et al.: Transition of deformation behavior in an ultrafine grained magnesium alloy. Mater. Sci. Eng. A **549**, 123–127 (2012)
22. Chen, W.Z., Wang, X., Wang, E.D., Liu, Z.Y., Hu, L.X.: Texture dependence of uniform elongation for a magnesium alloy. Scr. Mater. **67**, 858–861 (2012)
23. Le, Q., Jia, W., Ning, F.: Heat transfer analysis of magnesium alloy plate during transport process. J. Magnes. Alloy. **7**, 291–296 (2019)
24. Liu, C., et al.: Biodegradable Mg–Cu alloys with enhanced osteogenesis, angiogenesis, and long-lasting antibacterial effects. Sci. Rep. **6**, 27374 (2016)
25. Ali, M., Hussein, M.A., Al-Aqeeli, N.: Magnesium-based composites and alloys for medical applications: a review of mechanical and corrosion properties. J. Alloys Compd. **792**, 1162–1190 (2019)
26. Zheng, Y.F., Gu, X.N., Xi, Y.L., Chai, D.L.: In vitro degradation and cytotoxicity of Mg/Ca composites produced by powder metallurgy. Acta Biomater. **6**, 1783–1791 (2010)
27. Watanabe, H., Mukai, T., Mabuchi, M., Higashi, K.: High-strain-rate superplasticity at low temperature in a ZK61 magnesium alloy produced by powder metallurgy. Scr. Mater. **41**, 209–213 (1999)
28. Wong, W.L.E., Karthik, S., Gupta, M.: Development of high performance Mg–Al2O3 composites containing Al2O3 in submicron length scale using microwave assisted rapid sintering. Mater. Sci. Technol. **21**, 1063–1070 (2005)
29. Rashad, M., et al.: Effect of graphene nanoplatelets (GNPs) addition on strength and ductility of magnesium-titanium alloys. J. Magnes. Alloy **1**, 242–248 (2013)
30. Xu, N., Song, Q., Bao, Y., Fujii, H.: Investigation on microstructure and mechanical properties of cold source assistant friction stir processed AZ31B magnesium alloy. Mater. Sci. Eng. A **761**, 138027 (2019)
31. Ren, W., Li, J., Xin, R.: Texture dependent shifting behavior of neutral layer in bending of magnesium alloys. Scr. Mater. **170**, 6–10 (2019)
32. Zhou, J., et al.: Effect of laser surface melting with alternating magnetic field on wear and corrosion resistance of magnesium alloy. Surf. Coat. Technol. **309**, 212–219 (2017)
33. Barani, B., Lakshminarayanan, A.K., Subashini, R.: Microstructural characteristics of chitosan deposited az91 magnesium alloy. Mater. Today Proc. **16**, 456–462 (2019)
34. Kozina, I., Krawiec, H., Starowicz, M., Kawalec, M.: Corrosion resistance of MgZn alloy covered by chitosan-based coatings. Int. J. Mol. Sci. **22**, 8301 (2021)
35. Dargusch, M.S., et al.: Improved biodegradable magnesium alloys through advanced solidification processing. Scr. Mater. **177**, 234–240 (2020)

36. Seifiyan, H., Heydarzadeh Sohi, M., Ansari, M., Ahmadkhaniha, D., Saremi, M.: Influence of friction stir processing conditions on corrosion behavior of AZ31B magnesium alloy. J. Magnes. Alloy **7**, 605–616 (2019)
37. Xu, P., et al.: Investigation of the surface modification of magnesium particles with stannate on the corrosion resistance of a Mg-rich epoxy coating on AZ91D magnesium alloy. Prog. Org. Coat. **135**, 591–600 (2019)
38. Witte, F., et al.: Degradable biomaterials based on magnesium corrosion. Curr. Opin. Solid State Mater. Sci. **12**, 63–72 (2008)
39. Gunde, P., Hänzi, A.C., Sologubenko, A.S., Uggowitzer, P.J.: High-strength magnesium alloys for degradable implant applications. Mater. Sci. Eng. A **528**, 1047–1054 (2011)
40. Li, L., Gao, J., Wang, Y.: Evaluation of cyto-toxicity and corrosion behavior of alkali-heat-treated magnesium in simulated body fluid. Surf. Coat. Technol. **185**, 92–98 (2004)
41. Carboneras, M., Hernández, L.S., del Valle, J.A., García-Alonso, M.C., Escudero, M.L.: Corrosion protection of different environmentally friendly coatings on powder metallurgy magnesium. J. Alloys Compd. **496**, 442–448 (2010)
42. Xianhua, C., Yuxiao, G., Fusheng, P.: Research progress in magnesium alloys as functional materials. Rare Met. Mater. Eng. **45**, 2269–2274 (2016)
43. Yang, X., Patel, J.B., Huang, Y., Mendis, C.L., Fan, Z.: Towards directly formable thin gauge AZ31 Mg alloy sheet production by melt conditioned twin roll casting. Mater. Des. **179**, 107887 (2019)
44. Bai, S., Fang, G., Zhou, J.: Integrated physical and numerical simulations of weld seam formation during extrusion of magnesium alloy. J. Mater. Process. Technol. **266**, 82–95 (2019)
45. Wang, Y., Culbertson, D., Jiang, Y.: An experimental study of anisotropic fatigue behavior of rolled AZ31B magnesium alloy. Mater. Des. **186**, 108266 (2020)
46. Chen, J., et al.: Mechanical properties of magnesium alloys for medical application: a review. J. Mech. Behav. Biomed. Mater. **87**, 68–79 (2018)
47. Shuai, C., et al.: Mechanical alloying of immiscible metallic systems: process, microstructure, and mechanism. Adv. Eng. Mater. **23** (2021)
48. Deng, H., et al.: A high-Nb–TiAl alloy with ultrafine-grained structure fabricated by cryomilling and spark plasma sintering. Rare Met. **42**, 1678–1685 (2023)
49. Guyon, J., Hazotte, A., Monchoux, J.P., Bouzy, E.: Effect of powder state on spark plasma sintering of TiAl alloys. Intermetallics **34**, 94–100 (2013)
50. Shanmugasundaram, T., Guyon, J., Monchoux, J.P., Hazotte, A., Bouzy, E.: On grain refinement of a γ-TiAl alloy using cryo-milling followed by spark plasma sintering. Intermetallics **66**, 141–148 (2015)
51. Khan, M.U.F., Patil, A., Christudasjustus, J., Borkar, T., Gupta, R.K.: Spark plasma sintering of a high-energy ball milled Mg-10 wt% Al alloy. J. Magnes. Alloy **8**, 319–328 (2020)
52. Wang, N., et al.: AZ31 magnesium alloy with ultrafine grains as the anode for Mg-air battery. Electrochim. Acta **378**, 138135 (2021)
53. Seth, P.P., Singh, N., Singh, M., Prakash, O., Kumar, D.: Formation of fine Mg2Si phase in Mg–Si alloy via solid-state sintering using high energy ball milling. J. Alloys Compd. **821**, 153205 (2020)
54. Kumar, S.D., et al.: Mechanical properties of magnesium-silicon carbide composite fabricated through powder metallurgy route. Mater. Today Proc. **27**, 1137–1141 (2020)
55. Feng, J., et al.: Tensile flow and work hardening behaviors of ultrafine-grained Mg–3Al–Zn alloy at elevated temperatures. Mater. Sci. Eng. A **667**, 97–105 (2016)
56. Tejeda-Ochoa, A., et al.: Metastable FCC structure of Ti–Mg alloy synthesized by mechanical alloying. Microsc. Microanal. **26**, 2918–2920 (2020)
57. Luiggi A.N.J., Rondón, M., Valera, M.: Synthesis and characterization of Mg-In alloys (17.47% at. and 38.84% at.) prepared by mechanical alloying. Rev. LatinAm. Metal. Mat. **40**(1), 15–35 (2020). https://doi.org/10.5281/zenodo.7693707
58. Salleh, E.M., Ramakrishnan, S., Hussain, Z.: Synthesis of biodegradable Mg–Zn alloy by mechanical alloying: effect of milling time. Procedia Chem. **19**, 525–530 (2016)

59. Li, B., Li, J., Zhao, H., Yu, X., Shao, H.: Mg-based metastable nano alloys for hydrogen storage. Int. J. Hydrog. Energy **44**, 6007–6018 (2019)
60. Fang, C., et al.: Microstructures and mechanical properties of Mg2Sn-nanophase reinforced Mg–Mg2Sn composite. Mater. Sci. Eng. A **684**, 229–232 (2017)
61. Wang, R., Fang, C.F., Xu, Z.Y., Wang, Y.M.: Correlation of milling time with phase evolution and thermal stability of Mg-25 wt% Sn alloy. J. Alloys Compd. **891**, 162014 (2022)
62. Hrapkowicz, B., Lesz, S., Kremzer, M., Karolus, M., Pakieła, W.: Mechanical alloying of Mg–Zn–Ca–Er alloy. Bull. Polish Acad. Sci. Tech. Sci. **69**(e137587), 1–7 (2021)
63. Zhong, H.C., Xu, J.B.: Tuning the de/hydriding thermodynamics and kinetics of Mg by mechanical alloying with Sn and Zn. Int. J. Hydrog. Energy **44**, 2926–2933 (2019)
64. Gonzaga, S., et al.: Synthesis of magnesium-based alloys by mechanical alloying for implant applications. Coatings **13**, 260 (2023)
65. Maweja, K., Phasha, M., van der Berg, N.: Microstructure and crystal structure of an equimolar Mg–Ti alloy processed by Simoloyer high-energy ball mill. Powder Technol. **199**, 256–263 (2010)
66. Zhang, Y., et al.: Improved hydrogen storage dynamics of amorphous and nanocrystalline Ce–Mg–Ni-based CeMg12-type alloys synthesized by ball milling. Renew. Energy **132**, 167–175 (2019)
67. Zhang, Y., et al.: Structure and electrochemical hydrogen storage behaviors of Mg–Ce–Ni–Al-based alloys prepared by mechanical milling. J. Rare Earths **38**, 1093–1102 (2020)
68. Bu, W., et al.: Nanocrystalline structure and electrochemical hydrogen storage properties of the as-milled Mg–V–Ni–Fe–Zn-based materials. Int. J. Hydrog. Energy **48**, 6937–6946 (2023)
69. Hou, Z., Wei, X., Zhang, W., Yuan, Z., Ge, Q.: Electrochemical hydrogen storage performances of Mg–Ti–Ni–Co–Al-based alloys prepared by mechanical milling. J. Phys. Chem. Solids **149**, 109788 (2021)
70. Zhang, Y., et al.: Effect of graphite (GR) content on electrochemical hydrogen storage performances of nanocrystalline and amorphous La9Ce1Mg80Ni5–Ni–GR composites synthesized by mechanical milling. Int. J. Hydrog. Energy **45**, 29023–29033 (2020)
71. Sübütay, H., Şavklıyıldız, İ: Effect of high-energy ball milling in ternary material system of (Mg–Sn–Na). Crystals **13**, 1230 (2023)
72. Yong, H., et al.: Characterization of microstructure, hydrogen storage kinetics and thermodynamics of ball-milled Mg90Y1.5Ce1.5Ni7 alloy. Int. J. Hydrog. Energy**46**, 17802–17813 (2021)
73. Yuan, Z., et al.: Influence of CeO2 nanoparticles on microstructure and hydrogen storage performance of Mg–Ni–Zn alloy. Mater. Charact. **178**, 111248 (2021)
74. Song, M.Y., Kwak, Y.J., Shin, H.S., Lee, S.H., Kim, B.G.: Improvement of hydrogen-storage properties of MgH2 by Ni, LiBH4, and Ti addition. Int. J. Hydrog. Energy **38**, 1910–1917 (2013)
75. Xiao, F., Guo, Y., Yang, R., Li, J.: Hydrogen generation from hydrolysis of activated magnesium/low-melting-point metals alloys. Int. J. Hydrog. Energy **44**, 1366–1373 (2019)
76. Khan, M.U.F., Mirza, F., Gupta, R.K.: High hardness and thermal stability of nanocrystalline Mg–Al alloys synthesized by the high-energy ball milling. Materialia **4**, 406–416 (2018)
77. Youssef, K.M., Zaddach, A.J., Niu, C., Irving, D.L., Koch, C.C.: A novel low-density, high-hardness, high-entropy alloy with close-packed single-phase nanocrystalline structures. Mater. Res. Lett. **3**, 95–99 (2015)
78. Ferraz, M. de B., Botta, W.J., Zepon, G.: Synthesis, characterization and first hydrogen absorption/desorption of the Mg35Al15Ti25V10Zn15 high entropy alloy. Int. J. Hydrog. Energy**47**, 22881–22892 (2022)
79. Hashimoto, H., Isobe, S., Hashimoto, N., Oka, H.: Synthesis of Li–Mg–Al–Ti based lightweight high entropy alloys by mechanical alloying and investigation of conditions for solid solution formation. J. Alloy Metall. Syst. **4**, 100037 (2023)
80. Marques, F., et al.: Mg-containing multi-principal element alloys for hydrogen storage: a study of the MgTiNbCr0.5Mn0.5Ni0.5 and Mg0.68TiNbNi0.55 compositions. Int. J. Hydrog. Energy**45**, 19539–19552 (2020)

81. Cermak, J., Kral, L., Roupcova, P.: A new light-element multi-principal-elements alloy AlMg2TiZn and its potential for hydrogen storage. Renew. Energy **198**, 1186–1192 (2022)
82. Strozi, R.B., Leiva, D.R., Huot, J., Botta, W.J., Zepon, G.: Synthesis and hydrogen storage behavior of Mg–V–Al–Cr–Ni high entropy alloys. Int. J. Hydrog. Energy **46**, 2351–2361 (2021)
83. Stalin, B., Vidhya, V.S., Ravichandran, M., Kumar, A.N., Sudha, G.T.: Characterization and properties of Mg–TiO2 composites produced via ball milling and powder metallurgy. Metallofiz. Noveishie Tekhnol. **42**, 497–509 (2020). https://doi.org/10.15407/mfint.42.04.0497
84. Bemanifar, S., Rajabi, M., Hosseinipour, S.J.: Microstructural characterization of Mg–SiC nanocomposite powders fabricated by high energy mechanical milling. Silicon **9**, 823–827 (2017)
85. Soorya Prakash, K., Balasundar, P., Nagaraja, S., Gopal, P.M., Kavimani, V.: Mechanical and wear behaviour of Mg–SiC–Gr hybrid composites. J. Magnes. Alloy **4**, 197–206 (2016)
86. Shahin, M., Munir, K., Wen, C., Li, Y.: Magnesium-based composites reinforced with graphene nanoplatelets as biodegradable implant materials. J. Alloys Compd. **828**, 154461 (2020)
87. Kamrani, S., Penther, D., Ghasemi, A., Riedel, R., Fleck, C.: Microstructural characterization of Mg–SiC nanocomposite synthesized by high energy ball milling. Adv. Powder Technol. **29**, 1742–1748 (2018)
88. Li, P., Tan, W., Gao, M., Shi, K.: Strengthening of the magnesium matrix composites hybrid reinforced by chemically oxidized carbon nanotubes and in situ Mg2Sip. J. Alloys Compd. **858**, 157673 (2021)
89. Razzaghi, M., Kasiri-Asgarani, M., Bakhsheshi-Rad, H.R., Ghayour, H.: Microstructure, mechanical properties, and in-vitro biocompatibility of nano-NiTi reinforced Mg–3Zn–0.5Ag alloy: prepared by mechanical alloying for implant applications. Compos. Part B Eng. **190**, 107947 (2020)
90. Zhang, Y., et al.: Electrochemical hydrogen storage behaviors of as-milled Mg–Ce–Ni–Al-based alloys applied to Ni-MH battery. Appl. Surf. Sci. **494**, 170–178 (2019)
91. Sun, S., et al.: Microstructure and mechanical properties of AZ31 magnesium alloy reinforced with novel sub-micron vanadium particles by powder metallurgy. J. Mater. Res. Technol. **15**, 1789–1800 (2021)
92. Zhou, D., et al.: Hydrogen storage property improvement of ball-milled Mg2.3Y0.1Ni alloy with graphene. Int. J. Hydrog. Energy **50**, part D, 123–135 (2023)
93. Peng, C., Li, Y., Zhang, Q.: Enhanced hydrogen desorption properties of MgH2 by highly dispersed Ni: the role of in-situ hydrogenolysis of nickelocene in ball milling process. J. Alloys Compd. **900**, 163547 (2022)
94. Xie, G., Takada, H., Kanetaka, H.: Development of high performance MgFe alloy as potential biodegradable materials. Mater. Sci. Eng. A **671**, 48–53 (2016)
95. Park, K., Park, J., Kwon, H.: Effect of intermetallic compound on the Al-Mg composite materials fabricated by mechanical ball milling and spark plasma sintering. J. Alloys Compd. **739**, 311–318 (2018)
96. Lesz, S., Hrapkowicz, B., Karolus, M., Gołombek, K.: Characteristics of the Mg–Zn–Ca–Gd alloy after mechanical alloying. Materials (Basel) **14**, 226 (2021)
97. Zhou, H., et al.: Microstructures and mechanical properties of nanocrystalline AZ31 magnesium alloy powders with submicron TiB2 additions prepared by mechanical milling. Crystals **10**, 550 (2020)
98. Chaubey, A.K., et al.: High-strength ultrafine grain Mg–7.4%Al alloy synthesized by consolidation of mechanically alloyed powders. J. Alloys Compd. **610**, 456–461 (2014)
99. Feng, J., Li, X.W., Sun, H.F., Fang, W.B.: An ultra-high strength Mg–3Al–Zn alloy with low tension-compression yield asymmetry. Mater. Lett. **269**, 127489 (2020)
100. Zheng, B., et al.: High strength, nano-structured Mg–Al–Zn alloy. Mater. Sci. Eng. A **528**, 2180–2191 (2011)

101. Yu, H., Sun, Y., Hu, L., Wan, Z., Zhou, H.: Microstructure and properties of mechanically milled AZ61 powders dispersed with submicron/nanometer Ti particulates. Mater. Charact. **127**, 272–278 (2017)
102. Al Bacha, S., et al.: Hydrogen generation from ball milled Mg alloy waste by hydrolysis reaction. J. Power. Sour. **479**, 228711 (2020)
103. Al Bacha, S., et al.: Effect of ball milling strategy (milling device for scaling-up) on the hydrolysis performance of Mg alloy waste. Int. J. Hydrog. Energy**45**, 20883–20893 (2020)
104. Rattan Paul, D., et al.: Effect of ball milling and iron mixing on structural and morphological properties of magnesium for hydrogen storage application. Mater. Today Proc. **42**, 1673–1677 (2021)
105. Lyu, J., Lider, A., Kudiiarov, V.: Using ball milling for modification of the hydrogenation/dehydrogenation process in magnesium-based hydrogen storage materials: an overview. Metals **9**, 768 (2019)
106. Wang, Y., Wang, X., Li, C.M.: Electrochemical hydrogen storage of ball-milled MmMg12 alloy–Ni composites. Int. J. Hydrog. Energy **35**, 3550–3554 (2010)
107. Zhang, Y., et al.: A comparison study of hydrogen storage performances of SmMg11Ni alloys prepared by melt spinning and ball milling. J. Rare Earths **36**, 409–417 (2018)
108. Yang, Y., et al.: Research advances in magnesium and magnesium alloys worldwide in 2020. J. Magnes. Alloy **9**, 705–747 (2021)
109. Pighin, S.A., Urretavizcaya, G., Bobet, J.-L., Castro, F.J.: Nanostructured Mg for hydrogen production by hydrolysis obtained by MgH2 milling and dehydriding. J. Alloys Compd. **827**, 154000 (2020)
110. Al Bacha, S., Aubert, I., Zakhour, M., Nakhl, M., Bobet, J.-L.: Hydrolysis properties, corrosion behavior and microhardness of AZ91 "model" alloys. J. Alloys Compd. **845**, 156283 (2020)
111. Lobo, N., Takasaki, A., Mineo, K., Klimkowicz, A., Goc, K.: Stability investigation of the γ-MgH2 phase synthesized by high-energy ball milling. Int. J. Hydrog. Energy **44**, 29179–29188 (2019)
112. Fujiwara, K., et al.: New Mg–V–CrBCC alloys synthesized by high-pressure torsion and ball milling. Mater. Trans. **59**, 741–746 (2018)
113. Pang, X., Ran, L., Chen, Y., Luo, Y., Pan, F.: Enhancing hydrogen storage performance via optimizing Y and Ni element in magnesium alloy. J. Magnes. Alloy. **10**, 821–835 (2022)
114. Panwar, K., Srivastava, S.: Enhancement in hydrogenation properties of ball-milled AB5-type hydrogen storage alloy through catalyst. J. Phys. Conf. Ser. **2267**, 012052 (2022)
115. Liao, W., et al.: Enhancing (de)hydrogenation kinetics properties of the Mg/MgH2 system by adding ANi5 (A = Ce, Nd, Pr, Sm, and Y) alloys via ball milling. J. Rare Earths **39**, 1010–1016 (2021)
116. Mordike, B., Ebert, T.: Magnesium. Mater. Sci. Eng. A **302**, 37–45 (2001)
117. Gupta, M.: Utilizing magnesium based materials to reduce green house gas emissions in aerospace sector. Aeronaut. Aerosp. Open Access J. **1**(1), 41–46 (2017)
118. Gialanella, S., Malandruccolo, A.: Aerospace Alloys (2020). https://doi.org/10.1007/978-3-030-24440-8
119. Kurzynowski, T., Pawlak, A., Smolina, I.: The potential of SLM technology for processing magnesium alloys in aerospace industry. Arch. Civ. Mech. Eng. **20**, 23 (2020)
120. Luo, A.A.: Applications: aerospace, automotive and other structural applications of magnesium. In: Pekguleryuz, M.O., Kainer, K.U., Kaya, A.A. (eds.) Fundamentals of Magnesium Alloy Metallurgy, pp. 266–316. Woodhead Publishing (2013). https://doi.org/10.1533/978 0857097293.266
121. Tan, J., Ramakrishna, S.: Applications of magnesium and its alloys: a review. Appl. Sci. **11**, 6861 (2021)
122. Powell, B.R., Krajewski, P.E., Luo, A.A.: Chapter 4 - Magnesium alloys for lightweight powertrains and automotive structures. In: Mallick, P.K. (eds.) Materials, Design and Manufacturing for Lightweight Vehicles, 2nd edn, pp. 125–186. Woodhead Publishing (2021). https://doi.org/10.1016/B978-0-12-818712-8.00004-5

123. Kulekci, M.K.: Magnesium and its alloys applications in automotive industry. Int. J. Adv. Manuf. Technol. **39**, 851–865 (2008)
124. Sankaranarayanan, S., Gupta, M.: Emergence of god's favorite metallic element: magnesium based materials for engineering and biomedical applications. Mater. Today Proc. **39**, 311–316 (2021)
125. Tekumalla, S., Gupta, M.: Introductory chapter: an insight into fascinating potential of magnesium. In: Gupta, M. (ed.) Magnesium—The Wonder Element for Engineering/Biomedical Applications. IntechOpen (2020). https://doi.org/10.5772/intechopen.90866
126. Chinthamani, S., Kannan, G., George, G.D., Sreedharan, C.E.S., Rajagopal, K.S.: Effect of nano B4C on the tribological behaviour of magnesium alloy prepared through powder metallurgy. Mater. Sci. **26**, 392–400 (2020)
127. Xu, L., et al.: In vitro and in vivo evaluation of the surface bioactivity of a calcium phosphate coated magnesium alloy. Biomaterials **30**, 1512–1523 (2009)
128. Liu, C., Xin, Y., Tang, G., Chu, P.K.: Influence of heat treatment on degradation behavior of bio-degradable die-cast AZ63 magnesium alloy in simulated body fluid. Mater. Sci. Eng. A **456**, 350–357 (2007)
129. Xin, Y., et al.: Corrosion behavior of ZrN/Zr coated biomedical AZ91 magnesium alloy. Surf. Coat. Technol. **203**, 2554–2557 (2009)
130. Haghshenas, M.: Mechanical characteristics of biodegradable magnesium matrix composites: a review. J. Magnes. Alloy **5**, 189–201 (2017)
131. Gu, X.-N., Zheng, Y.-F.: A review on magnesium alloys as biodegradable materials. Front. Mater. Sci. China **4**, 111–115 (2010)
132. Feng, A., Han, Y.: The microstructure, mechanical and corrosion properties of calcium polyphosphate reinforced ZK60A magnesium alloy composites. J. Alloys Compd. **504**, 585–593 (2010)
133. Razavi, M., Fathi, M.H., Meratian, M.: Microstructure, mechanical properties and bio-corrosion evaluation of biodegradable AZ91-FA nanocomposites for biomedical applications. Mater. Sci. Eng. A **527**, 6938–6944 (2010)
134. Gu, X.N., et al.: Microstructure and characteristics of the metal–ceramic composite (MgCa-HA/TCP) fabricated by liquid metal infiltration. J. Biomed. Mater. Res. Part B Appl. Biomater. **99B**, 127–134 (2011)
135. Huan, Z.G., Leeflang, M.A., Zhou, J., Duszczyk, J.: ZK30-bioactive glass composites for orthopedic applications: a comparative study on fabrication method and characteristics. Mater. Sci. Eng. B **176**, 1644–1652 (2011)
136. Lei, T., Tang, W., Cai, S.-H., Feng, F.-F., Li, N.-F.: On the corrosion behaviour of newly developed biodegradable Mg-based metal matrix composites produced by in situ reaction. Corros. Sci. **54**, 270–277 (2012)
137. Salleh, E.M., Zuhailawati, H., Mohd Noor, S.N.F., Othman, N.K.: In vitro biodegradation and mechanical properties of Mg–Zn Alloy and Mg–Zn-hydroxyapatite composite produced by mechanical alloying for potential application in bone repair. Metall. Mater. Trans. A. **49**, 5888–5903 (2018)
138. Grewal, N.S., Sharma, G.K., Kumar, K., Batra, U.: Thermally and mechanically tuned interfaces of magnesium alloys for bioimplant applications. Surf. Interfaces **41**, 103284 (2023)
139. Sunil, B.R., Ganapathy, C., Sampath Kumar, T.S., Chakkingal, U.: Processing and mechanical behavior of lamellar structured degradable magnesium-hydroxyapatite implants. J. Mech. Behavior Biomed. Mater. **40**, 178–189 (2014)
140. Viswanathan, R., et al.: Plasma electrolytic oxidation and characterization of spark plasma sintered magnesium/hydroxyapatite composites. Mater. Sci. Forum **765**, 827–831 (2013)
141. Xiong, G., et al.: Characterization of biomedical hydroxyapatite/magnesium composites prepared by powder metallurgy assisted with microwave sintering. Curr. Appl. Phys. **16**, 830–836 (2016)
142. Iqbal, A.A., Ismail, N.B.: Mechanical properties and corrosion behavior of silica nanoparticle reinforced magnesium nanocomposite for bio-implant application. Materials (Basel) **15**, 8164 (2022)

143. Yartys, V.A., et al.: Magnesium based materials for hydrogen based energy storage: past, present and future. Int. J. Hydrog. Energy **44**, 7809–7859 (2019)
144. Ding, Z., Shaw, L.: Enhancement of hydrogen desorption from nanocomposite prepared by ball milling MgH2 with in situ aerosol spraying LiBH4. ACS Sustain. Chem. Eng. **7**, 15064–15072 (2019)
145. Thandalam, S.K., Ramanathan, S., Sundarrajan, S.: Synthesis, microstructural and mechanical properties of ex situ zircon particles (ZrSiO4) reinforced metal matrix composites (MMCs): a review. J. Mater. Res. Technol. **4**, 333–347 (2015)
146. Sankhla, A., Patil, A., Kamila, H., Yassen, M., Farah, N., Mueller, E., de Boor, J.: Mechanical alloying of optimited Mg(2Si, Sn) solid solutions: understanding phase evolution and tuning synthesis parameters for thermoelectric applications. ACS Appl. Energy Mater. **1**(2), 531–542 (2018)

Chapter 14
Fabrication of Intermetallic Alloys

R. Raghavendra Rao, B. N. Sharath, Puttaswamy Madhusudan, and S. Pradeep

Abstract Intermetallic alloys come from a family of metallic materials made up of two or more elements, at least one of which is a metal. The ordered atomic arrangement of intermetallic alloys' crystal structures sets them apart from normal alloys; in most metallic alloys, the arrangement is random. Intermetallic alloys have special and frequently advantageous qualities because of their ordered structures. It is noteworthy that the properties of intermetallic alloys vary depending on the particular elements present, their ratios, and the crystal structure that forms. Engineers and researchers are still investigating and creating novel intermetallic alloys with improved characteristics for a range of uses. In order to fabricate intermetallic alloys, materials must be processed to yield the appropriate intermetallic phases with particular characteristics. The alloy's composition, the desired qualities, and the intended uses all influence the production techniques. It is crucial to remember that the intermetallic alloy's unique specifications and the intended use determine which fabrication procedure is best. To enhance the qualities of intermetallic alloys for a range of industrial uses, scientists and engineers frequently combine different approaches.

Keywords Intermetallic alloys · Fabrication · Phase transformation

14.1 Introduction

The goal of materials science has always been to improve the properties and functionality of materials via scientific investigation [1]. Chemical modifications and conventional thermal, mechanical, and thermomechanical approaches have yielded notable

R. R. Rao (✉) · B. N. Sharath · S. Pradeep
Department of Mechanical Engineering, Malnad College of Engineering, Hassan, Affiliated to VTU, Belagavi, Karnataka, India
e-mail: rrr@mcehassan.ac.in

P. Madhusudan (✉)
Department of Civil and Environmental Engineering, Hanyang University, 222, Wangsimni-ro, Seongdong-gu, Seoul Campus, Seoul 04763, South Korea
e-mail: madhu2010@hanyang.ac.kr

S. Rajendrachari (ed.), *Mechanically Alloyed Novel Materials*, Advanced Structured Materials 220, https://doi.org/10.1007/978-981-97-6504-1_14

improvements in mechanical, chemical, and physical properties [2, 3]. Sophisticated materials have been developed as a result of the search for materials with higher strength, stiffness, and lower weight in comparison to ordinary materials. The high-tech sectors have provided additional impetus for the initiatives [4, 5]. Materials that prioritise the meticulous synthesis and control of their structure in order to produce a precisely tailored range of qualities appropriate for demanding applications are known as advanced materials [6]. Metallic compositions consisting of two or more elements, with at least one of the elements being a metal, are referred to as intermetallic alloys. Because the atomic arrangement in intermetallic alloys' crystal structures is organized, as opposed to random as in most metallic alloys, this sets them apart from normal alloys [7, 8]. Intermetallic alloys possess distinct and frequently appealing characteristics due to their ordered structures [9].

14.1.1 Characteristics of Intermetallic Alloys

Intermetallic alloys (Fig. 14.1) are characterized by a well-defined and ordered atomic arrangement in their crystal structures, in contrast to the typical disorganized or random patterns seen in ordinary alloys. At room temperature, many intermetallic alloys have a tendency to be brittle, which can restrict their use in some situations. On the other hand, certain intermetallic compounds can be customized for particular applications and show high-temperature ductility. Compared to traditional alloys, intermetallic compounds frequently have higher melting points, which make them appropriate for high-temperature applications. There are several phases that intermetallic alloys can exist, each with a distinct composition and set of features. The stoichiometry of the component elements determines the phases. A few intermetallic alloys have remarkable qualities, like great strength, superior resistance to oxidation, and distinct magnetic or electrical characteristics. They are appropriate for particular engineering applications because of their qualities. Intermetallic alloys are used in a wide range of industries, such as electronics, automotive, aerospace, and catalysis. Certain intermetallic alloys based on nickel, for instance, are employed in high-temperature settings, such as gas turbine engines. To increase the spectrum of uses for intermetallic alloys, ongoing research is concentrated on creating these materials with enhanced ductility, toughness, and other mechanical qualities. Materials with two or more metallic elements together, usually in organized crystal formations, are called intermetallic alloys. These alloys frequently have special qualities that fit them for use [10–13].

14.1.2 Applications of Intermetallic Alloys

In aircraft applications (Fig. 14.2), intermetallic alloys with high strength and corrosion resistance at high temperatures, including gamma titanium aluminides (TiAl),

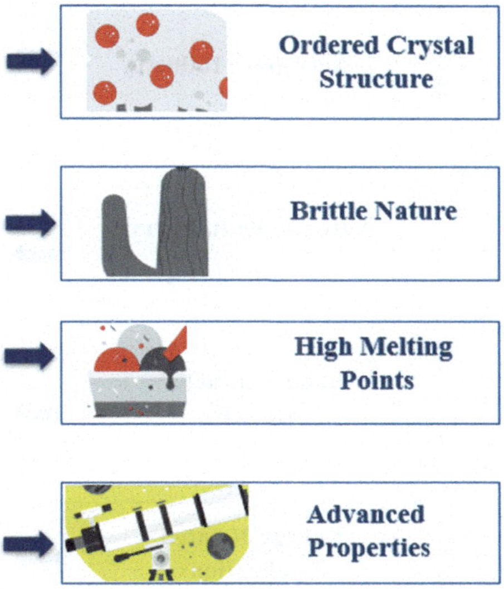

Fig. 14.1 Characteristics of intermetallic alloys

are utilized. Intermetallic alloys are a desirable option for lightweight vehicle components because they may be engineered to have high strength and low density. Take magnesium–aluminum intermetallic, for instance. In chemical processes, specific intermetallic compounds can act as catalysts. Palladium-aluminum alloys, for example, can be used in catalytic converters to remove pollutants from vehicle exhaust. High-frequency devices like microwave amplifiers and high-speed transistors are produced in the electronics industry using intermetallic compounds like gallium arsenide. The possibility of using intermetallic alloys as materials for hydrogen storage is being researched. The capacity of alloys with magnesium bases to reversibly absorb and release hydrogen is being investigated. Because intermetallic such as nickel aluminides (NiAl) can be employed in high-temperature settings, they are a good fit for gas turbine components. High-strength permanent magnets are used in electric motors and electronic equipment. These magnets are produced using intermetallic compounds, such as neodymium-iron-boron. Medical implants such as knee and hip replacements are made of biocompatible intermetallic alloys, such as titanium–aluminum–vanadium. Because of their mechanical qualities and resistance to corrosion, certain intermetallic alloys, particularly those based on zirconium, are utilized as cladding materials in nuclear fuel rods. The chemical processing sector uses nickel-based super alloys and other intermetallic alloys with exceptional corrosion resistance for equipment, pipes, and vessels. It is noteworthy that there is a continuous investigation into novel intermetallic alloys and their possible uses, broadening the spectrum of sectors that stand to gain from their distinct characteristics [14–17].

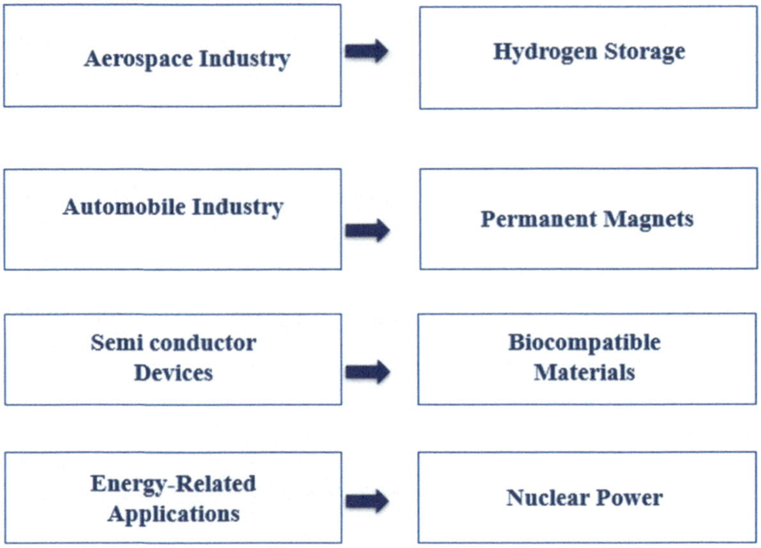

Fig. 14.2 Applications of intermetallic alloys

14.2 Phase Transformation in Intermetallic Alloys

Intermetallic alloys undergo phase transformations (Fig. 14.3), which are the transitions between stable states (phases) in the atom arrangement. Intermetallic alloys frequently display intricate phase diagrams and change in a variety of ways depending on the composition and temperature [18].

Order–disorder conversions can occur in some intermetallic alloys. Atoms are placed randomly in a disordered state, whereas they follow a regular and predictable pattern in an ordered state. The material's mechanical and thermal properties may be impacted by these changes. In intermetallic alloys, the order–disorder transformation is a phase shift that modifies the atoms' arrangement inside the alloy's crystal structure. Compounds created by two or more metallic elements in particular stoichiometry ratios are known as intermetallic alloys. The order–disorder transformation is linked to variations in the degree of atomic ordering within the crystal lattice and usually takes place at high temperatures. In an ordered intermetallic alloy, the atoms are arranged in a specific and regular manner within the crystal lattice. This ordered arrangement often leads to distinct crystallographic structures with well-defined unit cells. The disorder occurs when the atoms within the crystal lattice become less organized and more randomly arranged. This disordering process usually takes place at higher temperatures where thermal energy disrupts the ordered structure. Ordering at Low Temperatures: At low temperatures, many intermetallic alloys tend to exhibit ordered structures due to the minimization of free energy associated with specific atomic arrangements. As the temperature increases, thermal energy disrupts the ordered arrangement, leading to disorder within the crystal lattice. This can result

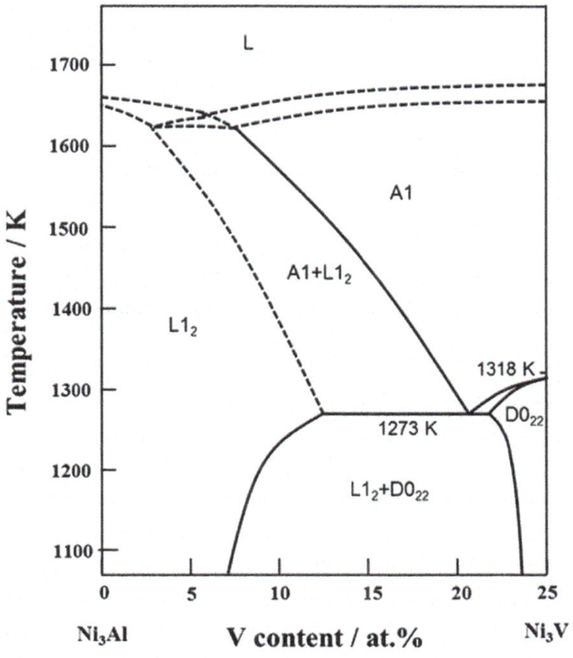

Fig. 14.3 Binary phase diagram [19]

in a transition from an ordered to a disordered state. The kinetics of the order–disorder transformation depend on factors such as temperature, time, and the activation energy for the movement of atoms within the crystal lattice. Diffusion of atoms is a crucial process in the transformation, and the rate of diffusion is influenced by temperature and the presence of defects in the crystal structure. The order–disorder transformation can significantly influence the mechanical properties of intermetallic alloys. Ordered structures often exhibit improved strength and hardness, while disordered structures may have better ductility and toughness. Understanding the order–disorder transformation is essential in designing and optimizing the properties of intermetallic alloys for specific applications, such as in aerospace, automotive, and electronic industries [20–22].

14.2.1 Alloy Systems

Common examples of intermetallic alloy systems undergoing order-disorder transformations include various phases in aluminum-based alloys, nickel-based superalloys, and titanium aluminides. The order-disorder transformation in intermetallic alloys is a dynamic process influenced by temperature and time, and it plays a crucial role in determining the alloy's mechanical and thermal properties. Researchers and engineers study these transformations to tailor the properties of intermetallic alloys for specific applications and improve their overall performance [21–24].

14.2.2 Precipitation Hardening

Precipitation hardening, sometimes referred to as age hardening or hardening by precipitation, is a heat treatment technique (Fig. 14.4) that is frequently applied to intermetallic alloys in order to enhance their mechanical characteristics, including hardness and strength. Through the creation of tiny precipitate particles inside the alloy matrix, this process results in an increase in strength by obstructing the migration of atoms and dislocations. Intermetallic alloys have the potential to precipitate harden, especially those with several phases. This results in an increase in hardness and strength due to the development of fine precipitates inside the alloy matrix. A common method for achieving this is heat treatment [25] (Figs. 14.5 and 14.6).

Alloys with particular compositions are frequently subjected to precipitation hardening; these alloys usually consist of a solid solution that can phase separate when heated. To produce a single-phase solid solution, the alloy must first be heated to a high temperature, usually over its solvus temperature [26–29]. The alloying components dissolve into the matrix as a consequence of this high-temperature treatment. The alloy is brought to room temperature quickly. In order to "freeze" the alloy in a supersaturated condition and stop equilibrium phases from forming, this rapid cooling is helpful. Following quenching, the alloy is exposed to an ageing temperature, which causes fine particle precipitation within the supersaturated solid solution. These precipitates typically provide efficient strengthening since they are tiny and coherent. During the ageing process, the precipitates keep expanding and dispersing throughout the alloy matrix. The resulting mechanical properties are

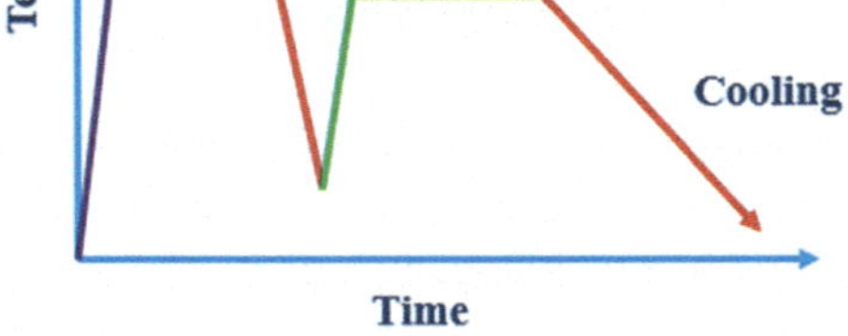

Fig. 14.4 Precipitation hardening [25]

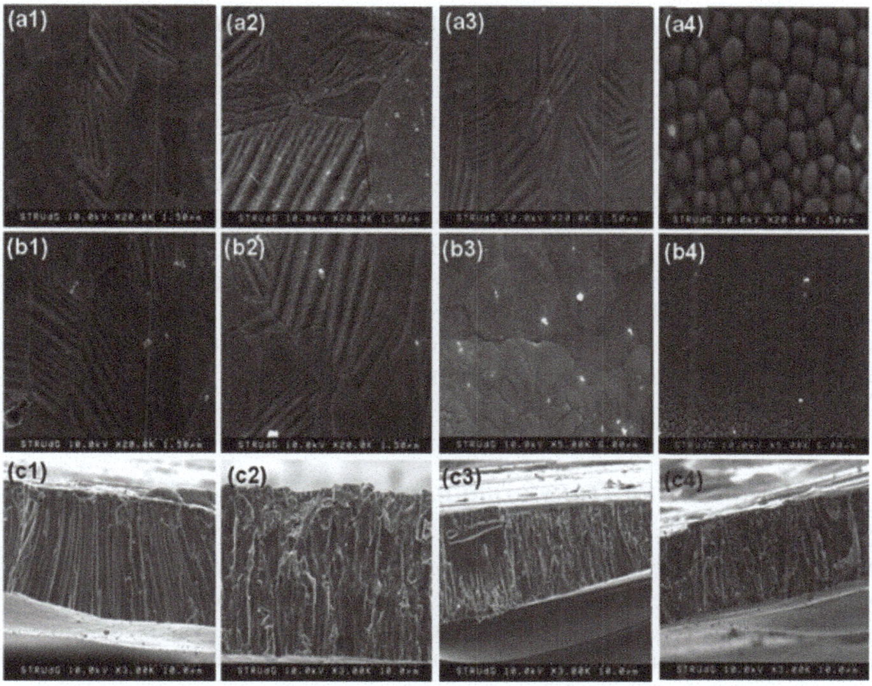

Fig. 14.5 Martensitic transformation and crystalline structure of $Ni_{50}Mn_{50-x}Sn_x$ melt-spun Heusler alloys [37]

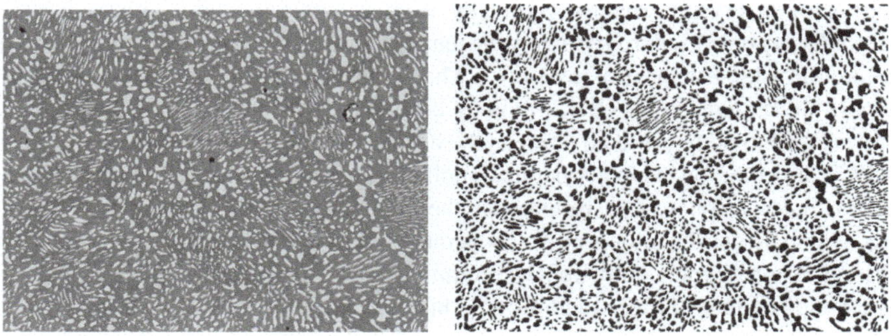

Fig. 14.6 Allotropic transformation [45]

strongly influenced by the size, distribution, and density of precipitates. Precipitation hardening improves strength by particle cutting, Orowan looping, and solid-solution hardening. This mechanism hinders the mobility of dislocations in the crystal lattice by dispersing precipitates, making it difficult for them to go through. The alloy's mechanical properties can be optimised for particular purposes by adjusting the precipitation hardening process's parameters, such as temperature and ageing time. Precipitation hardening of intermetallic alloys is frequently observed in some aluminium alloys (such as the 2000 series), nickel-based superalloys, and certain titanium alloys [30–32]. Applications for precipitation-hardened intermetallic alloys can be found in the aerospace, automotive, and other industries that demand a blend of corrosion resistance, high strength, and toughness [33]. To achieve the desired characteristics, optimisation of the heat treatment cycle is crucial, and precipitation hardening may need careful control of processing parameters. Excessive or insufficient ageing can impact the alloy's ultimate functionality [34, 35]. Precipitation hardening is an effective heat treatment technique for enhancing intermetallic alloys' mechanical qualities. Through the regulation of precipitate growth and formation inside the alloy matrix, engineers are able to customize the properties of the material to fulfill certain performance demands in a range of industrial settings.

14.2.3 Martensitic Transformation

A common solid-state phase transformation in intermetallic alloys that results in the production of a martensitic microstructure is called the martensitic transformation. A quick rearranging of the atoms within the crystal lattice causes this transformation, which changes the crystal structure. Diffusion lessness the absence of long-range atom diffusion is a defining characteristic of the martensitic transition. It is distinguished by the quick atom rearrangement that produces a martensitic phase. This change is frequent in certain intermetallic alloys and has a big impact on the material's mechanical characteristics [36].

A shift in crystal structure, typically from a high-temperature phase with a particular crystal structure (typically austenite) to a lower-temperature phase with a different crystal structure (martensite), is what is meant by the martensitic transformation. Body-centered cubic (BCC) to body-centered tetragonal (BCT), face-centered cubic (FCC) to body-centered tetragonal (BCT), and hexagonal close-packed (HCP) to BCT are common crystal structures implicated in martensitic transformations. In contrast to other phase transformations, the martensitic transformation is diffusion less, which means that atoms move cooperatively through displacements rather than by long-distance atomic diffusion. It is common for abrupt cooling or quenching from a high temperature to cause martensitic transition. The alloy cools down so fast that the atoms cannot have time to disperse and realign. Because there is not enough time for atomic diffusion, the quick cooling causes lattice distortions and stresses to occur inside the crystal structure as the atoms attempt to adapt to the altered crystal structure. A fine distribution of martensitic plates or needles throughout the alloy matrix characterizes the resultant microstructure. The distinct mechanical qualities are influenced by this microstructure. The shape memory effect is seen in certain intermetallic alloys that go through the martensitic transformation. This implies that they are capable of reversible deformation at specific temperatures and that, upon heating; they will regain their original shape. Due to the intricate structure of the martensitic plates and the lattice distortions, the martensitic transition frequently gives the alloy considerable hardness and strength. But it could also lead to more brittleness. The design of shape memory alloys, high-strength steels, and several nickel-based super alloys used in aerospace and other industries depends heavily on martensitic transformation. To customize the mechanical characteristics of intermetallic alloys, the martensitic transition must be managed. The transformation behaviour is dependent on various parameters, including the alloy composition, heat treatment settings, and rate of cooling. In conclusion, the development of distinct microstructures and mechanical characteristics in intermetallic alloys is a result of the notable phenomenon known as martensitic transformation. For the purpose of designing and optimizing alloys for particular applications in engineering and materials science, it is imperative to comprehend and manage this change [38–44].

14.2.4 Allotropic Transformation

In intermetallic alloys, the phrase "allotropic transformation" describes the change between distinct crystal forms, or "allotropes," under particular pressure and temperature circumstances. Allotropic transitions of intermetallic alloys can alter their microstructure and characteristics. The transition between various phases with differing crystal structures is the most prevalent example. At a particular temperature, certain intermetallic alloys can experience allotropic transformations, which cause them to transition from one crystal structure to another. For instance, an allotropic transformation is involved in the transition of titanium alloys from the beta phase to the alpha phase.

Intermetallic alloys frequently occur at different temperatures in a variety of crystal shapes. The transformation entails a modification of the atoms' configuration inside the crystal lattice.

The majority of allotropic transformations depend on temperature. An alteration in temperature can cause a phase transition in the intermetallic alloy, resulting in a distinct crystal structure. Certain phases of an intermetallic alloy are stable and energetically advantageous at particular pressures and temperatures. These stable phases are frequently dictated by the alloy composition and the state of the atmosphere. Phase diagrams are widely used to depict how intermetallic alloys behave in relation to allotropic transformations. These diagrams show the stable phases at various composition and temperature settings. Polymorphism is the existence of several crystal forms for a particular chemical makeup. Polymorphism is a property of intermetallic alloys, and the change from one polymorphic form to another is an allotropic transformation. Intermetallic alloys' mechanical, thermal, and magnetic properties can all be greatly impacted by allotropic transformations. For instance, the transformation might result in modifications to thermal, ductility, and hardness. Intermetallic alloys frequently undergo allotropic transformations, such as crystal structure transitions from face-centered cubic (FCC) to body-centered cubic (BCC) or from BCC to hexagonal close-packed (HCP). Solid–solid phase transitions are the result of allotropic transformations in intermetallic alloys, as opposed to liquid–gas (vaporization) or solid–liquid (melting) phase transitions. Reversible allotropic transformations cause some intermetallic alloys, including some nickel-titanium (Ni–Ti) alloys, to show the shape memory phenomenon. This enables the material to regain its pre-deformation shape. It is essential to comprehend and manage allotropic transformations in order to modify the properties of intermetallic alloys to satisfy certain engineering specifications. The optimization of transformation behaviour depends on various parameters, including alloy composition and heat treatment settings. In conclusion, crystal structures in intermetallic alloys undergo allotropic modifications in response to temperature changes. These changes are fundamental to the larger subject of materials science and metallurgy and have significant effects on the performance and design of materials in a variety of applications [46, 47].

14.2.5 Eutectoid and Peritectoid Reactions

One phase can simultaneously change into two additional phases in eutectoid processes. In peritectoid reactions, two phases combine to form a single phase. These kinds of reactions are typical in intricate systems of intermetallic alloys. Specific phase transitions that can take place in intermetallic alloys include eutectoid and peritectoid reactions. At a specific temperature, one phase can convert into two or more phases through solid-state processes. At a particular composition and temperature, a single phase in a eutectoid reaction splits into two new phases. One crystal structure changes into two distinct crystal structures as a result of this process. The eutectoid temperature is the name given to the transition temperature.

Commonly, the eutectoid reaction is denoted as $\alpha \rightarrow \beta + \gamma$, where α represents the initial single phase, and β and γ represent the two phases formed after the reaction. In the iron-carbon phase diagram, the eutectoid reaction occurs during the transformation of austenite (α) into ferrite (β) and cementite (γ) at the eutectoid temperature of about 727 °C [48].

At a particular composition and temperature, two phases merge to generate a new phase in a peritectoid reaction. There is a shift from two crystal structures to one crystal structure during this reaction. The peritectoid temperature is the name given to the transformation temperature. The peritectoid reaction is commonly represented by the formula $\alpha + \beta \rightarrow \gamma$, where α and β stand for the starting phases and γ for the new phase that is generated after the reaction. Example: The peritectoid reaction takes place at the peritectoid temperature of approximately 727 °C during the transition of austenite (α) and cementite (β) into pearlite (γ) in the iron-carbon phase diagram. These reactions are essential to comprehending the phase transitions that take place in alloy systems and are important to metal heat treatment procedures. The mechanical properties of the alloys are significantly impacted by the microstructures that emerge from eutectoid and peritectoid processes. In engineering and metallurgy, the design and optimization of materials for particular purposes depend heavily on the understanding and management of these reactions [49].

14.2.6 Diffusion-Less Phase Transformations

In intermetallic alloys, solid-state transitions known as "diffusion-less phase transformations" take place in the absence of atoms diffusing across vast distances. Rapid atom rearrangements inside the crystal lattice cause these alterations, which alter the material's architecture and characteristics. The martensitic transformation and the displacive transformation are two typical instances of diffusion less phase transformations. Phase transitions can happen in some intermetallic alloys without atoms diffusing. This is typical of martensitic transitions, in which atoms do not need to move in order for the crystal structure to change quickly. Let us discuss each of them. The martensitic transformation is a solid-state phase transformation that occurs rapidly and involves a change in crystal structure without diffusion. It occurs at low temperatures. It involves shear deformation. It leads to the development of martensitic, a fine microstructure with lath or plate-like features. Steel alloys are known to undergo the Martensitic transition, especially those with greater carbon contents. Applications requiring great strength and hardness, such as cutting tools, blades, and specific structural components, call for the usage of martensitic alloys. A type of solid-state transformation known as displacive transformation results in a change in crystal structure without the need for atoms to diffuse over great distances. This covers ferro elastic transitions and twinning processes, among others. In this atomic diffusion is absent from the coordinated deformation of the crystal lattice. It often entails the cooperative movement of atoms changing the crystal's form. Different intermetallic alloys and ceramics are susceptible to displacive changes. Unique mechanical, thermal, or electrical properties may be displayed by the resultant material, depending on the particular type of displacive transformation. Shape memory alloys experience a diffusion-free reversible solid-state phase transition. They display a shape memory effect, in which heating causes deformation at low temperatures to be restored. Some nickel-titanium (Ni–Ti) alloys are common SMAs. Actuators, medical devices (such as guide wires and stents), and other applications where a reversible form change is desired all use the shape memory alloys In materials science and engineering, these diffusion-less phase transitions are important because they make it possible to create alloys with certain microstructures and characteristics. For materials to be optimized for a variety of purposes, from high-strength structural components to electronics with shape memory capabilities, it is essential to comprehend the mechanisms underlying these transformations. In intermetallic alloys, the term "order–disorder transformation" describes a phase transition wherein the atomic arrangement inside the crystal structure shifts from a disordered to an ordered state, or vice versa. The atoms' distribution inside the crystal lattice is altered during this transformation, which results in noticeable modifications to the material's characteristics. For intermetallic alloys to be engineered to satisfy particular criteria, it is essential to comprehend these phase transitions. To achieve the desired mechanical, thermal, and chemical properties of intermetallic alloys, phase optimisation and control are frequently achieved using heat treatment procedures and alloy composition modifications. The degree of atomic ordering inside the crystal lattice varies during the order–disorder transition. Atoms are distributed more randomly in the disordered state than they are in the ordered state, where they are structured in a predictable and regular way. Usually, the order–disorder transition takes place at high temperatures. Higher temperatures cause disorder because thermal energy upsets the atoms' orderly arrangement. On the other hand, the atoms have a tendency to arrange themselves in an orderly fashion at lower temperatures. Temperature, time, and the activation energy of atomic mobility inside the crystal lattice are among the variables that affect the kinetics of the order–disorder transformation. Defects in the crystal structure like vacancies or interstitials may have an impact on the transformation. There are several intermetallic alloy systems where the order–disorder transition is seen. Typical instances comprise certain stages in alloys based on aluminium, superalloys based on nickel, and aluminides of titanium. In intermetallic alloys, the order–disorder transformation can significantly affect the mechanical, thermal, and electrical properties. While the disordered state may improve toughness and ductility, the ordered state frequently yields improved strength and hardness. It is essential to comprehend and manage the order–disorder transition in order to customize the characteristics of intermetallic alloys for particular uses. Industries like electronics, automobiles, and aerospace use this knowledge. A range of experimental methods are utilized to investigate and evaluate the level of atomic ordering in intermetallic alloys that are undergoing order–disorder transitions, such as transmission electron microscopy and X-ray diffraction. Annealing and quenching are two examples of heat treatment procedures that can have an impact on the order–disorder transformation. Elevated temperatures during quenching can result in a disorganized state, but lowering the temperature during annealing can encourage atomic arrangement.

In summary, temperature and time have a dynamic effect on the order–disorder transformation in intermetallic alloys. It is essential for figuring out how the atoms are arranged in the crystal lattice, which in turn affects the material's characteristics. This process is studied and manipulated by engineers and researchers to create alloys with desired properties for particular uses [50–57].

14.3 Microstructure in Intermetallic Alloys

The distribution and organization of phases, grains, and flaws at the microscopic level are referred to as the microstructure of intermetallic alloys (Fig. 14.7). It is essential to comprehend and manage the microstructure in order to customize the mechanical, thermal, and electrical characteristics of intermetallic alloys for particular uses.

Intermetallic alloys may have a grain structure, just like other metallic materials. The mechanical properties of an alloy are largely influenced by the size, shape, and distribution of its grains. Grain boundary strengthening mechanisms are responsible for the enhanced strength and hardness of fine-grained intermetallic alloys as compared to their coarse-grained counterparts. The microstructure of intermetallic alloys can contain a variety of phases. Different phases can have an impact on characteristics including strength, ductility, and resistance to corrosion. Comprehending

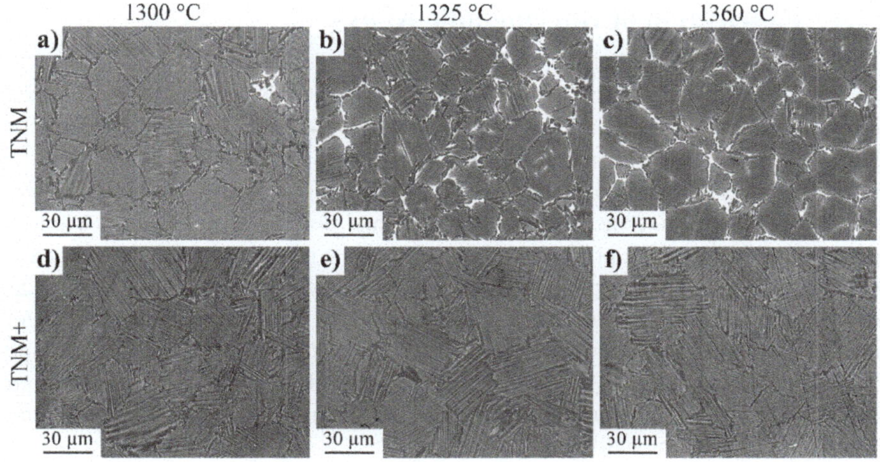

Fig. 14.7 Microstructure in intermetallic alloys [58]

the distribution and composition of phases is essential for customizing the alloy's characteristics for particular uses. Grain boundaries, vacancies, and other defects have a big impact on how an intermetallic alloy behaves mechanically. The strength, ductility, and deformation behaviour of the alloy can all be impacted by the interplay between dislocations and other microstructural characteristics. At high temperatures, some intermetallic alloys may experience disordering while others show ordered atomic configurations. Properties including electrical conductivity, diffusion behaviour, and thermal stability can all be impacted by the degree of ordering or disordering. During processing or heat treatment, intermetallic alloys may develop secondary phase precipitates. The mechanical qualities of these precipitates, such as strength, hardness, and creep resistance, can be significantly impacted. To maximize alloy performance, control over precipitate generation and dispersion is necessary. Diffusion processes or segregation during solidification might result in composition variations at the microscale. The mechanical characteristics, phase stability, and corrosion resistance of an alloy can be affected by composition gradients. Especially under stress or at high temperatures, the interfaces separating various phases, grains, or precipitates are critical in influencing the mechanical behaviour of the alloy. Properties like fracture toughness and fatigue resistance are influenced by the energy and structural makeup of these interactions. For intermetallic alloys to be tailored to specific performance needs in a variety of applications, such as the electronics, automotive, and aerospace sectors, their microstructure must be understood and controlled. High-precision characterisation methods including electron microscopy, X-ray diffraction, and atom probe tomography are frequently utilized to efficiently examine and modify the microstructure of these alloys [58–61].

14.4 Fabrication of Intermetallic Alloys

Intermetallic alloys are fabricated using a variety of techniques that are customized to the precise qualities required. An outline of the typical fabrication techniques for intermetallic alloys is shown in Fig. 14.8.

Casting is a commonly used process for creating huge or intricately shaped intermetallic alloys. Melting the component metals and pouring the molten alloy into a mould to solidify it is the process. For the appropriate microstructure and characteristics, mould design and cooling rates must be properly controlled. Another popular technique for creating intermetallic alloys is powder metallurgy. The constituent metals' elemental powders are combined, and crushed into the required shape (for example, via pressing), and then the compacted powder is sintered at high temperatures to fuse the particles together. Precise control over microstructure and composition is possible with this approach.

To create intermetallic alloys, mechanical alloying is a solid-state powder processing method. The process entails the ball milling of elemental powders at high energy, resulting in mechanical deformation, cold welding, and particle fracturing. This process produces alloyed powders that have a uniform composition.

Directly from computer-aided design (CAD) models, intermetallic alloys can be fabricated layer by layer using additive manufacturing processes such as selective laser melting (SLM) and electron beam melting (EBM). This technique allows for fine control over composition and microstructure, as well as the production of complicated geometries with design flexibility. Using methods like flame spraying or plasma spraying, thermal spray coating entails depositing molten or semi-molten intermetallic alloy particles onto a substrate surface. This technique offers improved resistance to oxidation, corrosion, and wear in surface modification or repair applications. These processes entail selectively melting and solidifying intermetallic alloys in order to control their composition and cleanse them. These techniques are very

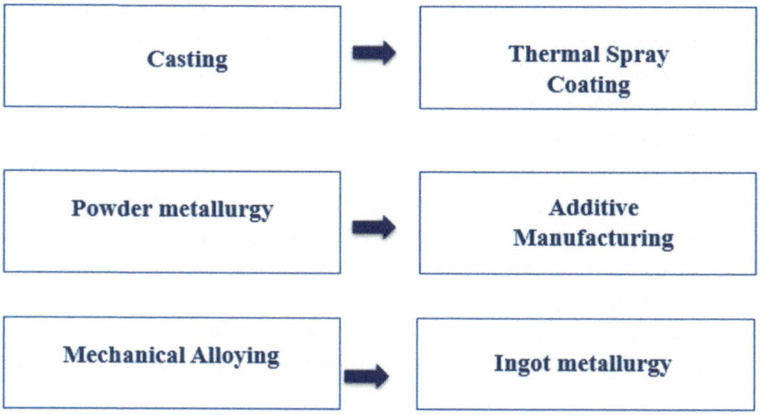

Fig. 14.8 Fabrication of intermetallic alloys

helpful for producing intermetallic alloys with high purity and exact control over impurity levels. Using a crucible to melt the component metals and a controlled cooling process to consolidate them into ingots is known as ingot metallurgy. After that, these ingots can be further treated to create intermetallic alloy components with certain mechanical qualities by using methods like hot forging, rolling, or extrusion. Every manufacturing technique has benefits and drawbacks based on the alloy's complexity, the required qualities, and cost concerns. The particular requirements of the application usually have an impact on the fabrication technique selection [62–92].

14.5 Industrial Applications of Intermetallic Alloys

Intermetallic alloys are used in many different industrial sectors because of their special set of qualities, which include high strength, resistance to oxidation and corrosion, and superior performance at high temperatures. Typical industrial uses for intermetallic alloys as shown in Fig. 14.9.

Because of their high thermal strength and lightweight nature, intermetallic alloys are used in aircraft applications. They are utilized in parts including gas turbine and aviation engine structural elements, exhaust systems, and turbine blades. To increase vehicle performance, lower emissions, and improve fuel efficiency, intermetallic alloys are being employed more and more in the automotive sector. They are used in parts such as lightweight structural parts, exhaust systems, turbocharger parts, and engine valves. Because of their high electrical conductivity, thermal stability, and corrosion resistance, intermetallic alloys are used in electronics and electrical devices. They are utilized in semiconductor devices and electrical circuits as soldering materials, connectors, and contacts. To handle corrosive fluids and high-temperature

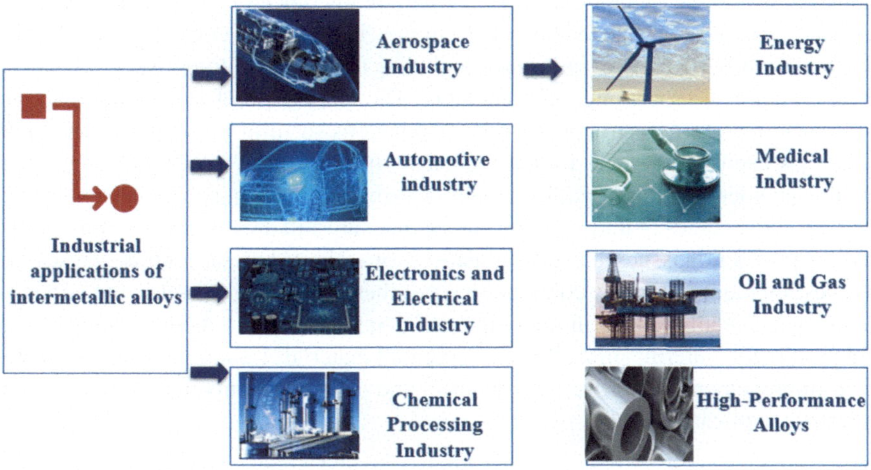

Fig. 14.9 Industrial applications of intermetallic alloys

conditions, chemical processing equipment uses intermetallic alloys with exceptional corrosion resistance. Chemical factories and petrochemical refineries employ them in heat exchangers, reactors, pumps, and valves.

Power generation and distribution are two areas in which intermetallic alloys are extremely important. They are utilized in parts including electrical transmission lines, high-temperature furnace elements, steam and gas turbine components, and turbine blades. Because of its mechanical qualities, biocompatibility, and resistance to corrosion, intermetallic alloys are used in implants and medical equipment. They find application in surgical instruments, dental implants, orthopaedic implants, and medical diagnostic devices. Some intermetallic alloys can catalyse reactions, which makes them helpful in chemical synthesis operations, environmental cleanup procedures, and catalytic converters for car exhaust systems. In the offshore and onshore oil and gas exploration and production processes, intermetallic alloys are used for drilling equipment, valves, pumps, and other components that are subjected to high temperatures and pressures. Advanced military gear, rocket engines, and aircraft propulsion systems all use intermetallic alloys with certain combinations of qualities, such as high strength, temperature resistance, and corrosion resistance [13, 14, 16, 17, 93–97].

These are only a few examples of the numerous industrial applications of intermetallic alloys. Ongoing research and development efforts continue to expand the range of applications for these materials across several sectors.

14.6 Conclusion

In conclusion, intermetallic alloys are a class of materials that have a variety of interesting and potentially useful features that make them useful for a large number of industrial uses. Due to their exceptional strength, tolerance to temperature changes, resistance to corrosion, and tendency to be lightweight, they are considered indispensable materials in a variety of industries, including aerospace, automotive, electronics, energy, chemical processing, and medicine. The fabrication of intermetallic alloys with customized microstructures and compositions to fulfil particular application needs has been made easier by advancements in fabrication techniques, including casting, powder metallurgy, additive manufacturing, and refining procedures. Despite the many advantages of intermetallic alloys, there are still drawbacks, including brittleness, production issues, and cost considerations. Nevertheless, continued research and development initiatives continue to solve these issues and increase the application of intermetallic alloys in many industrial applications. In modern engineering and production, intermetallic alloys are essential materials because they enable technical improvements, improve performance, and solve problems in a wide range of industrial applications.

References

1. Mordike, B.L., Ebert, T.: Magnesium: properties—Applications—Potential. Mater. Sci. Eng. A **302**(1), 37–45 (2001)
2. Schleder, G.R., Padilha, A.C., Acosta, C.M., Costa, M., Fazzio, A.: From DFT to machine learning: recent approaches to materials science—A review. J. Phys.: Mater. **2**(3), 032001 (2019)
3. Basu, B., Kalin, M.: Tribology of Ceramics and Composites: a Materials Science Perspective. Wiley (2011)
4. Sharath, B.N., Venkatesh, C.V., Afzal, A., Aslfattahi, N., Aabid, A., Baig, M., Saleh, B.: Multi ceramic particles inclusion in the aluminium matrix and wear characterization through experimental and response surface-artificial neural networks. Materials **14**(11), 2895 (2021)
5. Sharath, B.N., Venkatesh, C.V.: Study on Effect of boron carbide, aluminium oxide and graphite on dry sliding wear behaviour of aluminium based metal matrix composite at different temperature. Tribol.-Finnish J. Tribol. **38**(1–2), 35–46 (2021)
6. Madhu, K.S., Venkatesh, C.V., Sharath, B.N., Karthik, S.: Effect of boron carbide on wear resistance of graphite containing Al7029 based hybrid composites and its dry sliding wear characterization through experimental, response surface method and ANOVA. Tribol.-Finnish J. Tribol. **38**(3–4), 48–60 (2021)
7. Ferro, R., Saccone, A.: Intermetallic Chemistry. Elsevier (2011)
8. Zhou, M., Li, C., Fang, J.: Noble-metal based random alloy and intermetallic nanocrystals: syntheses and applications. Chem. Rev. **121**(2), 736–795 (2020)
9. Cinca, N., Lima, C.R., Guilemany, J.M.: An overview of intermetallics research and application: status of thermal spray coatings. J. Market. Res. **2**(1), 75–86 (2013)
10. Zamanzade, M., Barnoush, A., Motz, C.: A review on the properties of iron aluminide intermetallics. Crystals **6**(1), 10 (2016)
11. Bauer, J.C., Chen, X., Liu, Q., Phan, T.H., Schaak, R.E.: Converting nanocrystalline metals into alloys and intermetallic compounds for applications in catalysis. J. Mater. Chem. **18**(3), 275–282 (2008)
12. Mousavi, T., Karimzadeh, F., Abbasi, M.H.: Synthesis and characterization of nanocrystalline NiTi intermetallic by mechanical alloying. Mater. Sci. Eng. A **487**(1–2), 46–51 (2008)
13. Jozwik, P., Polkowski, W., Bojar, Z.: Applications of Ni3Al based intermetallic alloys—Current stage and potential perceptivities. Materials **8**(5), 2537–2568 (2015)
14. Sauthoff, G.: Multiphase intermetallic alloys for structural applications. Intermetallics **8**(9–11), 1101–1109 (2000)
15. Bochenek, K., Basista, M.: Advances in processing of NiAl intermetallic alloys and composites for high temperature aerospace applications. Prog. Aerosp. Sci. **79**, 136–146 (2015)
16. Stoloff, N.S., Liu, C.T., Deevi, S.C.: Emerging applications of intermetallics. Intermetallics **8**(9–11), 1313–1320 (2000)
17. Judkins, R.R., Rao, U.S.: Fossil energy applications of intermetallic alloys. Intermetallics **8**(9–11), 1347–1354 (2000)
18. Musi, M., Graf, G., Clemens, H., Spoerk-Erdely, P.: Alloying elements in intermetallic γ-TiAl based alloys—A review on their influence on phase equilibria and phase transformations. Adv. Eng. Mater. 2300610 (2023)
19. Semboshi, S., Sasaki, R., Kaneno, Y., Takasugi, T.: Age-induced precipitation and hardening behavior of Ni3Al intermetallic alloys containing vanadium. Metals **9**(2), 160 (2019)
20. Que, Z., Mendis, C.L.: Heterogeneous nucleation and phase transformation of Fe-rich intermetallic compounds in Al–Mg–Si alloys. J. Alloy. Compd. **836**, 155515 (2020)
21. Jabłońska, M., Jasik, A., Hanc-Kuczkowska, A.: Structures and phases transitions of the alloys on the bases of Fe–Al intermetallic phases
22. Pope, D.P., Ezz, S.S.: Mechanical properties of Ni3AI and nickel-base alloys with high volume fraction of γ'. Int. Metals Rev. **29**(1), 136–167 (1984)
23. Tung, C.C., Yeh, J.W., Shun, T.T., Chen, S.K., Huang, Y.S., Chen, H.C.: On the elemental effect of AlCoCrCuFeNi high-entropy alloy system. Mater. Lett. **61**(1), 1–5 (2007)

24. Tasan, C.C., Deng, Y., Pradeep, K.G., Yao, M.J., Springer, H., Raabe, D.: Composition dependence of phase stability, deformation mechanisms, and mechanical properties of the CoCrFeMnNi high-entropy alloy system. Jom **66**, 1993–2001 (2014)
25. Kong, H.J., Liu, C.T.: A review on nano-scale precipitation in steels. Technologies **6**(1), 36 (2018)
26. Chikkegouda, S.P., Gurudath, B., Sharath, B.N., Karthik, S., Mahale, R.S.: Mechanical and tribological characteristics of aluminium 2618 matrix composite reinforced with boron carbide. Biointerface Res. Appl. Chem. **12**(4) (2022)
27. Madhu, K.S., Sharath, B.N., Venkatesh, C.V., Pradeep, D.G.: Evaluation of mechanical properties of ceramic reinforced aluminium-7029 hybrid composite. In: IOP Conference Series: materials Science and Engineering 2021, vol. 1189, no. 1, p. 012019. IOP Publishing
28. Madhu, K.S., Venkatesh, C.V., Sharath, B.N., Karthik, S.: Characterization and evaluation of mechanical properties of Al–Zn based hybrid metal matrix composites. Appl. Sci. Eng. Progress **16**(1):5804 (2022)
29. Cinkilic, E., Yan, X., Luo, A.A.: Modeling Precipitation hardening and yield strength in cast Al–Si–Mg–Mn alloys. Metals **10**(10), 1356 (2020)
30. Ryen, Ø., Holmedal, B., Nijs, O., Nes, E., Sjölander, E., Ekström, H.E.: Strengthening mechanisms in solid solution aluminum alloys. Metall. Mater. Trans. A **37**, 1999–2006 (2006)
31. Wagner, R., Kampmann, R., Voorhees, P.W.: Homogeneous second-phase precipitation. Phase Transform. Mater. 309–407 (2001)
32. Myhr, O.R., Grong, Ø., Andersen, S.J.: Modelling of the age hardening behaviour of Al–Mg–Si alloys. Acta Mater. **49**(1), 65–75 (2001)
33. Sjölander, E., Seifeddine, S., Svensson, I.L.: Modelling yield strength of heat treated Al–Si–Mg casting alloys. Int. J. Cast Met. Res. **24**(6), 338–346 (2011)
34. Rao, R.R., Sharath, B.N., Madhu, K.S., Pradeep, S.: Comparison of wet and dry milling. In: Advancements in Powder Metallurgy: processing, Applications, and Properties, pp. 1–32. IGI Global (2024)
35. Sharath, B.N., Rao, R.R., Madhu, K.S., Pradeep, S.: Sintering of mechanically alloyed powders. In: Advancements in Powder Metallurgy: processing, Applications, and Properties, pp. 188–222. IGI Global (2024)
36. Jiang, H., Xi, R., Li, X., Kustov, S., Van Humbeeck, J., Wang, X.: Structure, martensitic transformation, and damping properties of functionally graded NiTi shape memory alloys fabricated by laser powder bed fusion. Materials **15**(14), 5073 (2022)
37. Ameur, R., Chemingui, M., Bachaga, T., Escoda, L., Khitouni, M., Suñol, J.J.: Martensitic transformation and crystalline structure of Ni50Mn50−xSnx melt-spun Heusler alloys. Crystals **10**(10), 853 (2020)
38. Auguet, C., Isalgue, A., Lovey, F., Pelegrina, J., Ruiz, S., Torra, V.: Metastable effects on martensitic transformation in SMA: part III. Tentative temperature effects in a NiTi alloy. J. Thermal Anal Calorim. **89**(2), 537–542 (2007)
39. Krenke, T., Acet, M., Wassermann, E.F., Moya, X., Mañosa, L., Planes, A.: Martensitic transitions and the nature of ferromagnetism in the austenitic and martensitic states of Ni−Mn−Sn alloys. Phys. Rev. B **72**(1), 014412 (2005)
40. Srinivas, A., Pavan, D., Venkatesha, B.K., Rao, R.R., Mohith, L.: Study on mechanical properties of AZ91 magnesium alloy. Mater. Today: Proc. **54**, 291–294 (2022)
41. Pradeep, N.B., Venkatesha, B.K., Parameshwara, S., Rao, R.R., Raviprakash, M.: Synthesis of nanostructured ternary Ti based alloy for bio-medical applications. Mater. Today: Proc. **54**, 372–377 (2022)
42. Shivashankara, B.S., Gopi, K.R., Pradeep, S., Rao, R.R.: Investigation of mechanical properties of ECAP processed AL7068 aluminium alloy. In: IOP Conference Series: materials Science and Engineering, vol. 1189, no. 1, p. 012027. IOP Publishing
43. Mukesh, Y.B., Naik, P.K., Vishwanatha, N.R., Prema, N.S., Girish, H.N., Laxmana, N.L., Madhusudan, P.: Impact of heat treatment on the mechanical performance of hot extruded Al6061-BN reinforced metal matrix composites. Bull. Polish Acad. Sci. Tech. Sci. **69**(3) (2021)

44. Sridhara, R., Bharadwaj, A.S., Tile, V.G., Rao, R.R.: Measurement of thermochemical properties of some metal hydrides-titanium (Ti), misch metal (Mm) and lanthanum (La) based alloys a-review. Indian J. Sci. Res. 126–133 (2015)
45. Dörries, K., Mukherji, D., Rösler, J., Esleben, K., Gorr, B., Christ, H.J.: Influence of σ phase on the allotropic transformation of the matrix in Co–Re–Cr-based alloys with Ni addition. Metals **8**(9), 706 (2018)
46. Mukherji, D., Strunz, P., Gilles, R., Hofmann, M., Schmitz, F., Rösler, J.: Investigation of phase transformations by in-situ neutron diffraction in a Co–Re-based high temperature alloy. Mater. Lett. **64**(23), 2608–2611 (2010)
47. Beran, P., Mukherji, D., Strunz, P., Gilles, R., Hölzel, M., Rösler, J.: Coexistence of two cubic-lattice Co matrices at high temperatures in Co–Re–Cr–Ni alloy studied by neutron diffraction. Adv. Mater. Sci. Eng. (2018)
48. Grosse, M. (2012). High-temperature oxidation in nuclear reactor systems. In: Nuclear Corrosion Science and Engineering, pp. 265–300. Woodhead Publishing
49. Kiatgamolchai, S., Sakulkalavek, A.: Distribution of elements in a Cu–Added FeSi2 Alloy under peritectoid and eutectoid reactions. J. Electron. Mater. **40**, 1029–1034 (2011)
50. Nagase, T., Yamauchi, I., Ohnaka, I.: Eutectoid decomposition in rapidly solidified α-Fe2Si5-based thermoelectric alloys. J. Alloy. Compd. **316**(1–2), 212–219 (2001)
51. Adorno, A.T., Carvalho, T.M., Magdalena, A.G., Dos Santos, C.M., Silva, R.A.: Activation energy for the reverse eutectoid reaction in hypo-eutectoid Cu–Al alloys. Thermochim. Acta **531**, 35–41 (2012)
52. Sathyanarayana, K., Puttegowda, M., Rangappa, S.M., Siengchin, S., Shivanna, P., Nagaraju, S.B., Somashekara, M.K., Girijashankar, P.B., Girijappa, Y.G.: Metallic lightweight materials: properties and their applications. In: Lightweight and Sustainable Composite Materials, pp. 47–67. Woodhead Publishing (2023)
53. Pradeep, D.G., Nithin, H.S., Sharath, B.N., Madhu, K.S.: Evaluation of dry sliding wear behavior of thermally sprayed and microwave post-processed TiO2 reinforced tungsten carbide composite coating. Welding World 1–3 (2023)
54. Sharath, B.N., Madhu, K.S., Pradeep, D.G., Madhu, P., Premkumar, B.G., Karthik, S.: High temperature tensile behaviour of ceramic-hybridized metal matrix composites for above-room-temperature applications. Silicon 1–2 (2023)
55. Karthik, S., Sharath, B.N., Madhu, P., Madhu, K.S., Prem Kumar, B.G., Verma, A.: Experimental and artificial neural network-based slurry erosion behavior evaluation of cast iron. Int. J. Interact. Des. Manuf. (IJIDeM) 1–1 (2023)
56. Jones, D.R., Ashby, M.F.: Engineering Materials 2: an Introduction to Microstructures and Processing. Butterworth-Heinemann (2012)
57. Dias, H.W., Medeiros, A.B., Binder, C., Rodrigue Neto, J.B., Klein, A.N., de Mello, J.D.: Tribological evaluation of turbostratic 2d graphite as oil additive. Lubricants **9**(4), 43 (2021)
58. Musi, M., Deshayes, C., Molénat, G., Toualbi, L., Galy, B., Spoerk-Erdely, P., Hantcherli, M., Monchoux, J.P., Thomas, M., Clemens, H., Couret, A.: Microstructure, plasticity and ductility of a TNM+ alloy densified by Spark Plasma Sintering. Metals **12**(11), 1915 (2022)
59. Sharath, B.N., Karthik, S., Madhu, P., Madhu, K.S., Pradeep, D.G.: Predictive analysis of slurry erosion behaviour in aluminium-based hybrid metal matrix composites: experimental and machine learning approach. J. Bio-and Tribol.-Corros. **9**(4), 70 (2023)
60. Nagaraju, S.B., Somashekara, M.K., Govindaswamy, P.D., Puttegowda, M., Shankar, P.B., Sathyanarayana, K.: Wear behaviour of hybrid (boron carbide-graphite) aluminium matrix composites under high temperature. J. Eng. Appl. Sci. **70**(1), 124 (2023)
61. Shivashankara, B.S., Gopi, K.R., Pradeep, S., Venkatesha, B.K.: Effect of processing route on grain refinement and mechanical properties in equal channel angular pressing of aluminium alloy AA7068. J Mines Metals Fuels **1**, 1–7 (2021)
62. Dzogbewu, T.C., du Preez, W.B.: Additive manufacturing of Ti-based intermetallic alloys: a review and conceptualization of a next-generation machine. Materials **14**(15), 4317 (2021)
63. Wimler, D., Lindemann, J., Reith, M., Kirchner, A., Allen, M., Vargas, W.G., Franke, M., Klöden, B., Weißgärber, T., Güther, V., Schloffer, M.: Designing advanced intermetallic titanium aluminide alloys for additive manufacturing. Intermetallics **131**, 107109 (2021)

64. Dilip, J.J., Miyanaji, H., Lassell, A., Starr, T.L., Stucker, B.: A novel method to fabricate TiAl intermetallic alloy 3D parts using additive manufacturing. Defence technology **13**(2), 72–76 (2017)
65. Katou, M., Oh, J., Miyamoto, Y., Matsuura, K., Kudoh, M.: Freeform fabrication of titanium metal and intermetallic alloys by three-dimensional micro welding. Mater. Des. **28**(7), 2093–2098 (2007)
66. Karczewski, K., Stępniowski, W.J., Chojnacki, M., Jóźwiak, S.: Crystalline oxalic acid aided FeAl intermetallic alloy sintering. Fabrication of intermetallic foam with porosity above 45%. Mater. Lett. **164**, 32–34
67. Pradeep, D.G., Nithin, H.S., Sharath, B.N., Madhu, K.S., Venkatesh, C.V.: Microstructure and wear behavior of microwave treated WC-10Co-4Cr composite coating on AISI 4140 alloy steel. In: IOP Conference Series: materials Science and Engineering, vol. 1189, no. 1, p. 012012. IOP Publishing (2021)
68. Sujan, G.K., Pan, Z., Li, H., Liang, D., Alam, N.: An overview on TiFe intermetallic for solid-state hydrogen storage: microstructure, hydrogenation and fabrication processes. Crit. Rev. Solid State Mater. Sci. **45**(5), 410–427 (2020)
69. Pradeep, D.G., Sharath, B.N., Madhu, K.S., Karthik, S., Venkatesh, C.V.: Investigating the adhesion strength of electrodeposited Ni–Al2O3 nano composite on Al-2618 substrate by using the scratch test technique. Mater. Today: Proc. **52**, 1702–1706 (2022)
70. Pradeep, D.G., Sharath, B.N., Afzal, A., Baig, M.A., Shanmugasundaram, M.: Study on scratch behavior of Ni–Al2O3 coating composition on Al-2219 substrate by electro deposited technique. Mater. Today: Proc. **46**, 8716–8722 (2021)
71. Matsuura, K., Kudoh, M., Oh, J.H., Kirihara, S., Miyamoto, Y.: Development of freeform fabrication of intermetallic compounds. Scripta Mater. **44**(3), 539–544 (2001)
72. Wang, L., Zhang, Y., Hua, X., Shen, C., Li, F., Huang, Y., Ding, Y.: Fabrication of γ-TiAl intermetallic alloy using the twin-wire plasma arc additive manufacturing process: microstructure evolution and mechanical properties. Mater. Sci. Eng. A **812**, 141056 (2021)
73. Shashanka, R., Chaira, D.: Optimization of milling parameters for the synthesis of nano-structured duplex and ferritic stainless steel powders by high energy planetary milling. Powder Technol. **278**, 35–45 (2015)
74. Chaira, D.: Development of nano-structured duplex and ferritic stainless steels by pulverisette planetary milling followed by pressureless sintering. Mater. Charact. **99**, 220–229 (2015)
75. Chaira, D.: Phase transformation and microstructure study of nano-structured austenitic and ferritic stainless steel powders prepared by planetary milling. Powder Technol. **259**, 125–136 (2014)
76. Rajendrachari, S., Kamacı, Y., Taş, R., Ceylan, Y., Bülbül, A.S., Uzun, O., Karaoğlanlı, A.C.: Antimicrobial investigation of CuO and ZnO nanoparticles prepared by a rapid combustion method. Phys. Chem. Res. **7**(4), 799–812 (2019)
77. Rajendrachari, S.: Effect of sintering temperature on the pitting corrosion of ball milled duplex stainless steel by using linear sweep voltammetry. Anal. Bioanal. Electrochem. **10**(3), 349–361 (2018)
78. Nayak, A.K., Shashanka, R., Chaira, D.: Effect of nanosize yittria and tungsten addition to duplex stainless steel during high energy planetary milling. In: IOP Conference Series: materials Science and Engineering, vol. 115, no. 1, p. 012008. IOP Publishing (2016)
79. Gupta, S., Shashanka, R., Chaira, D.: Synthesis of nano-structured duplex and ferritic stainless steel powders by planetary milling: An experimental and simulation study. In: IOP Conference Series: materials Science and Engineering, vol. 75, no. 1, p. 012033. IOP Publishing (2015)
80. Pavitra, V., Praveen, B.M., Nagaraju, G., Shashanka, R.: Energy storage, Photocatalytic and electrochemical nitrite sensing of ultrasound-assisted stable Ta2O5 nanoparticles. Top. Catal. 1–4 (2022)
81. Shashanka, R.: Synthesis of nano-structured stainless steel powder by mechanical alloying-an overview. Int. J. Sci. Eng. Res. **8**, 588–594 (2017)
82. Shashanka, R., Chaira, D.: Effects of nano-Y2O3 and sintering parameters on the fabrication of PM duplex and ferritic stainless steels. Acta Metall. Sinica (English Lett.) **29**, 58–71 (2016)

83. Shashanka, R., Chaira, D.: Effect of sintering temperature and atmosphere on nonlubricated sliding wear of nano-yttria-dispersed and yttria-free duplex and ferritic stainless steel fabricated by powder metallurgy. Tribol. Trans. **60**(2), 324–336 (2017)
84. Rajendrachari, S.: An overview of high-entropy alloys prepared by mechanical alloying followed by the characterization of their microstructure and various properties. Alloys **1**(2), 116–132 (2022)
85. Shashanka, R., Chaira, D., Chakravarty, D.: Fabrication of nano-yttria dispersed duplex and ferritic stainless steels by planetary milling followed by spark plasma sintering and non-lubricated sliding wear behaviour study. J. Mater. Sci. Eng. B **6**, 111–125 (2016)
86. Shashanka, R., Uzun, O., Chaira, D.: Synthesis of nano-structured duplex and ferritic stainless steel powders by dry milling and its comparison with wet milling. Archiv. Metall. Mater. **65**(1)
87. Pradeep, N.B., Hegde, M.R., Rajendrachari, S., Surendranathan, A.O.: Investigation of microstructure and mechanical properties of microwave consolidated TiMgSr alloy prepared by high energy ball milling. Powder Technol. **408**, 117715 (2022)
88. Shashanka, R., Chaira, D.: Ball Milled Nano-structured Stainless Steel Powders: fabrication and Characterization. Educreation Publishing (2017)
89. Mahale, R.S, Shashanka, R.: Mechanical testing of spark plasma sintered materials: a review. In: AIP Conference Proceedings, vol. 2469, no. 1. AIP Publishing (2022)
90. Adimule, V., Keri, R., Rajendrachari, S., Kendrekar, P., Kulkarni, C.V.: Highly photoluminescence and wide band gap insulating metal hybrid nanoparticles array of samarium-doped SrO: CoO: synthesis, characterizations and sensor characteristics. J. Mater. Sci.: Mater. Electron. **34**(5), 442 (2023)
91. Rajendrachari, S., Adimule, V.M., Jayaprakash, G.K., Pandith, A.: Electrochemical oxidation of methylene blue dye in wastewater using mechanically alloyed high entropy alloy modified carbon paste electrode using cyclic voltammetry. Mater. Res. Express **10**(5), 054003 (2023)
92. Rajendrachari, S., Basavegowda, N., Vinaykumar, R., Narsimhachary, D., Somu, P., Lee, M.J.: Electrocatalytic determination of methyl orange dye using mechanically alloyed novel metallic glass modified carbon paste electrode by cyclic voltammetry. Inorg. Chem. Commun. **30**, 111010 (2023)
93. Sampath, S., Ravi, V.P., Sundararajan, S.: An overview on synthesis, processing and applications of nickel aluminides: from fundamentals to current prospects. Crystals **13**(3), 435 (2023)
94. Sharath, B.N., Madhu, K.S., Pradeep, D.G., Madhu, P., Premkumar, B.G., Karthik, S.: Effects of tertiary ceramic additives on the micro hardness and wear characteristics of Al2618+Si3N4–B4C–Gr hybrid composites for automotive applications. J. Alloys Metall. Syst. 100014 (2023)
95. Sikka, V.K., Deevi, S.C., Viswanathan, S., Swindeman, R.W., Santella, M.L.: Advances in processing of Ni3Al-based intermetallics and applications. Intermetallics **8**(9–11), 1329–1337 (2000)
96. Varin, R.A.: Structural and functional intermetallics-an overview. Inżynieria Materiałowa **22**(1), 11–18 (2001)
97. David, S.A., Deevi, S.C.: Welding of unique and advanced ductile intermetallic alloys for high-temperature applications. Sci. Technol. Weld. Join. **8**, 681–705 (2017)

Chapter 15
Mechanical Alloying of Super Alloys Like High-Entropy Alloys and Smart Materials

Emre Altas

Abstract The acceleration of technological developments in today's industrial fields causes researchers working on engineering materials to engage in different searches. It is desirable to obtain features such as improving traditional materials, working at higher temperatures to obtain materials with higher wear resistance, and ensuring that the obtained materials have higher corrosion resistance. Today, where technological developments continue rapidly, the need for superior materials is increasing day by day. Therefore, in order to meet the need for materials with superior properties, nano-sized material synthesis methods have been developed. Mechanical alloying, one of these methods, stands out as one of the most preferred synthesis methods among production methods, thanks to its superior properties. High-entropy alloys and smart materials have high strength, hardness, high temperature resistance, toughness, structure stability, wear and high. It is known to have material properties needed by today's industry, such as corrosion resistance. Due to these superior features, it has a wide range of uses. In this study, mechanical alloying of superalloys such as high-entropy alloys and smart materials, which is a production technique of powder metallurgy, is discussed.

Keywords High-entropy alloys · Smart materials · Mechanical alloying

List of Abbreviations

XRD X-Ray Diffractometer
DTA Differential Thermal Analysis
DSC Differential Scanning Calorimetry
SEM Scanning Electron Microscopy
TEM Transmission Electron Microscopy

E. Altas (✉)
Department of Mechanical Engineering, Faculty of Engineering, Architecture and Design, Bartın University, Bartın, Turkey
e-mail: emrealtas@bartin.edu.tr

S. Rajendrachari (ed.), *Mechanically Alloyed Novel Materials*, Advanced Structured Materials 220, https://doi.org/10.1007/978-981-97-6504-1_15

CNC Computer Numerical Control
EDM Electrical Discharge Machining

15.1 High-Entropy Alloys

The acceleration of technological developments in today's industrial fields causes researchers working on engineering materials to engage in different searches. It is desirable to obtain features such as improving traditional materials, working at higher temperatures to obtain materials with higher wear resistance, and ensuring that the obtained materials have higher corrosion resistance [1–3]. In addition, it is to ensure that the materials work under dynamic load and mechanical load and to raise the working load to higher levels when necessary. One of the methods of giving these properties to materials is to use pure materials to make alloys with superior properties compared to their components [4]. By making an alloy, many useful properties can be achieved, such as obtaining a material with the desired qualities, changing the physical properties of the material, changing the mechanical properties of the material, making the material suitable for heat treatments and reducing the cost of the material. Conventional alloys contain a main solvent element with some additional additives. These additional additives function to improve the working properties of the main solvent element, but some additional additives may cause the brittleness of the main solvent element and the emergence of some undesirable functions during the processing of the alloy [5, 6]. For this reason, multi-element (component) otherwise known as high-entropy alloys, have attracted the attention of researchers in recent years in order to obtain properties that cannot be achieved with traditional alloys. These alloys, which were extremely perfect and still maintain their perfection and continue to develop today, were defined as high-entropy alloys, especially by Jien-Wei Yeh and his team in 2004 [7, 8].

High-entropy alloys contain alloys with superior mechanical and chemical properties than other materials used in today's technology because they contain various basic alloying elements. These properties can be classified as high strength, high hardness value, high wear resistance, high temperature resistance, high structural stability, high corrosion resistance and high oxidation resistance [9]. The fact that many of these properties cannot be obtained simultaneously in conventional alloys makes high-entropy alloys interesting. High-entropy alloys attract the attention of scientific researchers due to their simple configurations and superior performance. Yeh et al. [10] in their study conducted, defined high-entropy alloys as multicomponent alloys containing more than one main element in the form of a solid melt with equal or near-equal atomic ratios (between 5 and 35%). In a very small part of the studies conducted on high-entropy alloys, the non-major element is below 5%. However, since the solid-molten state of these alloys has a higher mixing entropy than conventional alloys, they are also in the group of high-entropy alloys [11, 12].

The main feature of high-entropy alloys is that they have high mixing entropy. Due to this feature, solid melt phases are more stable than intermetallics even at high temperatures. Therefore, they can be easily synthesized, analyzed and produced. High-entropy alloys can present solutions as intermetallic components as well as simple solid phases such as face-centered cubic and body-centered cubic [13]. High-entropy alloys emerge as a result of the mixing of main elements as well as the formation of complex compounds and phases. As high-entropy alloys are mixed in their chemical structures according to the atomic weights of the alloying elements, the Gibbs free energy decreases and in some, a solution with the simplest crystal structure is formed. As can be seen from this situation, high-entropy alloys have developed as a result of acquiring a wide variety of properties [14]. They are used in different industrial areas and fields. Although there are areas where more than 300 high-entropy alloys have been developed, these advanced alloys lead to the formation of both other metal materials and intermetallic materials. These studies help find the relationship between mechanical properties, microstructure and phases. There are many reasons why high-entropy alloys are popular in the markets, but a few of them are: It has features such as being able to work at high temperatures, high strength properties, high creep resistance, high hardness and magnetic properties, as well as being resistant to corrosion [15]. In addition, the above-mentioned properties of high-entropy alloys make them both available in the markets and still maintain their importance in the markets. Due to these properties, high-entropy alloys are in a more advantageous position against their competitors. At the same time, these properties also show the advantages of high-entropy alloys [16].

Purpose of High-Entropy Alloys

Generally, alloying elements are added to materials to gain, develop and improve their desired properties. Thus, properties that cannot be achieved with traditional alloys can be achieved more easily and in a time-saving manner with high-entropy alloys. The main superior features achieved are [17]:

- High toughness and strength,
- High wear resistance,
- High resistance to high temperatures,
- Opportunity to work in different environments,
- Low weight,
- Easy formability,
- Magnetism,
- High-entropy alloys are produced in order to obtain superior properties such as less harmful effects on the environment.

15.1.1 Uses of High-Entropy Alloys

Although traditional alloys are indispensable materials of our age, the search for new generation materials continues with the developing technology. Traditional alloys began to have difficulty keeping up with changing technologies and began to fail to meet the desired performance criteria, and this eventually led to the acceleration of research on more specialized materials. With developing technology, it is preferred that the materials used have one or more features such as greater corrosion and wear resistance, more affordable production costs, longer life in high temperature and pressure environments, and lower density. In addition to all these, low-density materials contribute to fuel savings, especially in the automotive and aviation industries, and cause fewer environmental problems [18].

High-entropy alloys can be nominated as new-generation alloy systems that can meet all these superior properties. Thanks to their wide range of design possibilities, high-entropy alloys can contain properties such as low density, high strength, high corrosion resistance, wear resistance and chemical stability. Apart from all these, they can be used as coating materials thanks to their high wear and corrosion resistance. High-entropy alloys can be converted into powder form and coated on part surfaces using methods such as the thermal spray coating technique. High-entropy alloys are also candidates for use in the coating of nuclear fuel tanks and high-pressure tanks, thanks to their resistance to radiation and corrosion resistance. High-entropy alloys can be used as refractory metals in thermal barrier coating applications due to their stability at high temperatures. High-entropy alloys can be easily used in the food industry because they are resistant to high oxidation and corrosion [19, 20].

High-entropy alloys can also exhibit very good paramagnetic, ferromagnetic and soft magnetic properties. Due to these features, they have the potential to be used in the electronics industry. Due to their high hardness resistance and very low diffusion coefficients, it is possible to use high-entropy alloys in high-speed steels and cutting tool steels to increase surface hardness and as diffusion barriers. As a result of the research, some high-entropy alloy systems have been produced and are ready to be used in the market. The $Al_5Cr_{12}Fe_{35}Mn_{28}Ni_{20}$ high-entropy alloy system is the best example of this. The forming ability of this high-entropy alloy system is good. It has hardness values of approximately 360 HV. This easily shaped alloy can be used as a flexible substrate material for some electronic devices, solar cells and displays [21, 22].

15.1.2 The Future Place of High-Entropy Alloys

Recently, many articles, theses and journals have been written on high-entropy alloys. However, we are still at the beginning of understanding high-entropy alloys. Since traditional alloys have been used as references in the studies so far, we are experiencing some problems in understanding high-entropy alloys. Matrices dissolved in high-entropy alloys, the effect factor of mixing entropy at different rates, the interaction between incompatible atom pairs, lattice distortion effect, phase transformations, thermal and electrical conductivity values, diffusion coefficients, dislocation defect energy, grain boundary energy, strength enhancement, brittleness, ductility, some behavioral parameters such as fatigue, creep, toughness, corrosion and oxidation need to be discussed and understood in detail [23, 24].

Apart from high-entropy alloys, another type of alloy that should be considered is medium entropy alloys. Not enough studies have been done on medium entropy alloys. Medium entropy alloys are an alloy system that falls between traditional alloys and high-entropy alloys. They can contain properties specific to both alloy systems. It is anticipated that studies on medium entropy alloys will intensify in the coming years. The issue of ceramic reinforcement using high-entropy alloys as matrix also needs to be investigated. It is envisaged that such composite materials can achieve superior properties that cannot be achieved by traditional composite materials. In addition, this type of composite material that can be used in many special areas can be produced [25].

It is anticipated that high-entropy alloys will be used in many applications in the coming years and will serve as an alternative material to replace traditional alloys. Considering these developments, high and medium-entropy alloys are expected to be used in many areas in the future [26].

15.1.3 Production of High-Entropy Alloys

Production of high-entropy alloys is most commonly carried out by arc melting, a liquid casting method. Arc melting is a method that gives very good results for the production of high-entropy alloys, in which more than five elements are blended in appropriate proportions, melted at high temperatures to achieve good solubility, and then cast. However, it is not always easy or possible to distribute the components homogeneously with this method, and this method also has serious limitations in shaping [27]. In order to ensure a more homogeneous distribution of the particles in the structure, the mechanical alloying solid-state production method, which is one of the fields of powder metallurgy, is being studied as an alternative. With powder metallurgy, compositions very close to the theoretical composition can be produced by reducing the possibility of segregation. In addition, it is being studied as a method that will enable easier control of the microstructure and the easy production of high-entropy alloys with nanocrystalline structures and metal matrix composites [27].

A high-entropy alloy system generally contains a minimum of 5 alloying elements and each alloying element in equal atomic ratios or elements in close atomic ratios. Generally speaking, high-entropy alloys each have an atomic percentage between 5 and 35%. It is also emphasized that the atomic percentage of the element or elements that are not the main elements but whose doping effect is desired to be examined should not be less than 5%. However, although it is seen in a very small part of the literature studies that the percentage of the non-main element is less than 5%, the reason why it is considered a high-entropy alloy is because the mixing entropy is higher than traditional alloys. Production phases of high-entropy alloys: They are classified into three main groups: solid, liquid and gas phases. Methods such as mechanical alloying and derivatives of mechanical alloying under the title of solid phase production can be used in the production of high-entropy alloys. This process has several advantages. Thanks to this process, almost all kinds of materials are produced, including intermetallic compounds with brittle structures and composite ductile metal alloys. As a disadvantage, contamination problems may occur due to air or the device during grinding. However, these contaminations can be reduced or eliminated by choosing the appropriate grinding environment and grinding in inert gases with high purity rates or in a vacuum environment [28, 29].

High-entropy alloys, which have attracted much attention in recent years, can be easily produced by developing mechanical alloying methods. High-entropy alloys are the synthesis of materials with higher properties in terms of many physical and chemical properties, such as temperature, hardness, and melting point, by allowing at least four elements to react at equal molar ratios of each element. It is very important to compare the properties of the materials produced following these existing manufacturing methods. X-ray diffractometer (XRD) analysis, differential thermal analysis (DTA), differential scanning calorimetry (DSC), scanning electron microscopy (SEM), transmission electron microscopy (TEM), high-entropy alloys synthesized by mechanical alloying method at nanoscale these are methods performed with the help of an optical microscope to examine particle sizes and microstructures. Microstructural analysis is performed with the help of an optical microscope. Following microstructural analysis, the synthesized samples are physically examined by subjecting them to hardness and wear tests. In high-entropy alloys, the concentration of each major alloying element is between 5 and 35%. High-entropy alloys may contain less than 5% minor elements in addition to the major elements. These alloys are called high-entropy alloys because their mixing entropy in the molten or solid state is significantly higher than conventional alloys. According to the existing metallurgical knowledge, it is known that alloys containing many (binary or ternary) phase diagrams have a more complex and brittle structure. Characterization of alloys with such brittle microstructures indicates the formation of intermetallic compounds that are difficult and have limited application. Contrary to this expectation, experimental studies have shown that the high mixing entropy of High-entropy alloys promotes the formation of solid melt phases but also produces a simple structure, thereby reducing the number of phases. In order to make the mixture powders formed after mechanical alloying is integrated, they must first be pressed and then subjected to the sintering process. The point that should be taken into consideration during the sintering process is that grain growth may occur by keeping the powders to be sintered at high temperatures for a long time [14, 30].

15.1.4 Mechanical Alloying

With developing technology, the need for different material properties is increasing day by day. The fact that existing manufacturing methods in the industry cannot fully meet the desired material properties has led to an increased interest in different types of materials. One of these production methods is the mechanical alloying method. In this study, the mechanical alloying method developed in the 1960s was applied as the production method. The mechanical alloying method is the technique of processing solid powders with the help of balls. The solid elemental powders prepared in mechanical alloying are placed in a mortar and balls, usually ten times the weight of the alloy powder, are placed and connected to the grinding device and the rotation process is started. As the mechanical energy is transferred to the elemental powders during rotation, the particle size of the elemental powders becomes smaller and blended, and their diffusion distances decrease. As the temperature increases during grinding, the elemental powders disperse homogeneously among each other, causing the powders to alloy. Here, the rotation speed of the container is adjusted to maximize the speed of the balls hitting the powder. Therefore, the impact of the balls on the powder increases the microstructural strains of the powders and an increase in the mechanical properties of the powders is observed. Thus, the powders produced by the mechanical alloying method have an irregular lattice structure and high-entropy alloys are prepared. The prominent parameters in mechanical alloying are The structure of the powders used initially, the type of grinding material, the material of the container in which the powders are placed, the grinding time and the atmosphere used [31, 32].

The mechanical alloying method is a method in which the powders are agglomerated or welded and then broken as a process. As the particle size decreases with the mechanical energy transmitted by the device to the crucible, chemical reactions and structural changes are observed. With the rotational movement in mechanical alloying, the powders move away from their stable state and their bonds are broken. The advantages of this method are: many advantages can be listed, such as the formation of amorphous phases, the production of finely dispersed secondary phases, which are more abundant in oxides, and the obtaining of powders close to nano sizes. This method comes to the fore because it is possible to combine and alloy elements that cannot be achieved with the known classical Hume-Routery rules and to obtain a homogeneous alloy. During the alloying process, while the powders are alloyed, the fracture process continues in a cycle due to the cold welding effect. As a result of the balls hitting the powder in the crucible, the powder particles undergo deformation and change shape. Agglomeration and breakage continue until the particle size decreases and stabilizes. During the process, the temperature increases, resulting from the balls beating the powders in a way that can also involve diffusion and cause the emergence of new phases. In the final stage of the process, a homogeneous structure is obtained when the balance between this welding and breaking event is achieved [11, 33].

In mechanical alloying, where the microstructure can be controlled as a result of constantly repeated welding and fractures, a balance must be achieved between fracture and welding for successful alloying. It is recommended to use a suitable process control agent to prevent cold welding from taking place. In the presence of a process control agent, the interaction of metal surfaces with each other can be reduced, which will reduce welding. Additionally, the mill should be kept cold to accelerate breakage. Thus, a more balanced process will occur. Mechanical alloying is a method that can enable the production of various alloys and compositions that are difficult and impossible to produce with traditional production methods such as casting. Mechanical alloying can be used in the production of various material groups such as amorphous, nitride and hydrate in various structures such as composites and nanocrystals with good performance [34].

15.1.5 High-Entropy Alloys Produced by Mechanical Alloying

The superior materials needed by today's developing industry cannot be met with conventional alloy groups and production techniques, so the use of new material groups and production techniques has become inevitable. High-entropy alloys, which exhibit superior properties compared to ordinary material groups, are among the materials needed by today's industry. The fact that high-entropy alloys contain many properties that are difficult or impossible to find together in other materials in a single alloy makes these materials an important position in the scientific World [35].

High-entropy alloys are among the most interesting material groups today because they have high hardness, toughness, strength, corrosion, fatigue, creep resistance and excellent soft magnetic properties. High-entropy alloys with soft magnetic properties are among the most interesting magnetic materials thanks to their near-perfect magnetic properties. This material group, which has managed to become the center of attention, is produced using production methods such as arc melting, casting and mechanical alloying. Thanks to its superior properties, the mechanical alloying method stands out as one of the most preferred synthesis methods among production methods. This method enables the synthesis of alloys of the desired quality thanks to the controllable ball-to-powder ratio, grinding speed, grinding mill type, grinding time, grinding chamber and ball material, grinding environment and a wide variety of alloy compositions. It is also a widely used method for synthesizing materials such as amorphous, nanocrystalline, intermetallic, high-entropy alloys and composites. By using this method, soft magnetic materials, which are among the most needed materials in today's technology and industry, can be synthesized in the desired quality. The fact that this material group is used in an increasingly wider area means that the need for these materials is increasing. High-entropy alloys synthesized by mechanical alloying method are summarized below with their results [36, 37].

The effects of mill speed, type, material, alloying time and atmosphere on the formation of high-entropy alloys are studied. Especially in the formation of high-entropy alloys, the mechanical alloying time must be determined very well. In the production of high-entropy alloys, studies are carried out for various periods of time, generally lasting 15–40 h. If grinding is not done at sufficient speed and time, the formation of high-entropy alloys with a simple structure will not be possible. In a study conducted with AlCoCuNiZn, at the end of 50 h of grinding at 300 rpm, the formation of high-entropy alloys was not observed according to the XRD results, but when worked at 350 rpm, the expected simple structured solid solution was formed [38]. While higher grinding times provide a more homogeneous alloy formation, they are not preferred because they will also increase the risk of unwanted impurity contamination from the environment to the powders. Therefore, consolidation processes are applied to ensure the formation of the alloy after a relatively short grinding process [16].

CoCrFeNiCuAl alloy was synthesized as high-entropy alloys by Zhang et al. [39] using the mechanical alloying method. It has been observed that grinding the alloy powders for more than 42 h results in the formation of a supersaturated alloy. Grinding the alloy powders for 60 h causes dense agglomerations and the particle size reaches five μm. It was observed that the powders milled for 60 h by mechanical alloying method in the form of high-entropy alloys were exposed to high lattice strains and therefore had excellent chemical homogeneity. Face-centered cubic phases were transformed into body-centered phases by annealing the solid solution alloy at 600 °C for one hour. The remaining phases with the face-centered structure were precipitated after annealing at 1000 °C.

When the literature is examined, it can be said that there has been increasing scientific interest, especially in recent years, on the producibility of high-entropy alloys from starting metal powders by mechanical alloying. However, when the relevant studies are examined in detail, it is seen that the appropriate conditions may vary for each system. It is seen from the literature that high-energy grinders such as planetary, SpexTM and shaker are generally used in the mechanical alloying of high-entropy alloys. Depending on the target composition, studies are carried out with various containers and balls suitable for the chemistry of the starting powders such as WC, hardened steel, zircon, and stainless steel. As process control agents, stearic acid, cyclohexane, methanol, ethanol and heptane are preferred to be used according to the test system. In order to prevent oxidation of powders in a dry environment, mechanical alloying is often preferred to be carried out in an Ar atmosphere. In addition, it is seen that mechanical alloying is used not only in wet or dry environments but also in successive dry and wet alloying [40, 41].

15.2 Smart Materials

While the search for alternative materials with superior performance continues unabated on different platforms, simultaneous developments in material science and technology continue to appear in innovative product forms that make our lives easier. Some materials exhibit different physical reactions under heat and stress due to their properties arising from their atomic structure. These reactions sometimes correspond exactly to the behaviors we want to achieve with complex hydraulic and pneumatic systems. Therefore, there are solutions that do not require complex engineering circuits with a simple mechanism. Materials that offer these solutions are called smart materials. These materials can be metallic, ceramic, polymer, composite or even gel. They can be used in the aviation industry, automotive industry, medical field and many other sectors. One of these materials is shape memory alloys and is an intermetallic compound. These alloys are divided among themselves as copper-based, iron-based and NiTi-based alloys. Among these alloys, the ones that attract the most attention from scientists and are the most researched are NiTi-based alloys [42, 43].

The properties of shape memory alloys are entirely related to their microstructure. In these alloys, the phase transformation that occurs at microscopic dimensions is observed macroscopically as a shape change. The transformations in question take place between the low-temperature phase called martensite and the high-temperature phase called austenite. These transformations are divided into thermoelastic and non-thermoelastic martensitic transformations. Shape memory alloys show thermoelastic martensitic transformation characteristics that occur under the influence of stress and temperature. The transformation that occurs is a diffusion-free transformation, and after the transformation, the atoms preserve their original neighborhoods [44].

Shape memory alloys, called nitinol, consist of a chemical composition of approximately equal elements Ni and Ti. Thanks to their unique shape memory property and good damping effect, excellent biocompatibility, and wear and corrosion resistance, these alloys are widely used in numerous industrial and medical application fields. Shape memory alloys consisting of the NiTi alloy system are generally obtained using the casting method. However, properties such as satisfactory ductility, fracture resistance and superelasticity cannot be achieved in samples produced by the melting method. These properties may be affected by production conditions and environmental conditions. The hydrogen that is likely to form in the structure directly affects the determination of parameters and environmental conditions in the production of parts with the NiTi alloy system. For this reason, new technologies and application areas are needed to eliminate the hydrogen content resulting from environmental conditions. The powder metallurgy method is one of the leading fields with its different methods and methods in the production and shaping of NiTi alloys. Among the known powder metallurgy methods, Application areas such as metal injection molding, spark plasma sintering and hot isostatic pressing are among these methods [45, 46]

15.2.1 Usage Areas of Smart Materials

Materials that exhibit superelastic behavior and shape memory materials are widely preferred in many practical and advanced applications today. These materials are widely used in many industrial and medical applications such as machinery and building materials, and medical devices. It is also used in advanced applications such as electronic devices and spacecraft and in the production of many tools and equipment that make daily life easier, such as superelastic eyeglass frames and telephone antennas. Recently, many applications in the field of robotics have been used extensively. Among these alloys, Nitinol alloy is used in heart stents, a medical device, due to its high biocompatibility and superelastic behavior. 30 °C, show a superelastic behavior above the Af temperature inside the body. It is frequently preferred due to its flexible behavior within the vein and its ease during operation. In addition, materials are frequently used for catheters and guide cables used in various operations. Shape memory alloys broke new ground in the medical industry in the 2000s when they were used as stents to open vascular occlusions in the human body. The areas of use of shape memory alloys in aviation include unmanned aerial vehicles. By using shape memory alloys, compatible wings can be designed for unmanned aerial vehicles, and these wings can change shape according to changing weather conditions and provide aerodynamic optimization. In addition, it has been observed that wings designed with shape memory alloys can withstand different aerodynamic pressures and there is no loss in their stiffness properties with changing weather conditions, unlike traditional actuators [47, 48].

15.2.2 Production Methods of Smart Materials

Today, with the developing technology, the products produced are also changing and transforming accordingly. Different production methods have been developed to produce higher quality and more precise products, not to mention making the products less costly. With these production methods, materials with higher strength are obtained. Combining two or more metal powders and creating new products with micron-sized powders by alloying creates different usage areas. The powder metallurgy method is one of these methods and is the process of producing metal powders and transforming these powders into the desired piece shape. In this method, powders are produced by different powder production methods and the desired part is obtained by mixing the powders, pressing and sintering processes [49].

Recently, smart materials with superior properties have been preferred in various application areas, especially in many fields such as engineering, agriculture, space and aviation. There are many application areas and different types of smart materials for these application areas. In particular, shape memory alloys are alloys in which the material shape and properties can be recovered through temperature change or stress under the influence of large applied deformations. Casting and powder metallurgy technologies can mostly be used in the production of these alloys. However, in casting technology, the cost and manufacturing of shape memory parts with small piece sizes and complex geometry can be difficult [50]. In addition to these situations, the combination of molten NiTi alloy and oxygen in the air during production can cause the formation of undesirable precipitate phases in the material by creating a precipitate phase with the Ti element. Unlike the casting method, composition control of porous NiTi alloys produced by powder metallurgy can be achieved. Previous studies have revealed that the chemical composition of the NiTi alloy is important. Because the martensite $\leftrightarrow$ austenite transformation temperature depends on the nickel ratio in the alloy. However, with this method, subsequent machining operations are largely eliminated, mass production of parts with controllable porous structures in complex shapes or close to the final shape is ensured, and errors that may occur due to segregation during casting are prevented. This is the reason why powder metallurgy technology is preferred as well as casting technology in the production of NiTi alloys [47].

The widely used mechanical alloying method allows two or more powders to be alloyed together. With this method, materials with superior technological features are produced. Different types of grinders (horizontal, spex, planetary) are used in the mechanical alloying method. Composite powder structures are formed as a result of ongoing fracture and cold welding, thanks to the collisions between the powder-surface-ball during the alloying process. In the mechanical alloying method, powders of different sizes are used to increase the pressing and sintering capabilities of the materials. In this way, powders can reach higher packaging density. In the mechanical alloying method, powder is obtained with the help of the friction movement of abrasive materials on each other. Particle-reinforced composites, such as materials with increased strength with oxides, have been used for many years due to their high-temperature friction resistance. When producing these composites, it is difficult to obtain a homogeneous structure in every region of the main structure. Milling techniques have been developed to facilitate the production of composite materials [51].

15.2.3 Smart Materials Produced by Mechanical Alloying

Thermo-mechanical heat treatments are important in controlling the structural properties of NiTi products obtained by the mechanical alloying method. Especially after cold working, the NiTi alloy must be annealed in order to show a good shape memory effect and superelasticity. The principle purpose of this annealing is to allow martensitic transformation by reducing residual errors during shaping. Control of the annealing process is very important. If the annealing time and temperature are too low, the martensitic transformation will occur and be completed in a wide temperature range. Otherwise, if the annealing temperature and time are too high, the martensitic transformation will occur normally, but the mechanical behavior of the matrix will be low and its shape memory properties will be poor. In fact, each application must be heat treated according to its own requirements. If the application requires a high number of cycles and less shape memory amplitude is desired, low heat treatment should be applied. Conversely, if a high amount of shape memory amplitude is desired, suitably high annealing temperatures are required. The mechanical alloying method, one of the powder metallurgy production methods, is one of the most suitable methods used to add a different alloy powder into the existing alloy system and change the composition [46, 52].

The mechanical alloying method is widely used in NiTi shape memory alloys due to its unique efficiency in changing the Ni and Ti ratios in the alloy system and synthesizing different alloying elements into the structure. The main reasons why this technique comes to the fore are the ease of obtaining a homogeneous structure with the applied mechanical alloying, the ease of material synthesis in terms of the desired material properties and the convenience provided by the production processes. By using this technique, the desired particle size can be obtained and even unique physical, chemical and mechanical properties can be obtained in materials thanks to the production of nanoparticles [51, 53].

Chen et al. [54]. synthesized and characterized nanocrystalline NiTi by mechanical alloying method. They used titanium and nickel powders of 99.9% purity as powder. They subjected the 50%Ni–50Ti composition by weight to mechanical alloying with a 40:1 Ball/Powder ratio. In order to determine the recrystallization temperature of mechanically alloyed powders, they performed DSC analysis with a heating rate of 40 K/min and found that the crystallization temperatures of 32 h mechanically alloyed powders were 752.6 and 873 K, according to the DSC analysis. As a result of their study, they obtained amorphous NiTi powders after 24 h of grinding and determined the recrystallization temperature of this amorphous structure to be 828 K. As a result of heat treatment at 873 K, Ti_2Ni, NiTi and Ni_3Ti intermetallic compounds were obtained.

In their study to obtain a very dense NiTi alloy with a very thin β phase by mechanical alloying and sintering, Mariarz et al. [55], used starting powders with 99.5% purity and average particle size of 100 μm and processed them with a high-energy mechanical alloying device in a planetary type Fritsch Pulverist. 4, 8, 16, 32 h of mechanical alloying. With the XRD analysis, they found that there was a decrease in the intensity and width of the peaks after 8 h. They stated that the reason for this was the high stresses released by the collision of the balls during grinding. They noted that this situation caused a decrease in crystal size and an increase in lattice distortion. As a result of these studies; they obtained a completely amorphous structure from Ni and Ti powders after 32 h of milling time. By hot pressing, they obtained a density reaching 98% of the theoretical density and a structure in which martensitic transformation was observed with β-NiTi and B19 phases, with Ni_3Ti and Ti_2Ni dominant phases.

References

1. Tsai, M.-H., Yeh, J.-W.: High-entropy alloys: a critical review. Mater. Res. Lett. **2**(3), 107–123 (2014)
2. George, E.P., Curtin, W., Tasan, C.C.: High entropy alloys: a focused review of mechanical properties and deformation mechanisms. Acta Mater. **188**, 435–474 (2020)
3. Shashanka, R., Chaira, D.: Optimization of milling parameters for the synthesis of nano-structured duplex and ferritic stainless steel powders by high energy planetary milling. Powder Technol. **278**, 35–45 (2015)
4. Li, W., Xie, D., Li, D., Zhang, Y., Gao, Y., Liaw, P.K.: Mechanical behavior of high-entropy alloys. Prog. Mater Sci. **118**, 100777 (2021)
5. Chen, J., Zhou, X., Wang, W., Liu, B., Lv, Y., Yang, W., Xu, D., Liu, Y.: A review on fundamental of high entropy alloys with promising high–temperature properties. J. Alloy. Compd. **760**, 15–30 (2018)
6. Nayak, A., Shashanka, R., Chaira, D.: Effect of nanosize yittria and tungsten addition to duplex stainless steel during high energy planetary milling. In: IOP Conference Series: materials Science and Engineering, p. 012008. IOP Publishing (2016)

7. Murty, B.S., Yeh, J.-W., Ranganathan, S., Bhattacharjee, P.P.: High-Entropy Alloys. Elsevier (2019)
8. Zhang, Y., Yang, X., Liaw, P.: Alloy design and properties optimization of high-entropy alloys. Jom **64**, 830–838 (2012)
9. Rajendrachari, S.: An overview of high-entropy alloys prepared by mechanical alloying followed by the characterization of their microstructure and various properties. Alloys **1**(2), 116–132 (2022)
10. Yeh, J.W., Chen, S.K., Lin, S.J., Gan, J.Y., Chin, T.S., Shun, T.T., Tsau, C.H., Chang, S.Y.: Nanostructured high-entropy alloys with multiple principal elements: novel alloy design concepts and outcomes. Adv. Eng. Mater. **6**(5), 299–303 (2004)
11. Vaidya, M., Muralikrishna, G.M., Murty, B.S.: High-entropy alloys by mechanical alloying: a review. J. Mater. Res. **34**(5), 664–686 (2019)
12. Moghaddam, A.O., Shaburova, N.A., Samodurova, M.N., Abdollahzadeh, A., Trofimov, E.A.: Additive manufacturing of high entropy alloys: a practical review. J. Mater. Sci. Technol. **77**, 131–162 (2021)
13. Li, W., Liaw, P.K., Gao, Y.: Fracture resistance of high entropy alloys: a review. Intermetallics **99**, 69–83 (2018)
14. Chen, S., Tong, Y., Liaw, P.K.: Additive manufacturing of high-entropy alloys: a review. Entropy **20**(12), 937 (2018)
15. Sharma, P., Dwivedi, V., Dwivedi, S.P.: Development of high entropy alloys: a review. Mater. Today: Proc. **43**, 502–509 (2021)
16. Pradeep, N., Hegde, M.R., Rajendrachari, S., Surendranathan, A.: Investigation of microstructure and mechanical properties of microwave consolidated TiMgSr alloy prepared by high energy ball milling. Powder Technol. **408**, 117715 (2022)
17. Pickering, E.J., Carruthers, A.W., Barron, P.J., Middleburgh, S.C., Armstrong, D.E., Gandy, A.S.: High-entropy alloys for advanced nuclear applications. Entropy **23**(1), 98 (2021)
18. Li, W., Liu, P., Liaw, P.K.: Microstructures and properties of high-entropy alloy films and coatings: a review. Mater. Res. Lett. **6**(4), 199–229 (2018)
19. Shang, Y., Brechtl, J., Pistidda, C., Liaw, P.K.: Mechanical behavior of high-entropy alloys: a review. In: High-Entropy Materials: theory, Experiments, and Applications, pp. 435–522 (2021)
20. Zhang, Y., Li, R.: New Advances in High-Entropy Alloys, ed., MDPI, p. 1158 (2020)
21. Xia, S.-Q., Zhen, W., Yang, T.-F., Zhang, Y.: Irradiation behavior in high entropy alloys. J. Iron. Steel Res. Int. **22**(10), 879–884 (2015)
22. Dewangan, S.K., Mangish, A., Kumar, S., Sharma, A., Ahn, B., Kumar, V.: A review on high-temperature applicability: A milestone for high entropy alloys. Eng. Sci. Technol. Int. J. 101211 (2022)
23. Rajendrachari, S., Adimule, V., Gulen, M., Khosravi, F., Somashekharappa, K.K.: Synthesis and characterization of high entropy alloy 23Fe–21Cr–18Ni–20Ti–18Mn for electrochemical sensor applications. Materials **15**(21), 7591 (2022)
24. Sathiyamoorthi, P., Kim, H.S.: High-entropy alloys with heterogeneous microstructure: processing and mechanical properties. Prog. Mater Sci. **123**, 100709 (2022)
25. Manzoni, A.M., Glatzel, U.: New multiphase compositionally complex alloys driven by the high entropy alloy approach. Mater. Charact. **147**, 512–532 (2019)
26. Li, R., Xie, L., Wang, W.Y., Liaw, P.K., Zhang, Y.: High-throughput calculations for high-entropy alloys: a brief review. Front. Mater. **7**, 290 (2020)
27. Taheriniya, S., Davani, F.A., Hilke, S., Hepp, M., Gadelmeier, C., Chellali, M.R., Boll, T., Rösner, H., Peterlechner, M., Gammer, C.: High entropy alloy nanocomposites produced by high pressure torsion. Acta Mater. **208**, 116714 (2021)
28. Yeh, J.-W.: Overview of high-entropy alloys. In: High-Entropy Alloys: fundamentals and Applications, pp. 1–19 (2016)
29. Dada, M., Popoola, P., Adeosun, S., Mathe, N.: High entropy alloys for aerospace applications. Aerodynamicsed., IntechOpen (2019)

30. Cai, Y., Wang, G., Ma, Y., Cao, Z., Meng, X.: High hardness dual-phase high entropy alloy thin films produced by interface alloying. Scripta Mater. **162**, 281–285 (2019)
31. Suryanarayana, C.: Mechanical alloying and milling. Prog. Mater Sci. **46**(1–2), 1–184 (2001)
32. Zhu, M., Gao, Y., Chung, C., Che, Z., Luo, K., Li, B.: Improvement of the wear behaviour of Al–Pb alloys by mechanical alloying. Wear **242**(1–2), 47–53 (2000)
33. Nagesha, K., Rajanish, M., Shivappa, D.: A review on mechanical alloying. Change **3** (2013)
34. Arik, H., Ozcatalbas, Y., Turker, M.: Dry sliding wear behavior of in situ Al–Al4C3 metal matrix composite produced by mechanical alloying technique. Mater. Des. **27**(9), 799–804 (2006)
35. Suryanarayana, C., Froes, F.H.: Mechanical alloying of titanium-base alloys. Adv. Mater. **5**(2), 96–106 (1993)
36. Addepalli, S.N., Joladarashi, S., Ramesh, M., Arya, S.: Effect of mechanical alloying on the microstructure of CoCrNiTiMox high entropy alloy. J. Therm. Spray Technol. **31**(4), 1045–1055 (2022)
37. El-Eskandarany, M.S.: Mechanical Alloying: energy Storage, Protective Coatings, and Medical Applications. William Andrew (2020)
38. Cardoso, K.R., Izaias, B.d.S., Vieira, L.d.S., Bepe, A.M.: Mechanical alloying and spark plasma sintering of AlCrCuFeZn high entropy alloy. Mater. Sci. Technol. **36**(17), 1861–1869 (2020)
39. Zhang, K., Fu, Z., Zhang, J., Shi, J., Wang, W., Wang, H., Wang, Y., Zhang, Q.: Annealing on the structure and properties evolution of the CoCrFeNiCuAl high-entropy alloy. J. Alloy. Compd. **502**(2), 295–299 (2010)
40. Basavarajappa, P.N., Hegde, M.M.R., Rajendrachari, S., Surendranathan, A.O.: Investigation of structural and mechanical properties of nanostructured TiMgSr alloy for biomedical applications
41. Zhang, Y., Xing, Q.: High entropy alloys: manufacturing routes. In: Reference Module in Materials Science and Materials Engineering. Elsevier, Amsterdam, The Netherlands
42. Altas, E., Erkan, O., Ozkan, D., Gokkaya, H.: Optimization of cutting conditions, parameters, and cryogenic heat treatment for surface roughness in milling of NiTi shape memory alloy. J. Mater. Eng. Perform. 1–13 (2022)
43. Altas, E., Gokkaya, H., Karatas, M.A., Ozkan, D.: Analysis of surface roughness and flank wear using the taguchi method in milling of NiTi shape memory alloy with uncoated tools. Coatings **10**(12), 1259 (2020)
44. Altas, E., Khosravi, F., Gokkaya, H., Maleki, V.A., Akınay, Y., Ozdemir, O., Bayraktar, O., Kandas, H.: Finite element simulation and experimental investigation on the effect of temperature on pseudoelastic behavior of perforated Ni–Ti shape memory alloy strips. Smart Mater. Struct. **31**(2), 025031 (2022)
45. Altas, E., Altin Karatas, M., Gokkaya, H., Akinay, Y.: Surface integrity of NiTi shape memory alloy in milling with cryogenic heat treated cutting tools under different cutting conditions. J. Mater. Eng. Perform. (2021)
46. Bahl, S., Nagar, H., Singh, I., Sehgal, S.: Smart materials types, properties and applications: a review. Mater. Today: Proc. **28**, 1302–1306 (2020)
47. Drossel, W.-G., Kunze, H., Bucht, A., Weisheit, L., Pagel, K.: Smart3–Smart materials for smart applications. Procedia Cirp **36**, 211–216 (2015)
48. Yin, Y., Rogers, J.A.: Introduction: smart Materials, ed., pp. 4885–4886. ACS Publications (2022)
49. Drossel, W.-G., Meinel, F., Bucht, A., Kunze, H.: Smart materials for smart production—A cross-disciplinary innovation network in the field of smart materials. Procedia Manuf. **21**, 197–204 (2018)
50. Nguyen, P.Q., Courchesne, N.M.D., Duraj-Thatte, A., Praveschotinunt, P., Joshi, N.S.: Engineered living materials: prospects and challenges for using biological systems to direct the assembly of smart materials. Adv. Mater. **30**(19), 1704847 (2018)
51. Bengisu, M., Ferrara, M.: Materials that Move: smart Materials, Intelligent Design. Springer (2018)

52. Farber, E., Zhu, J.-N., Popovich, A., Popovich, V.: A review of NiTi shape memory alloy as a smart material produced by additive manufacturing. Mater. Today: Proc. **30**, 761–767 (2020)
53. Ashima, R., Haleem, A., Bahl, S., Javaid, M., Mahla, S.K., Singh, S.: Automation and manufacturing of smart materials in Additive Manufacturing technologies using internet of things towards the adoption of industry 4.0. Mater. Today: Proc.
54. Shenqi, C., Yanchun, Z. Yiyi, L., Shouhua, H.: Synthesis and characterization of nanocrystalline Ni–Ti and Ni–Cr powders by mechanical alloying. 材料科学与技术 **13**(2), 86 (1997)
55. Maziarz, W., Dutkiewicz, J., Van Humbeeck, J., Czeppe, T.: Mechanically alloyed and hot pressed Ni–49.7 Ti alloy showing martensitic transformation. Mater. Sci. Eng. A **375**, 844–848 (2004)

Chapter 16
Automobile Applications of Mechanically Alloyed Magnesium and Titanium Material

Gangadhar M. Kanaginahal, M. C. Kiran, Kiran Shahapurkar, Rayappa Srinivas Mahale, and Prashant Kakkamari

Abstract This chapter intricately examines the pivotal role played by magnesium and titanium alloys in the sphere of automotive part fabrication. It meticulously dissects their distinctive attributes, spotlighting their innate lightweight composition and unparalleled robustness, attributes that fundamentally elevate vehicular performance and operational efficacy. Through an in-depth exploration of the singular merits of these alloys and the symbiotic advantages derived from their amalgamation, the chapter accentuates their formidable resistance to corrosion, thermal steadfastness, and mechanical prowess, all pivotal facets in the realm of sustainable automotive design. Furthermore, the chapter critically evaluates the impediments inherent in manufacturing these alloys and the pioneering methodologies devised to optimize their efficacy within the automotive domain. By spotlighting tangible case studies, it elucidates how these alloys seamlessly fortify vehicular structures while substantially alleviating weight burdens, thereby significantly augmenting fuel efficiency. Additionally, the chapter contemplates the economic and environmental ramifications attendant upon the deployment of these cutting-edge materials, underscored

G. M. Kanaginahal · R. S. Mahale
Department of Mechanical Engineering, Jain College of Engineering and Research, Belagavi, Karnataka 590008, India

M. C. Kiran
Department of Mechanical Engineering, NITTE Meenakshi Institute of Technology, Bengaluru, Govindapura, Karnataka 560064, India

K. Shahapurkar (✉)
Department of Mechanical Engineering, School of Mechanical, Chemical and Materials Engineering, Adama Science and Technology University, 1888 Adama, Ethiopia
e-mail: kiranhs1588@astu.edu.et

R. S. Mahale
School of Mechanical Engineering, REVA University, Bengaluru, Karnataka 560064, India

P. Kakkamari
Department of Mechanical Engineering, KLS Gogte Institute of Technology Belagavi, Belgaum, Karnataka 590006, India
e-mail: ppkakamari@git.edu

S. Rajendrachari (ed.), *Mechanically Alloyed Novel Materials*, Advanced Structured Materials 220, https://doi.org/10.1007/978-981-97-6504-1_16

by their potential to revolutionize the industry, fostering sustainability and resource maximization. This comprehensive elucidation of magnesium and titanium alloyed materials serves as an imperatival conduit for future exploration and innovation, proffering inspiration for extensive integration within automotive frameworks, thereby steering the industry towards pioneering, efficient, and ecologically cognizant design paradigms.

Keywords Magnesium alloys · Titanium materials · Automobile parts · Mechanical alloying · Power train

16.1 Introduction

Automotive requisites dictate the imperative of commendable ductility for myriad components, especially in scenarios necessitating energy dissipation during untoward incidents. In this crucible, alloy and process refinement for wrought alloys is steered towards optimizing the material's energy absorption potential. Nonetheless, certain components mandate a predilection for heightened strength over ductility. Thus, alloy progression aligns itself with diverse imperatives, delineating discernible alloy cohorts tailored to proffer specific material traits. Magnesium, with its ethereal density of 1.74 g per cubic centimeter. Nevertheless, the pervasive deployment of magnesium and its alloys in pivotal engineering applications remains circumscribed by its modest absolute strength, ductility, and resistance to deformation under sustained load—a phenomenon known as creep. Despite the spectrum of magnesium lightweight alloys already in existence, there persists a need for alloys boasting augmented structural attributes, imperative to satiate contemporary and impending material exigencies [1, 2]. The Mg has shown better specific strength and good castability with corrosion resistance. They have found applications in transmission castings, instrument panels, cylinder cover head, steering components, seat frame, engine blocks etc. Magnesium emerges as an alluring contender for automotive application, primarily owing to its enviable lightness—a staggering 36% less dense per unit volume than its aluminum counterpart and an impressive 78% lighter than iron. When judiciously alloyed, magnesium boasts the loftiest strength-to-weight ratio among all structural metals, rendering it an exemplar of material efficiency. Abundant by nature, magnesium claims its place as the eighth most ubiquitous element. Seawater, the principal fount of supply, harbors 0.13% of this element, thereby offering a near-boundless reservoir. The amalgam of its feathery mass, superlative specific strength, and commendable castability positions magnesium alloys as a beacon of promise in the domains of automotive and aerospace engineering.

16.1.1 Importance of Mechanical Alloying on Mg Materials

In its as-cast state, magnesium exhibits a notable susceptibility to rapid degradation coupled with a concomitant weakness in structural integrity. Consequently, the material's inherent characteristics can be substantially enhanced through the judicious application of alloying techniques and advanced processing methodologies. Reactivity inherent to magnesium facilitates a chemical interplay with alloying agents, thereby engendering the formation of intermetallic compounds. These discernible intermetallic phases, pervasive within magnesium alloys, serve as the principal agents influencing the microstructural configuration, and by extension, the material's overall properties. The solubility dynamics are primarily governed by two fundamental factors: the valence of the alloying element and its atomic dimensions relative to magnesium. Magnesium's crystalline lattice, characterized by a hexagonal close-packed (HCP) structure, along with its precisely measured atomic diameter, facilitates the formation of composite solid solutions encompassing a diverse array of elements. Magnesium alloys are prevalent in various compositional formats, including binary, ternary, and other intricate formulations. However, a critical prerequisite lies in the meticulous selection of the alloying element, a choice that profoundly impacts the augmentation of mechanical properties and the acquisition of superior corrosion resistance in magnesium alloys. This selection must be harmoniously aligned with the imperative of achieving impeccable biocompatibility in implanted materials, achieved through the inherently non-toxic nature of the chosen alloy [3]. The domain of cast magnesium (Mg) alloys bifurcates along chemical composition lines, distinguishing between Mg–Aluminum (Al) alloys meticulously calibrated with Al content and supplemented with adjunct elements to facilitate cost-effective production, and Al-deficient alloys featuring Zinc (Zn), Rare Earth (RE) elements, Yttrium (Y), and Zirconium (Zr) to cater to elevated-temperature exigencies. For wrought Mg alloys, a dichotomy emerges between those amenable to heat treatment and those that eschew such modification, owing to the constraining influence of the hexagonally arranged close-packed crystal lattice of Mg. Pioneering processing methodologies present a tantalizing avenue for refining microstructural attributes. It is notable that cast Mg alloys reign supreme, commanding approximately 90% of the utilization landscape. The infusion of 1–9% Al into Mg alloys augments tensile potency, ductility, and the capacity for facile casting operations at temperatures below 120 °C. However, once the Al content exceeds the 2% threshold, an enhancement in corrosion resistance is offset by a concomitant diminution in mechanical performance, particularly at elevated thermal regimes [4] (Fig. 16.1).

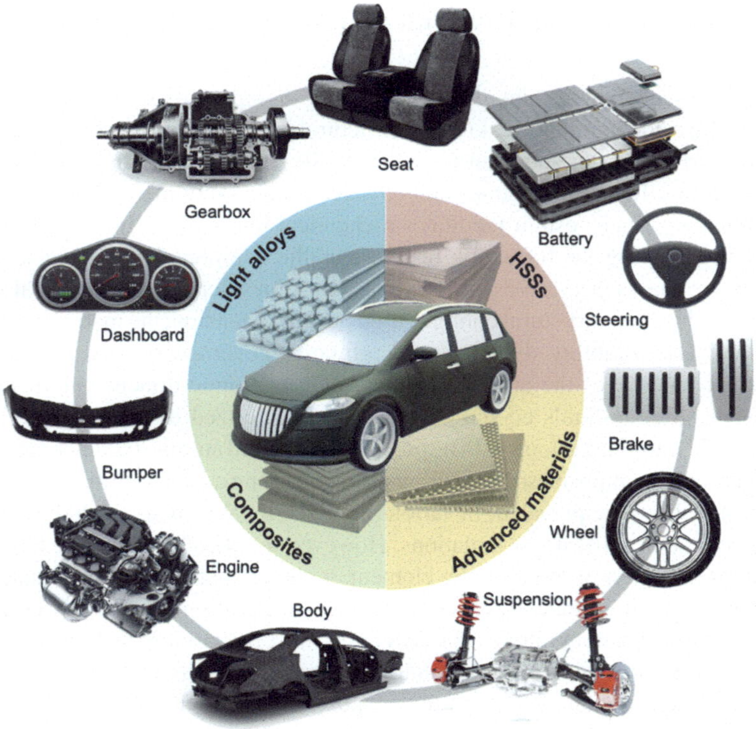

Fig. 16.1 Materials utilized in automobile parts [5]

16.1.2 Other Materials Used in Automotive Manufacturing

The significant advancements in metallurgy have afforded a diverse array of mechanical and physical properties in both steel and aluminum alloys, broadening the spectrum of available options. Beyond the expanded metallic repertoire, due consideration is warranted for plastics, encompassing twenty distinct types viable for application in motor vehicles. The integration of computer-aided design (CAD) systems has revolutionized the approach to materials selection. Design engineers now have access to a database of pre-established material options, streamlining the decision-making process. This shift empowers engineers to make informed choices without relying solely on past experiences or the expertise of dedicated materials engineers (Table 16.1).

Table 16.1 Elements utilized as in Mg-alloys for fabrication [3]. It is open access

Sl. No	Elements	Improvement in mechanical properties
1	Calcium	Enhances grain refinement and strengthens MMCs
2	Lithium	Increases ductility while reducing strength
3	Zinc	Strengthens composite, optimal up to 5 wt.%
4	Strontium	3 wt.% enhances strength, refines grain size
5	Manganese	Improves grain refinement and tensile strength
6	Tin	Enhances both tensile and compressive strength
7	Zirconium	Provides high damping, tensile strength, and ductility
8	Aluminum	Improves hardness and tensile strength
9	Titanium	Increases elongation
10	Copper	Reduces ductility
11	Silicon	Enhances fluidity
12	Iron	Not suitable for corrosion resistance
13	Neodymium	Strength improvement observed
14	Molybdenum	Enhances Young's modulus, hardness, and elongation
15	Cerium	Reduces yield strength

16.1.3 Mechanical Alloying of Magnesium

Unearthed in 1774, magnesium holds the position of being the eighth most plentiful element in the Earth's crust, making up around 2% of its overall composition. Due to its classification in the alkaline earth metals group, magnesium is naturally found in compound forms rather than as a pure element. Notable sources of magnesium extraction include magnesite, dolomite, carnallite, and seawater. With an approximate density of 1.738 g/cm^3, magnesium stands as the lightest structural metal, yet it boasts remarkable strength and stiffness. To enhance its properties, magnesium is alloyed with various elements, and these alloys are identified using an ASTM-based naming convention. In the realm of material processing, powder metallurgy plays a vital role in enhancing magnesium's mechanical characteristics and corrosion resistance. This method involves blending powders and compacting them under pressure to form green compacts. Subsequently, sintering is employed to promote uniformity and facilitate interdiffusion between particles, thereby improving the material's performance [6, 7]. The realm of cast magnesium (Mg) alloys can be divided into two groups based on chemical composition: Mg–Aluminium (Al) alloys with controlled Al content and additional elements for cost-effective manufacturing, and Al-free alloys with Zn, RE, Y, and Zr for high-temperature applications. Wrought Mg alloys have heat-treatable and non-heat-treatable variants, limited by the HCP crystal structure of Mg. Advanced processing techniques offer potential for microstructure refinement. Cast Mg alloys dominate with around 90% usage. Incorporating 1–9% Al in Mg alloys enhances tensile strength, ductility, and castability below 120 °C, but beyond 2% Al, corrosion resistance improves while mechanical properties decline at higher temperatures. Adding Zinc (Zn) improves tensile properties but affects weldability [8]. Mg–Zn-based alloys exhibit strength and ductility, with research focusing on optimizing properties through alloying elements like the Mg–Zn–Zirconium (Zr) system. Mn enhances corrosion resistance but has limited strengthening effects. Adding Mn to Mg–Gd alloys increases strength but reduces ductility. The incorporation of calcium (Ca) in magnesium (Mg) alloys enhances both mechanical properties and corrosion resistance, making them well-suited for biomedical applications. Mg–Zn–Calcium (Ca) alloys are being considered as potential materials for biodegradable implants. Yttrium (Y) is introduced into Mg alloys to improve their performance at elevated temperatures, enhancing plasticity and creep resistance. In the context of Mg alloys for automotive engineering, rare-earth (RE) elements, such as Gd (gadolinium), are employed to refine crystals, thereby improving both creep and corrosion resistance. Gd, in particular, demonstrates significant solubility in Mg, allowing for solid solution and precipitation strengthening [4, 9] (Fig. 16.2).

Researchers have developed magnesium alloys with the Mg_2Si phase, exhibiting excellent mechanical properties. Al–Mg–Silicon (Si) alloys, including Mg_2Si, show improved characteristics, making them attractive in the automotive and aerospace industries. The abundance and low cost of magnesium, aluminum, and silicon further enhance their appeal. Structural modifications of Mg_2Si through various techniques hold promise for enhancing mechanical performance. AS series alloys, like AS21, developed by Volkswagen, demonstrate stress and creep resistance up to 150 °C [10]. These alloys feature Mg-rich solid solution, $Mg_{17}Al_{12}$, and a beneficial Mg_2Si phase, making them desirable for critical applications at elevated temperatures [11, 12]. A novel method using mechanical alloying (MA) to produce a biodegradable Mg–Zn alloy and investigates the influence of milling time on its properties. Extended milling time has been found to elevate the density and microhardness of sintered Mg–Zn alloy.

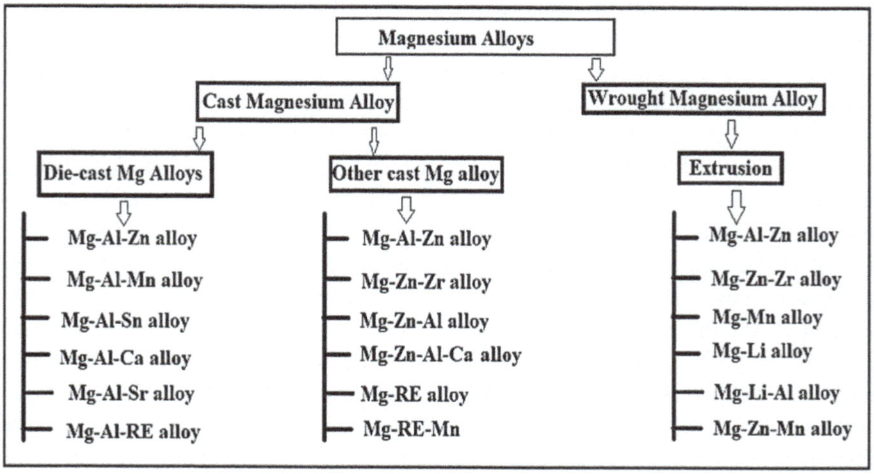

Fig. 16.2 Classification of Mg alloys [3]

This research underscores the promise of Mg-based materials for use as load-bearing implants in the medical field, attributed to their commendable biocompatibility and biodegradability. The process of mechanical alloying emerges as a key factor in enhancing the mechanical properties and corrosion resistance of Mg-based materials [13, 14]. Spark plasma sintering (SPS) of Zn–1Mg powder Zn–6Mg and Zn–16Mg alloys ensured the preservation of a refined and uniform microstructure from the precursor powder, resulting in the emergence of new intermetallic phases [15]. Notably, the compressive properties of the sintered samples are primarily governed by the concentration of alloying elements, with increased concentration leading to improved strength but reduced ductility. Among the examined alloys, the Zn–1Mg alloy demonstrates the most favorable properties. The remarkable fine-grained and homogenous microstructure exhibited by the Zn–1Mg alloy holds great promise for enhancing the mechanical attributes of biodegradable medical devices, particularly in applications requiring exceptional compression-loaded performance. Furthermore, the high tensile strength values attained by the Zn–1Mg alloy underscore its potential suitability for engineering applications demanding robust mechanical performance [16, 17] (Fig. 16.3)

Mechanical alloying holds paramount significance in the realm of biomedical applications as it enables the fabrication of sophisticated materials with precisely tailored properties. This technique empowers the production of nanocrystalline or ultrafine-grained biomaterials, elevating their mechanical robustness, hardness, and wear resistance. Biodegradable constructs can be ingeniously devised through

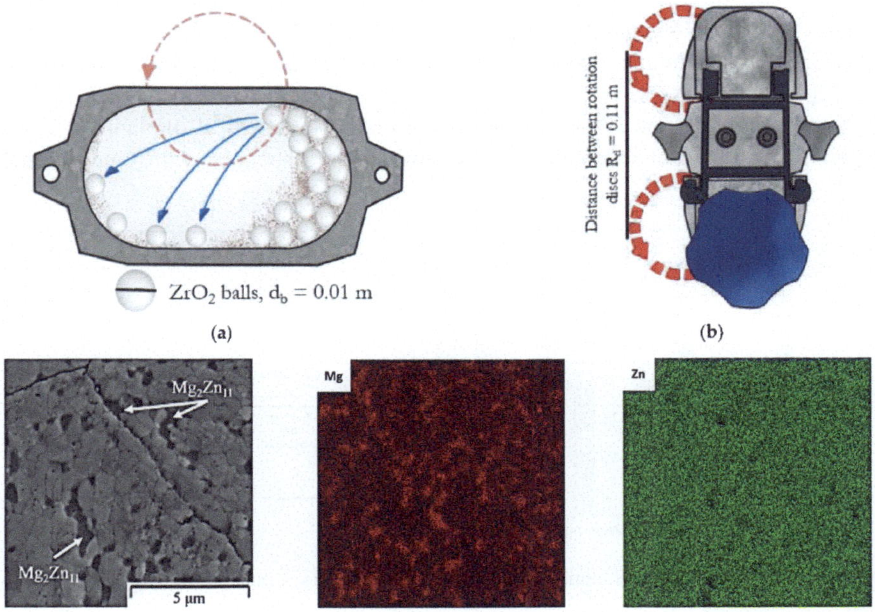

Fig. 16.3 Zn–1Mg alloy compacted by spark plasma sintering and their distribution [16]

mechanical alloying, offering tunable degradation rates in alignment with tissue healing kinetics, thereby mitigating complications associated with implants [18]. Integration of bioactive constituents, such as calcium and phosphorus, augments osteointegration and fosters osseous bonding. Moreover, antibacterial agents can be judiciously dispersed within the alloy, mitigating infection risks. The versatility and precise control afforded by mechanical alloying bestow paramount value in crafting biocompatible and high-performance biomaterials, propelling medical devices and implants towards enhanced patient outcomes in the biomedical domain [19].

An extensive investigation into the influence of various alloying elements on the mechanical attributes of Mg-based materials tailored for biomedical applications encompassed diverse binary and ternary Mg alloys, wherein ZEK100, WE43, and EW62 emerge as highly promising candidates, showcasing commendable biodegradability, biocompatibility, and reduced corrosion rates. Additionally, the study explores the implementation of surface modifications to effectively govern degradation rates, rendering Mg-based materials exceptionally suitable for biomedical implants. The noteworthy findings furnish invaluable insights for researchers and manufacturers in the biomedical industry, facilitating judicious material selection and fostering the development of cutting-edge implants meeting critical biocompatibility and mechanical prerequisites [3]. The work on ternary Mg–Zn–Ca alloys, with a particular emphasis on their mechanical properties, corrosion resistance, and biocompatibility results shows that the MA process yields alloys with a fine microstructure, enhancing their mechanical characteristics. Additionally, the synthesized alloys exhibit improved corrosion resistance compared to pure magnesium. The biocompatibility of these alloys makes them promising candidates as temporary implants for bone regeneration applications. Notably, cytotoxicity tests reveal low adverse effects on human fibroblast cells, supporting their potential for safe clinical use [20] (Fig. 16.4)

Significantly, the utilization of magnesium alloys as medical implants presents a promising prospect to counteract the limitations associated with conventional

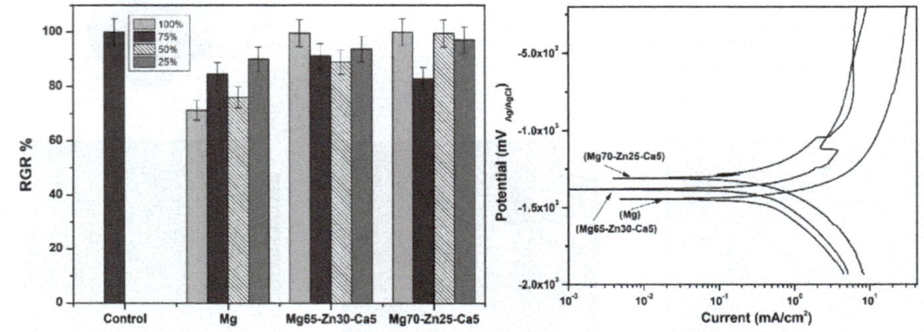

Fig. 16.4 Cytotoxicity and potentiodynamic polarization results of pure and alloyed Mg materials [20]

implant materials, including the potential release of toxic metallic ions and particles through corrosion, necessitating secondary surgical procedures for permanent fixture removal, resulting in escalated costs and patient morbidity [21].

The insights into the comprehensive investigation of the microstructure and mechanical properties of as-cast Mg–Zn–Mn–Ca alloys, specifically exploring their potential suitability as medical implants found that incorporation of Zn significantly refines the grain size of the alloys, thereby resulting in substantial enhancements in their mechanical characteristics. Notably, the main secondary phase observed in these alloys is identified as the Mg–Zn–Ca phase, with its volume fraction exhibiting an incremental trend alongside the Zn content [21]. Furthermore, the application of solution treatment facilitates the dissolution of select secondary phases into the matrix, culminating in a marked augmentation of the alloy's plasticity. The findings of this investigation offer valuable avenues for the development of novel biodegradable magnesium alloys for medical implantation, catering to bone fractures and vascular tissue treatments due to their inherent degradability, biocompatibility, and well-suited elastic modulus (40–45 GPa) [22, 23]. The mechanical alloying of Mg65–Zn30–Ca5 and Mg70–Zn25–Ca5 alloys, specifically engineered for biomedical implant applications, manifests heightened corrosion resistance and attenuated cytotoxicity relative to pristine Mg. X-ray diffraction analysis substantiates the formation of MgZn intermetallic phases during the alloying process. Implementation of a low-energy mechanical alloying regimen underscores its cost-effectiveness. Potentiodynamic polarization curves elucidate the pivotal role of augmented Zn content in instigating a protective film, thereby amplifying corrosion resistance. Employing the Sulforhodamine B (SRB) assay, cell viability assessments validate the non-cytotoxic nature of sintered Mg65–Zn30–Ca5 and Mg70–Zn25–Ca5 alloys towards human fibroblast cells [20]. The Mg–0.3Sr–0.5Y alloy exhibits a homogeneous degradation pattern characterized by a subdued corrosion rate and a well-balanced interplay between mechanical strength and bio-degradation properties, making it a compelling prospect for bio-applications. The study also sheds light on the osteogenic prowess of Sr, as it exerts a positive influence on bone growth while hindering bone resorption, rendering Mg–Sr–Y alloys a promising material choice for bone implants in the medical realm. The incorporation of Y in both the α-Mg phase and $Mg_{17}Sr_2$ phase results in a mitigated micro-galvanic corrosion effect, favoring a uniform degradation profile for the alloys. The formation of a compact composite film comprising $Mg(OH)_2$, Y_2O_3, and Sr-substituted phosphate (Sr-HA) further bolsters matrix protection, ensuring consistent degradation [24, 25].

The impact of various alloying elements on these alloys, revealing the potential to enhance their mechanical characteristics shows the use of powder metallurgy technique for fabricating magnesium-based alloys, with a detailed analysis of parameters and reinforcing materials that significantly influence mechanical properties. Notably, the selection of appropriate parameters and reinforcement materials emerges as a crucial factor in achieving desired mechanical attributes. The promising potential of alumina particles, silicon carbide, fullerene, and yttria as alloying elements to elevate the microhardness and mechanical performance of these alloys [26]. As a result, magnesium-based alloys demonstrate considerable applicability in aerospace, automotive, and biomedical industries, owing to their favorable combination of lightweight and high-strength properties. Researchers and engineers can utilize these insights to advance the development of lightweight, high-strength materials with diverse applications [27, 28].

This potentiality of Mechanical Alloying (MA) methodology to attain a refined microstructure in magnesium alloys accentuates the paramount significance of magnesium alloys in multifarious domains, encompassing automotive, aerospace, and medical applications. Augmenting the mechanical potency and corrosion resistance of magnesium alloys, the incorporation of Zn, Ca, and rare earth elements (REE) garners particular attention. Elucidating the unique applicability of Erbium (Er), in biomaterials, this research showcases how MA expedites the preparation of bespoke alloys, proving more expeditious, cost-effective, and specialized vis-à-vis conventional techniques [29]. Delving into the interactions between erbium and magnesium alloys under varied milling durations, scrutinizing phase composition, microstructure, and hardness, the findings reveal an upsurge in hardness and a homogeneous dispersion of fine particulates with prolonged milling. Paving the way for future considerations, the $Mg_{65}Zn_{30}Ca_4Er_1$ alloy powder emerges as a promising candidate for sintering materials [30, 31]. Moreover, the inquiry explores the ramifications of producing Al–Mg alloys via both direct current and pulse electrodeposition methodologies within the AlCl3–LiAlH4–benzene-tetrahydrofuran (THF) system. Notably, the pulse electrodeposition method exhibits superior efficiency and efficacy in yielding Al–Mg alloy coatings characterized by finer grain sizes and elevated magnesium contents. This breakthrough paves the way for the production of high-purity, uniform-quality nanometer alloys, signifying their potential applicability in diverse industries like aerospace, automotive, and electronics. Conversely, the mixing and mechanical alloying methods are beset by certain limitations, encompassing heightened heat consumption, metal oxidation loss, and inadvertent introduction of impurities. In stark contrast, the electrodeposition method emerges as a more promising alternative, proficiently generating high-quality Al–Mg alloy coatings [32, 33].

The exhaustive and comprehensive literature survey, elucidating the profound potential of biodegradable magnesium alloys as cutting-edge bone repair materials meticulously summarizes recent investigations, with a particular focus on alloy design, surface modification strategies, and the intricate biological performance of magnesium alloys. Distinguishing itself from broader reviews, this study furnishes targeted insights, catering to the specific niche of magnesium alloys for bone repair applications. The practical implications of this seminal work are substantial, heralding a promising path for the advancement of biodegradable magnesium alloys in orthopedic surgery. The inherent biocompatibility, osteoconductivity, and bone-inducing properties of these alloys elevate their standing as promising contenders for regenerative therapies [34, 35].

The use of mechanical milling to create nanocrystalline AZ31 magnesium alloy with titanium (Ti) addition revealed that Ti addition refines the Mg matrix grains, leading to improved mechanical properties. This has practical implications for the development of nanocrystalline AZ31 Mg alloys with enhanced performance, making them potential candidates for aerospace, automotive, and biomedical applications. Mechanical milling effectively reduces crystallite size, achieving nanocrystalline dimensions for Ti additions of 18 wt% and 27 wt%. The alloy exhibits increased microhardness and yield strength, reaching 147 Hv and 293 MPa at room temperature, respectively [36, 37].

Fabricating an advanced Mg-based nanocomposite with exceptional strength was accomplished through the incorporation of thermally stable and coherent second phases. Employing the powder metallurgy technique, the Mg–Ni–Gd alloy demonstrated the formation of Mg_5Gd, Mg_2Ni, Gd_2O_3, and MgO phases, accompanied by the emergence of ultra-fine grains and uniform dispersion of second-phase particles in the 100–200 nm range, ultimately yielding an impressive strength of 804 MPa. The attainment of fine crystallite size following milling and the establishment of solid solution contribute to the elevated mechanical properties, resilient even under high-temperature conditions [2, 38]. A novel lightweight Mg-based multicomponent alloy was fabricated by disintegrated melt deposition technique, followed by hot extrusion, the alloy manifests remarkable attributes of hardness, compressive strength, and ductility. This innovative multicomponent alloying strategy, incorporating Al, Cu, Mn, and Zn (each 5 at %), culminates in the integration of Al_6Mn and Al_2CuMg dispersed hard phases within the pliant Mg base hcp solid solution phase, culminating in extraordinary strength and ductility. The $Mg_{80}Al_5Cu_5Mn_5Zn_5$ alloy remarkably exhibits a substantial hardness of 196 HV, a significant advancement over commercially utilized alloys. It demonstrates a remarkable combination of compressive strength (616 MPa) and ductility (16.9%), outperforming conventional Mg alloys. Additionally, in its as-extruded state, the alloy showcases higher ultimate tensile strength (318 MPa) and commendable ductility (8.2%), obviating the need for supplementary heat treatment while maintaining symmetrical tensile/compression yielding behavior, underscoring its suitability for weight-critical industrial applications [1].

16.1.4 Mechanical Alloying of Magnesium for Automobile Parts

Magnesium (Mg) showcases a hexagonally dense-packed crystalline structure, diverging from the typical configuration of most structural metals. At room temperature, only a limited number of slip systems, can be activated, significantly constraining its deformation capacity. Non-basal slipping initiates only at temperatures around a minimum of 200 °C and a maximum of 300 °C, concurrent with multi-basal slipping that becomes effectively plasticized. Mg alloy materials are primarily classified into cast and deformed categories based on forming processes. Die casting, including methods like sand casting and investment casting, emerges as the most widely employed technique, imparting dimensional stability and high

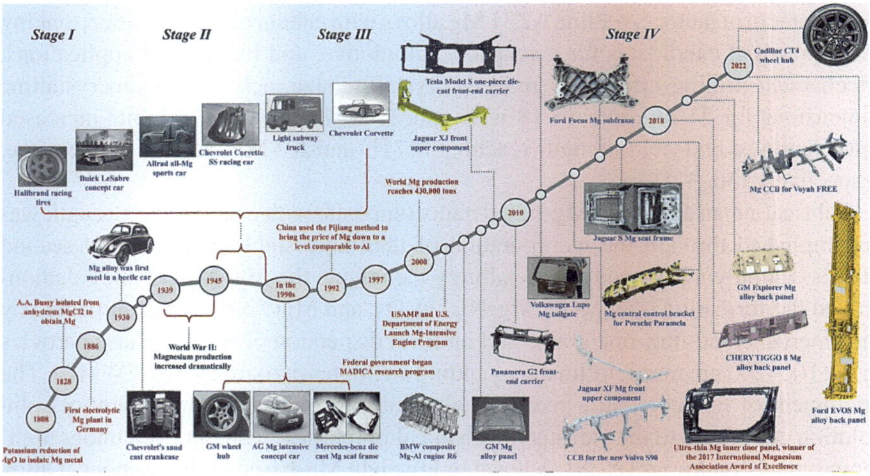

Fig. 16.5 Evolution of automobile parts [39]

productivity. Deformation processing enhances properties by meticulously controlling material organization and employing heat treatment, eliminating tissue defects, and refining grain structure. While forming performance is lackluster at low temperatures, Mg exhibits remarkable plasticity in the superplastic state, making it suitable for fabricating intricately shaped components [39] (Fig. 16.5)

Car Door

Magnesium (Mg) alloys revolutionize inner door panels, drastically cutting weight in esteemed vehicles like the Aston Martin DB9 and Mercedes S-Class coupe—43% and a mere 4.56 kg, respectively. Noteworthy cases include the 2009 Mercedes E-Class T-Model and the 2012 General Motors Corporation project, achieving a 50% mass reduction with integrated door speakers. Fiat Chrysler also employs Mg die castings in the 2018 Jeep Wrangler's rear doors. In 2010, the Lincoln MKT introduced a pioneering die-cast Mg inner door panel, a result of astute engineering. The 2017 Chrysler Pacifica augmented performance with Mg die castings in the tailgate, achieving a 50% weight reduction by integrating components. Mercedes-Benz's 2009 E-Class Wagon exemplifies the advantageous combination of Mg and Al alloys for outer panels, optimizing weight and structural integrity through varying wall thicknesses and reinforcing ribs.

Front End Carrier

In 2012, the Tesla Model S introduced a revolutionary one-piece die-cast Mg front-end carrier, impressively weighing a mere 6.49 kg. Porsche's Panamera G2 employs AM50A for its front-end carrier, achieving a 3.5 kg reduction. The Range Rover's front-end carrier material, AM60B, trims down weight by 7.1 kg. Notably, the 2009 Jaguar XJ utilized Mg alloy AM60B for its front upper component, a substantial 30% weight decrease from prior hydraulically expanded Al-tube solutions. In 2015, the Mercedes-Benz AMG GT employed Mg alloy AM60B for its front upper component, totaling 3.5 kg with a bolted connection. The 2017 fourth-generation Audi A8 mounts its cabin with Mg alloy castings, achieving a commendable 28% weight reduction.

Wheel

Wang's 2014 patented method revolutionized wheel production using forward and reverse extrusion, employing Mg and Al alloys. Wang then licensed Linzhou Dingxin Magnesium Technology Co., Ltd. and Dewei Co., Ltd. to pioneer a superplastic forming process, merging forging and extrusion in one step, streamlining production, and reducing costs. The process, involves blank cutting, unified extrusion post-homogenization, machining, and surface coating. Wheels made of AZ80 Mg alloy, though produced in limited quantities, have shown heightened efficiency, offering substantial market potential. The Chery AM50A Mg alloy steering wheel skeleton employs cold chamber die casting, resulting in an impressive 25% reduction in weight. The Mg alloy steering wheel for passenger cars by Dongfeng Nissan and Shen Long weighs merely 876.5 g. Shenyang University of Technology has ingeniously amalgamated traditional welding techniques into a high-toughness, one-piece die-cast Mg alloy steering wheel skeleton.

Subframe

In 1985, the Volvo LCP 2000 project embraced a pioneering concept: accounting for energy consumption throughout a vehicle's lifespan. This approach incorporated approximately 50 kg of Mg alloy, encompassing critical components like the transmission and clutch housings, wheels, rear suspension, subframe, and cylinder block. In 2005, Corvette opted for Mg front beams, displacing Al brackets to fulfill objectives of weight reduction and fuel efficiency. Audi R8's rear subframe material, AZ91, achieved weight reduction via high-pressure die casting. The 2016 Ford Focus introduced a cast Mg AE44 subframe, delivering a notable 32% weight reduction compared to its steel counterpart, with all joints undergoing rigorous bolt load retention tests [40].

The rigorous examination of the microstructural and mechanical attributes of the magnesium AZ80 alloy control arm, affirms its suitability for automotive use. Findings advocate warm forging at 623 K for optimal structural integrity. At

523 K, dynamic recrystallization reduces grain size, enhancing mechanical properties. Despite a slight yield strength decrease at 623 K, ultimate tensile strength and elongation improve. Overall, warm forging of the AZ80 alloy proves highly effective in crafting automotive control arms, promising reduced vehicle mass and heightened energy efficiency. This research underscores the alloy's potential, endorsing 623 K as the prime forging temperature for robust control arm production [41].

Power Train

In the 1930s, Louris Chevrolet made motoring history by integrating magnesium (Mg) pistons into a racing car. The pioneering Mg alloy crankcase of Chevrolet, epitomized innovation. Volkswagen's iconic Beetle powertrain, also in the 1930s, harnessed about 20 kg of Mg castings, marking a zenith in Mg consumption. VW's Passat, Audi's A4 and A6, along with Ford's truck gearbox housing, all incorporate AZ91D Mg alloy. The primary source of noise at low and moderate vehicle speeds is the engine's radiation, where components like the oil pan, cylinder head cover, and timing gear cover play a significant role. The oil pan, especially, requires specific mechanical properties, notably high tensile strength. However, the quest for reduced vibration and noise has led to an interest in lighter materials, with magnesium alloys emerging as a preferred choice due to their inherent advantages. Crafted from a heat-resistant Mg alloy, the oil pan presents a substantial 33% reduction in weight compared to its aluminum counterpart. This pioneering Mg alloy oil pan, utilized in both the Honda Insight and CA20 models, showcases a notable decrease in weight percentage. Tailored for FAW's latest-generation passenger car engines, operating at temperatures nearing 150 °C, road tests have attested to the oil pan's adherence to temperature standards.

Internal System

Mercedes-Benz pioneered a Mg die casting method, integrating complex components into two parts, weighing a total of 8.5 kg. AM50 and AM20 materials perfectly meet Mg seat requirements, enabling cars with robust, ductile frame components. Recent advancements include Alfa Romeo's Mg alloy front seat skeleton, reducing weight by 2.8 kg compared to steel. Changan's Mg alloy seat skeleton for the Oushang E01 model achieved a 25–30% weight reduction. Jaguar's upgraded model features a 3.08 kg die-cast Mg seat skeleton. Rear seats, made of die-cast Mg alloy, offer effective weight reduction. Chrysler's Stow'n Go seats use Al for the second-row seats and Mg castings for the third row. Though efficient, Mg seats face challenges of cost and limited size, potentially affecting their appeal in the Chinese market focused on comfort. GMC revolutionized automotive design in 1961 with the introduction of the first die-cast Mg alloy dashboard Core Carrier Board (CCB), reducing weight by approximately 4 kg despite its slender 4–5 mm thickness. Presently, models like the Buick LaCrosse boast GMC-produced die-cast Mg alloy CCBs weighing around

6.9 kg. Notably, senior vehicles like Chery Ruiqi G6 and SAIC Rongwei 550 utilize bulk Mg alloy CCBs, with lightweight single-piece structures constituting over 55%. Recent advancements, such as the implementation in Voyah's FREE model, have employed AM50 material and advanced Mg-alloy die-casting technology, resulting in a 45% weight reduction from 12.1 to 6.7 kg, showcasing a significant leap in efficiency and performance enhancement [39].

BMW stands as an exemplar in this realm, having engineered the R6, an engine comprised of a composite magnesium–aluminium alloy, which stands as the most lightweight 3.0 L in-line six-cylinder gasoline engine globally. The germane emphasis, according to the company, in adopting magnesium for this pioneering engine concept, was the substantial reduction in weight. BMW's aspiration was to attain an engine capable of enhanced power output and greater torque, all while curbing fuel consumption and CO_2 emissions. The composite crankcase, boasting a magnesium alloy housing enveloping an aluminium insert, renders it the most lightweight 3.0 L in-line six-cylinder gasoline engine worldwide, weighing in at a mere 161 kg. This resultant magnesium–aluminium alloy engine is 24% lighter than its conventional aluminium counterpart, concurrently elevating power performance and fuel efficiency. This environmentally conscious engine attains the finest power-to-weight ratio and lowest specific fuel consumption, necessitating fewer components and less laborious engine assembly. The development of Mg–Al–Sr magnesium alloy systems and a high-pressure die-casting (HPDC) process progressed hand in hand with the engine's design evolution. Mercedes-Benz, in a similar vein, pioneered a novel 7G-Tronic seven-speed automatic transmission sans any increase in weight, thanks to the integration of magnesium [42].

Titanium Alloys

16.2 Introduction

Around 1966, Mechanical alloying emerged as development at The International Nickel Company's (INCO's) Paul D Merica Research Laboratory. This advancement was a component of a program aimed at creating an alloy that combined oxide dispersion strengthening with gamma prime precipitation hardening. The primary purpose was to design a nickel-based super-alloy suitable for use in gas turbine applications [43]. INCO's exhaustive quest to amalgamate the elevated temperature strength characteristic of oxide dispersion with the intermediate temperature strength offered by a gamma prime precipitate in a nickel-based superalloy constituted the impetus for the innovation of mechanical alloying. Additionally, it was intended to include the essential corrosion and oxidation resistance. Modern mechanical alloying (MA) techniques have enabled sophisticated turbine-blading materials like the INCONEL alloy MA 6000 to accomplish this desired result [44, 45]. Beyond the formation of intricate oxide dispersion-strengthened (ODS) alloys, mechanical alloying is not just a process; it is a methodology for generating composite metal powders endowed with

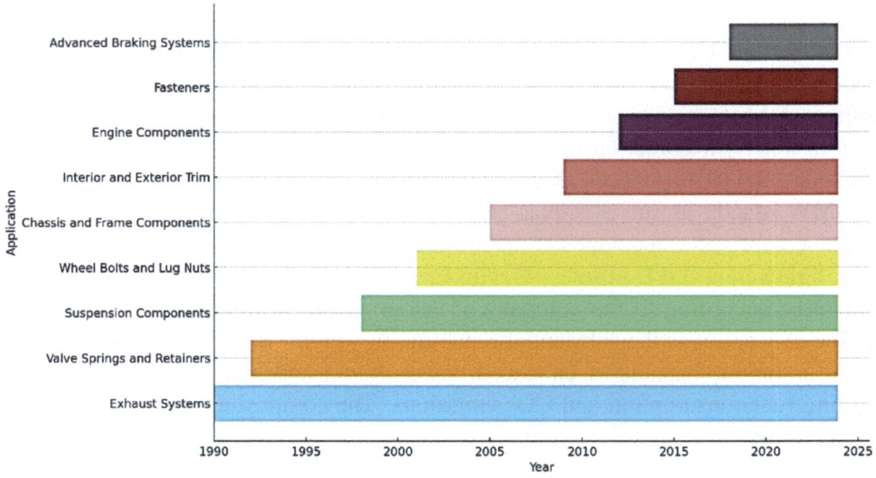

Fig. 16.6 Evolution of titanium in automobile parts

meticulously controlled, exceptionally fine microstructures. Consequently, it evolved into an invaluable tool for crafting alloys that are challenging or even unfeasible to produce through conventional melting and casting techniques [44] (Fig. 16.6).

Over 200 years ago, in 1791, titanium was initially isolated and named by Reverend William Gregor (1761–1817), an amateur chemist and the Bishop of Creed in Cornwall, UK. In Stockholm, Sweden, Jöns Jacob Berzelius first isolated titanium metal in an impure form in 1825. Subsequently, in 1887, Nilson and Pettersson also accomplished the isolation of titanium metal. However, the first large quantity of pure titanium metal was first produced at General Electric by Mathew A. Hunter in 1910. The US Bureau of Mines plant in Albany, Oregon, showcased the first commercial production of titanium in 1946. Since then, the process has remained largely unchanged from the method developed by DuPont in 1948 to produce titanium sponges commercially [46]. Due to its remarkable qualities, titanium and titanium alloys have been widely used in industry since the early 1950s. High strength-to-weight ratios, high melting points, and exceptional corrosion resistance are a few of these, especially in chloride-containing settings [47–49]. Furthermore, titanium alloys find extensive application in chemical processing industries, encompassing heat exchangers, electrochemical reactors, desalination plants, the paper industry, and hydrocarbon processing. Moreover, these alloys are also utilized in various consumer goods, including sports equipment, jewelry, architecture, decoration, and eyeglasses [50]. Since the 1950s, pure titanium has been extensively utilized as a material for surgical implants and medical instruments [51–54]. Brånemark made the ground-breaking discovery that titanium could achieve permanent integration with bone, coining the term "osseointegration" to describe the stable bonding between titanium

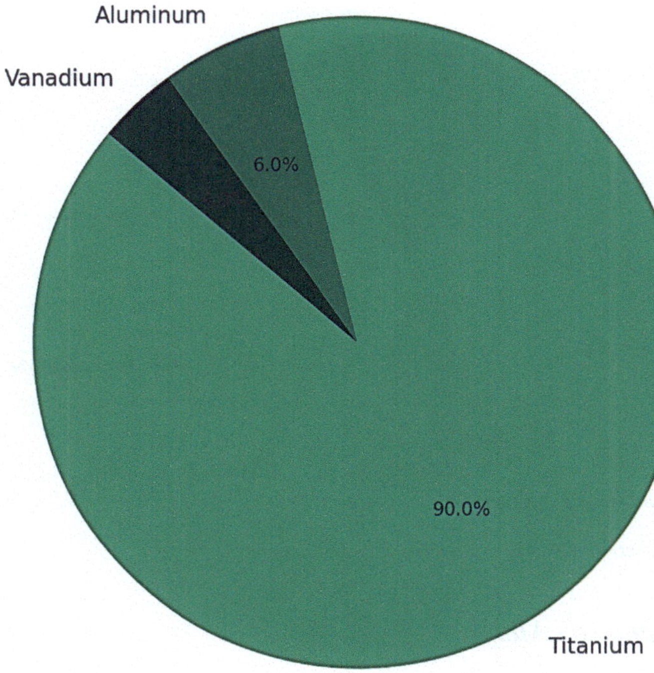

Fig. 16.7 Composition of titanium alloys in automotive applications

and bone tissues [55]. In addition to its exceptional corrosion resistance in biological settings, titanium exhibits non-cytotoxic properties, making it highly compatible with living tissues [56] (Fig. 16.7).

An alloy refers to a material comprising two or more elements (which can be either metals or nonmetals) that are thoroughly blended through processes like fusion or electro-deposition. Utilizing this principle, titanium alloys are created by introducing elements like aluminum, vanadium, molybdenum, niobium, zirconium, and various others. This combination results in the formation of diverse alloys [57]. These alloys exhibit remarkable characteristics, as demonstrated in the description below. Based on their impact on heat treating temperature and the alloying components, titanium alloys can be categorized into the subsequent three types:

1. Alpha (α) alloys
2. Alpha-beta (α-β) alloys
3. Beta (β) alloys.

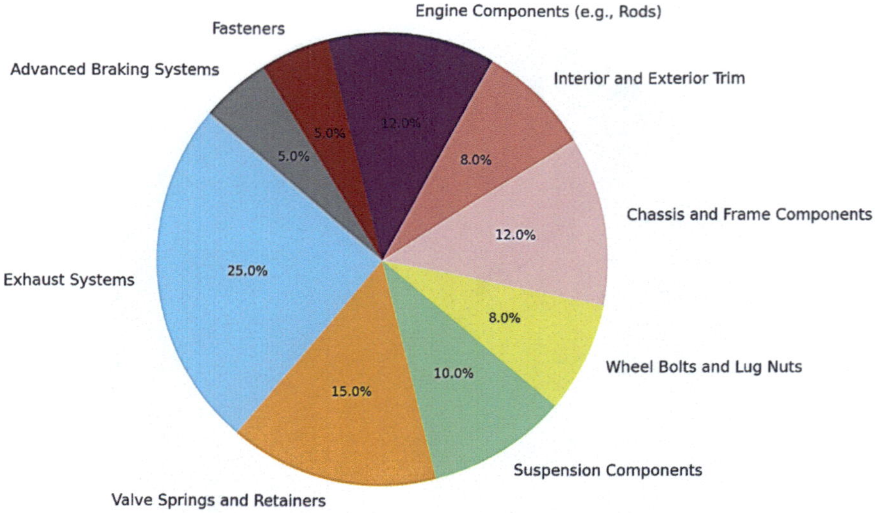

Fig. 16.8 Application of titanium alloys in automobile parts

16.2.1 Alpha (α) Alloys

Unalloyed titanium, alpha alloys, and near-alpha alloys are the three main classes of alloys that include alpha titanium alloys. Under usual temperatures, unalloyed titanium and alpha alloys that contain alpha stabilizers like aluminium and tin exhibit a solitary-phase, hexagonal close-packed (hcp) crystal structure. These alloys have only moderate tensile strengths due to their monophase composition. Nevertheless, their notable thermal stability contributes to acceptable resistance to creep and toughness. Near-alpha alloys can have up to 0.5% Si added to create silicide precipitates and up to 2% beta stabilizers, which add trace amounts of the beta phase to the microstructure. These alloys, as opposed to unalloyed titanium and alpha alloys, exhibit the highest tensile strength and the best creep resistance of all titanium alloys when subjected to temperatures above 400 °C [58]. The frequently employed α-phase alloys encompass Ti–3Al–2.5 V (designated as Ti-3-2.5) and Ti–5Al–2.5Sn (referred to as Ti-5-2.5) [59]. High strength-to-weight ratio, corrosion resistance, biocompatibility, good ductility and toughness, high temperature performance, ease of fabrication, creep resistance, low thermal expansion, high fracture toughness, and non-magnetic are the advantages of the alpha titanium alloys [60, 61] (Fig. 16.8)

16.2.2 Alpha-Beta (α-β) Alloys

The α-β alloys stand out as the most extensively utilized titanium alloys for structural purposes. These alloys contain higher quantities of β-stabilizing elements, typically ranging between 4 and 6% (by weight), including substances like molybdenum, vanadium, tungsten, tantalum, and silicon. As a result, their β-phase content surpasses that of near-α alloys, allowing for processing methods such as heat treatment to achieve enhanced strengths. The primary mechanisms for bolstering involve maintaining some metastable β-phase at room temperature or the formation of martensite from the preceding β-phase upon rapid cooling to ambient conditions. Approximately fifty percent of all titanium alloys manufactured fall into this category. Within these alloys, aluminum serves as both an α-phase stabilizer and a hardening agent due to its capacity for reinforcing solutions. An illustrative instance of an α-β titanium alloy is Ti–6Al–4V, featuring 6% aluminum and 4% vanadium by weight [57, 59]. The α-β titanium alloys—display important properties such as notable creep resistance, excellent corrosion resistance, strong fatigue strength, and favourable hot formability [62].

16.2.3 Beta (β) Alloys

Beta alloys are characterized as alloys that maintain an entirely beta structure when rapidly cooled from the beta phase range. These alloys include enough beta-stabilizing components to avoid martensite formation by keeping them from crossing the martensite start line (Ms). The composition threshold for what is referred to as metastable alloys are determined by the minimal content of beta stabilizers (βs) that identifies stable beta alloys [63]. These alloys serve as examples of the strongest titanium alloys and are referred to as "β -alloys" because they can quickly cool from the β -phase range while maintaining a totally metastable β-phase at ambient temperature. Later ageing causes the α-phase to precipitate, which then triggers reinforcing. For β-alloys, the ageing temperature is normally high enough to relieve any remaining stresses brought on by the solution treatment's quenching process. Conversely, stable beta alloys lack the capability for age hardening [59, 63]. These β-titanium alloys have great biological properties, including amazing characteristics like an extraordinarily high strength-to-weight ratio, a lowered elastic modulus, super-elasticity, increased elastic deflections, and less toxicity [62, 64, 65].

16.2.4 Titanium Alloys Production Processes

The titanium alloys can be produced by different methods;

1. Powder metallurgy
2. Atomization

    2.1 Gas atomisation process
    2.2 Plasma atomization process

3. Cold spray process
4. Electron beam melting
5. Selective laser melting
6. 3D printing.

Powder Metallurgy

The creation, processing, and compacting of finely split metal particles to produce useful engineering components is known as powder metallurgy (PM) [66]. Powder metallurgy (PM) techniques have been used to produce a wide variety of structural parts, self-lubricating bearings, and cutting tools since the 1920s. Compared to competing metalworking methods, PM processing has advantages such as form and material flexibility, application versatility, near-net-shape production capability, and recognition as an ecologically benign technology [67].

Conventional sintering, as shown in Fig. 16.9, is a popular process in the field of powder metallurgy (PM) for creating bone implants because of its capacity to successfully consolidate powder particles into a dense and strong material. In this method, feedstock powders are initially blended, then compacted under significant pressure, and ultimately subjected to sintering. The sintering step involves subjecting the materials to high temperature and pressure, inducing particle cohesion with minimal alteration to their shape. This process also enables controlled temperature regulation to facilitate the creation of porosity within the final product [69].

Atomization

Currently, the predominant method for producing spherical titanium powder is through atomization techniques, specifically gas atomization (GA), plasma atomization (PA), and plasma rotating electrode process (PREP). In these processes, three interconnected stages—melting, atomization, and solidification—are executed. Melting employs methods like vacuum induction melting or plasma arc melting, while atomization fragments liquid metal into droplets that solidify in a cooling chamber, shielded by inert gas. Commercial atomization utilizes high-pressure gas or rotating liquid streams, creating molten droplets that solidify during flight. These processes, conducted in ultra-high-purity inert gases, ensure oxygen pickup is minimal, ranging between 100 and 500 ppm [70].

Gas Atomization (GA)

The Crucible Materials Corporation invented the titanium gas atomization process in the 1980s. In this process, pre-alloyed titanium alloy ingots or bars are induction skull-melted in a copper crucible that has been water-cooled while being conducted in a vacuum. The molten metal is poured into a tundish through a refractory metal

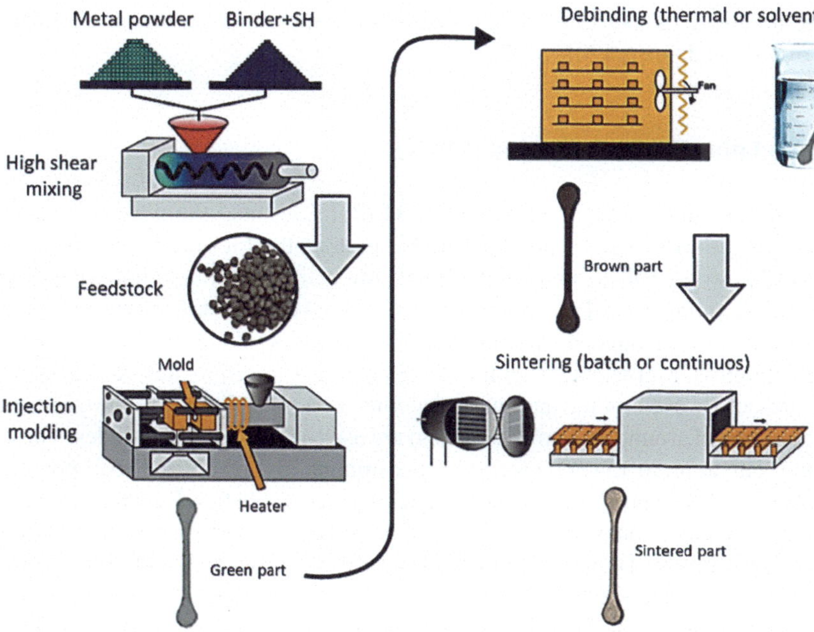

Fig. 16.9 Powder metallurgy process [68]

nozzle once the alloy composition has become homogeneous after spending enough time in the molten state. It is then atomized by streams of high-pressure input gas (typically, Argon is preferred over Helium due to cost savings). The molten metal stream descends under the influence of gravity for a predetermined distance prior to atomization, and this process is known as free-fall gas atomization (FFGA) [70].

Plasma Atomization (PA)

The Plasma Atomization (PA) technique was developed in 1996 with the aim of producing finely spherical powders, where the plasma atomization process involves introducing a pre-alloyed wire (for instance, 1/16″ or 1/8″) into a high-temperature region (approximately 10,000 K) that is heated by plasma torches. This causes the wire to melt and break into droplets that cool rapidly. The cooling rate typically falls within the range of 10–1000 °C/s (10–1000 °C/s). A notable distinction between plasma atomization and gas atomization lies in the method by which the wire is handled. In PA, the wire is simultaneously melted and atomized by an extremely high-temperature plasma, whereas in GA, the metal is melted using an induction coil or another heat source before being atomized by cold, high-pressure gas. Because the liquid metal is kept separate from any refractory metals or solid materials that could contaminate the powder before it hardens, plasma-atomized titanium powder boasts exceptional purity [70].

Plasma Rotating Electrode Process (PREP)

Nuclear Metals/Starmet (now a division of TIMET) invented the plasma rotating electrode process (PREP), a centrifugal atomization method for producing titanium pre-alloyed powder. This technique involves using helium plasma to liquefy the tip of a quickly spinning centerless ground bar. The molten droplets are hurled away and solidify in the air, surrounded by helium. Figure 2.6 shows the procedure schematically. A stainless steel chamber with a diameter of 2.4 m (96 inches) and positive helium pressure is used for the melting and atomization procedures. Pre-alloyed bars with a diameter of around 64 mm (2.5 inches) are used as the initial electrodes. These bars can rotate at up to 15,000 revolutions per minute. The resulting PREP powder has a spherical shape and superior flow and packing characteristics. Depending on the particular alloy, electrode diameter, and rotational speed, the particle size distribution changes. For instance, Ti–6Al–4V powder typically has a size range of 100–300 m, with a median particle size (d50) of roughly 175 m. Less than 1% of the powder produced during this manufacturing method has a particle size less than 50 m [71].

Cold Spray Process

The most recent innovation in surface coating techniques is the cold spray (CS) process. It involves coating metallic or dielectric substrates with metal, alloy, or composite powders. This is accomplished by launching a high-velocity jet of small particles—sized between 5 and 50 m at speeds between 300 and 1200 m/s through a specially designed nozzle into a stream of compressed and preheated gas. Initially, the compressed gas is heated by a gas heater; then powder is added to the gas stream; and the gas-powder mixture is accelerated through a nozzle of the de Laval type and applied to a substrate. To improve the core system, additional machinery for in-process tasks like milling, laser treatment, and helium recycling has been developed. Gas conditions, nozzle location and movement, and powder feed rates are the key process variables. It is possible to create compact coatings with deposition efficiency greater than 95% (for example, for CP-Ti) when these parameters are fine-tuned. This deposition technique produces coatings that have a number of benefits while operating at temperatures that are much below the melting point of the majority of feedstock powders. These benefits include low oxidation, outstanding repeatability, great flexibility, rapid deposition rates, and cost savings [69, 72].

Electron Beam Melting

One method for producing titanium coatings using additive manufacturing is electron beam melting (EBM). This process includes using an electron beam that is magnetically directed to melt and deposit metal particles one layer at a time. Although this method has been successful, it has higher initial setup costs because a high vacuum environment is required [57].

Selective Laser Melting

Another additive manufacturing method used to produce titanium alloy coatings is selective laser melting (SLM). In this procedure, a strong laser beam is used to completely melt the powder. Due to the requirement for complex quick cooling systems, SLM has substantial costs, just like the approach discussed earlier. Furthermore, the quality of the finished products may suffer as a result of temperature changes that occur during processing [57, 69].

16.2.5 Applications of Titanium Alloys in Automobile Parts

Incorporating TiC particulates has notably mitigated the wear rate of the MMC brake disc compared to grey cast iron. Wear assessments depict a consistent reduction in wear rate corresponding to increased TiC content, illustrating a clear influence of load and sliding speed. Consequently, the wear observed in the MMC stands substantially lower than that experienced by the cast iron counterpart. Interestingly, when automotive brake shoe materials interface with the MMC, the observed wear is relatively higher compared to the interaction with grey cast iron. This suggests a distinctive interplay between the brake shoe material and the MMC surface, hinting at potential disparities in material properties and frictional dynamics [73]. Titanium and its diverse alloys, encompassing commercial pure titanium grades like 1 and 2, Ti–6Al–4V, Ti–4.5Fe–6.8Mo–1.5Al, among others, epitomize foundational materials within the automotive manufacturing domain. Their assimilation stems from their unparalleled attributes: superior thermal resilience and formidable resistance to corrosion. These inherent qualities substantiate their indispensability in crafting engine components and accompanying paraphernalia pivotal to vehicular functionality. The infusion of titanium dioxide nanoparticles into steel alloys utilized in automotive fabrication presents a promising avenue for ameliorating component weight. This innovation unlocks the potential for employing titanium alloys in critical lightweight applications encompassing valves, steering knuckles, wheel hubs, brackets, suspension springs, and links, thereby heralding prospects for heightened fuel efficiency. Moreover, within the domain of braking systems, titanium alloys eclipse their aluminum counterparts in the context of disc brake rotors. Notably, the laser refinement of Ti–6Al–4V conspicuously enhances its hardness, positioning it as a quintessential material for brake disc rotors, thus amplifying braking efficacy and durability. Beyond these applications, the dual functionality of titanium dioxide nanoparticles as tire wall pigments and synergistic agents with silane in filler modifications augments interactions within rubber compositions. This strategic augmentation curtails dynamic hysteresis losses within tires, potentially yielding a 2% decrement in fuel consumption for every 10% reduction in hysteresis loss [74].

Conclusion

In summary, the fusion of magnesium and titanium alloys represents the zenith of automotive engineering, embodying the pursuit of efficiency, environmental stewardship, and heightened vehicular performance. This chapter meticulously dissected the intrinsic attributes of these alloys, showcasing their individual merits and the unparalleled synergy they manifest when coalesced. From their corrosion resistance to thermal stability and mechanical resilience, these materials epitomize a transformative frontier in sustainable automotive design. Notwithstanding the challenges in fabrication and processing, their assimilation has yielded tangible benefits, bolstering vehicle structures while significantly diminishing weight, thereby elevating fuel efficiency. The consequential economic and environmental implications of their assimilation are profound, heralding a paradigm shift towards eco-conscious practices and optimal resource utilization within the automotive sector.

References

1. Tun, K.S., Kumar, A., Gupta, M.: Introducing a high performance Mg-based multicomponent alloy as an alternative to Al-alloys. Front Mater **6**, 1–6 (2019). https://doi.org/10.3389/fmats.2019.00215
2. Ravikanth Reddy, C., Satya Prasad, K., Srinivasarao, B.: Microstructure and mechanical properties of Mg–Ni–Gd alloy synthesised by powder metallurgy. Powder Metall. **66**, 270–277 (2023). https://doi.org/10.1080/00325899.2022.2163553
3. Sharma, S.K., Saxena, K.K., Malik, V., Mohammed, K.A., Prakash, C., Buddhi, D., Dixit, S.: Significance of alloying elements on the mechanical characteristics of Mg-based materials for biomedical applications. Crystals **12**, 1–40 (2022). https://doi.org/10.3390/cryst12081138
4. Loukil, N.: Alloying elements of magnesium alloys: A literature review. In: Magnesium Alloys, pp 1–11 (2016)
5. Zhang, W., Xu, J.: Advanced lightweight materials for automobiles: a review. Mater. Des. **221**, 1–20 (2022). https://doi.org/10.1016/j.matdes.2022.110994
6. Guan, D.: Processing, microstructure and mechanical behavior of bulk nanostructured Mg alloy (2015)
7. Giridasappa, A., Kiran, K.K., Rajendrachari, S.: Nanotechnology and their evaluation of Bi-functional applications. J. Metastable Nanocrystalline Mater. **37**, 33–43 (2023). https://doi.org/10.4028/p-dKORv2
8. Rajendrachari, S.: An overview of high-entropy alloys prepared by mechanical alloying followed by the characterization of their microstructure and various properties. Alloys **1**, 116–132 (2022). https://doi.org/10.3390/alloys1020008
9. Rajendrachari, S., Shatokha, V., Avar, B.: New Advances in Powder Technology (2023)

10. Rajendrachari, S., Adimule, V., Gulen, M., Khosravi, F., Somashekharappa, K.K.: Synthesis and characterization of high entropy alloy 23Fe–21Cr–18Ni–20Ti–18Mn for electrochemical sensor applications. Materials (Basel) **15**, 1–12 (2022). https://doi.org/10.3390/ma15217591
11. Seth, P.P., Parkash, O., Kumar, D.: Structure and mechanical behavior of in situ developed Mg2Si phase in magnesium and aluminum alloys—A review. RSC Adv. **10**, 37327–37345 (2020). https://doi.org/10.1039/d0ra02744h
12. Kanaginahal, G.M., Mahale, R.S., Vasanth, S., Sharath, P.C., Shashanka, R.: Studies on dielectric, super-capacitive and photoluminescence properties of metal organic frameworks subjected to varying temperature and frequencies. Nano Hybrids Compos. **39**, 17–25 (2023). https://doi.org/10.4028/p-si6964
13. Salleh, E.M., Ramakrishnan, S., Hussain, Z.: Synthesis of biodegradable Mg–Zn alloy by mechanical alloying: effect of milling time. Procedia Chem. **19**, 525–530 (2016). https://doi.org/10.1016/j.proche.2016.03.048
14. Mahale, R.S., Kanaginahal, G.M., Vasanth, S., Tiwary, V.K., Shashanka, R., Sharath, P.C., Patil, A.: Applications of fused deposition modeling in dentistry. In: Keshavamurthy, R., Tambrallimath, V., Davim, J.P. (eds.) Development, Properties, and Industrial Applications of 3D Printed Polymer Composites, pp. 211–219. IGI Global, Hershey, Pennsylvania (2023)
15. Shashanka, R.: Synthesis of nano-structured stainless steel powder by mechanical alloying-an overview. Int. J. Sci. Eng. Res. **8**, 588–594 (2017)
16. Nečas, D., Marek, I., Pinc, J., Vojtěch, D., Kubásek, J.: Advanced Zinc–Magnesium alloys prepared by mechanical alloying and spark plasma sintering. Materials (Basel) **15**, 1–17 (2022). https://doi.org/10.3390/ma15155272
17. Rajendrachari, S., Adimule, V.M., Jayaprakash, G.K., Pandith, A.: Electrochemical oxidation of methylene blue dye in wastewater using mechanically alloyed high entropy alloy modified carbon paste electrode using cyclic voltammetry. Mater. Res. Express **10**, 1–13 (2023). https://doi.org/10.1088/2053-1591/acd3d7
18. Pradeep, N., Hegde, M.R., Rajendrachari, S., Surendranathan, A.: Investigation of microstructure and mechanical properties of microwave consolidated TiMgSr alloy prepared by high energy ball milling. Powder Technol. **408**, 1–15 (2022). https://doi.org/10.1016/j.powtec.2022.117715
19. Basavarajappa, P.N., Hegde, M.M.R., Rajendrachari, S., Surendranathan, A.O.: Investigation of structural and mechanical properties of nanostructured TiMgSr alloy for biomedical applications. Biointerface Res. Appl. Chem. **13**, 1–18 (2023). https://doi.org/10.33263/BRIAC132.118
20. Gonzaga, S., Molina, A., Guardián, R., Martínez, H., Vélez, E.V., Tapia, J.S.O.: Synthesis of Magnesium-based alloys by mechanical alloying for implant applications. Coatings **13**, 1–15 (2023). https://doi.org/10.3390/coatings13020260
21. Rajendrachari, S., Basavegowda, N., Vinaykumar, R., Narsimhachary, D., Somu, P., Lee, M.-J.: Electrocatalytic determination of methyl orange dye using mechanically alloyed novel metallic glass modified carbon paste electrode by cyclic voltammetry. Inorg. Chem. Commun. **155**, 1–13 (2023). https://doi.org/10.1016/j.inoche.2023.111010
22. Kane, S.N., Mishra, A., Dutta, A.K.: Microstructure and mechanical properties of the as-cast Mg–Zn–Mn–Ca alloys. J. Phys. Conf. Ser. **755** (2016). https://doi.org/10.1088/1742-6596/755/1/011001
23. Ramakrishna, D., Saravu, S., Rajendrachari, S.: A palladium complex dispersed in ionic liquid as an efficient catalytic combination for the synthesis of Benzazoles. Top. Catal. **65**, 1–12 (2022). https://doi.org/10.1007/s11244-021-01554-6
24. Li, Y., Yuan, Y., Wang, J., Wu, L., Cao, F., Zhang, L., Pan, F.: Controllable degradation behavior of Mg–Sr–Y alloys for the bio-applications. NPJ Mater. Degrad. **7** (2023). https://doi.org/10.1038/s41529-023-00366-1
25. Mahale, R.S., Vasanth, S., Chikkegouda, S.P., Rajendrachari, S., Narsimhachary, D., Basavegowda, N.: Electrochemical determination of ascorbic acid by mechanically alloyed super duplex stainless steel powders. Metals (Basel) **13**, 1–14 (2023). https://doi.org/10.3390/met13081430

26. Mahale, R.S., Shaygan, S., Attaeyan, A., Vasanth, S., Krishna, H., Sharath, P., Shashanka, R., Malekipour, M., Ghorbani, A., Prasad, H., Ghaffari, Y.: A Brief review of the structure designed using metallic 3D printing for biomechanics applications. In: Additive Manufacturing for Biocomposites and Synthetic Composites, pp. 120–137 (2023)
27. Esamael, S.K., Fatalla, A.A.: Analysis of effect of alloying Magnesium by powder metallurgy on mechanical properties: review. J. Res. Med. Dent. Sci. **9**, 100–106 (2021)
28. Adimule, V., Keri, R., Rajendrachari, S., Kendrekar, P., Kulkarni, C.V.: Highly photoluminescence and wide band gap insulating metal hybrid nanoparticles array of samarium-doped SrO:CoO: synthesis, characterizations and sensor characteristics. J. Mater. Sci. Mater. Electron. **34** (2023). https://doi.org/10.1007/s10854-023-09899-2
29. Halfadji, A., Naous, M., Rajendrachari, S., Ceylan, Y., Ceylan, K.B., Shekar, P.R.: Effective investigation of electro-catalytic, photocatalytic, and antimicrobial properties of porous CuO nanoparticles green synthesized using leaves of Cupressocyparis leylandii. J. Mol. Struct. **1–14** (2023). https://doi.org/10.1016/j.molstruc.2023.137318
30. Hrapkowiczk, B., Lesz, S., Kremzer, M., Karolus, M., Pakieła, W.: Mechanical alloying of Mg–Zn–Ca–Er alloy. Bull. Polish Acad. Sci. Tech. Sci. **69**, 1–8 (2021). https://doi.org/10.24425/bpasts.2021.137587
31. Adimule, V., Shashanka, R.: Dielectric, Supercapacitive, Photoluminescent Properties of Hybrid Metal Organic Frameworks (2023)
32. Zhu, S.-S., Kan, H.-M., Feng, X.-J., Zhang, N., Sun, H.: The preparation method of aluminum-magnesium alloys. 299–305 (2016). https://doi.org/10.2991/ame-16.2016.50
33. Patil, A., Nithin, S., Krishna, H., Vasanth, S., Shrinivas, R., Mahale, S.P., Rajendrachari, S.: Investigation of wear behavior for three-body made of CA-40 steel and nickel hard cast iron by abrasion method. AIP Conf. Proc. **2901**, 1–15 (2023). https://doi.org/10.1063/5.0178634
34. Liu, C., Ren, Z., Xu, Y., Pang, S., Zhao, X., Zhao, Y.: Biodegradable magnesium alloys developed as bone repair materials: a review. Scanning **2018** (2018). https://doi.org/10.1155/2018/9216314
35. Halfadji, A., Naous, M., nadia Kharroubi, K., el zahraà Belmehdi, F., Rajendrachari, S.: An ultrasonic-assisted synthesis, characterization, and application of Nano-Fe3O4/TiO2 as nano-catalyst for the removal of organic dye by Like-Photo-Fenton reactions. Inorg. Chem. Commun. **158**, 1–15 (2023). https://doi.org/10.1016/j.inoche.2023.111686
36. Zhou, H., Hu, L., Sun, Y., Zhang, H., Duan, C., Yu, H.: Synthesis of nanocrystalline AZ31 magnesium alloy with titanium addition by mechanical milling. Mater. Charact. **113**, 108–116 (2016). https://doi.org/10.1016/j.matchar.2016.01.014
37. Rajendrachari, S., Arslanoglu, H., Yaras, A., Golabhanvi, S.M.: Electrochemical detection of uric acid based on a carbon paste electrode modified with Ta2O5 recovered from ore by a novel method. ACS Omega **8**, 46946–46954 (2023). https://doi.org/10.1021/acsomega.3c06749
38. Halfadji, A., Chougui, A., Djeradi, R., Ouabad, F.Z., Aoudia, H., Rajendrachari, S.: TiO2-decorated by nano-γ-Fe2O3 as a catalyst for efficient photocatalytic degradation of orange G dye under eco-friendly white LED irradiation. ACS Omega **8**, 39907–39916 (2023). https://doi.org/10.1021/acsomega.3c06420
39. Liu, B., Yang, J., Zhang, X., Yang, Q., Zhang, J., Li, X.: Development and application of magnesium alloy parts for automotive OEMs: a review. J. Magnes. Alloy **11**, 15–47 (2023). https://doi.org/10.1016/j.jma.2022.12.015
40. Thiagarajan, C., Lakshminarayanan, N., Anand, A., Nikhil Santhosh, M., Anderson, N.J.: Investigation and analysis of properties of magnesium alloy for suitability to electric vehicle components. IOP Conf. Ser. Mater. Sci. Eng. **993** (2020). https://doi.org/10.1088/1757-899X/993/1/012007
41. Yoon, J., Lee, S.I.: Warm forging of magnesium AZ80 alloy for the control arm in an automobile. Proc. Inst. Mech. Eng. Part D J. Automob. Eng. **229**, 1732–1738 (2015). https://doi.org/10.1177/0954407014567909
42. Kulekci, M.K.: Magnesium and its alloys applications in automotive industry. Int. J. Adv. Manuf. Technol. **39**, 851–865 (2008). https://doi.org/10.1007/s00170-007-1279-2

43. Benjamin, J.S., Co, A.: Mechanical alloying—A perspective. Met. Powder Rep. **45**, 122–127 (1990). https://doi.org/10.1016/S0026-0657(10)80124-9
44. Gilman, P.S., Benjamin, J.S.: Mechanical alloying. Annu. Rev. Mater. Sci. **13**, 279–300 (1983)
45. Kim, Y.G., Merrick, H.F.: Characterization of an oxide dispersion strengthened superalloy, MA-6000E, for turbine blade applications (1979)
46. Cardarelli, F.: Less Common Nonferrous Metals. Springer, Cham, Switzerland (2008)
47. Balazic, M., Kopac, J., Jackson, M.J., Ahmed, W.: Review: titanium and titanium alloy applications in medicine. Int. J. Nano Biomater. **1**, 3–34 (2007). https://doi.org/10.1504/IJNBM.2007.016517
48. Eylon, D., Fujishiro, S., Postans, P.J., Froes, F.H.: High-temperature titanium alloys—A review. JOM 55–62 (1984)
49. Peters, B.M., Kumpfert, J., Ward, C.H., Leyens, C.: Titanium alloys for aerospace applications. Adv. Eng. Mater. **5**, 419–427 (2003). https://doi.org/10.1002/adem.200310095
50. Pio, E., Medina, A., Mart, C., Manuel, F., Cerda, C.: Evolution of face-centered cubic Ti alloys transformation by X-ray diffraction profile analysis in mechanical alloying. Metals (Basel) **11**, 1–17 (2021). https://doi.org/10.3390/met11111841
51. Zhou, Y., Niinomi, M., Akahori, T., Nakai, M., Fukui, H.: Comparison of various properties between titanium-tantalum alloy and pure titanium for biomedical applications. Mater. Trans. **48**, 380–384 (2007). https://doi.org/10.2320/matertrans.48.380
52. Zhang, L., Chen, L.: A review on biomedical titanium alloys: recent progress and prospect. Adv. Eng. Mater. **21**, 1–29 (2019). https://doi.org/10.1002/adem.201801215
53. Zhou, Y.L., Niinomi, M., Akahori, T.: Effects of Ta content on Young's modulus and tensile properties of binary Ti–Ta alloys for biomedical applications. Mater. Sci. Eng. **371**, 283–290 (2004). https://doi.org/10.1016/j.msea.2003.12.011
54. Geetha, M., Singh, A.K., Asokamani, R., Gogia, A.K.: Ti based biomaterials, the ultimate choice for orthopaedic implants—A review. Prog. Mater. Sci. **54**, 397–425 (2009). https://doi.org/10.1016/j.pmatsci.2008.06.004
55. Alipal, J., Lee, T.C., Koshy, P., Abdullah, H.Z., Idris, M.I.: Evolution of anodised titanium for implant applications. Heliyon **7**, e07408 (2021). https://doi.org/10.1016/j.heliyon.2021.e07408
56. Sarraf, M., Rezvani, E., Saeid, G., Seeram, A., Nazatul, R., Sukiman, L.: A state of the art review of the fabrication and characteristics of titanium and its alloys for biomedical applications. Bio-Des. Manuf. **5**, 371–395 (2022). https://doi.org/10.1007/s42242-021-00170-3
57. Hamweendo, A., Botef, I.: Modern production methods for titanium alloys: a review. In: Titanium Alloys-Novel Aspects of their Manufacturing and Processing, pp. 1–17 (2019)
58. Weiss, I., Semiatin, S.L.: Thermomechanical processing of alpha titanium alloys—An overview. Mater. Sci. Eng. **263**, 243–256 (1999)
59. Williams, J.C., Boyer, R.R.: Opportunities and issues in the application of titanium alloys for aerospace components. Metals (Basel) **10**, 705 (2020). https://doi.org/10.3390/met10060705
60. Tabie, V.M., Li, C., Saifu, W., Li, J., Xu, X.: Mechanical properties of near alpha titanium alloys for high-temperature applications—A review. Aircr. Eng. Aerosp. Technol. **92**, 521–540 (2020). https://doi.org/10.1108/AEAT-04-2019-0086
61. Been, J., Faller, K., Eli, T.: Using Ti-5111 for marine fastener applications. JOM **51**, 21–24 (1999)
62. Elahinia, M.H., Hashemi, M., Tabesh, M.: Progress in materials science manufacturing and processing of NiTi implants: a review. Prog. Mater. Sci. **57**, 911–946 (2012). https://doi.org/10.1016/j.pmatsci.2011.11.001
63. Weiss, I., Semiatin, S.L.: Thermomechanical processing of beta titanium alloys—An overview. Mater. Sci. Eng. **243**, 46–65 (1998)
64. Sidambe, A.T.: Biocompatibility of advanced manufactured titanium implants—A review. Materials (Basel) **7**, 8168–8188 (2014). https://doi.org/10.3390/ma7128168
65. Yang, R., Hao, Y., Li, S.: Development and application of low-modulus biomedical titanium alloy Ti2448. In: Biomedical Engineering: trends in Materials Science, pp. 1–25 (2014)
66. Froes, F.H., Eylon, D.: Powder-metallurgy of titanium alloys. Int. Mater. Rev. **35**, 162–184 (1990). https://doi.org/10.1179/095066090790323984

67. Alexandra, B., Hidalgo, A., Frykholm, R., Ebel, T., Pyczak, F.: Powder metallurgy strategies to improve properties and processing of titanium alloys: a review. Advanced **1–14** (2017). https://doi.org/10.1002/adem.201600743
68. Neto, F.C., Giaretton, M.V., Neves, G.O., Aguilar, C., Tramontin Souza, M., Binder, C., Klein, A.N.: An overview of highly porous titanium processed via metal injection molding in combination with the space holder method. Metals (Basel) **12** (2022). https://doi.org/10.3390/met12050783
69. Hamweendo, A., Malama, T., Botef, I.: Titanium-nickel alloys for bone tissue engineering application via cold titanium-nickel alloys for bone tissue engineering application. In: International Conference on Competitive Manufacturing COMA16, pp. 273–279 (2016)
70. Sun, P., Fang, Z.Z., Zhang, Y., Xia, Y.: Review of the methods for production of spherical Ti and Ti alloy powder. JOM **69**, 1853–1860 (2017). https://doi.org/10.1007/s11837-017-2513-5
71. Yolton, C.F., Froes, F.H.S.: Conventional Titanium Powder Production. Elsevier Inc. (2015)
72. Vo, P., Goldbaum, D., Wong, W., Irissou, E., Legoux, J., Chromik, R.R., Yue, S.: Cold-Spray Processing of Titanium and Titanium Alloys. Elsevier Inc. (2015)
73. Sivananth, V., Karuppusamy, P., Lingadurai, K.: Wear and corrosion behaviour of titanium carbide reinforced metal matrix composites for automobile brake disc application. Int. J. Mater. Eng. Innov. **10**, 246–267 (2019). https://doi.org/10.1504/IJMATEI.2019.101970
74. Akinlabi, S.A., Ogbonna, O.S., Mashinini, P.M., Adediran, A.A., Fatoba, O.S., Akinlabi, E.T.: Titanium and epoxy for automobile application: a review. In: Proceedings of the Eighth International Conference on Advances Civil, Structural and Mechanical Engineering CSM 2019, pp. 48–58 (2019)

Chapter 17
Energy Storage Applications of Mechanically Alloyed Materials Supercapacitors, Battery Applications

R. Raghavendra Rao, B. N. Sharath, Puttaswamy Madhusudan, and S. Pradeep

Abstract The ability of any society throughout human history to develop and instigate novel materials that suit their needs has resulted in the advancement of their performance and ranking in the world. The differences in the quality is measured by the progress made on health, education, industry, economic, culture, etc., between a country to country and region to another are always attributed to man's ability for developing materials and manufacturing equipment. Amorphous and metallic glassy alloys, carbon nanotubes, nanostructured materials, nanoparticles, nanocomposites, and nanostructured materials are some of the newly engineered materials that can be effectively produced using this room-temperature method of production. Mechanical alloying, also known as ball milling or mechanical alloying, is the process of mixing and grinding pre-alloyed or elemental powders in a high-energy ball mill. These materials can be beneficial for a variety of energy storage applications because of the distinctive characteristics and microstructures produced by this method. The revolution in Electric vehicle technology nowadays is leading to a focus on the inventions of battery storage applications. This has led many scientists and researchers to conduct research on energy storage applications to suit their needs for different industrial applications. In that context, the supercapacitors have emerged to be the ray of hope for the EV revolution. Due to their unique features supercapacitors synthesised by mechanically alloyed powders have emerged as the latest trend in this area.

Keywords Energy storage · Supercapacitors · Battery applications

R. Raghavendra Rao (✉) · B. N. Sharath · S. Pradeep
Department of Mechanical Engineering, Malnad College of Engineering, Hassan, Affiliated to VTU, Belagavi, Karnataka, India
e-mail: rrr@mcehassan.ac.in

B. N. Sharath
e-mail: sbn@mcehassan.ac.in

P. Madhusudan (✉)
Department of Civil and Environmental Engineering, Hanyang University, 222, Wangsimni-ro, Seongdong-gu, Seoul Campus, Seoul 04763, South Korea
e-mail: madhu2010@hanyang.ac.kr

S. Rajendrachari (ed.), *Mechanically Alloyed Novel Materials*, Advanced Structured Materials 220, https://doi.org/10.1007/978-981-97-6504-1_17

17.1 Introduction

Materials scientists have always focused their scientific inquiries on finding ways to enhance the characteristics and performance of materials. Notable enhancements in mechanical, chemical, and physical characteristics have been attained via chemical alterations and traditional techniques of thermal, mechanical, and thermomechanical techniques [1–4]. The pursuit of materials that possess greater strength, stiffness, and reduced weight compared to conventional materials has resulted in the creation and advancement of sophisticated materials. The endeavours have been further stimulated by the high-technology sectors. Advanced materials can be defined as materials that prioritise the methodical synthesis and regulation of their structure, with the aim of producing a precisely customised range of properties that are suitable for challenging applications [5–8]. Energy storage technologies have been in use for many years and have undergone constant development to reach their current levels of maturity, which are for numerous storage kinds. There are numerous types of energy storage systems, and they can be divided into numerous categories [9].

17.1.1 Battery Electrodes

Batteries and other electrochemical energy storage technologies depend heavily on their battery electrodes. Batteries use a process called redox (reduction-oxidation) to change chemical energy into electrical energy [10]. The cathode and the anode are the two primary electrode types found in batteries. Cathode is the electrode that experiences reduction, or the gain of electrons, during discharge known as the cathode (when the battery is generating power) [11]. For instance, the cathode material in a typical lithium-ion battery is frequently a metal oxide, such as lithium iron phosphate ($LiFePO_4$), lithium manganese oxide ($LiMn_2O_4$), or lithium cobalt oxide ($LiCoO_2$). During the discharge process, these minerals serve as a source of positively charged lithium ions. Anode: During discharge, oxidation, or the loss of electrons, takes place at the anode electrode. Graphite and other carbon-based materials are typically used as the anode material in lithium-ion batteries. Lithium ions transfer electrical energy from the anode to the cathode during discharge by passing through an electrolyte [12, 13]. The battery's total performance is largely dependent on the performance of its electrodes. Electrode composition, porosity, surface area, and conductivity are important factors. To improve the energy density, cycle life, and safety of batteries, researchers are always creating new electrode materials. Different materials may be used for the electrodes of other battery types, such as nickel-metal hydride (NiMH), lead-acid, and future technologies like solid-state batteries, in addition to conventional lithium-ion batteries. Based on the materials used for the electrodes and general construction, each type of battery has a unique combination of benefits and drawbacks. Advanced electrode materials for batteries can be produced using mechanically alloyed materials. These materials may function electrochemically more effectively, have greater capacity, and have longer cycle lives. For instance, mechanical alloying of nanoscale alloys or composites can improve the performance of the electrode in lithium-ion batteries [14, 15]

17.1.2 Supercapacitors

Materials having a large surface area and good electrical conductivity are necessary for supercapacitors, also known as ultra-capacitors. The energy and power density of supercapacitors can be increased by using high-surface-area electrode materials made from mechanically alloyed materials. Supercapacitors are energy storage devices that fill the void left by conventional capacitors and rechargeable batteries. They are sometimes referred to as ultra-capacitors or electrochemical capacitors. They work similarly to traditional capacitors in that they use electrochemical principles to obtain significantly greater capacitance values while still storing electrical energy through the separation of charges [16–20].

Capacitance

Compared to conventional capacitors, supercapacitors have a substantially greater capacitance. They are able to store and distribute more electrical energy as a result.

Energy Density

Supercapacitors store more energy per unit mass or volume than batteries do, but their power density—the rate at which energy can be provided or absorbed—is higher in supercapcitors than in batteries. They are therefore appropriate for uses requiring quick energy discharge and recharging.

Charge and Discharge Rate

Supercapacitors are well suited for applications requiring fast bursts of power because of their rapid charge and discharge rates. Batteries, on the other hand, normally have slower rates of charge and discharge.

Lifespan

Compared to batteries, supercapacitors have a longer cycle life. They don't significantly lose performance even after many cycles of charging and discharging.

Voltage

A single supercapacitor cell often has a low voltage. A series connection of many cells is frequently used to reach larger voltage levels.

Applications

Supercapacitors are used in consumer electronics, industrial applications, transportation (including electric and hybrid cars), renewable energy systems, and consumer electronics.

Construction

Supercapacitors are made up of two electrodes that are spaced apart by an electrolyte. Activated carbon is commonly used to make the electrodes, and organic or aqueous electrolyte can be used. At the interface where the electrodes and electrolyte meet, charges separate.

Types

Supercapacitors come in a variety of forms, such as electrochemical double-layer capacitors (EDLCs) and pseudocapacitors. While pseudocapacitors employ a combination of electrochemical and electrostatic processes, EDLCs store energy by the electrostatic separation of charges.

Research and Development

To make supercapacitors more competitive with batteries in terms of total energy storage, ongoing research aims to increase their energy density. In this sense, material and design advancements are essential. In certain situations, supercapacitors offer advantages over more conventional energy storage devices due to their special combination of properties, making them a promising technological advancement [21–26].

17.1.3 Hydrogen Storage

Hydrogen is seen as a viable energy source, and materials for storing hydrogen are essential for the advancement of hydrogen-based energy systems. Due to their better kinetics and thermodynamics, mechanically alloyed materials, such as metal hydrides, can be employed for hydrogen storage applications. One essential component of using hydrogen as a clean, renewable energy source is hydrogen storage. Despite having a large energy content by weight, hydrogen is difficult to store due to its low density. Hydrogen can be stored via a variety of techniques, each with pros and cons [27].

Compressed Hydrogen Gas (CHG)

High pressures are used to compress and store hydrogen gas, usually between 350 and 700 bar (5,000 and 10,000 psi). A well-established technology, compressed hydrogen can be used in both stationary and mobile applications. The energy needed for compression and the requirement for sturdy, large storage tanks provide the biggest obstacles [28].

Liquid Hydrogen (LH2)

To turn hydrogen into a liquid, it is cooled to a very low temperature (around −253 °C or −423°F). Compared to compressed gas, liquid hydrogen has a higher energy density; however, its liquefaction process requires a lot of energy, and cryogenic storage tanks are necessary.

Metal Hydrides

Metal hydrides are created when specific metal alloys absorb hydrogen. Hydrogen may be stored at lower temperatures and pressures thanks to this technique. Although they have a relatively high volumetric energy density, metal hydrides can be costly and hefty. Hydrogen may also leak slowly [29].

Chemical Hydrogen Storage

Liquid organic hydrogen carriers (LOHCs) and ammonia are two examples of substances to which hydrogen can form chemical bonds. The benefit of chemical storage is that reactions may be reversible, enabling the controlled release of hydrogen when needed. Developing effective and affordable catalysts for the release of hydrogen and the regeneration of the storage material are two challenges.

Carbon-Based Materials

By physisorption, materials such as porous carbon, graphene, and carbon nanotubes can adsorb hydrogen. Although this approach is lightweight, it is still difficult to achieve high storage capacity and quick-release kinetics.

Hydrogen Storage in Salt Caverns

Large underground salt caverns can be used to store hydrogen. High-pressure hydrogen is injected into the caverns and can be released when needed. This method is suitable for large-scale storage but depends on the location. The choice of storage technology is often determined by the particular application and requirements. Researchers are always looking for new materials and technologies to improve the efficiency, safety, and affordability of hydrogen storage [30].

17.1.4 Flywheels

Materials with great tensile strength and exceptional energy density are used in flywheel energy storage devices. Advanced composite materials for flywheel rotors, which are utilised to efficiently store and release energy, can be made from mechanically alloyed materials. One kind of energy storage device that stores energy as kinetic energy is a flywheel. They are comprised of a motor-generator coupled to a rotating mass (the flywheel). The motor turns electrical energy into kinetic energy by accelerating the flywheel when there is extra energy available. The rotating flywheel is connected to the generator so that it may transform kinetic energy back into electrical energy when the stored energy is needed [31–34].

Kinetic Energy Storage

The process of storing energy by transforming mechanical or electrical energy into kinetic energy is known as kinetic energy storage. When needed, this stored kinetic energy can be transformed back into electrical energy. A flywheel or rotating mass that stores energy in its rotational motion is the typical component of kinetic energy storage devices. A flywheel is a mechanical device that revolves around an axis and is the central component of kinetic energy storage systems. Kinetic energy that has been stored is represented by the flywheel's rotation. The system employs electricity to speed up the flywheel during times when there is excess energy, increasing rotational speed and converting energy into kinetic energy. By speeding up a mass's spin, flywheels accumulate energy. Kinetic energy is the form that the energy is kept in.

Working Principle

An electric motor is used to accelerate the flywheel to a high speed during times of low demand or when electricity is readily accessible. The rotating mass stores the kinetic energy. The flywheel's stored kinetic energy is transformed back into electricity when electrical energy is needed. Usually, to do this, a generator is attached to the flywheel, which rotates and generates electricity when it slows down.

Advantages

Systems utilising kinetic energy storage have the ability to respond quickly, which makes them appropriate for uses requiring short bursts of power. The flywheel's quick rotation allows them to produce significant power outputs.

Research and Development

The goals of on-going research are to decrease the size, increase overall performance, and improve the efficiency of kinetic energy storage devices. Kinetic energy storage has unique benefits in terms of quick response time and high power density for specific applications, even though it may not be as common as other storage technologies like batteries or pumped hydro. Future developments in materials science and technology could result in even higher levels of practicality and efficiency for kinetic energy storage devices [35–38].

Applications

By offering short-duration energy storage to mitigate variations in the supply and demand for power, kinetic energy storage can be utilised for grid stabilisation. Kinetic energy storage devices are occasionally utilised as a temporary backup power source in establishments like data centres and hospitals where a momentary power outage could be disastrous.

Problems/Challenges

Kinetic energy storage systems typically have a lower energy density than some other energy storage technologies, which means they might not store as much energy for a given mass or volume. To reduce energy losses from friction in the rotating flywheel, effective bearings and low-friction solutions are crucial. High rotational speed often known as fast rotational rates, describes the speed at which a shaft rotates around its axis. Units of measurement for angular velocity include revolutions per minute (RPM) and radians per second (rad/s). High rotational rates are a highly pertinent notion in physics, engineering, and technology, among other subjects. In order to extract as much energy as possible from stored energy, flywheels usually rotate at high rates. The square of the rotating speed determines the kinetic energy. The flywheel's performance is greatly dependent on the materials used in it. Flywheels made of advanced materials, like carbon fibre composites, are frequently robust and lightweight. Air resistance and friction reduction are two more important considerations in the design. Flywheel energy storage devices have a comparatively high efficiency, requiring little energy to be lost during the charging and discharging process. Flywheels have the benefit of being able to react swiftly to variations in demand. They are appropriate for uses requiring quick response times, including grid stabilisation, since they can quickly release stored energy. Flywheels can be utilised to supply sudden energy bursts when needed, which can help stabilise electrical grids. In the event of an outage, they can be used to supply temporary backup power. Flywheels have occasionally been investigated as potential energy storage components for hybrid automobiles. Flywheels undergo some degree of self-discharge, which is the progressive loss of energy over time as a result of friction and air resistance, much like any other energy storage device. One factor that may prevent high-performance flywheels from being widely adopted is their manufacturing cost. Research is still being done to increase flywheel energy storage devices' energy density, effectiveness, and affordability. Their performance keeps getting better because of design and material advancements. All things considered, flywheel energy storage offers a peculiar combination of qualities that make it ideal for particular uses, particularly those that call for quick reaction times and brief energy storage [39–43].

17.1.5 Thermoelectric Materials

Materials that have been mechanically alloyed can be modified to have certain thermoelectric properties. Thermoelectric materials are useful in waste heat recovery systems and thermoelectric generators because they can transform heat into electricity. The Seebeck effect allows materials that are thermoelectric to change thermal energy into electrical energy and vice versa. The creation of an electromotive force (EMF) in a circuit when there is a temperature differential between its endpoints is known as the Seebeck effect. Because of this reversible process, a material can experience a temperature differential—a phenomenon called the Peltier effect—when an electric current is supplied to it. A high electrical conductivity, low thermal conductivity, and a high Seebeck coefficient—a measure of the Seebeck effect's magnitude—are essential characteristics of a good thermoelectric material. The material's efficiency in converting heat into electricity is determined by the combination of these properties. The material's efficiency in converting heat into electricity is determined by the combination of these qualities. Furthermore, a dimensionless figure of merit known as "ZT," which is the product of the product of electrical conductivity and thermal conductivity and the square of the Seebeck coefficient, is frequently used to identify materials with strong thermoelectric qualities [44, 45].

Inorganic Semiconductors

Because of their comparatively high ZT values at ambient temperature, a number of inorganic materials, including bismuth telluride (Bi_2Te_3) and lead telluride (PbTe), are commonly utilised in commercial thermoelectric devices.

Organic Materials

Conducting polymers are one type of organic thermoelectric material that can provide flexibility and low weight. However, as compared to inorganic materials, their efficiency is frequently lower.

Nanomaterials

Materials' thermoelectric performance can be improved by nanostructuring them.

Complex Intermetallic Compounds

For high-temperature applications, certain complex intermetallic compounds, such as catharses and skutterudites, have shown considerable potential. To increase the efficiency of thermoelectric materials for a variety of applications, such as waste heat recovery, power generation, and cooling in electronic devices, researchers are constantly investigating new materials and fabrication processes. For thermoelectric materials to be more useful and competitive in real-world applications, their efficiency must be increased [45–48].

17.1.6 Phase Change Materials (PCMs)

Substances that go through a phase transition—typically from a solid to a liquid and vice versa—at a specific temperature are known as phase change materials, or PCMs. These materials have the ability to absorb or release large amounts of latent heat during this phase transition without experiencing a major temperature shift. PCMs are useful for many applications because of this feature, especially in the management and storage of thermal energy [49].

Different Phase Transition Types

The most frequent phase shift in PCMs is the heating-induced conversion from a solid to a liquid. During cooling, the opposite process takes place, going from liquid to solid.

Latent Heat Storage

Thermal energy storage systems make use of phase change materials (PCMs). PCMs with improved thermal properties, such as increased heat storage capacity and higher thermal conductivity, can be made using mechanically alloyed materials. During phase transitions, PCMs store and release thermal energy as latent heat, offering a high heat storage density. The energy needed to change a material's condition without causing a change in temperature is known as the latent heat of fusion, and it determines how much heat is absorbed or emitted.

Applications

PCMs are used to store thermal energy in a variety of applications, including controlling interior temperature and energy utilisation in buildings. By absorbing and dispersing heat, PCMs can be used in electronic devices to provide more effective cooling. PCMs incorporated into textiles can provide clothes with the ability to regulate temperature.

Types of Phase Change Materials

Fatty acids and paraffins are examples of organic PCMs. Their low toxicity, biodegradability, and versatility in many applications are generally the reasons behind their selection. Metallic alloys and salt hydrates are two examples of inorganic PCMs. Although inorganic PCMs can have problems like corrosion, they may have higher thermal conductivities than organic ones. Bio-based PCMs are known for their environmental sustainability, bio-based PCMs are made from renewable resources.

Challenges

The ideal operating temperature range for a given application determines which PCM to use. A PCM's overall performance is influenced by its capacity to conduct heat. Long-term dependability depends on the system's capacity to go through several phase transitions without suffering noticeably from degradation. PCMs efficiently manage thermal energy in a variety of systems, promoting sustainability and energy efficiency. The main goals of ongoing research include creating novel PCM formulations, enhancing their functionality, and investigating creative usage in many sectors [50–54].

17.1.7 Solar Thermal Storage

Materials that have been mechanically alloyed can be employed in concentrated solar power systems as high-temperature thermal storage materials to effectively store and release thermal energy. A technique known as solar thermal storage makes it possible to collect and store heat produced by sunshine for later use. This holds special significance in scenarios where the need for energy is not directly correlated with solar radiation, like at night or on overcast days. When needed, the heat from solar energy collection can be transformed into electricity or used for heating thanks to solar thermal storage devices [55].

Sensible Heat Storage

One popular method of storing solar heat is with water. Hot water can be kept in large, insulated tanks for use as home hot water or for space heating. Another option for a storage medium is a rock with high heat conductivity. During the day, the heat is retained in the rocks, which release it when needed. PCMs have been employed in solar thermal storage, as was previously mentioned. The PCM absorbs heat over the day and changes phases, usually melting, storing a significant quantity of thermal energy. As the PCM solidifies, this energy is released [56].

Thermochemical Storage

Chemical Reactions: In order to store and release heat, certain solar thermal storage systems use chemical reactions. Reversible chemical transformations that absorb and release heat are often involved in these reactions. Molten Salt Storage: Molten salt is utilised as a heat transmission fluid and a storage medium in certain large-scale solar thermal power facilities. Daytime heat from the sun causes the salt to retain heat, which it then releases to create steam and power when needed.

Advantages of Solar Thermal Storage

By overcoming the intermittent nature of solar energy, solar thermal storage enables a more constant and dependable energy source. Even in the absence of sunlight, solar thermal power plants possess the ability to provide electricity on demand through efficient storage. Solar thermal storage can be combined with current energy systems to store extra energy produced during sunny hours for usage at times when demand is highest. The expense of storage systems, the effectiveness of the storage procedure, and the selection of appropriate storage materials are obstacles, though. The goal of ongoing research is to lower the costs and increase the efficiency of solar thermal storage devices [57].

17.1.8 Magnetic Energy Storage

One kind of energy storage technology is magnetic energy storage, which uses magnetic fields to store energy. In comparison to other energy storage technologies like batteries or pumped hydro storage, this technology is still in its early stages of development and is not as commonly used. On the other hand, because of its high energy density, extended cycle life, and quick response times, it might have certain benefits. While there are other methods for storing magnetic energy, one prominent idea is based on the use of superconducting materials. Superconductors are substances that, when cooled to extremely low temperatures, have zero resistance to electrical conductivity. Electrical current generates a magnetic field that is stored in the superconducting coil of a magnetic energy storage device that uses superconductors. The magnetic field is stored and released as needed, causing a nearby coil to experience a current that can be converted into electrical power. Magnetic energy storage devices have the capacity to store a significant amount of energy in a comparatively small amount of area thanks to their high energy density. Superconducting materials may withstand multiple cycles of charging and discharging without experiencing appreciable degradation if they are properly maintained. Magnetic energy storage devices are well suited for applications requiring quick modifications to power output because they can react swiftly to variations in the demand for electricity. A lot of superconductors work at very low temperatures, which can make maintenance difficult and costly. The overall cost-effectiveness of magnetic energy storage devices may be impacted by the high cost of superconducting materials. Before the technology is widely used, a number of technical issues must be resolved. It is still in its early phases of development. Magnetic energy storage is still a subject of research and development, so it will be interesting to watch how this technology advances in the years to come. High-magnetic coercive materials are used in magnetic energy storage systems to store energy in the form of magnetic fields. For these purposes, materials that have been mechanically alloyed can be created to have the necessary magnetic characteristics [58–65].

17.1.9 Electrochemical Capacitors

Advanced materials for electrochemical capacitors can be created using mechanically alloyed materials, increasing their energy and power density as well as their cycling stability. Supercapacitors and ultra-capacitors, other names for electrochemical capacitors, are energy-storage devices that store electrical energy by separating charges electrostatically. They provide a distinct set of features that enable them to be utilised in a variety of applications, bridging the gap between conventional capacitors and batteries. The double-layer capacitor, which stores energy at the interface between an electrolyte and a conducting electrode (often activated carbon), is the most popular kind of supercapacitor [66]. A double layer of electric charge is produced by the separation of charges at the electrode–electrolyte contact. Reversible faradaic redox processes at the electrode–electrolyte interface store energy in pseudo capacitors, a form of supercapacitor. Transition metal oxides and conducting polymers are frequently used as electrode materials in pseudocapacitors. Supercapacitors' high power density is one of their key benefits. Because of their short charging and discharging times, they are appropriate for uses needing brief power spikes, including regenerative braking in electric cars or balancing out variations in renewable energy systems. In general, supercapacitors outlast conventional batteries in terms of cycle life. They don't significantly deteriorate after a lot of charge–discharge cycles. Supercapacitors have a lower energy density than batteries, despite their superior power density. As a result, they work better in applications that need short bursts of power as opposed to long-term energy storage. Supercapacitors are more suitable for some applications in difficult situations because they can perform at low temperatures longer than many batteries can. Supercapacitors are used in many different industries and sectors, such as consumer electronics, renewable energy systems, transportation (such as electric cars, buses, and trains), and industrial applications. They are frequently used in tandem with batteries to enhance each other's advantages. In order to make supercapacitors more competitive with conventional batteries in terms of overall energy storage capacity, ongoing research intends to improve the energy density of supercapacitors. Supercapacitors are especially well-suited for specialised applications where their quick charge–discharge times, extended cycle life, and high power density are useful. Engineers and researchers are still looking for ways to advance and broaden the use of a range of applications [67, 68].

17.1.10 Grid Energy Storage

Mechanically alloyed materials can be used in redox flow batteries and other grid energy storage technologies to improve the performance and longevity of electrodes and electrolytes. The term "grid energy storage" describes the range of devices and techniques that are employed within an electrical grid to store electrical energy in huge quantities. This storage is necessary to manage sporadic renewable energy sources, balance the supply and demand for electricity, and improve the overall efficiency and dependability of the grid. Grid energy storage solutions come in a variety of forms, each with unique benefits and uses [69–73].

17.1.11 Batteries

Lithium-Ion Batteries: Because of its high energy density and falling costs, lithium-ion batteries—which are also used in consumer gadgets and electric cars—are being employed more and more for grid energy storage [74].

Flow batteries: These batteries have the capacity to store energy in large-scale, long-duration applications and are made of liquid electrolytes. Vanadium redox flow batteries are one example [75].

17.1.12 Pumped Hydro Storage

Powered Reserve Hydroelectricity

This is the process of pumping water from a lower reservoir to a higher one by using extra electricity generated during periods of low demand. The stored water is released to create power during times of high demand [76].

Compressed Air Energy Storage (CAES)

To produce power, compressed air is subterranean stored and then inflated using a turbine. The efficiency of adiabatic CAES is increased by storing and reusing the heat produced during compression.

Energy Storage via Flywheels

Flywheels are a mechanical means of storing energy in the form of kinetic energy. A rotor that has been spinning on excess electricity is linked to a generator to generate electricity when needed.

Thermal Energy Storage

Excess heat is utilised in concentrated solar power facilities to melt and store salt. Later on, the heat that has been stored can be used to create steam and power. During off-peak hours, electrical energy is used to create ice. By using the frozen ice for cooling during moments of high demand, the electrical system is not as burdened.

Hybrid and Advanced Storage Technologies

Research is constantly being done to create novel and more effective grid storage technologies, like batteries of the future and creative materials. In order to maintain a stable and dependable electrical grid, grid energy storage is essential, especially as the proportion of renewable energy sources rises. These storage solutions promote grid stability, mitigate supply and demand fluctuations, and offer backup power in an emergency. A number of criteria, including the amount of capacity needed, how long it will be stored, geographical concerns, and economic viability, influence the choice of storage technology. In these applications, mechanically alloyed materials' distinct microstructures and specially crafted features can result in better performance, a longer lifespan, and higher energy storage efficiency, helping to create more effective and sustainable energy storage systems. However, the individual specifications and limitations of each application will determine the material selection and alloy composition [76–80].

17.2 Supercapacitors and Its Battery Applications

Supercapacitors, also known as ultra-capacitors, are high-capacity capacitors that have lower voltage restrictions than solid-state capacitors but far larger capacitance values. It fills the void left by rechargeable batteries and electrolytic capacitors. It can receive and distribute charge far faster than batteries, stores 10–100 times more energy per unit volume or mass than electrolytic capacitors, and can withstand many more cycles of charge and discharge than rechargeable batteries [81]. Supercapacitors are employed for regenerative braking, short-term energy storage, or burst-mode power supply in vehicles, buses, trains, cranes, and lifts. They are not used for long-term compact energy storage, but rather for applications needing several fast charge/discharge cycles. Static random-access memory uses smaller units as power backup (SRAM). Supercapacitors are not like regular capacitors. Batteries can use supercapacitors, often referred to as ultra-capacitors or electrochemical capacitors, in a variety of ways. Supercapacitors are not conventional batteries, but they have special features that enhance and complement a variety of battery applications [82] (Fig. 17.1).

17.2.1 Battery Applications for Supercapacitors

Energy Storage and Backup

Supercapacitors can offer quick and powerful energy storage options to supplement batteries for energy storage and backup. They are excellent at supplying brief bursts of power and are frequently utilised in situations where immediate energy delivery

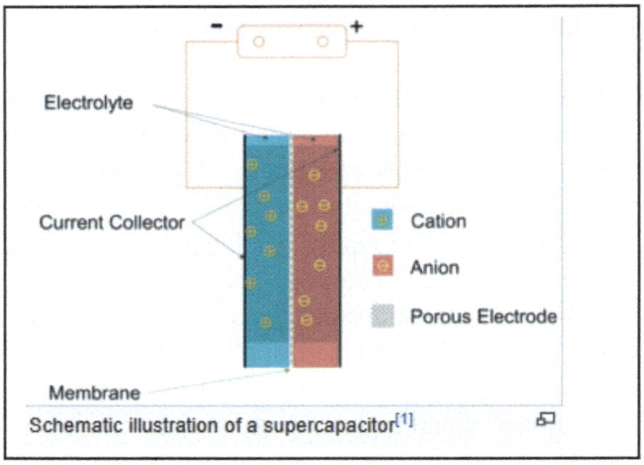

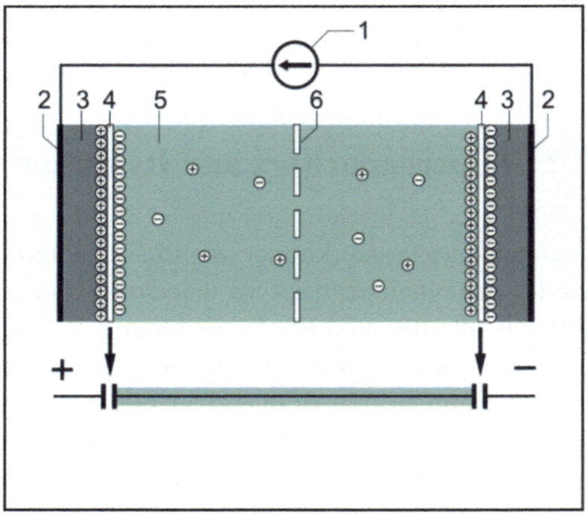

Electric double-layer capacitor model (#1)
No Text 1.Charger (Current source)
2.Collector 3.Polarization electrode
4.Electric double-layer 5.Electrolytic
solution 6.Separator

Fig. 17.1 Supercapacitor and battery

is required to assist batteries. For instance, supercapacitors in uninterruptible power supply (UPS) systems can fill the void during power outages until the principal batteries are operational [83].

Peak Power Shaving

The term "peak power shaving by supercapacitors" refers to the process of using supercapacitors, sometimes referred to as electric double-layer capacitors or ultracapacitors, to reduce the peaks and valleys of electrical system power consumption. To increase system performance and efficiency, this technology is frequently utilised in conjunction with batteries and other energy storage options. To handle peak power consumption, supercapacitors can be incorporated into battery systems. Supercapacitors decrease the burden on the battery by providing high-power bursts when necessary, resulting in longer battery life and greater system efficiency. Unlike conventional batteries, which rely on chemical reactions to store energy, supercapacitors are energy storage devices that store electrical energy through the electrostatic separation of charges. They are ideal for applications that need fast bursts of power since they can charge and discharge quickly. The power requirements for many electrical systems, including electric cars, heavy equipment, and renewable energy systems, are frequently fluctuating or intermittent. In order to achieve this high load needs during peak hours, additional energy input is needed. To meet these peak power requirements, supercapacitors can be employed as power buffers. When there is a surplus of energy compared to consumption during times of low energy demand, they are charged. Supercapacitors release their stored energy during times of peak power demand, giving the system a rapid power boost and helping to close the supply-demand gap. This can stop voltage sags, maintain system performance, and stabilise the power supply [84, 85].

Advantages of Supercapacitors for Power Shaving

Supercapacitors are capable of charging and discharging quickly, which makes them ideal for supplying power in brief spurts. Compared to some battery technologies, supercapacitors have a longer cycle life, making them resistant to repeated charge and discharge cycles. Supercapacitors typically require little maintenance, making them maintenance-free. Supercapacitors can increase the system's overall energy efficiency by minimising the requirement for frequent, high-power draws from the primary energy source [86, 87].

17.2.2 Applications of Supercapacitors

A wide variety of applications, including the use supercapacitors to reduce peak power, to offer more power for acceleration in electric automobiles. Systems using renewable energy to balance out erratic energy production from sources like wind and sun. Backup power during grid outages will be provided by an uninterruptible power supply (UPS) Industrial equipment to control peak power use while operating. Integration with Other Energy Storage Solutions: Supercapacitors and batteries are frequently used in tandem, with the former handling short-term energy storage while the latter handles high-power assistance at peak loads. The requirements for power and energy are optimised by this combination. Overall, by effectively handling high-power demands and lessening stress on primary energy sources, peak power shaving with supercapacitors can assist improve the stability and effectiveness of electrical systems [88–90].

Applications for Supercapacitors in Electric Vehicles (EV)

Supercapacitors are used in EVs to recover and store energy during regenerative braking, which is then utilised to deliver quick acceleration when required. Regenerative braking contributes to an overall improvement in electric vehicle energy efficiency. Supercapacitors work effectively in systems that use regenerative braking. The electric motor acts as a generator when the car brakes, transforming the kinetic energy back into electrical energy. This energy can be effectively captured and stored for later use using supercapacitors. Supercapacitors may produce sudden bursts of energy and have a high power density. Because of this, they are useful for applications that need to accelerate or decelerate quickly, including supplying extra power during acceleration or absorbing energy during abrupt braking. In start-stop systems, the internal combustion engine is swiftly restarted when necessary and switched off for extended periods of time when not in use. Supercapacitors can be employed in these systems. Supercapacitors are a good fit for this application because of their quick charge and discharge times. While there are unexpected spikes in power use, like while accelerating or descending a steep grade, supercapacitors can help by offering peak power assistance [91]. This can lessen the strain on the primary battery, which might not be able to deliver high power quickly enough. Supercapacitors enhance the life of an electric vehicle's primary battery by managing high-power tasks. With the supercapacitor's ability to withstand quick cycles of charging and discharging, the battery can concentrate on longer-duration, higher-energy jobs. Supercapacitors are renowned for their high performance in low temperatures, an environment where batteries may lose some of their efficiency. In order to retain high power performance in cold climates, supercapacitors can be used in conjunction with batteries. Batteries and supercapacitors are combined in hybrid energy storage systems found in certain automobiles. The goal of this hybrid strategy is to maximise overall energy storage and delivery by utilising the high power density of supercapacitors and the energy density of batteries. In general, supercapacitors are less heavy than batteries. This feature is useful in situations where reducing the weight of the vehicle is essential for better performance and efficiency. It's vital to remember that supercapacitors are not usually utilised as the main energy storage option for electric vehicles, even though they do provide unique benefits for some applications. Rather, they are incorporated into the energy management system as a whole to enhance the functions of conventional batteries. The goal of ongoing research and development is to increase supercapacitors' energy density so that they can play a bigger part in the designs of future electric vehicles [92].

17.2.3 Hybrid Electric Vehicles (HEVs)

Supercapacitors can be used in hybrid electric vehicles (HEVs) to store energy and supply power during acceleration, lowering the load on the internal combustion engine and increasing fuel efficiency. In hybrid electric vehicles (HEVs), supercapacitors are an important component of the entire energy management system. Internal combustion engines (ICEs) and electric propulsion systems are combined in hybrid electric vehicles (HEVs) to increase fuel economy and lower pollutants. HEV systems incorporate supercapacitors to offer particular benefits under particular driving conditions. Supercapacitors are included in hybrid electric vehicles' energy management systems to solve particular issues with power delivery and quick energy absorption. Through integration, the capabilities of the internal combustion engine and the electric propulsion system are better balanced, which enhances the hybrid vehicle's overall performance and efficiency [93, 94].

Start-Stop Systems

Start-stop systems, sometimes referred to as idle-stop, micro-hybrid, or stop-start systems, are automotive technologies that automatically turn off and restart the internal combustion engine when the car stops, for example at a stop sign or during traffic congestion, in an effort to improve fuel economy and lower emissions. The fundamental idea behind start-stop systems is to cut down on the amount of time the engine is left idle, which saves fuel and lowers emissions from the exhaust. Supercapacitors are occasionally employed in start-stop systems, which are common in contemporary automobiles. When the car stops, they immediately supply power to restart the engine, cutting down on fuel use and pollution.

Supercapacitors are essential to automotive start-stop systems because they improve the system's dependability and efficiency. As was already noted, the purpose of start-stop systems is to increase fuel efficiency by automatically turning off and on the internal combustion engine when there is no movement. These systems frequently incorporate supercapacitors to handle the high power needs brought on by frequent engine restarts. Because supercapacitors are excellent at quick cycles of charging and discharging, they are a good fit for the high power requirements of engine restarts. Supercapacitors can effectively store and release energy in brief bursts, in contrast to conventional lead-acid batteries, which may find it difficult to deliver the required power rapidly. The battery in a start-stop system is in charge of supplying electricity when the engine resumes. By managing the high power requirements of repeated engine starts, supercapacitors can help the battery by lessening the strain and wear on it. This may help to increase the battery's overall lifespan. Supercapacitors provide energy in short bursts, which makes engine restarts smooth and swift. The engine must be able to release energy quickly in order for the restart to go smoothly and without any obvious lag. Supercapacitors improve start-stop systems' overall dependability. Their high power density and capacity to manage many charge–discharge cycles make them a reliable answer to the frequent and dynamic needs of the start-stop operation. The electrical system may experience voltage decreases as a result of engine restarts. Supercapacitors offer rapid power support during the restart process, which helps to offset these voltage reductions. Keeping the electrical system steady, helps avoid problems with other car electronics. Supercapacitors are used by certain start-stop systems with regenerative braking capability to store the energy recovered during braking. Subsequent engine restarts or the operation of electrical systems can be facilitated by the stored energy. Supercapacitors are a good choice for start-stop systems in a range of environmental circumstances since they typically function well in cold temperatures. Automobile manufacturers may optimise the energy management of their vehicles and guarantee dependable and efficient operation by including supercapacitors in their start-stop systems. Although a battery normally serves as the primary energy storage for the car's electrical systems, the inclusion of supercapacitors enables instantaneous, high-power support in crucial situations, such as engine restarts, hence enhancing the overall efficacy of start-stop technology [95, 96].

17.2.4 Energy Harvesting

Supercapacitors may store energy obtained from renewable sources, such as solar panels or motion-based kinetic energy. They guarantee effective energy collecting and storage when used in conjunction with batteries. Because of their quick charge and discharge abilities, supercapacitors are frequently utilised in energy harvesting applications. This makes them ideal for storing and releasing energy from intermittent energy sources. Because of their quick charge and discharge times, supercapacitors are frequently utilised in energy harvesting applications. This makes them ideal for storing and releasing energy from sporadic energy sources. Supercapacitors should be used with the following important factors in mind when harvesting energy [97, 98]

17.2.5 Energy Source Compatibility

Applications that gather energy from short-duration, intermittent sources—such as solar cells, vibration, or thermoelectric generators—are a good fit for supercapacitors. They are perfect for storing sudden energy bursts because of their rapid charging and discharging times.

Matching of Voltage and Capacity

Make sure that the voltage and capacity of the supercapacitor are suitable for the particular energy harvesting system. Optimising these parameters maximises the efficiency of energy storage and recovery.

Low Rate of Self-discharge

In general, supercapacitors self-discharge at a slower pace than conventional batteries. In energy harvesting systems, this is crucial, particularly if the energy being harvested is irregular. Longer periods of energy storage are made possible by a low self-discharge rate.

Taking Temperature into Account

The operating temperature ranges of supercapacitors must be in line with the energy harvesting application's ambient parameters. Extreme temperatures may restrict the performance of some supercapacitors.

Handling Power and Current

High-power burst applications are a perfect fit for supercapacitors. Make sure the supercapacitor can support the energy harvesting system's maximal current and power requirements.

Balancing and Protection Circuits

To ensure that the voltage across every supercapacitor cell remains within a safe range, implement balancing circuits. In order to avoid overcharging or overcharging, which might shorten the supercapacitor's lifespan, protection circuits could be required.

Integration with Energy Harvesting Electronics

Supercapacitors must be incorporated into the energy harvesting system as a whole. Power management electronics are usually included in this system to control the charging and discharging procedures. Effective energy transfer between the supercapacitor and the energy harvester is ensured by these electronics.

Life Cycle

Comparing supercapacitors to conventional batteries, the former have a far shorter cycle life. It is imperative to take into account the anticipated quantity of charge and discharge cycles throughout the device's lifespan and select a supercapacitor with a suitable cycle life.

Size and Cost

Think about the supercapacitor's size and cost in relation to the whole requirements of the energy harvesting system. A key consideration in the choosing process is cost-effectiveness, as supercapacitors might be more costly than certain conventional batteries. Supercapacitors, which effectively store and distribute captured energy when needed, can improve the efficiency and dependability of energy harvesting systems when they are chosen and integrated correctly.

Applications for Pulsed Load

When instantaneous and reliable power is essential, such as in some military and aerospace systems, supercapacitors are useful in applications with intermittent, high-power demands.

Devices for IoT and Remote Monitoring

Supercapacitors can store energy and supply power for data transmission or sensor readings in low-power, battery-operated remote monitoring and IoT devices, prolonging the battery life and system dependability. Supercapacitors are perfect for applications that call for speedy charging and discharging since they can provide and take energy quickly.

Extended Battery Life

Supercapacitors are frequently combined with conventional batteries (such lithium-ion batteries) in various battery applications to develop hybrid energy storage systems that provide the benefits of both energy density and power density. Supercapacitors and batteries work together to improve energy efficiency, extend battery life, and boost performance across a range of applications.

Fast Charging and Discharging

Supercapacitors are perfect for applications that call for speedy charging and discharging since they can provide and take energy quickly. Supercapacitors are frequently combined with conventional batteries (such as lithium-ion batteries) in various battery applications to develop hybrid energy storage systems that provide the benefits of both energy density and power density. Supercapacitors and batteries work together to improve energy efficiency, extend battery life, and boost performance across a range of applications [99, 100].

17.3 Comparison of Supercapacitors with Batteries

17.3.1 Energy-Storage System

Supercapacitors: By dividing positive and negative charges across an electrolyte, they store energy electrostatically. At the electrode–electrolyte interface, they create an electric double layer by storing energy through the passage of ions.

Batteries: During the charging and discharging phases, chemical reactions take place that store energy electrochemically.

17.3.2 Energy Density

Supercapacitors: Compared to batteries, they have a lower energy density. Less energy may be stored by them per mass or volume unit. They are able to deliver powerful blasts, though.

Batteries: Batteries may store more energy for longer periods of time than supercapacitors because they typically have a higher energy density. However, batteries may not discharge as quickly or be able to handle sudden power demands as well.

17.3.3 Power Density

Supercapacitors: Provide a high power density that makes quick cycles of charging and discharging possible. They are capable of managing high-power applications and brief energy bursts. Suitable for applications needing a consistent, sustained power output, batteries often have a lower power density than supercapacitors due to their slower rate of discharge and charge.

17.3.4 Life Cycle

Supercapacitors: Offer a longer cycle life benefit. They don't significantly lose performance even after being charged and discharged hundreds of thousands to millions of times.

Batteries: The majority of battery types have a short cycle life, notwithstanding advancements in technology. Over time, frequent charging and draining cause capacity deterioration.

17.3.5 Sensitivity to Temperature

Supercapacitors: Generally speaking, they are more suited for locations with extreme temperatures since they are less susceptible to temperature changes than some batteries.

Batteries: The performance and durability of certain battery chemistries can be impacted by temperature fluctuations.

17.3.6 Applications

Supercapacitors: Perfect for applications requiring short bursts of power, like energy harvesting, some backup power systems, and regenerative braking in cars.

Batteries: Suitable for grid energy storage, portable gadgets, and electric vehicles that need a more consistent, long-term power supply.

17.3.7 Cost

Supercapacitors: In the past, supercapacitors cost more than batteries. However, as production and technology have advanced, their costs have been steadily declining.

Batteries: Because of improvements in production and material science, battery costs have been declining over time, however, they still vary depending on the kind and size. Batteries and supercapacitors each have advantages and disadvantages of their own. The exact needs of the application will determine which option is best, taking into account balance aspects such as cycle life, cost, power density, and energy density. In certain situations, hybrid energy storage systems which combine batteries and supercapacitors may be employed to take advantage of each technology's advantages and achieve better overall performance [90, 101–103].

17.4 Conclusion

The widespread use of mechanically alloyed materials in the field of energy storage, namely in supercapacitors and batteries, represents a notable advancement in addressing the increasing need for effective and environmentally friendly energy storage options. The distinctive characteristics resulting from the process of mechanical alloying, including increased surface area, greater conductivity, and customised structures, play a significant role in the exceptional performance of these materials inside energy storage systems. Mechanically alloyed materials provide noteworthy capacitance values and rapid charge–discharge rates within the domain of supercapacitors, rendering them very intriguing contenders for high-power applications. Supercapacitors play a vital role in applications necessitating rapid energy storage and release, such as electric cars and grid stabilisation, due to their exceptional capability in delivering brief bursts of power. Furthermore, materials produced by mechanical alloying have significant promise in the field of battery technology, highlighting progress in terms of energy density, cycle life, and overall performance. These materials provide a means to surpass conventional constraints commonly associated with batteries, hence facilitating the advancement of energy storage technologies that are both more effective and have a longer lifespan. The utilisation of mechanical alloying techniques facilitates the attainment of customised structures, which in turn allows for improved regulation of electrochemical phenomena. This capability serves to address concerns such as electrode deterioration and ultimately enhances the dependability of batteries.

References

1. Raabe, D.: The materials science behind sustainable metals and alloys. Chem. Rev. **123**(5), 2436–2608 (2023)
2. Shashanka, R., Chaira, D.: Optimization of milling parameters for the synthesis of nano-structured duplex and ferritic stainless steel powders by high energy planetary milling. Powder Technol. **1**(278), 35–45 (2015)
3. Rajendrachari, S.: Effect of sintering temperature on the pitting corrosion of ball milled duplex stainless steel by using linear sweep voltammetry. Anal. Bioanal. Electrochem. **10**(3), 349–361 (2018)
4. Sharath, B.N., Madhu, K.S., Venkatesh, C.V.: Experimental study on dry sliding wear behaviour of Al-B4C-Gr metal matrix composite at different temperatures. Appl. Mech. Mater. **2**(895), 96–101 (2019)
5. Rajendrachari, S., Kamacı, Y., Taş, R., Ceylan, Y., Bülbül, A.S., Uzun, O., Karaoğlanlı, A.C.: Antimicrobial investigation of CuO and ZnO nanoparticles prepared by a rapid combustion method. Phys. Chem. Res. **7**(4), 799–812 (2019)
6. Sharath, B.N., Venkatesh, C.V.: Study on Effect of boron carbide, aluminium oxide and graphite on dry sliding wear behaviour of aluminium based metal matrix composite at different temperature. Tribol.-Finnish J. Tribol. **38**(1–2), 35–46 (2021)
7. Shashanka, R., Chaira, D.: Effects of nano-Y2O3 and sintering parameters on the fabrication of PM duplex and ferritic stainless steels. Acta Metall. Sinica (English Lett.) **29**, 58–71 (2016)

8. Guney, M.S., Tepe, Y.: Classification and assessment of energy storage systems. Renew. Sustain. Energy Rev. **1**(75), 1187–1197 (2017)
9. Aneke, M., Wang, M.: Energy storage technologies and real life applications—A state of the art review. Appl. Energy **1**(179), 350–377 (2016)
10. Luo, X., Wang, J., Dooner, M., Clarke, J.: Overview of current development in electrical energy storage technologies and the application potential in power system operation. Appl. Energy **1**(137), 511–536 (2015)
11. Hall, P.J., Bain, E.J.: Energy-storage technologies and electricity generation. Energy Policy **36**(12), 4352–4355 (2008)
12. Baker, J.: New technology and possible advances in energy storage. Energy Policy **36**(12), 4368–4373 (2008)
13. Nitta, N., Wu, F., Lee, J.T., Yushin, G.: Li-ion battery materials: present and future. Mater. Today **18**(5), 252–264 (2015)
14. Verma, P., Maire, P., Novák, P.: A review of the features and analyses of the solid electrolyte interphase in Li-ion batteries. Electrochim. Acta **55**(22), 6332–6341 (2010)
15. Sharath, B.N., Venkatesh, C.V., Afzal, A., Aslfattahi, N., Aabid, A., Baig, M., Saleh, B.: Multi ceramic particles inclusion in the aluminium matrix and wear characterization through experimental and response surface-artificial neural networks. Materials **14**(11), 2895 (2021)
16. Pandolfo, A.G., Hollenkamp, A.F.: Carbon properties and their role in supercapacitors. J. Power Sour. **157**(1), 11–27 (2006)
17. Wang, Y., Xia, Y.: Recent progress in supercapacitors: from materials design to system construction. Adv. Mater. **25**(37), 5336–5342 (2013)
18. Zhang, L., Zhang, F., Yang, X., Long, G., Wu, Y., Zhang, T., Leng, K., Huang, Y., Ma, Y., Yu, A., Chen, Y.: Porous 3D graphene-based bulk materials with exceptional high surface area and excellent conductivity for supercapacitors. Sci. Rep. **3**(1), 1408 (2013)
19. Obreja, V.V.: On the performance of supercapacitors with electrodes based on carbon nanotubes and carbon activated material—A review. Phys. E **40**(7), 2596–2605 (2008)
20. Nayak, A.K., Shashanka, R., Chaira, D.: Effect of nanosizeyittria and tungsten addition to duplex stainless steel during high energy planetary milling. In: IOP Conference Series: materials Science and Engineering, vol. 115, no. 1, p. 012008. IOP Publishing (2016)
21. Gupta, S., Shashanka, R, Chaira, D.: Synthesis of nano-structured duplex and ferritic stainless steel powders by planetary milling: an experimental and simulation study. In: IOP Conference Series: materials Science and Engineering, vol. 75, no. 1, p. 012033. IOP Publishing (2015)
22. Pavitra, V., Praveen, B.M., Nagaraju, G., Shashanka, R.: Energy storage, photocatalytic and electrochemical nitrite sensing of ultrasound-assisted stable Ta2O5 nanoparticles. Top. Catal. **8**, 1–4 (2022)
23. Shashanka, R.: Synthesis of nano-structured stainless steel powder by mechanical alloying-an overview. Int. J. Sci. Eng. Res. **8**, 588–594 (2017)
24. Sharath, B.N., Madhu, K.S., Pradeep, D.G., Venkatesh, C.V.: Tribological Suitability of aluminium hybrid composite above atmospheric temperature. In: IOP Conference Series: materials Science and Engineering, vol. 1189, no. 1, p. 012018. IOP Publishing (2021)
25. Ballupete Nagaraju, S., Kodigarahalli Somashekara, M., Puttegowda, M., Manjulaiah, H., Kini, C.R., Channarayapattana Venkataramaiah, V.: Effect of B4C/Gr on hardness and wear behavior of Al2618 based hybrid composites through Taguchi and artificial neural network analysis. Catalysts **12**(12), 1654 (2022)
26. Madhu, K.S., Venkatesh, C.V., Sharath, B.N., Karthik, S.: Characterization and evaluation of mechanical properties of Al–Zn based hybrid metal matrix composites. Appl. Sci. Eng. Prog. **16**(1), 5804 (2022)
27. Züttel, A.: Hydrogen storage methods. Naturwissenschaften **91**, 157–172 (2004)
28. Graetz, J.: New approaches to hydrogen storage. Chem. Soc. Rev. **38**(1), 73–82 (2009)
29. Niaz, S., Manzoor, T., Pandith, A.H.: Hydrogen storage: materials, methods and perspectives. Renew. Sustain. Energy Rev. **1**(50), 457–469 (2015)
30. Züttel, A.: Materials for hydrogen storage. Mater. Today **6**(9), 24–33 (2003)

31. Bai, Y., Shen, B., Zhang, S., Zhu, Z., Sun, S., Gao, J., Li, B., Wang, Y., Zhang, R., Wei, F.: Storage of mechanical energy based on carbon nanotubes with high energy density and power density. Adv. Mater. **31**(9), 1800680 (2019)
32. DeTeresa, S.J.: Materials for advanced flywheel energy-storage devices. MRS Bull. **24**(11), 51–56 (1999)
33. Zhang, R., Wen, Q., Qian, W., Su, D.S., Zhang, Q., Wei, F.: Superstrong ultralong carbon nanotubes for mechanical energy storage. Adv. Mater. **23**(30), 3387–3391 (2011)
34. Li, X., Palazzolo, A.: A review of flywheel energy storage systems: state of the art and opportunities. J. Energy Storage **1**(46), 103576 (2022)
35. Olabi, A.G., Wilberforce, T., Abdelkareem, M.A., Ramadan, M.: Critical review of flywheel energy storage system. Energies **14**(8), 2159 (2021)
36. DeTeresa, S.J., Groves, S.E.: Properties of fiber composites for advanced flywheel energy storage devices. Lawrence Livermore National Lab. (LLNL), Livermore, CA (United States) (2001)
37. Bolund, B., Bernhoff, H., Leijon, M.: Flywheel energy and power storage systems. Renew. Sustain. Energy Rev. **11**(2), 235–258 (2007)
38. Shashanka, R., Chaira, D.: Effect of sintering temperature and atmosphere on nonlubricated sliding wear of nano-yttria-dispersed and yttria-free duplex and ferritic stainless steel fabricated by powder metallurgy. Tribol. Trans. **60**(2), 324–336 (2017)
39. Rajendrachari, S.: An overview of high-entropy alloys prepared by mechanical alloying followed by the characterization of their microstructure and various properties. Alloys **1**(2), 116–132 (2022)
40. Shashanka, R., Chaira, D., Chakravarty, D.: Fabrication of nano-yttria dispersed duplex and ferritic stainless steels by planetary milling followed by spark plasma sintering and nonlubricated sliding wear behaviour study. J. Mater. Sci. Eng. B **6**, 111–125 (2016)
41. Sharath, B.N., Madhu, K.S., Pradeep, D.G., Madhu, P., Premkumar, B.G., Karthik, S.: Effects of tertiary ceramic additives on the micro hardness and wear characteristics of Al2618+Si3N4–B4C–Gr hybrid composites for automotive applications. J. Alloys Metall. Syst. **31**, 100014 (2023)
42. Nagaraju, S.B., Sathyanarayana, K., Somashekara, M.K., Pradeep, D.G., Puttegowda, M., Verma, A.: Artificial neural networks for predicting mechanical properties of Al2219-B4C-Gr composites with multireinforcements. Proc. Inst. Mech. Eng. Part C: J. Mech. Eng. Sci. 09544062231196038 (2023)
43. Nagaraju, S.B., Somashekara, M.K, Govindaswamy, P.D., Puttegowda, M., Girijashankar, P.B., Sathyanarayana, K.: Mechanical characterization of B4C-Gr Al2618 based composites synthesized by stir casting method. Appl. Sci. Eng. Prog. **16**(3), 6579 (2023)
44. LeBlanc, S.: Thermoelectric generators: linking material properties and systems engineering for waste heat recovery applications. Sustain. Mater. Technol. **1**(1), 26–35 (2014)
45. Zoui, M.A., Bentouba, S., Stocholm, J.G., Bourouis, M.: A review on thermoelectric generators: progress and applications. Energies **13**(14), 3606 (2020)
46. Ovik, R., Long, B.D., Barma, M.C., Riaz, M., Sabri, M.F., Said, S.M., Saidur, R.: A review on nanostructures of high-temperature thermoelectric materials for waste heat recovery. Renew. Sustain. Energy Rev. **1**(64), 635–659 (2016)
47. Patil, D., Arakerimath, R.R.: A review of thermoelectric generator for waste heat recovery from engine exhaust. Int. J. Res. Aeronaut. Mech. Eng. **1**(8), 1–9 (2013)
48. Patil, D.S., Arakerimath, R.R., Walke, P.V.: Thermoelectric materials and heat exchangers for power generation—A review. Renew. Sustain. Energy Rev. **1**(95), 1–22 (2018)
49. Dutil, Y., Rousse, D.R., Salah, N.B., Lassue, S., Zalewski, L.: A review on phase-change materials: mathematical modeling and simulations. Renew. Sustain. Energy Rev. **15**(1), 112–130 (2011)
50. Raj, C.R., Suresh, S., Bhavsar, R.R., Singh, V.K.: Recent developments in thermo-physical property enhancement and applications of solid solid phase change materials: a review. J. Therm. Anal. Calorim. **139**, 3023–3049 (2020)

51. Socaciu, L.G.: Thermal energy storage with phase change material. Leonardo Electron. J. Pract. Technol. **20**, 75–98 (2012)
52. Nazir, H., Batool, M., Osorio, F.J., Isaza-Ruiz, M., Xu, X., Vignarooban, K., Phelan, P., Kannan, A.M.: Recent developments in phase change materials for energy storage applications: a review. Int. J. Heat Mass Transf. **1**(129), 491–523 (2019)
53. Kalnæs, S.E., Jelle, B.P.: Phase change materials and products for building applications: a state-of-the-art review and future research opportunities. Energy Build. **1**(94), 150–176 (2015)
54. Socaciu, L., Pleşa, A., Ungureşan, P., Giurgiu, O.: Review on phase change materials for building applications. Leonardo Electron. J. Pract. Technol. **25**, 179–194 (2014)
55. Tian, Y., Zhao, C.Y.: A review of solar collectors and thermal energy storage in solar thermal applications. Appl. Energy **1**(104), 538–553 (2013)
56. Shukla, R., Sumathy, K., Erickson, P., Gong, J.: Recent advances in the solar water heating systems: a review. Renew. Sustain. Energy Rev. **1**(19), 173–190 (2013)
57. Abedin, H.A, Rosen, M.A.: A critical review of thermochemical energy storage systems. Open Renew. Energy J. **4**(1) (2011)
58. Molina, M.G., Mercado, P.E.: Power flow stabilization and control of microgrid with wind generation by superconducting magnetic energy storage. IEEE Trans. Power Electron. **26**(3), 910–922 (2010)
59. Pradeep, D.G., Nithin, H.S., Sharath BN, Madhu KS, Venkatesh CV. Microstructure and wear behavior of microwave treated WC-10Co-4Cr composite coating on AISI 4140 alloy steel. In: IOP Conference Series: materials Science and Engineering, vol. 1189, no. 1, p. 012012. IOP Publishing (2021)
60. Shashanka, R., Chaira, D.: Ball Milled Nano-Structured Stainless Steel Powders: fabrication and Characterization. Educreation Publishing (2017)
61. Mahale, R.S., Shashanka, R.: Mechanical testing of spark plasma sintered materials: a review. In: AIP Conference Proceedings, vol. 2469, no. 1. AIP Publishing (2022)
62. Adimule, V., Keri, R., Rajendrachari, S., Kendrekar, P., Kulkarni, C.V.: Highly photoluminescence and wide band gap insulating metal hybrid nanoparticles array of samarium-doped SrO: CoO: synthesis, characterizations and sensor characteristics. J. Mater. Sci.: Mater. Electron. **34**(5), 442 (2023)
63. Nagaraju, S.B., Somashekara, M.K., Govindaswamy, P.D., Puttegowda, M., Shankar, P.B., Sathyanarayana, K.: Wear behaviour of hybrid (boron carbide-graphite) aluminium matrix composites under high temperature. J. Eng. Appl. Sci. **70**(1), 1–23 (2023)
64. Sharath, B.N., Karthik, S., Madhu, P., Madhu, K.S., Pradeep, D.G.: Predictive analysis of slurry erosion behaviour in aluminium-based hybrid metal matrix composites: experimental and machine learning approach. J. Bio- Tribo-Corros. **9**(4), 70 (2023)
65. Sharath, B.N., Madhu, P., Verma, A.: Enhancing tribological performance: a review of ceramic reinforced aluminium hybrid composites for high-temperature engineering applications. Hybrid Adv. **1**, 100094 (2023)
66. Kötz, R., Carlen, M.J.: Principles and applications of electrochemical capacitors. Electrochimica Acta **45**(15–16), 2483–2498 (2000)
67. Wang, Y., Song, Y., Xia, Y.: Electrochemical capacitors: mechanism, materials, systems, characterization and applications. Chem. Soc. Rev. **45**(21), 5925–5950 (2016)
68. Lewandowski, A., Galinski, M.: Practical and theoretical limits for electrochemical double-layer capacitors. J. Power Sour. **173**(2), 822–828 (2007)
69. Gür, T.M.: Review of electrical energy storage technologies, materials and systems: challenges and prospects for large-scale grid storage. Energy Environ. Sci. **11**(10), 2696–2767 (2018)
70. Leung, P., Li, X., De León, C.P., Berlouis, L., Low, C.J., Walsh, F.C.: Progress in redox flow batteries, remaining challenges and their applications in energy storage. RSC Adv. **2**(27), 10125–10156 (2012)
71. Chakrabarti, M.H., Mjalli, F.S., AlNashef, I.M., Hashim, M.A., Hussain, M.A., Bahadori, L., Low, C.T.: Prospects of applying ionic liquids and deep eutectic solvents for renewable energy storage by means of redox flow batteries. Renew. Sustain. Energy Rev. **1**(30), 254–270 (2014)

72. Rajendrachari, S., Adimule, V.M., Jayaprakash, G.K., Pandith, A.: Electrochemical oxidation of methylene blue dye in wastewater using mechanically alloyed high entropy alloy modified carbon paste electrode using cyclic voltammetry. Mater. Res. Express **10**(5), 054003 (2023)
73. Rajendrachari, S., Basavegowda, N., Vinaykumar, R., Narsimhachary, D., Somu, P., Lee, M.J.: Electrocatalytic determination of methyl orange dye using mechanically alloyed novel metallic glass modified carbon paste electrode by cyclic voltammetry. Inorg. Chem. Commun. **30**, 111010 (2023)
74. Kim, T., Song, W., Son, D.Y., Ono, L.K., Qi, Y.: Lithium-ion batteries: outlook on present, future, and hybridized technologies. J. Mater. Chem. A **7**(7), 2942–2964 (2019)
75. Chen, H., Zhang, X., Wu, S., Chen, F., Xu, J.: A comparative study of iron-vanadium and all-vanadium flow battery for large scale energy storage. Chem. Eng. J. **1**(429), 132403 (2022)
76. Rehman, S., Al-Hadhrami, L.M., Alam, M.M.: Pumped hydro energy storage system: a technological review. Renew. Sustain. Energy Rev. **1**(44), 586–598 (2015)
77. Blakers, A., Stocks, M., Lu, B., Cheng, C.: A review of pumped hydro energy storage. Prog. Energy **3**(2), 022003 (2021)
78. Yang, C.J.: Pumped hydroelectric storage. In: Storing Energy, pp. 25–38. Elsevier
79. Ding, Z., Wu, W., Leung, M.: Advanced/hybrid thermal energy storage technology: material, cycle, system and perspective. Renew. Sustain. Energy Rev. **1**(145), 111088 (2021)
80. Afif, A., Rahman, S.M., Azad, A.T., Zaini, J., Islan, M.A., Azad, A.K.: Advanced materials and technologies for hybrid supercapacitors for energy storage—A review. J. Energy Storage **1**(25), 100852 (2019)
81. Häggström, F., Delsing, J.: IoT energy storage—A forecast. Energy Harvest. Syst. **5**(3–4), 43–51 (2018). https://doi.org/10.1515/ehs-2018-0010. S2CID 64526195. Accessed 30 Oct. 2020
82. Tehrani, Z., Thomas, D.J., Korochkina, T., Phillips, C.O., Lupo, D., Lehtimäki, S., O'Mahony, J., Gethin, D.T.: Large-area printed supercapacitor technology for low-cost domestic green energy storage. (PDF). Energy **118**, 1313–1321 (2017). https://doi.org/10.1016/j.energy.2016.11.019. ISSN 0360-5442. S2CID 55090490
83. Gidwani, M., Bhagwani, A., Rohra, N.: Supercapacitors: the near future of batteries. Int. J. Eng. Invent. **4**(5), 22–27 (2014)
84. Chen, X., Huang, L., Liu, J., Song, D., Yang, S.: Peak shaving benefit assessment considering the joint operation of nuclear and battery energy storage power stations: Hainan case study. Energy **15**(239), 121897 (2022)
85. Zheng, W., Ma, K., Wang, X.: Hybrid energy storage with supercapacitor for cost-efficient data center power shaving and capping. IEEE Trans. Parallel Distrib. Syst. **28**(4), 1105–1118 (2016)
86. Wang, C., Zhang, L., Zhang, Z., Zhao, R., Zhao, D., Ma, R., Yin, L.: Layered materials for supercapacitors and batteries: applications and challenges. Prog. Mater. Sci. **1**(118), 100763 (2021)
87. Libich, J., Máca, J., Vondrák, J., Čech, O., Sedlaříková, M.: Supercapacitors: properties and applications. J. Energy Storage **1**(17), 224–227 (2018)
88. Smith, T.A., Mars, J.P., Turner, G.A.: Using supercapacitors to improve battery performance. In: 2002 IEEE 33rd Annual IEEE Power Electronics Specialists Conference. Proceedings (Cat. No. 02CH37289), vol. 1, pp. 124–128. IEEE (2002)
89. Uno, M., Tanaka, K.: Accelerated charge–discharge cycling test and cycle life prediction model for supercapacitors in alternative battery applications. IEEE Trans. Industr. Electron. **59**(12), 4704–4712 (2011)
90. Shukla, A.K., Sampath, S., Vijayamohanan, K.: Electrochemical supercapacitors: energy storage beyond batteries. Curr. Sci. **79**(12), 1656–1661 (2000)
91. Burke, A., Zhao, H.: Applications of Supercapacitors in Electric and Hybrid Vehicles. Institute of Transportation Studies. University of California, Davis (2015)
92. Faggioli, E., Rena, P., Danel, V., Andrieu, X., Mallant, R., Kahlen, H.: Supercapacitors for the energy management of electric vehicles. J. Power Sour. **84**(2), 261–269 (1999)

93. Horn, M., MacLeod, J., Liu, M., Webb, J., Motta, N.: Supercapacitors: a new source of power for electric cars? Econ. Anal. Policy **1**(61), 93–103 (2019)
94. Zhang, Z., Zhang, X., Chen, W., Rasim, Y., Salman, W., Pan, H., Yuan, Y., Wang, C.: A high-efficiency energy regenerative shock absorber using supercapacitors for renewable energy applications in range extended electric vehicle. Appl. Energy **15**(178), 177–188 (2016)
95. Asekar, A.K.: STOP-START system using microhybrid technology for increasing fuel efficiency. Int. J. Mech. Prod. Eng. **1**(6), 20–26 (2013)
96. Dinh, T.Q., Marco, J., Niu, H., Greenwood, D., Harper, L., Corrochano, D.: A novel method for idle-stop-start control of micro hybrid construction equipment—Part A: fundamental concepts and design. Energies **10**(7), 962 (2017)
97. Cao, H., Kong, L., Tang, M., Zhang, Z., Wu, X., Lu, L., Li, D.: An electromagnetic energy harvester for applications in a high-speed rail pavement system. Int. J. Mech. Sci. **1**(243), 108018 (2023)
98. Becker, T., Borjesson, V., Cetinkaya, O., Baoxing, C., Colomer-Farrarons, J., Maeve, D., Elefsiniotis, A., Govoni, L., Hadas, Z., Hayes, M., Holmes, A.S.: Energy harvesting for a green internet of things
99. Afonso, T.L., Marques, A.C., Fuinhas, J.A.: Strategies to make renewable energy sources compatible with economic growth. Energ. Strat. Rev. **1**(18), 121–126 (2017)
100. Harte, J., Jassby, A.: Energy technologies and natural environments: the search for compatibility. Annual Rev. Energy **3**(1), 101–146 (1978)
101. Chu, A., Braatz, P.: Comparison of commercial supercapacitors and high-power lithium-ion batteries for power-assist applications in hybrid electric vehicles: I. Initial characterization. J. Power Sour. **112**(1), 236–246 (2002)
102. Musolino, V., Tironi, E., di Milano, P.: A comparison of supercapacitor and high-power lithium batteries. In: Electrical Systems for Aircraft, Railway and Ship Propulsion, pp. 1–6. IEEE (2010)
103. Castaings, A., Lhomme, W., Trigui, R., Bouscayrol, A.: Comparison of energy management strategies of a battery/supercapacitors system for electric vehicle under real-time constraints. Appl. Energy **1**(163), 190–200 (2016)

Chapter 18
Electrochemical Sensor Applications of Mechanically Alloyed Powders

Halfadji Ahmed and Naous Mohamed

Abstract Electrochemical redox reactions are fundamental processes involving electron transfer that occur at the interface between electrodes and electrolytes, playing a crucial role in various scientific fields. This chapter explores the significance of these reactions, particularly in the context of electrochemical sensors, and discusses their applications in chemistry, materials science, and energy conversion. Various electrochemical techniques, such as cyclic voltammetry and chronoamperometry, are introduced to analyze redox reactions and develop practical applications. The study of bioactive molecules in electrochemistry is of paramount importance, given its wide-ranging applications. These molecules, including neurotransmitters, antioxidants, metabolites, and biomarkers, are pivotal in biological processes, healthcare, disease diagnosis, and environmental monitoring. Understanding the electrochemical behavior of bioactive molecules enables the development of diagnostic tools, neuroscience research, antioxidant assessment, metabolism monitoring, environmental protection, and food safety. Furthermore, it contributes to drug development and biotechnology applications, underscoring its interdisciplinary significance. Mechanically alloyed modifiers play a crucial role in enhancing the electrochemical performance of materials and electrodes. Mechanical alloying, a process involving powder milling and blending, creates alloyed or composite materials with improved properties. These modifiers enhance conductivity, increase surface area, catalyze reactions, improve stability, and enable selective sensing. They find applications in energy storage, environmental sensing, biomedical devices, and more, making them invaluable for tailoring electrode materials and advancing electrochemical systems. This chapter sheds light on the importance of understanding electrochemical redox reactions, especially concerning bioactive molecules, and highlights the

H. Ahmed (✉) · N. Mohamed
Department of Sciences and Technology, Faculty of Applied Sciences, Ibn Khaldoun University of Tiaret, 14000 Tiaret, Algeria
e-mail: ahmedhalfadji@gmail.com; ahmed.halfadji@univ-tiaret.dz

H. Ahmed
Synthesis and Catalysis Laboratory, Ibn Khaldoun University of Tiaret, 14000 Tiaret, Algeria

N. Mohamed
LCPM Laboratory, Faculty of Sciences, Chemistry Department, University of Oran 1 Ahmed Ben Bella, Oran, Algeria

S. Rajendrachari (ed.), *Mechanically Alloyed Novel Materials*, Advanced Structured Materials 220, https://doi.org/10.1007/978-981-97-6504-1_18

role of mechanically alloyed modifiers in advancing sensor technology and a diverse range of applications. It underscores the interdisciplinary nature of electrochemistry and its pivotal role in addressing critical challenges in healthcare, environmental protection, and materials science.

Keywords Cyclic voltammetry · Electrochemical sensors · Redox reactions · Mechanically alloyed modifiers · Carbon paste electrodes

18.1 Introduction

Electrochemical redox reactions are fundamental processes that involve the transfer of electrons between chemical species [1]. These reactions occur at the interface between an electrode and an electrolyte solution and are crucial in various fields, including chemistry, electrochemistry, materials science, and energy conversion. In an electrochemical cell, there are typically two half-reactions: the oxidation half-reaction (loss of electrons) and the reduction half-reaction (gain of electrons). The overall reaction is known as a redox (reduction-oxidation) reaction. Electrons flow from the species being oxidized (the reducing agent) to the species being reduced (the oxidizing agent) through an external circuit, producing an electrical current [2]. Electrochemical redox reactions can be described using various electrochemical techniques, including cyclic voltammetry, chronoamperometry, and potentiostatic/galvanostatic methods. These techniques allow scientists and engineers to study the kinetics, thermodynamics, and mechanisms of redox reactions, as well as to develop practical applications such as electrochemical sensors, fuel cells, batteries, and corrosion protection [3, 4].

The study of bioactive molecules in electrochemistry holds significant importance due to its wide-ranging applications in various fields. Bioactive molecules, which include neurotransmitters, antioxidants, metabolites, and biomarkers, play crucial roles in biological processes, health, disease diagnosis, and environmental monitoring. We can cite some key reasons why studying bioactive molecules in electrochemistry is significant, the electrochemical analysis of bioactive molecules allows for the development of diagnostic tools and biosensors for monitoring biomarkers associated with diseases. For example, glucose sensors are used for diabetes management by measuring blood glucose levels [5]. Also, in neuroscience research, the diagnostic of neurotransmitters such as dopamine and serotonin are essential for neural communication. Electrochemical sensors help researchers understand brain function and disorders like Parkinson's and depression [6]. Electrochemical methods are employed to assess antioxidant levels like ascorbic acid (vitamin C) and glutathione to protect cells from oxidative stress in biological samples [7]. Metabolism monitoring via electrochemical sensors are used to monitor metabolites like lactate, uric acid, and cholesterol in clinical and research settings to gain insights into metabolic processes [8]. Moreover, Bioactive molecules, including environmental contaminants and pollutants, can be detected using electrochemical sensors for environmental monitoring and assessing water and soil quality [9]. To food safety electrochemical sensors are applied in food analysis to detect foodborne pathogens and toxins, ensuring food safety and quality [10]. Pharmaceuticals as emergence contaminate groups as bioactive molecules are essential in drug development, as electrochemical methods can be used to screen potential drug candidates and assess drug efficacy [11]. In the field of biotechnology, bioactive molecules are crucial in biotechnology for applications such as biocatalysis and enzyme-based biosensors [12]. Overall, understanding and studying bioactive molecules in electrochemistry have broad implications for improving healthcare, advancing scientific research, and addressing critical issues related to health, the environment, and biotechnology.

Mechanically alloyed modifiers play a pivotal role in enhancing the electrochemical performance of materials and electrodes. Mechanical alloying is a process that involves the milling and blending of powders to create alloyed or composite materials with improved properties. These modified materials can significantly impact various aspects of electrochemical systems, including their sensitivity, stability, and efficiency. Here are some key roles of mechanically alloyed modifiers in enhancing electrochemical performance: which can improve conductivity, the mechanically alloyed modifiers often contain conductive elements or materials, such as nanoparticles of metals or conductive polymers. These modifiers can enhance the electrical conductivity of electrodes, allowing for more efficient charge transfer during electrochemical reactions [13]. Also, can enhance surface area, mechanical alloying can create nanoscale structures and increase the specific surface area of electrode materials. This increased surface area provides more active sites for electrochemical reactions and improves the overall electrode performance [14]. Mechanically alloyed modifiers can introduce catalytic elements or nanoparticles into electrode materials to enhance catalytic activity. These modifiers can catalyze specific electrochemical reactions, reducing overpotential and improving the overall efficiency of the process [15]. The incorporation of mechanically alloyed modifiers can enhance the structural stability of electrode materials (improved stability). This increased stability reduces electrode degradation and improves the long-term performance of electrochemical devices [16]. In the context of sensors, mechanically alloyed modifiers can improve selectivity by enhancing the electrode's ability to discriminate between analytes (selective sensing). This is achieved by modifying the electrode surface to selectively adsorb or react with specific target molecules [15]. Mechanically alloyed modifiers are used in energy storage and conversion devices such as batteries and supercapacitors (energy storage and conversion). They can improve the electrochemical performance of these devices by optimizing charge–discharge kinetics and capacity [17]. In environmental sensors, mechanically alloyed modifiers can enhance the sensitivity and selectivity of electrodes for detecting pollutants, heavy metals, and other analytes in environmental samples (Environmental Sensing) [18]. In the field of biomedical applications, mechanically alloyed modifiers are utilized in biosensors and bioelectrodes for detecting biomolecules in medical diagnostics. They improve the electrochemical response and reliability of these devices [19].

In summary, mechanically alloyed modifiers serve as valuable tools for tailoring the properties of electrode materials, thereby enhancing their electrochemical performance. Their versatility and ability to improve conductivity, catalytic activity, stability, and selectivity make them essential components in various electrochemical applications.

18.2 Mechanical Alloying: Fundamentals and Applications

18.2.1 Explanation of the Mechanical Alloying Process

Mechanical alloying is a solid-state powder processing technique used to synthesize alloy materials and composite powders with finely dispersed phases. It involves the repeated mechanical deformation, fracturing, and welding of powder particles in a high-energy ball mill or attritor mill. The process is often used to create materials with improved properties, such as enhanced hardness, wear resistance, and electrical conductivity. Here is a step-by-step explanation of the mechanical alloying process:

Selection of Powder Materials: Mechanical alloying begins with selecting high-purity elemental or pre-alloyed powders as starting materials. These powders are typically mixed in precise proportions to achieve the desired alloy composition.

Loading the Mill: The selected powders are loaded into a grinding container, such as a ball mill, along with grinding media (usually balls or beads) made from materials like steel, tungsten carbide, or ceramic. The ratio of powder to grinding media is critical to the process.

Mill Operation: The mill is sealed, and the milling process begins. The grinding media and powders inside the mill jar are subjected to high mechanical stresses caused by the collision and friction between particles and the balls. These interactions lead to particle deformation, fracturing, and cold welding.

Deformation and Fracture: During milling, the kinetic energy of the grinding media is transferred to the powder particles. This results in the deformation and fracturing of the particles. Deformation promotes the formation of alloy phases, while fracturing reduces the particle size.

Cold Welding: As particles collide, they also experience cold welding due to the high-pressure conditions within the mill. Cold welding leads to the formation of bonds between particles, facilitating the consolidation of the alloy.

Intermediate Steps: The milling process is often carried out in several intermediate steps with intermittent periods of rest to avoid excessive heat generation. Cooling systems may be employed to control temperature rise.

Particle Size Reduction: Over time, the milling process reduces the particle size of the powders, resulting in fine, homogenized alloy particles. The degree of reduction and the final particle size depend on the milling time, milling speed, and the properties of the starting materials.

Final Alloy Formation: Mechanical alloying continues until the desired alloy composition and particle size distribution are achieved. The resulting alloy powder can be used as-is or further processed (e.g., through consolidation techniques like hot pressing) to form bulk materials.

Mechanical alloying offers advantages such as the ability to create metastable alloys, tailor material properties, and produce fine-grained materials with enhanced mechanical properties. It is widely used in the development of advanced materials for applications in aerospace, electronics, automotive, and more.

18.2.2 Advantages and Applications of Mechanically Alloyed Materials in Electrochemistry

Mechanically alloyed materials offer unique advantages that make them highly valuable in electrochemical applications. The mechanical alloying process allows for precise control over material composition, particle size, and structure, leading to enhanced electrochemical performance. These materials exhibit improved electrical conductivity, increased surface area, and tailored catalytic properties, making them ideal for various electrochemical applications. They are extensively used in the development of high-performance electrodes for fuel cells, batteries, and supercapacitors [20]. Additionally, mechanically alloyed modifiers play a pivotal role in enhancing the sensitivity and selectivity of electrochemical sensors for detecting bioactive molecules, environmental pollutants, and heavy metals [21, 22]. Their application in corrosion protection coatings and advanced materials for energy storage and conversion further underscores their significance in the field of electrochemistry [23]. Overall, mechanically alloyed materials have revolutionized the design and performance of electrochemical devices, enabling advancements in energy technology, environmental monitoring, and analytical chemistry.

18.2.3 Tailoring Material Properties Through Mechanical Alloying

Mechanical alloying is a versatile powder processing technique that enables precise control over material properties by manipulating the composition, microstructure, and phase evolution of materials. Through the mechanical alloying process, it is possible to tailor material properties to meet specific application requirements. This includes the ability to create nanoscale structures, finely disperse phases, and control grain size, resulting in enhanced mechanical, thermal, and electrical properties. The process can also induce changes in lattice structure and crystal defects, leading to improved functional characteristics. Additionally, mechanical alloying facilitates the formation of metastable phases that can exhibit unique properties not attainable through conventional methods [24]. These tailored material properties find applications in a wide range of fields, including aerospace, automotive, electronics, and energy storage, where advanced materials with superior performance are in high demand [25, 26]

18.3 Bioactive Molecules in Electrochemistry

18.3.1 Introduction to Bioactive Molecules

Bioactive molecules are a diverse group of chemical compounds that play essential roles in biological systems, influencing various physiological and biochemical processes within living organisms. These molecules can have profound effects on cellular functions, signaling pathways, and overall health. Bioactive molecules encompass a wide range of compounds, including neurotransmitters like dopamine and serotonin, antioxidants such as ascorbic acid (vitamin C), metabolites like glucose and lactate, hormones, enzymes, and many others [27–29]. They are involved in processes such as neural transmission, immune response, metabolism, and cell signaling. The study of bioactive molecules is integral to understanding biological functions, diseases, and the development of diagnostic tools and therapeutic interventions [30]. Moreover, bioactive molecules (bio-molecules) also find applications in fields such as clinical diagnostics, pharmaceuticals, and environmental monitoring [31].

18.3.2 Categories of Bioactive Molecules (e.g., Neurotransmitters, Antioxidants, Metabolites)

Bioactive molecules are a diverse group of chemical compounds that play essential roles in biological systems, influencing various physiological and biochemical processes within living organisms. These molecules can have profound effects on cellular functions, signaling pathways, and overall health. Bioactive molecules encompass a wide range of compounds, including neurotransmitters like dopamine and serotonin, antioxidants such as ascorbic acid (vitamin C), metabolites like glucose and lactate, hormones, enzymes, and many others. They are involved in processes such as neural transmission, immune response, metabolism, and cell signaling. The study of bioactive molecules is integral to understanding biological functions, diseases, and the development of diagnostic tools and therapeutic interventions [30]. Moreover, bioactive molecules also find applications in fields such as clinical diagnostics, pharmaceuticals, and environmental monitoring [6, 32].

18.3.3 Relevance of Studying Bioactive Molecules in Health, Diagnostics, and Environmental Monitoring

The study of bioactive molecules holds immense relevance in the fields of health, diagnostics, and environmental monitoring. In health, bioactive molecules such as neurotransmitters, hormones, and metabolites play crucial roles in maintaining physiological balance and health. Understanding these molecules is vital for diagnosing and managing various medical conditions, including neurological disorders, endocrine diseases, and metabolic disorders [33]. In diagnostics, bioactive molecules serve as valuable biomarkers for disease detection and monitoring, providing insights into disease progression and treatment efficacy. For instance, the measurement of glucose levels in diabetes management exemplifies the diagnostic significance of bioactive molecules [34, 35]. In environmental monitoring, bioactive molecules are indicators of pollution, contamination, and ecosystem health. Detecting and quantifying these molecules help assess water and soil quality, ensuring the safety of the environment and human health [36]. The study of bioactive molecules bridges the gap between basic science and practical applications, ultimately improving healthcare, diagnostics, and environmental sustainability.

18.4 Electrochemical Techniques for Bioactive Molecule Analysis

18.4.1 Overview of Common Electrochemical Techniques (e.g., Cyclic Voltammetry, Chronoamperometry)

Electrochemical techniques are essential tools in analytical chemistry, materials science, and electrocatalysis for studying redox reactions and characterizing electrode processes. Two widely used electrochemical techniques are cyclic voltammetry (CV) and chronoamperometry (CA). Cyclic voltammetry involves applying a voltage waveform to an electrochemical cell and measuring the resulting current response. This technique provides information about the electrochemical behavior of analytes, including redox potential, reaction kinetics, and diffusion coefficients. Chronoamperometry, on the other hand, measures the current as a function of time at a constant applied potential. It is useful for studying reaction rates, electrode kinetics, and diffusion-controlled processes. These techniques, along with others like potentiostatic/galvanostatic methods and impedance spectroscopy, play crucial roles in electrochemical analysis, sensor development, and the study of energy storage and conversion devices [37, 38].

18.4.2 Principles of Electrochemical Sensing

Electrochemical sensing is founded on the principles of electrochemistry, leveraging the interaction between electrical signals and chemical processes to detect and quantify target analytes. At the core of electrochemical sensing are three primary components: the working electrode, reference electrode, and counter electrode, immersed in an electrolyte solution. When a potential is applied across the electrodes, redox reactions occur at the working electrode's surface, generating a current that is proportional to the concentration of the analyte of interest. The key electrochemical techniques used in sensing include amperometry, potentiometry, and impedance spectroscopy, which provide insights into various aspects of the electrochemical process, such as reaction kinetics and analyte concentration. Electrochemical sensors offer advantages like high sensitivity, selectivity, and rapid response times, making them indispensable tools in applications ranging from clinical diagnostics to environmental monitoring [39–41].

18.4.3 Importance of Electrode Modifiers in Enhancing Sensor Performance

Electrode modifiers play a pivotal role in enhancing the performance of electrochemical sensors, making them critical components in sensor design and development. These modifiers are materials or coatings applied to the electrode surface, and they serve various functions to improve sensor sensitivity, selectivity, and stability. Common electrode modifiers include nanomaterials like nanoparticles and nanotubes, conductive polymers, and enzyme immobilization matrices [42, 43]. They can facilitate electron transfer, increase the active surface area, and enhance catalytic activity, resulting in improved signal response and lower detection limits [44]. Additionally, modifiers can impart selectivity by promoting specific interactions with target analytes or suppressing interference from other substances [45]. By tailoring the properties of electrode modifiers, sensors can be customized for applications in clinical diagnostics, environmental monitoring, and food analysis, contributing significantly to the advancement of analytical chemistry and sensor technology [46, 47].

18.5 Mechanically Alloyed Modifiers for Electrochemical Sensors

18.5.1 Types of Mechanically Alloyed Modifiers (e.g., Nanoparticles, Nanocomposites)

Mechanically alloyed modifiers encompass a diverse range of materials engineered to enhance the properties of base materials for various applications. Nanoparticles, including metal nanoparticles, metal oxide nanoparticles, and carbon-based nanoparticles like carbon nanotubes and graphene, are commonly employed as modifiers [48]. These nanoparticles can improve electrical conductivity, catalytic activity, and mechanical strength when incorporated into materials. Nanocomposites, on the other hand, consist of a combination of materials at the nanoscale, often with a matrix material and reinforcing nanoparticles. These nanocomposites offer tailored properties, such as enhanced hardness, wear resistance, and thermal stability [49]. Both nanoparticles and nanocomposites are used as modifiers in various fields, including aerospace, automotive, electronics, and energy storage, where the goal is to achieve superior material performance through mechanical alloying [50].

18.5.2 Properties and Advantages of Mechanically Alloyed Materials in Sensor Design

Mechanically alloyed materials exhibit a unique set of properties that make them highly advantageous for sensor design and development. The mechanical alloying process allows for precise control over material composition, particle size, and microstructure. These materials often possess nanoscale grains, high surface area, and improved electrical conductivity, making them ideal for sensor applications [51]. The enhanced surface area provides more active sites for sensing reactions, leading to higher sensitivity and lower detection limits. Additionally, the tailored composition of mechanically alloyed materials can facilitate catalytic activity, improving the sensor's selectivity and response time. The ability to create metastable phases through mechanical alloying enables the development of sensors with unique properties and enhanced stability. These advantages make mechanically alloyed materials invaluable in the design of advanced sensors for a wide range of applications, from environmental monitoring to biomedical diagnostics [52, 53].

18.5.3 Examples of Mechanically Alloyed Modifiers Used in Electrochemical Sensors

Mechanically alloyed modifiers have been extensively utilized in the development of high-performance electrochemical sensors. One common example is the incorporation of metal nanoparticles, such as platinum (Pt), gold (Au), or silver (Ag), into sensor electrodes. These nanoparticles, when mechanically alloyed with electrode materials, significantly enhance the electrode's catalytic activity for various electrochemical reactions, leading to improved sensor sensitivity and selectivity [54]. Additionally, carbon-based nanomaterials like carbon nanotubes (CNTs) and graphene, often modified through mechanical alloying, serve as excellent modifiers due to their high electrical conductivity and large surface area. These modifiers have been employed to enhance the electron transfer kinetics and immobilize biomolecules in biosensors [55]. Moreover, mechanically alloyed nanocomposites, such as metal oxide–carbon nanocomposites, offer synergistic properties, combining the conductivity of carbon and the catalytic activity of metal oxides. These modifiers find applications in the development of sensors for various analytes, including neurotransmitters, biomarkers, and environmental pollutants [56].

18.6 Electrochemical Redox Reactions of Neurotransmitters

18.6.1 In-Depth Discussion of Neurotransmitters (e.g., Dopamine, Serotonin)

Neurotransmitters are crucial chemical messengers that play a central role in communication between nerve cells, or neurons, within the nervous system. Two prominent neurotransmitters, dopamine and serotonin, exert profound effects on mood, behavior, and various physiological functions.

Dopamine

Dopamine is well-known for its role in the brain's reward and pleasure pathways. It is involved in regulating mood, motivation, and the experience of pleasure. Dysfunction in the dopaminergic system has been implicated in conditions such as Parkinson's disease, schizophrenia, and addiction. In addition to its role in the brain, dopamine also functions as a peripheral neurotransmitter, impacting processes like blood pressure regulation and digestion [57–59].

Serotonin

Serotonin, often referred to as the "feel-good" neurotransmitter, is closely associated with mood regulation and emotional well-being. It plays a critical role in controlling mood, sleep, appetite, and pain perception. Serotonin is also involved in various psychiatric and neurological disorders, including depression, anxiety disorders, and migraines. Medications that target the serotonin system, such as selective serotonin reuptake inhibitors (SSRIs), are widely used in the treatment of depression and anxiety [60, 61].

Understanding the intricate functions of neurotransmitters like dopamine and serotonin is essential in neuroscience and psychiatry, as it sheds light on the pathophysiology of numerous mental health disorders and informs the development of therapeutic interventions [60].

18.6.2 Mechanisms of Redox Reactions of Neurotransmitters on Mechanically Alloyed Modifiers

The redox reactions of neurotransmitters on mechanically alloyed modifiers involve intricate electrochemical processes that underpin the operation of electrochemical sensors and biosensors [39]. Dopamine and serotonin, as prominent neurotransmitters, often serve as analytes of interest in these systems [59, 62]. Mechanically alloyed modifiers, such as nanocomposites or metal nanoparticles, play a vital role in facilitating these reactions. During the sensing process, neurotransmitters are oxidized or reduced at the electrode surface, leading to the transfer of electrons [63]. Mechanically alloyed modifiers enhance this electron transfer process by providing a high surface area and promoting catalytic activity. The modifiers can facilitate the redox reactions of neurotransmitters by serving as active sites for electron transfer and enhancing the electrocatalytic behavior of the electrode material [39, 64]. Understanding these mechanisms is essential for the development of highly sensitive and selective electrochemical sensors for neurotransmitter detection in various applications, including neuroscience research and clinical diagnostics [65].

18.6.3 Applications in Neuroscience Research and Disease Diagnosis

Electrochemical sensors have found extensive applications in the field of neuroscience research and disease diagnosis, offering invaluable tools for understanding neural processes and detecting neurochemical markers associated with various disorders [66]. In neuroscience research, electrochemical sensors are employed to monitor neurotransmitter dynamics in real time, providing insights into neural communication, behavior, and the effects of neuropharmacological interventions [67]. These sensors enable the study of neurotransmitters like dopamine, serotonin, and glutamate, aiding in the exploration of brain function and disorders such as Parkinson's disease, depression, and addiction [67]. Furthermore, electrochemical sensors have significant potential in the diagnosis and monitoring of neurological and neuropsychiatric disorders [68]. They can detect biomarkers like amyloid-beta and tau proteins associated with Alzheimer's disease and contribute to the early diagnosis and management of conditions like epilepsy and multiple sclerosis [69]. Overall, electrochemical sensors play a pivotal role in advancing our understanding of neurobiology and improving the diagnosis and treatment of neurological disorders [70].

18.7 Electrochemical Detection of Antioxidants

18.7.1 Overview of Antioxidants and Their Importance in Health

Antioxidants are a diverse group of compounds that play a critical role in safeguarding health by combating oxidative stress and protecting cells from damage caused by reactive oxygen species (ROS) and free radicals [71]. These harmful molecules are produced during normal metabolic processes and can accumulate due to various factors such as pollution, smoking, and UV radiation exposure. Antioxidants include vitamins (e.g., vitamin C and vitamin E), minerals (e.g., selenium), phytochemicals (e.g., flavonoids and carotenoids), and enzymes (e.g., superoxide dismutase) [71]. They work by neutralizing ROS and free radicals, thereby reducing the risk of oxidative damage to DNA, proteins, and lipids. The importance of electrochemical detection of antioxidants in health is underscored by their potential to prevent chronic diseases, including cardiovascular disease, cancer, and neurodegenerative disorders [72]. A diet rich in antioxidant-rich foods, such as fruits, vegetables, and nuts, is associated with reduced inflammation and improved overall well-being, highlighting the significance of antioxidants in maintaining health.

18.7.2 Redox Reactions of Antioxidants (e.g., Ascorbic Acid, Glutathione) on Modified Electrodes

The redox reactions of antioxidants, such as ascorbic acid (vitamin C) and glutathione, on modified electrodes are of great interest in electrochemical research and sensor development. Ascorbic acid is a potent water-soluble antioxidant that undergoes reversible redox reactions involving the conversion between its reduced form (ascorbate) and oxidized form (dehydroascorbic acid) at appropriate electrode surfaces. Modified electrodes, often functionalized with nanomaterials like carbon nanotubes or metal nanoparticles, provide a favorable environment for enhancing the electrochemical detection of ascorbic acid due to improved electron transfer kinetics and increased surface area. Similarly, glutathione, a tripeptide antioxidant, plays a vital role in maintaining cellular redox balance. Modified electrodes facilitate the sensitive detection of glutathione through redox reactions that involve the reversible conversion between its reduced and oxidized forms. These electrochemical sensors find applications in studying antioxidant behavior, assessing oxidative stress levels, and monitoring antioxidant concentrations in biological and environmental samples.

18.7.3 Implications for Monitoring Antioxidant Levels in Biological Samples

The monitoring of antioxidant levels in biological samples holds significant implications for both clinical and research applications. Antioxidants, such as ascorbic acid (vitamin C), glutathione, and tocopherols, play a pivotal role in protecting cells from oxidative damage and maintaining redox homeostasis in living organisms. Assessing antioxidant levels in biological samples, including blood, urine, and tissues, provides valuable insights into an individual's oxidative stress status and overall health. High levels of antioxidants are associated with a reduced risk of chronic diseases, including cardiovascular disease and cancer, whereas low levels may indicate an increased susceptibility to oxidative stress-related disorders [73, 74]. Monitoring antioxidant levels is particularly relevant in clinical practice for assessing nutritional deficiencies, oxidative stress in chronic diseases, and the effectiveness of antioxidant therapies. Additionally, in research settings, understanding antioxidant dynamics aids in elucidating disease mechanisms and developing targeted interventions to mitigate oxidative damage [75].

18.8 Metabolite Sensing via Electrochemical Methods

18.8.1 Introduction to Metabolites and Their Role in Metabolism

Metabolites are the small molecules involved in the myriad chemical reactions that collectively constitute metabolism, the complex set of biochemical processes that occur within living organisms. These molecules serve as the building blocks, intermediates, and end products of metabolic pathways. Metabolites can encompass a wide range of compounds, including sugars, amino acids, lipids, nucleotides, and various organic acids. They are vital for energy production, growth, and the maintenance of cellular functions. Metabolites are not only the substrates and products of metabolic reactions but also play crucial roles as signaling molecules and regulators of gene expression, influencing cellular responses to environmental changes and nutrient availability. Studying the diverse array of metabolites and their interactions provides valuable insights into the biochemical intricacies of living systems and is central to fields such as biochemistry, systems biology, and metabolomics [76, 77].

18.8.2 Electrochemical Detection of Metabolites (e.g., Glucose, Lactate, Uric Acid) on Mechanically Alloyed Modifiers

The electrochemical detection of metabolites, including glucose, lactate, and uric acid, has garnered significant attention in various applications, ranging from clinical diagnostics to food analysis. Mechanically alloyed modifiers have emerged as promising platforms for enhancing the sensitivity and selectivity of electrochemical sensors targeting these metabolites. Glucose, as a fundamental biomarker for diabetes management, has been extensively studied. Modified electrodes, incorporating nanomaterials like carbon nanotubes or metal nanoparticles through mechanical alloying, offer improved catalytic activity, increased surface area, and enhanced electron transfer kinetics for glucose oxidation [78, 79]. Similarly, lactate and uric acid sensors benefit from mechanically alloyed modifiers, which facilitate their electrochemical detection by providing optimized electrode surfaces and enhancing the electrocatalytic behavior [6]. These sensors hold great potential for real-time monitoring of metabolite concentrations in biological fluids and are crucial for understanding metabolic processes, disease diagnosis, and personalized medicine [79].

18.8.3 Clinical and Environmental Applications

Metabolite sensing using electrochemical methods has found extensive applications in both clinical and environmental contexts. In clinical settings, the detection of metabolites such as glucose, lactate, and uric acid is of paramount importance for disease diagnosis and management [80]. Electrochemical sensors provide rapid, sensitive, and cost-effective means for monitoring these metabolites in biological fluids, offering critical insights into conditions like diabetes, cardiovascular diseases, and metabolic disorders [81]. Moreover, electrochemical sensors enable real-time monitoring of biomarkers, facilitating personalized medicine and enhancing patient care. In the environmental domain, electrochemical sensors play a crucial role in detecting and quantifying various metabolites and pollutants in water, air, and soil. Metabolite sensing via electrochemical methods aids in assessing water quality, detecting contaminants, and monitoring environmental changes. For instance, the detection of organic metabolites like phenols and nitrates is essential for assessing water pollution and ensuring safe drinking water. Furthermore, electrochemical sensors are employed in environmental monitoring to detect volatile organic compounds (VOCs), heavy metals, and other pollutants, contributing to pollution control and the preservation of ecosystems [82]. These versatile electrochemical sensors offer practical solutions for a wide range of applications, bridging the gap between clinical and environmental monitoring by providing efficient and accurate metabolite detection [83].

18.9 Challenges and Future Directions

18.9.1 Current Challenges in Electrochemical Sensing of Bioactive Molecules

While electrochemical sensing of bioactive molecules has made significant advancements, several challenges persist in this field. One prominent challenge is achieving high selectivity, especially in complex biological matrices containing numerous interfering species. Developing sensors with specific recognition elements and advanced signal-processing techniques is essential to address this issue [84]. Additionally, enhancing the sensitivity and detection limits of electrochemical sensors remains a continuous goal. Strategies involving the design of novel nanomaterial-based modifiers and the optimization of electrode configurations are being explored [85]. Ensuring sensor stability and reproducibility over extended periods and under varying environmental conditions is another challenge. Efforts are being made to improve sensor robustness and reliability for practical applications. Finally, the translation of lab-based research into commercially available, user-friendly sensors for point-of-care diagnostics and environmental monitoring presents challenges related to standardization and cost-effective manufacturing [85, 86]. Overcoming these challenges will further advance the field of electrochemical sensing and its widespread adoption in various applications.

18.9.2 Potential Improvements and Innovations in Mechanically Alloyed Modifiers

Mechanically alloyed modifiers have demonstrated their effectiveness in enhancing the performance of electrochemical sensors. However, ongoing research is focused on achieving further improvements and innovations in this field. One area of exploration is the development of novel nanomaterial-based modifiers with tailored properties, such as enhanced catalytic activity, selectivity, and stability. Advances in nanotechnology offer the possibility of creating finely tuned modifiers that can address specific sensing challenges [87]. Another avenue for improvement lies in optimizing the synthesis and fabrication processes of mechanically alloyed modifiers to ensure consistency and reproducibility [88]. Innovations in electrode design, including the integration of 3D-printing technologies and microfabrication techniques, hold promise for creating sensors with enhanced performance and miniaturization [88, 89]. Furthermore, the exploration of new materials, such as 2D materials and hybrid nanocomposites, opens up exciting possibilities for tailoring the properties of modifiers and expanding their applications in diverse sensing scenarios. Continued research and development in these areas are expected to lead to breakthroughs in the design and utilization of mechanically alloyed modifiers for electrochemical sensors [90, 91].

18.9.3 Emerging Trends and Future Prospects in the Field

The field of electrochemical sensing continues to evolve, driven by emerging trends and promising future prospects. One significant trend is the integration of artificial intelligence (AI) and machine learning algorithms into electrochemical sensor data analysis, enabling advanced pattern recognition, real-time monitoring, and improved accuracy in analyte quantification [92]. Additionally, the development of flexible and wearable sensors has gained momentum, offering non-invasive and continuous monitoring of various biomarkers, paving the way for personalized healthcare applications [93]. The miniaturization of sensors and the advent of lab-on-a-chip technologies hold promise for point-of-care diagnostics and on-site environmental monitoring [94]. Furthermore, the utilization of 3D printing and additive manufacturing techniques allows for the rapid prototyping of custom-designed sensors, making sensor fabrication more accessible [95]. The incorporation of new nanomaterials, such as metal–organic frameworks (MOFs) and 2D materials, as modifiers, expands the range of detectable analytes and enhances sensor performance [96]. As the field continues to advance, electrochemical sensors are expected to play an increasingly pivotal role in health care, environmental protection, and various industries.

18.10 Conclusion

In this chapter, we have delved into the fascinating world of electrochemical redox reactions of bioactive molecules on mechanically alloyed modifiers. We began with an overview of electrochemical redox reactions, emphasizing their significance in understanding the behavior of various bioactive molecules. The role of mechanically alloyed modifiers in enhancing electrochemical performance was highlighted, showcasing their potential to revolutionize sensor technology.

Mechanical alloying, a fundamental process, was discussed in detail, showcasing its advantages and diverse applications in electrochemistry. The tailoring of material properties through mechanical alloying was explored, underscoring its role in creating advanced modifiers for electrochemical sensors. We introduced various categories of bioactive molecules, including neurotransmitters, antioxidants, and metabolites, emphasizing the relevance of studying these molecules in the context of health, diagnostics, and environmental monitoring.

Our exploration extended to electrochemical techniques commonly used for bioactive molecule analysis, providing insights into their principles and the importance of electrode modifiers in enhancing sensor performance. Mechanically alloyed modifiers, ranging from nanoparticles to nanocomposites, were showcased for their unique properties and advantages in sensor design. We presented examples of their utilization in electrochemical sensors, highlighting their impact on sensor sensitivity and selectivity.

A significant portion of this chapter was devoted to the electrochemical redox reactions of neurotransmitters and the electrochemical detection of antioxidants and metabolites. We discussed the mechanisms of redox reactions, applications in neuroscience research, disease diagnosis, and the implications for monitoring antioxidant levels in biological samples. Finally, we touched upon current challenges in electrochemical sensing, potential innovations in mechanically alloyed modifiers, and emerging trends that promise to shape the future of this field. Altogether, this chapter has provided a comprehensive understanding of electrochemical redox reactions of bioactive molecules on mechanically alloyed modifiers, emphasizing their importance in advancing sensor technology and expanding applications in healthcare and environmental monitoring.

References

1. Iwuozor, K.O., et al.: Adsorption of methyl orange: a review on adsorbent performance. Curr. Res. Green Sustain. Chem. **4**, 100179 (2021)
2. Zhang, C., et al.: Metal oxide resistive sensors for carbon dioxide detection. Coord. Chem. Rev. **472**, 214758 (2022)
3. Sopha, H., et al.: In-situ plated antimony film electrode for adsorptive cathodic stripping voltammetric measurement of trace nickel. Electrochem. Commun. **20**, 23–25 (2012)
4. Skoog, D., Holler, F., Crouch, S.: Principles of instrumental analysis: Cengage learning. Cengage Learning (2017)
5. Ahmad, T., et al.: Recent advances in electrochemical sensing of hydrogen peroxide (H2O2) released from cancer cells. Nanomaterials **12**(9), 1475 (2022)
6. Teymourian, H., Barfidokht, A., Wang, J.: Electrochemical glucose sensors in diabetes management: an updated review (2010–2020). Chem. Soc. Rev. **49**(21), 7671–7709 (2020)
7. Fan, Z., et al.: Highly sensitive photoelectrochemical sensing of bisphenol a based on zinc phthalocyanine/TiO2 nanorod arrays. Talanta **189**, 16–23 (2018)
8. Islam, M.S., Banik, S., Collinson, M.M.: Recent advances in Bimetallic Nanoporous gold electrodes for electrochemical sensing. Nanomaterials **13**(18), 2515 (2023)
9. Okpara, E.C., et al.: Optical and electrochemical techniques for point-of-care water quality monitoring: a review. Results Chem. 100710 (2022)

10. Reta, N., et al.: Nanostructured electrochemical biosensors for label-free detection of water-and food-borne pathogens. ACS Appl. Mater. Interfaces **10**(7), 6055–6072 (2018)
11. Sang, W., et al.: Recent advances in nanomaterial-based synergistic combination cancer immunotherapy. Chem. Soc. Rev. **48**(14), 3771–3810 (2019)
12. Dong, H., et al.: Engineering bioscaffolds for enzyme assembly. Biotechnol. Adv. **53**, 107721 (2021)
13. Chang, H., et al.: In situ fabrication of (Sr, La) FeO 4 with CoFe alloy nanoparticles as an independent catalyst layer for direct methane-based solid oxide fuel cells with a nickel cermet anode. J. Mater. Chem. A **4**(36), 13997–14007 (2016)
14. Lyu, H., et al.: Ball-milled carbon nanomaterials for energy and environmental applications. ACS Sustain. Chem. Eng. **5**(11), 9568–9585 (2017)
15. Li, G., et al.: A label-free electrochemical aptasensor based on platinum@ palladium nanoparticles decorated with hemin-reduced graphene oxide as a signal amplifier for glypican-3 determination. Biomater. Sci. **10**(23), 6804–6817 (2022)
16. Jabeen, S., et al.: Synthesis of In2O3/GNPs nanocomposites with integrated approaches to tune overall performance of electrochemical devices. Ceram. Int. **47**(16), 22345–22355 (2021)
17. Islam, M.S., Shudo, Y., Hayami, S.: Energy conversion and storage in fuel cells and supercapacitors from chemical modifications of carbon allotropes: state-of-art and prospect. Bull. Chem. Soc. Jpn. **95**(1), 1–25 (2022)
18. Tajik, S., et al.: Recent developments in polymer nanocomposite-based electrochemical sensors for detecting environmental pollutants. Ind. Eng. Chem. Res. **60**(3), 1112–1136 (2021)
19. Iftikhar, F.J., et al.: Advancements in nanofiber-based electrochemical biosensors for diagnostic applications. Biosensors **13**(4), 416 (2023)
20. Yu, L., Chen, G.Z.: Supercapatteries as high-performance electrochemical energy storage devices. Electrochem. Energy Rev. **3**, 271–285 (2020)
21. Ali, Z., et al.: Colorimetric sensing of heavy metals on metal doped metal oxide nanocomposites: a review. Trends Environ. Anal. Chem. **37**, e00187 (2023)
22. Nemiwal, M., Kumar, D.: Recent progress on electrochemical sensing strategies as comprehensive point-care method. Monatshefte für Chemie-Chem. Monthly **152**, 1–18 (2021)
23. Thakur, A., Rasyotra, A., Jasuja, K.: Review and perspectives of electrochemical energy conversion and storage in metal diborides and XBenes. Energy Fuels (2023)
24. Suryanarayana, C.: Mechanical alloying: a novel technique to synthesize advanced materials. Research (2019)
25. Fragassa, C., Lesiuk, G., Epp, J.: High-performance applications of metals and alloys: material properties, behaviour modeling, optimal design and advanced processes, MDPI, p. 1485 (2023)
26. Vaidya, M., Muralikrishna, G.M., Murty, B.S.: High-entropy alloys by mechanical alloying: a review. J. Mater. Res. **34**(5), 664–686 (2019)
27. Sidhu, J.S., Zafar, T.A.: Bioactive compounds in banana fruits and their health benefits. Food Quality Safety **2**(4), 183–188 (2018)
28. Arazi, H., Eghbali, E.: Possible effects of beetroot supplementation on physical performance through metabolic, neuroendocrine, and antioxidant mechanisms: a narrative review of the literature. Front. Nutr. **8**, 660150 (2021)
29. Santhi, V.P., et al.: Fruits: a potential source of vitamin c as essential human nutrition and immunity development: a review (2020)
30. Kussmann, M., Cunha, D.H.A., Berciano, S.: Bioactive compounds for human and planetary health. Front. Nutr. **10** (2023)
31. Chang, M., et al.: Cellulose-based biosensor for bio-molecules detection in medical diagnosis: a mini-review. Curr. Med. Chem. **27**(28), 4593–4612 (2020)
32. Nemčeková, K., Labuda, J.: Advanced materials-integrated electrochemical sensors as promising medical diagnostics tools: a review. Mater. Sci. Eng. C **120**, 111751 (2021)
33. Marchev, A.S., et al.: Metabolomics and health: from nutritional crops and plant-based pharmaceuticals to profiling of human biofluids. Cell. Mol. Life Sci. **78**(19–20), 6487–6503 (2021)

34. Soleimany, A.P., Bhatia, S.N.: Activity-based diagnostics: an emerging paradigm for disease detection and monitoring. Trends Mol. Med. **26**(5), 450–468 (2020)
35. Kerry, R.G., et al.: Molecular prospect of type-2 diabetes: nanotechnology based diagnostics and therapeutic intervention. Rev. Endocr. Metab. Disord. **22**, 421–451 (2021)
36. Zaghloul, A., et al.: Biological indicators for pollution detection in terrestrial and aquatic ecosystems. Bull. National Res. Centre **44**(1), 1–11 (2020)
37. Kaur, J., Malik, R., Gangwar, D.: A reliable cyclic voltammetry Technique for the degradation of Salicylaldehyde: electrode kinetics. Nat. Environ. Pollut. Technol. **20**(5), 1967–1972 (2021)
38. Puthongkham, P., Venton, B.J.: Recent advances in fast-scan cyclic voltammetry. Analyst **145**(4), 1087–1102 (2020)
39. Fazio, E., et al.: Metal-oxide based nanomaterials: synthesis, characterization and their applications in electrical and electrochemical sensors. Sensors **21**(7), 2494 (2021)
40. Baranwal, J., et al.: Electrochemical sensors and their applications: a review. Chemosensors **10**(9), 363 (2022)
41. Hernández-Rodríguez, J.F., Rojas, D., Escarpa, A.: Electrochemical sensing directions for next-generation healthcare: trends, challenges, and frontiers. Anal. Chem. **93**(1), 167–183 (2020)
42. Beitollahi, H., et al.: A review on the effects of introducing CNTs in the modification process of electrochemical sensors. Electroanalysis **31**(7), 1195–1203 (2019)
43. Ambaye, A.D., et al.: Recent developments in nanotechnology-based printing electrode systems for electrochemical sensors. Talanta **225**, 121951 (2021)
44. Tajik, S., et al.: Performance of metal–organic frameworks in the electrochemical sensing of environmental pollutants. J. Mater. Chem. A **9**(13), 8195–8220 (2021)
45. Cardoso, R.M., et al.: Additive-manufactured (3D-printed) electrochemical sensors: a critical review. Anal. Chim. Acta **1118**, 73–91 (2020)
46. Cheng, W., et al.: Applications of metal-organic framework (MOF)-based sensors for food safety: enhancing mechanisms and recent advances. Trends Food Sci. Technol. **112**, 268–282 (2021)
47. Pérez-Fernández, B., Costa-García, A., Muñiz, A.d.l.E.: Electrochemical (bio)sensors for pesticides detection using screen-printed electrodes. Biosensors **10**(4), 32 (2020)
48. Mishra, R.R., et al.: Research progress on nano-metal matrix composite (NMMC) fabrication method: a comprehensive review. Mater. Today: Proc. **56**, 2104–2109 (2022)
49. Aziz, S.S.A., Abulkhair, H., Moustafa, E.B.: Role of hybrid nanoparticles on thermal, electrical conductivity, microstructure, and hardness behavior of nanocomposite matrix. J. Market. Res. **13**, 1275–1284 (2021)
50. Iqbal, A.A., et al.: Graphene-based nanocomposites and their fabrication, mechanical properties and applications. Materialia **12**, 100815 (2020)
51. Kumar, A., Singh, A., Suhane, A.: Mechanically alloyed high entropy alloys: existing challenges and opportunities. J. Market. Res. **17**, 2431–2456 (2022)
52. Yadav, S., et al.: GLAD based advanced nanostructures for diversified biosensing applications: recent progress. Biosensors **12**(12), 1115 (2022)
53. Wan, H., et al.: Biomedical sensors. In: Biomedical Information Technology, pp. 51–79. Elsevier (2020)
54. Rajendrachari, S., et al.: Synthesis and characterization of high entropy alloy 23Fe–21Cr–18Ni–20Ti–18Mn for electrochemical sensor applications. Materials **15**(21), 7591 (2022)
55. Mahale, R.S., et al.: Electrochemical determination of ascorbic acid by mechanically alloyed super duplex stainless steel powders. Metals **13**(8), 1430 (2023)
56. Rajendrachari, S., et al.: Electrocatalytic determination of methyl orange dye using mechanically alloyed novel metallic glass modified carbon paste electrode by cyclic voltammetry. Inorg. Chem. Commun. **155**, 111010 (2023)
57. Speranza, L., et al.: Dopamine: the neuromodulator of long-term synaptic plasticity, reward and movement control. Cells **10**(4), 735 (2021)
58. Berke, J.D.: What does dopamine mean? Nat. Neurosci. **21**(6), 787–793 (2018)

59. Rajendrachari, S., et al.: Electrocatalytic investigation by improving the charge kinetics between carbon electrodes and dopamine using bio-synthesized CuO nanoparticles. Catalysts **12**(9), 994 (2022)
60. June, M.S., Serotonin versus Dopamine: why do we need them & what are they? Omega **4**, 59 (2023)
61. Breuning, L.: The neurochemistry of science bias. In: Groupthink in Science: greed, Pathological Altruism, Ideology, Competition, and Culture, pp. 3–14 (2020)
62. Banerjee, S., et al.: Electrochemical detection of neurotransmitters. Biosensors **10**(8), 101 (2020)
63. Khoshnevisan, K., et al.: Nanomaterial based electrochemical sensing of the biomarker serotonin: a comprehensive review. Microchim. Acta **186**, 1–21 (2019)
64. Thomas, J., et al.: The influence of B-site cation in LaBO3 (B= Fe Co, Ni) perovskites on the nanomolar sensing of neurotransmitters. Sens. Actuat. B Chem. **332**, 129362 (2021)
65. Jeon, J., Hwang, I., Chung, T.D.: Electrochemical detection of neurotransmitters: toward synapse-based neural interfaces. Biomed. Eng. Lett. **6**, 123–133 (2016)
66. Maduraiveeran, G., Sasidharan, M., Ganesan, V.: Electrochemical sensor and biosensor platforms based on advanced nanomaterials for biological and biomedical applications. Biosens. Bioelectron. **103**, 113–129 (2018)
67. Xu, T., et al.: Electrochemical sensors for nitric oxide detection in biological applications. Electroanalysis **26**(3), 449–468 (2014)
68. Ou, Y., et al.: Frontiers in electrochemical sensors for neurotransmitter detection: towards measuring neurotransmitters as chemical diagnostics for brain disorders. Anal. Methods **11**(21), 2738–2755 (2019)
69. Brazaca, L.C., et al.: Applications of biosensors in Alzheimer's disease diagnosis. Talanta **210**, 120644 (2020)
70. Karaboğa, M.N.S., Sezgintürk, M.K.: Biosensor approaches on the diagnosis of neurodegenerative diseases: sensing the past to the future. J. Pharm. Biomed. Anal. **209**, 114479 (2022)
71. Gulcin, İ: Antioxidants and antioxidant methods: an updated overview. Arch. Toxicol. **94**(3), 651–715 (2020)
72. Nejadmansouri, M., et al.: An overview of optical and electrochemical sensors and biosensors for analysis of antioxidants in food during the last 5 years. Sensors **21**(4), 1176 (2021)
73. Olmos Moya, P.M., et al.: Simultaneous electrochemical speciation of oxidized and reduced glutathione. Redox profiling of oxidative stress in biological fluids with a modified carbon electrode. Anal. Chem. **89**(20), 10726–10733 (2017)
74. Liu, J., et al.: Antioxidant redox sensors based on DNA modified carbon screen-printed electrodes. Anal. Chem. **78**(19), 6879–6884 (2006)
75. Mello, L.D., Kubota, L.T.: Biosensors as a tool for the antioxidant status evaluation. Talanta **72**(2), 335–348 (2007)
76. Thirumurugan, D., et al.: An introductory chapter: secondary metabolites. In: Secondary Metabolites-Sources and Applications, pp. 3–21 (2018)
77. Pagare, S., et al.: Secondary metabolites of plants and their role: overview. Curr. Trends Biotechnol. Pharm. **9**(3), 293–304 (2015)
78. Moteshakeri, M.: Development of PEDOT-modified electrochemical sensors for uric acid analysis in Milk, ResearchSpace@ Auckland (2019)
79. Khaleque, M., et al.: Nanostructured wearable electrochemical and biosensor towards healthcare management: a review. RSC Adv. **13**(33), 22973–22997 (2023)
80. Jeerapan, I., Sonsa-Ard, T., Nacapricha, D.: Applying nanomaterials to modern biomedical electrochemical detection of metabolites, electrolytes, and pathogens. Chemosensors **8**(3), 71 (2020)
81. Xiao, F., Wang, L., Duan, H.: Nanomaterial based electrochemical sensors for in vitro detection of small molecule metabolites. Biotechnol. Adv. **34**(3), 234–249 (2016)
82. Kumar, P., et al.: Progress and challenges in electrochemical sensing of volatile organic compounds using metal-organic frameworks. Crit. Rev. Environ. Sci. Technol. **49**(21), 2016–2048 (2019)

83. Kumar, Y.R., et al.: A systematic review on 2D materials for volatile organic compound sensing. Coord. Chem. Rev. **461**, 214502 (2022)
84. Ramachandran, R., et al.: A review of the advanced developments of electrochemical sensors for the detection of toxic and bioactive molecules. Inorg. Chem. Front. **6**(12), 3418–3439 (2019)
85. Zhang, H., et al.: A review of innovative electrochemical strategies for bioactive molecule detection and cell imaging: CURRENT advances and challenges. Analytica Chimica Acta 341920 (2023)
86. Romanholo, P.V., et al.: Biomimetic electrochemical sensors: new horizons and challenges in biosensing applications. Biosens. Bioelectron. **185**, 113242 (2021)
87. Baskaran, J., Raghuvaran, P., Ashwin, S.: Experimental investigation on the effect of microstructure modifiers and heat treatment influence on A356 alloy. Mater. Today: Proc. **37**, 3007–3010 (2021)
88. Rajendrachari, S.: Practical Implementations of Additive Manufacturing Technologies. Springer Nature (2023)
89. Shashanka, R.: Fabrication of nano-structured duplex and ferritic stainless steel by planetary milling followed by consolidation (2016)
90. Mahale, R.S., et al.: Electrochemical sensor applications of nanoparticle modified carbon paste electrodes to detect various neurotransmitters: a review. Appl. Mech. Mater. **908**, 69–88 (2022)
91. Moses, O.A., et al.: 2D materials inks toward smart flexible electronics. Mater. Today **50**, 116–148 (2021)
92. Pouyanfar, N., et al.: Artificial intelligence-based microfluidic platforms for the sensitive detection of environmental pollutants: recent advances and prospects. Trends Environ. Anal. Chem. **34**, e00160 (2022)
93. Banitaba, S.N., et al.: Recent progress of bio-based smart wearable sensors for healthcare applications. Mater. Today Electron. **5**, 100055 (2023)
94. da Silva, E.T., et al.: Electrochemical biosensors in point-of-care devices: recent advances and future trends. ChemElectroChem **4**(4), 778–794 (2017)
95. Xu, Z., et al.: Additive manufacturing of self-sensing carbon fiber composites. Adv. Eng. Mater. (2023)
96. Liu, C.-S., Li, J., Pang, H.: Metal-organic framework-based materials as an emerging platform for advanced electrochemical sensing. Coord. Chem. Rev. **410**, 213222 (2020)

Chapter 19
Present and the Future of Mechanically Alloyed Materials

M. Kalyan Phani and D. Narsimhachary

Abstract Mechanically alloyed materials have emerged as a transformative technology in materials science, revolutionizing the landscape of advanced materials with their unique and tailorable properties. The current chapter delves into the present applications of mechanically alloyed materials, emphasizing their widespread use in industries such as aerospace, automotive, electronics, and energy. These materials exhibit exceptional mechanical, thermal, and electrical characteristics, making them invaluable for numerous high-performance applications. The present research is focused on designing the microstructure of the alloys by optimizing the mechanical alloying process parameters to achieve the desired mechanical properties, such as increased hardness, wear resistance, and corrosion resistance. The chapter also explores the exciting prospects for mechanically alloyed materials in the future. One of the primary directions of research is the development of lightweight, high-strength materials for the transportation industry. These materials have the potential to significantly reduce fuel consumption and carbon emissions in various modes of transportation. Furthermore, the integration of mechanically alloyed materials in 3D printing and additive manufacturing processes is presented as a pivotal advancement, allowing for the fabrication of complex components with precisely tailored properties, thus revolutionizing the manufacturing sector. In the energy sector, the book chapter discusses the role of mechanically alloyed materials in advanced battery and supercapacitor applications. Their enhanced electrical conductivity and durability are instrumental in improving the efficiency and lifespan of energy storage systems. Additionally, the exploration of high-temperature-resistant materials for gas turbine engines and nuclear reactors demonstrates their potential to reshape the energy landscape. These materials, with their versatile properties and wide-ranging applications, are a focal point of materials science research. As technology advances, the

M. K. Phani (✉)
Department of Metallurgical Engineering, O.P. Jindal University, Raigarh, Chattisgarh 496109, India
e-mail: kalyan.makkuva@opju.ac.in

D. Narsimhachary
Department of Mechanical Engineering, G.H. Raisoni University, Saikheda, Madhya Pradesh, India

S. Rajendrachari (ed.), *Mechanically Alloyed Novel Materials*, Advanced Structured Materials 220, https://doi.org/10.1007/978-981-97-6504-1_19

synthesis and characterization of these materials are expected to become more efficient, enabling the creation of innovative solutions to address the evolving challenges of the twenty-first century.

Keywords Mechanical alloying · Solid state · Industry 4.0 · Additive manufacturing · Biomaterials

19.1 Introduction

Mechanical alloying is a solid-state powder processing technique that plays a crucial role in creating advanced materials with tailored properties for various applications [1–3]. It is a versatile and innovative solid-state powder processing method that has gained significant prominence in materials science and engineering. This technique involves the repetitive milling and blending of elemental or pre-alloyed powders to create new alloys or materials with tailored properties. This process leads to the formation of nanocrystalline and amorphous structures, offering a wide range of unique properties—physical, mechanical, and microstructural—that are not attainable through conventional metallurgical methods [4].

Mechanical alloying has emerged as a key research area with transformative potential and diverse applications across various industries. Research in the field is multifaceted and continually evolving, encompassing materials development, process optimization, microstructural analysis, and exploration of novel applications [5–9]. Researchers focus on tailoring mechanical alloying process parameters, such as milling time, milling speed, ball size, and the choice of starting materials, to achieve specific material properties [10–12]. The goal is to enhance properties such as hardness, wear resistance, and corrosion resistance while simultaneously reducing grain size and improving homogeneity.

Mechanical alloying research also extends to the exploration of advanced materials for aerospace, automotive, electronics, energy, and medical applications. The development of high-strength, lightweight materials for the transportation sector and the enhancement of energy storage materials are just a few examples of its impact [13, 14]. As technology continues to advance, mechanical alloying research holds the promise of addressing emerging challenges, offering sustainable solutions, and revolutionizing the field of materials science [15–18] This introduction highlights the fundamental role of mechanical alloying research in shaping the future of materials engineering and its potential to provide innovative solutions to meet the evolving needs of various industries. This chapter delves into the current status and the future prospects of mechanically alloyed materials, which have the potential to revolutionize industries such as aerospace, automotive, energy, and electronics. To comprehend the evolution of mechanically alloyed materials, we begin with a historical overview.

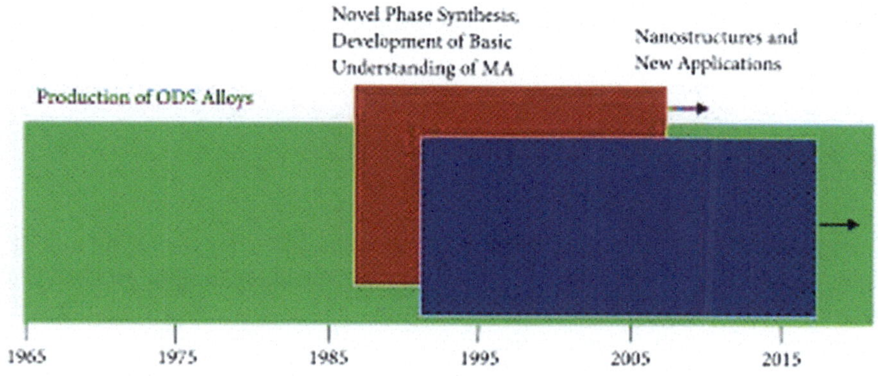

Fig. 19.1 Historical development of MA [5]

19.2 Historical Development of Mechanical Alloying

19.2.1 Emergence of Mechanical Alloying

The history of mechanical alloying can be traced back to the early 1960s when Benjamin and Volin conducted pioneering work on powder metallurgy [2]. They introduced the concept of high-energy ball milling, a key process in mechanical alloying. The technique gained recognition due to its ability to create homogeneous alloys of elements that were otherwise immiscible in the liquid state.

19.2.2 Evolution of Mechanical Alloying Techniques

Over the decades, mechanical alloying techniques have evolved significantly. Initial methods predominantly employed high-energy ball mills, but more advanced equipment such as attritors, planetary ball mills, and vibratory mills have been developed. These innovations have improved the efficiency and precision of mechanical alloying [1, 2] (Fig. 19.1).

19.3 Principles of Mechanical Alloying

19.3.1 Process Overview

Mechanical alloying involves the repeated cold welding, fracturing, and rewelding of powder particles inside a milling container. The process is initiated by mechanical impacts, leading to a decrease in grain size and the creation of a homogeneous alloy.

Understanding the process's fundamental principles is crucial to advancing the field [5].

19.3.2 Key Parameters

Several parameters influence the success of mechanical alloying, including milling time, milling speed, ball-to-powder ratio, and the choice of milling media. The optimal combination of these parameters varies with the materials being alloyed, making process optimization a critical step in material development [2].

19.4 Recent Advancements

19.4.1 Nanostructured Materials

One of the significant advancements in mechanical alloying is the production of nanostructured materials. The reduction in powder size which in turn reduced the grain size and further increased defect density offered unique mechanical, electrical, and thermal properties [8]. Nanostructured materials have found applications in areas such as catalysis, energy storage, and sensors [15].

19.4.2 Additive Manufacturing

Mechanically alloyed powders have become increasingly significant in the realm of additive manufacturing, particularly in 3D printing and rapid prototyping. The process of mechanical alloying involves the repeated welding, fracturing, and rewelding of powder particles in a high-energy ball mill, resulting in a homogeneous mixture of elements at the atomic level.

The benefits of using mechanically alloyed powders in additive manufacturing are multifold. One of the key advantages is the ability to create intricate and complex shapes that might be challenging or impossible with traditional manufacturing methods. The fine and uniform powder particles obtained through mechanical alloying contribute to the production of materials with enhanced mechanical properties.

Despite the promising potential of mechanically alloyed powders, research in this area is still in its early stages. The complexity arises from the multitude of parameters that must be considered to achieve a desired component. These parameters include additive manufacturing machine parameters (speed, overlap distance, energy density, powder flow rate, shielding gas, focal length, height of the build area) milling time, milling speed, ball-to-powder ratio, and the type of milling media, among others. Adjusting these parameters can influence the final microstructure, composition, and properties of the alloyed powders, making the optimization process intricate.

To advance the use of mechanically alloyed powders in additive manufacturing, ongoing research is essential. Scientists and engineers are working to better understand the interplay of different parameters and their effects on the final product. This involves experimentation, characterization, and modeling to develop a comprehensive understanding of the mechanical alloying process.

As the research progresses, it is anticipated that the use of mechanically alloyed powders will become more refined, enabling the production of tailored materials with optimized properties for specific applications. This could lead to ground-breaking advancements in additive manufacturing, opening up new possibilities for creating components with superior performance and unique characteristics [55–58].

19.4.3 Bulk Metallic Glasses

The development of bulk metallic glasses (BMGs) contributes to the advancement of materials with extraordinary properties. BMGs are unique in that they lack the crystalline structure found in traditional metals; instead, they exhibit an amorphous, or glassy, structure. This amorphous nature contributes to their exceptional mechanical properties, including high strength, elasticity, and impressive corrosion resistance. The process of mechanical alloying is particularly effective in producing BMGs because it allows for the creation of alloys with compositions that would be challenging or impossible to achieve through conventional methods. The high-energy ball milling involved in mechanical alloying facilitates the mixing and homogenization of elements on a nanoscale, leading to the formation of metallic glasses.

The outstanding mechanical properties of BMGs make them suitable for a wide range of applications across various industries. In aerospace, for example, BMGs can be utilized in the manufacturing of lightweight and strong components, contributing to fuel efficiency and overall performance. In the medical field, the biocompatibility and corrosion resistance of BMGs make them potential candidates for certain medical implants. Consumer electronics also benefit from the unique properties of BMGs, where their strength and elasticity can enhance the durability and performance of electronic devices.

While the application of BMGs in various industries is promising, challenges such as processing and forming methods, and potential brittleness in certain compositions, still need to be addressed. Ongoing research in the field of mechanical alloying and BMGs aims to overcome these challenges and unlock even more opportunities for these remarkable materials in diverse applications. The combination of mechanical alloying and bulk metallic glasses represents a fascinating intersection of materials science and advanced manufacturing, with the potential to revolutionize the way we design and produce high-performance materials [17].

19.5 Applications of Mechanically Alloyed Materials

19.5.1 Aerospace Industry

Mechanically alloyed materials have indeed played a crucial role in meeting the demands of modern aircraft design, where a balance between strength, weight, and durability is essential. The aerospace industry's adoption of these materials, particularly titanium-based alloys produced through mechanical alloying, has been a key contributor to the success of aircraft like the Boeing 787 Dreamliner and Airbus A350 XWB.

The lightweight yet strong characteristics of these alloys contribute significantly to fuel efficiency and overall structural integrity. The use of titanium alloys in critical components, such as the fuselage structure and landing gear, allows for a high strength-to-weight ratio, reducing the overall weight of the aircraft. This not only improves fuel efficiency, leading to reduced emissions, but also aligns with the sustainability goals of the aviation sector.

The corrosion resistance inherent in mechanically alloyed materials is another crucial factor. This property ensures the longevity of critical components, reducing the need for frequent maintenance. This not only contributes to the sustainability profile of aircraft but also helps airlines optimize operational costs. The Boeing 777X, as you mentioned, is an excellent example of how mechanically alloyed sustainable alloys are integral in achieving these goals [38].

Continued research and innovation in this field are indeed essential. Exploring new alloy compositions, refining manufacturing processes, and addressing specific challenges related to different aircraft components will drive advancements further. This ongoing pursuit of improvement is vital for staying at the forefront of innovation and meeting the ever-evolving demands of the aerospace industry.

19.5.2 Automotive Sector

The automotive sector, particularly in the context of sustainable alloys, has significant implications for vehicle design, performance, and sustainability. Let's delve deeper into the key aspects you mentioned [39].

Weight Reduction and Fuel Efficiency:

- Mechanically alloyed materials, such as aluminum alloys, contribute to lightweight yet high-strength components in vehicles.
- The use of these alloys in automotive bodies reduces overall weight, leading to improved fuel efficiency.

Enhanced Safety:

- The high-strength properties of mechanically alloyed materials can enhance the safety features of vehicles.
- Lightweight yet strong materials can contribute to better crash performance and overall safety.

Application in Electric Vehicles (EVs):

- As the automotive industry shifts towards sustainability, electric vehicles are becoming increasingly popular.
- The use of mechanically alloyed sustainable alloys in EVs, as seen in the Tesla Model S, contributes to the lightweight design crucial for electric vehicle efficiency.

Energy Efficiency in EVs:

- The lightweight nature of mechanically alloyed materials directly impacts the energy consumption of electric vehicles.
- By reducing weight, these alloys help extend the range of electric vehicles, making them more practical for everyday use.

Role in Green Solutions:

- The automotive industry's focus on greener solutions aligns with the use of mechanically alloyed sustainable alloys.
- These materials play a pivotal role in achieving sustainability goals by reducing the environmental impact of vehicles.

19.5.3 Energy Sector

Mechanical alloying significantly contributes to the energy sector, particularly in renewable energy and electronic devices. In wind energy, advanced alloys with enhanced fatigue and corrosion resistance are integral to the longevity and reliability of turbine blades. Vestas, a prominent wind turbine manufacturer, employs these alloys to enhance the durability and efficiency of their turbines. Solar energy applications benefit from mechanically alloyed sustainable alloys in structural components, providing corrosion resistance and lightweight properties that contribute to system durability and ease of installation [45–47]. The overall sustainability of renewable energy systems is bolstered by the reduced maintenance requirements and extended lifespan enabled by these alloys. Beyond renewables, the electronics industry, exemplified by companies like Apple, utilizes mechanically alloyed materials in smartphones for improved thermal conductivity, supporting the development of energy-efficient and sustainable electronic products. In both energy and electronics, the role of mechanically alloyed sustainable alloys is pivotal in advancing technologies for a more sustainable future.

19.6 Future Prospects

19.6.1 Integration with Industry 4.0

The integration of mechanical alloying processes with Industry 4.0 technologies, such as artificial intelligence, the Internet of Things (IoT), and automation, can enhance efficiency, quality control, and real-time monitoring in materials manufacturing.

19.6.2 Bio-inspired Materials

Drawing inspiration from natural materials, biomimicry presents a promising avenue for developing mechanically alloyed materials with exceptional properties. Materials inspired by the structure and properties of biological materials, such as nacre or spider silk, could revolutionize various industries [25].

19.6.3 Sustainable Alloys

The development of environmentally friendly alloys with reduced reliance on critical raw materials and toxic elements is vital. Research into sustainable alloys, such as those based on earth-abundant elements and recyclable materials, can address both economic and environmental challenges [35].

19.7 Conclusions

Mechanical alloying has come a long way since its inception, with numerous breakthroughs in materials science and engineering. The ability to create novel materials with tailored properties has opened up exciting opportunities in various industries.

The future of mechanically alloyed materials holds even more promise, as advances in computational techniques and sustainability considerations drive innovation. As we continue to explore this field, the development of cutting-edge materials that meet the demands of our rapidly evolving world remains at the forefront of scientific research.

In conclusion, the future of mechanical alloying holds great promise and significance in materials science and engineering. This innovative solid-state powder processing technique is poised to continue revolutionizing materials development across a spectrum of industries, from aerospace and automotive to energy and healthcare.

Researchers are increasingly focused on optimizing the process to create materials with exceptional properties, while also emphasizing sustainability. With ongoing advances in technology, mechanical alloying will enable the synthesis of materials with tailored properties, contributing to enhanced performance, cost-effectiveness, and sustainability. As the demand for advanced materials grows, mechanical alloying is well-positioned to address the evolving challenges of the twenty-first century and provide innovative solutions for a wide range of applications.

References

1. Benjamin, J.S., Volin, T.E.: The fabrication of aluminum-transition metal alloys by mechanical alloying. J. Appl. Phys. **37**(7), 2980–2982 (1966)
2. Suryanarayana, C.: Mechanical alloying and milling. Prog. Mater Sci. **46**(1–2), 1–184 (2001)
3. Gleiter, H.: Nanocrystalline materials. Prog. Mater Sci. **33**(4), 223–315 (1989)
4. Löffler, J.F.: Bulk metallic glasses. Mater. Sci. Eng., A **485**(1–2), 1–14 (2008)
5. Ma, E., Atzmon, M.: Nanostructured Materials: processing, Properties and Potential Applications. Elsevier (2010)
6. Suryanarayana, C., Ivanov, E.: Future directions of mechanical alloying. Mater. Sci. Eng. A **573**, 55–59 (2013)
7. Bala, H., Leparoux, M.: Mechanical alloying: a novel technique to synthesize advanced materials for aerospace applications. Procedia Struct. Integr. **2**, 1365–1372 (2016)
8. Simoes, S., Santos, T.: Emerging trends in mechanical alloying and milling. Mater. Today: Proc. **17**, 508–515 (2019)
9. Zarei-Hanzaki, A., et al.: Recent advances in mechanical alloying, its applications and challenges: a review. J. Alloy. Compd. **717**, 271–284 (2017)
10. Rajendrachari, S., Shatokha, V., Avar, B. (eds.).: New Advances in Powder Technology. IntechOpen (2023). https://doi.org/10.5772/intechopen.102106
11. Chookaew, S., Ngamjarurojana, A.: Advances in mechanical alloying and milling. Mater. Sci. Forum **783–786**, 1791–1796 (2014)
12. Mittemeijer, E.J., Somers, M.A.J.: Thermochemical Surface Engineering of Steels: improving Materials Performance. Elsevier (2013)
13. Gu, L., Tan, Y.: Mechanical alloying and milling. In: Encyclopedia of Renewable and Sustainable Materials, vol. 4, pp. 580–590. Elsevier (2017)
14. Volek, A., Kudlacek, J.: Advances in mechanical alloying: technological process and perspective. Mater. Today: Proc. **4**(11), 11858–11866 (2017)
15. Oghbaei, M., Emami, Z.: Mechanical alloying as a solid state processing technique for the synthesis of advanced materials. J. Alloy. Compd. **629**, 238–249 (2015)
16. Leszczyńska-Sejda, K.: The future of mechanical alloying. Arch. Mater. Sci. Eng. **74**(2), 90–99 (2015)
17. Bala, H., et al.: Mechanical alloying: a promising route to produce high-strength aluminum alloys for aerospace applications. Mater. Des. **142**, 34–45 (2018)
18. Adimule, V., Wei, P.S., Tanaka, M., Shashanka, R.: Synthesis and Modifications of Materials and its Properties, pp. 1–134 (2022)
19. Zarei-Hanzaki, A., Eskandari, M.: Mechanical alloying and milling: a review of powder and grain surface effects. Powder Technol. **287**, 3–13 (2015)
20. Zhang, Y., Yu, H.: Mechanical alloying: recent developments and future prospects. J. Mater. Res. **30**(11), 1501–1518 (2015)
21. Amiri, S., et al.: Mechanical alloying: a versatile technique to synthesize advanced materials for energy and environmental applications. Sustain. Mater. Technol. **15**, 1–12 (2018)

22. Talha, M., Ghasemi, A.: Mechanical alloying and advanced materials in aerospace industry. In: Proceedings of the International Conference on Mechanical, Aeronautical and Automotive Engineering (ICMAAE), pp. 1–7 (2017)
23. Bala, H., Maleque, M.A.: Mechanical alloying: a promising route to synthesize high-performance aerospace materials. Mater. Des. **83**, 768–779 (2015)
24. Guo, X., et al.: Recent advancements in aerospace materials and manufacturing processes. Mater. Des. **159**, 1–23 (2018)
25. Ghasemi, A., Talha, M.: Advanced materials in aerospace industry: a comprehensive review. J. Aerosp. Eng. **32**(3), 04019045 (2019)
26. Leszczyńska-Sejda, K., Bojar, Z.: Advanced materials for aerospace applications: a review. Mater. Sci. Eng. A **701**, 140–149 (2017)
27. Simões, S., Santos, T.: Recent advances in the application of mechanical alloying for aerospace materials. J. Aerosp. Technol. Manag. **7**(2), 127–136 (2015)
28. Chookaew, S., Ngamjarurojana, A.: Mechanical alloying of advanced aerospace materials. Procedia CIRP **74**, 429–433 (2018)
29. Volek, A., Kudlacek, J.: High-strength alloys for aerospace applications synthesized by mechanical alloying. Mater. Des. **169**, 107685 (2019)
30. Bhadeshia, H.K.D.H.: Nanotechnology in aerospace. Mater. Sci. Technol. **32**(7), 743–749 (2016)
31. Bala, H., Maleque, M.A.: Advanced materials in aerospace industry: challenges and solutions. Mater. Today: Proc. **3**(2), 2404–2412 (2016)
32. Goh, C.Y., Yap, D.A.: Application of advanced materials in the aerospace industry. Procedia Eng. **184**, 10–17 (2017)
33. Amiri, S., et al.: Aerospace materials and mechanical alloying: a review of recent advancements. Aerosp. Sci. Technol. **98**, 105739 (2020)
34. Zarei-Hanzaki, A., Eskandari, M.: Aerospace alloys and mechanical alloying: a comprehensive review. Mater. Des. **142**, 220–236 (2018)
35. Chen, C., et al.: Recent developments in advanced materials for aerospace applications. Prog. Aerosp. Sci. **106**, 100539 (2019)
36. Nene, S.S., Ogale, A.A.: Recent developments in aerospace materials. Mater. Today: Proc. **3**(2), 2404–2412 (2016)
37. Bala, H., Maleque, M.A.: Mechanical alloying: a promising route to synthesize high-performance materials for industrial applications. Mater. Des. **111**, 522–533 (2016)
38. Zarei-Hanzaki, A., Eskandari, M.: Industrial applications of mechanically alloyed materials: a comprehensive review. J. Alloy. Compd. **767**, 186–205 (2018)
39. Simões, S., Santos, T.: Mechanical alloying and its industrial applications. Mater. Today: Proc. **4**(11), 11216–11223 (2017)
40. Chookaew, S., Ngamjarurojana, A.: Advanced materials produced by mechanical alloying for industrial applications. Procedia Eng. **138**, 229–236 (2016)
41. Talha, M., Ghasemi, A.: Mechanical alloying and industrial materials: a comprehensive review. Mater. Des. **183**, 108146 (2019)
42. Leszczyńska-Sejda, K., Bojar, Z.: Industrial applications of mechanically alloyed materials: challenges and solutions. Mater. Sci. Eng. A **711**, 105–116 (2018)
43. Bala, H., et al.: Advanced materials synthesized by mechanical alloying for industrial use: a review. Procedia Manuf. **8**, 477–484 (2017)
44. Goh, C.Y., Yap, D.A.: Mechanical alloying for industrial materials: progress and prospects. Mater. Today: Proc. **5**(6), 13376–13382 (2018)
45. Amiri, S., et al.: Mechanical alloying and its industrial applications in the 21st century: a comprehensive review. Mater. Today: Proc. **28**, 1651–1656 (2020)
46. Volek, A., Kudlacek, J.: Industrial applications of mechanically alloyed materials: recent developments and future outlook. Procedia CIRP **62**, 30–35 (2017)
47. Amiri, S., et al.: Sustainable mechanical alloying: a review of environmental and economic implications. J. Clean. Prod. **214**, 151–160 (2019)

48. Simões, S., Santos, T.: Sustainability aspects of mechanical alloying in materials production. J. Mater. Process. Technol. **252**, 621–629 (2018)
49. Goh, C.Y., Yap, D.A.: Sustainable materials synthesis by mechanical alloying: challenges and prospects. Procedia CIRP **62**, 23–29 (2017)
50. Zarei-Hanzaki, A., Eskandari, M.: Sustainability in mechanical alloying: a comprehensive overview. J. Market. Res. **9**(1), 2105–2123 (2020)
51. Chookaew, S., Ngamjarurojana, A.: Sustainable practices in mechanical alloying for materials production. Procedia CIRP **81**, 1487–1492 (2019)
52. Nene, S.S., Ogale, A.A.: Environmental sustainability and mechanical alloying: a holistic review. Procedia CIRP **68**, 87–92 (2017)
53. Leszczyńska-Sejda, K., Bojar, Z.: Sustainability in materials synthesis by mechanical alloying: a critical analysis. J. Clean. Prod. **139**, 1102–1111 (2016)
54. Volek, A., Kudlacek, J.: Sustainable mechanical alloying for materials development. Mater. Des. **154**, 94–104 (2018)
55. Talha, M., Ghasemi, A.: Sustainable practices in mechanical alloying: environmental impact and solutions. J. Sustain. Metall. **7**(1), 54–67 (2021)
56. Mahale, R.S., Vasanth, S., Sharath, P.C., Shashanka, R., Tambrallimath, V., Badari, A.: Effect of ball to powder ratios on the phase transformation of austenite and ferrite during ball milling of SAF-2507 super duplex stainless steel powders. J. Inst. Eng. (India): Ser. D 1–14
57. Rajendrachari, S. (ed.).: Practical Implementations of Additive Manufacturing Technologies. Springer Nature (2023)
58. Samantray, R., Priyadarshinee, M., Rajendrachari, S.: General job opportunities for the graduates, Post Graduates and Ph.D. graduates in additive manufacturing industries. In: Practical Implementations of Additive Manufacturing Technologies, pp. 325–332. Springer Nature Singapore, Singapore (2023)
59. Rajendrachari, S., Basavegowda, N., Vinaykumar, R., Narsimhachary, D., Somu, P., Lee, M.J.: Electrocatalytic determination of methyl orange dye using mechanically alloyed novel metallic glass modified carbon paste electrode by cyclic voltammetry. Inorg. Chem. Commun. **155**, 111010 (2023)
60. Narsimhachary, D., Kalyan Phani, M.: Additive manufacturing: environmental impact, and future perspective. In: Practical Implementations of Additive Manufacturing Technologies, pp. 295–308. Springer Nature Singapore, Singapore (2023)
61. Pradeep, N.B., Hegde, M.R., Rajendrachari, S., Surendranathan, A.O.: Investigation of microstructure and mechanical properties of microwave consolidated TiMgSr alloy prepared by high energy ball milling. Powder Technol. **408**, 117715 (2022)
62. Mahale, R.S., Vasanth, S., Krishna, H., Chikkegouda, S.P., Rajendrachari, S., Patil, A., Rathod, B.S.: Sensor-based additive manufacturing technologies (2022)
63. Shetty, D.P., Mahale, R.S., Shamanth, V., Sharath, P.C., Shashanka, R.: A review on metal additive manufacturing processes. AIJR Abstracts, p. 80 (2021)

The manufacturer's authorised representative in the EU is Springer Nature Customer Service Centre GmbH, Europaplatz 3, 69115 Heidelberg, Germany. If you have any concerns regarding our products, please contact ProductSafety@springernature.com

Printed and bound by CPI Group (UK) Ltd, Croydon, CR0 4YY
01/10/2026
02225936-0016